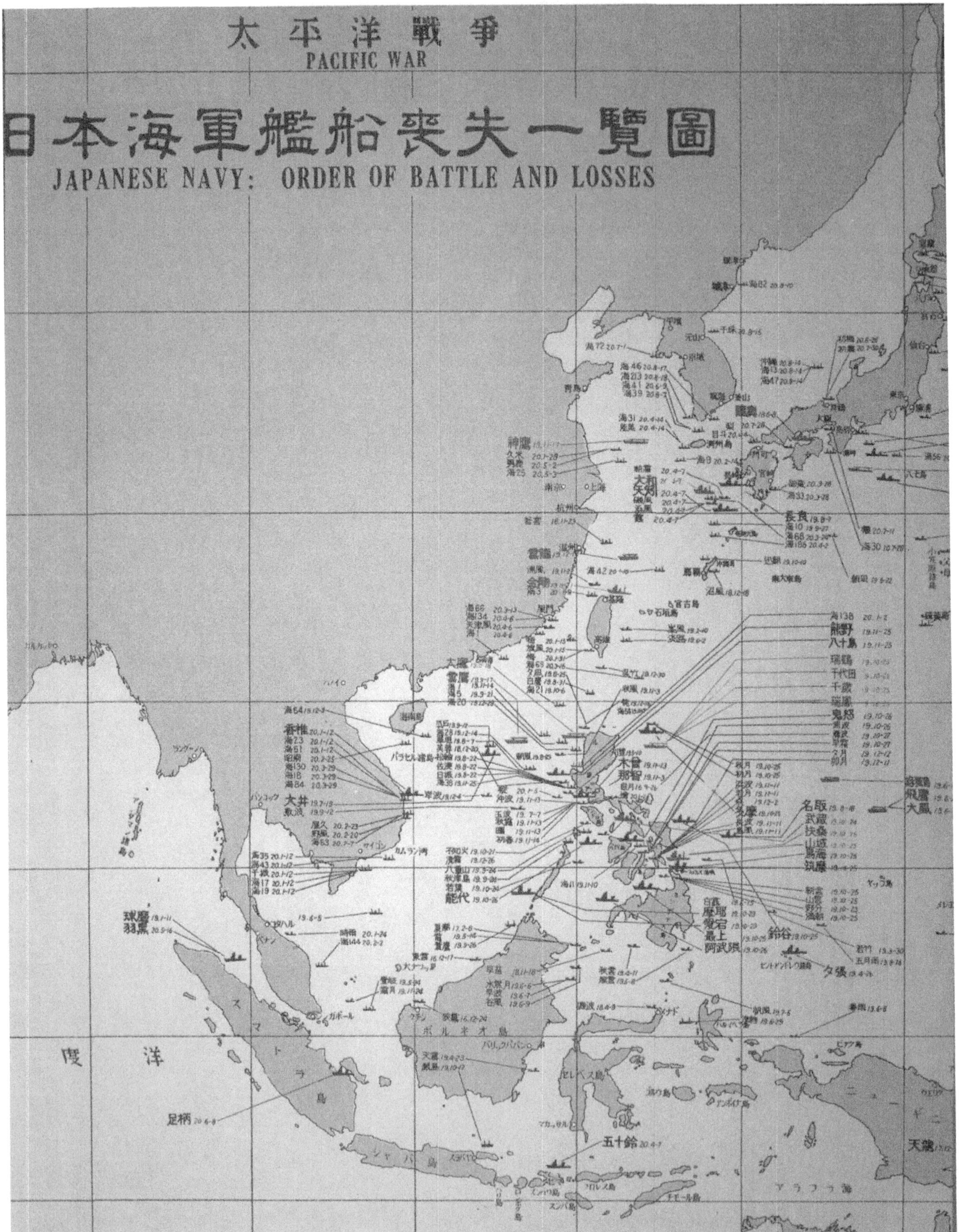

太平洋戰爭
PACIFIC WAR
日本海軍艦船喪失一覽圖
JAPANESE NAVY: ORDER OF BATTLE AND LOSSES

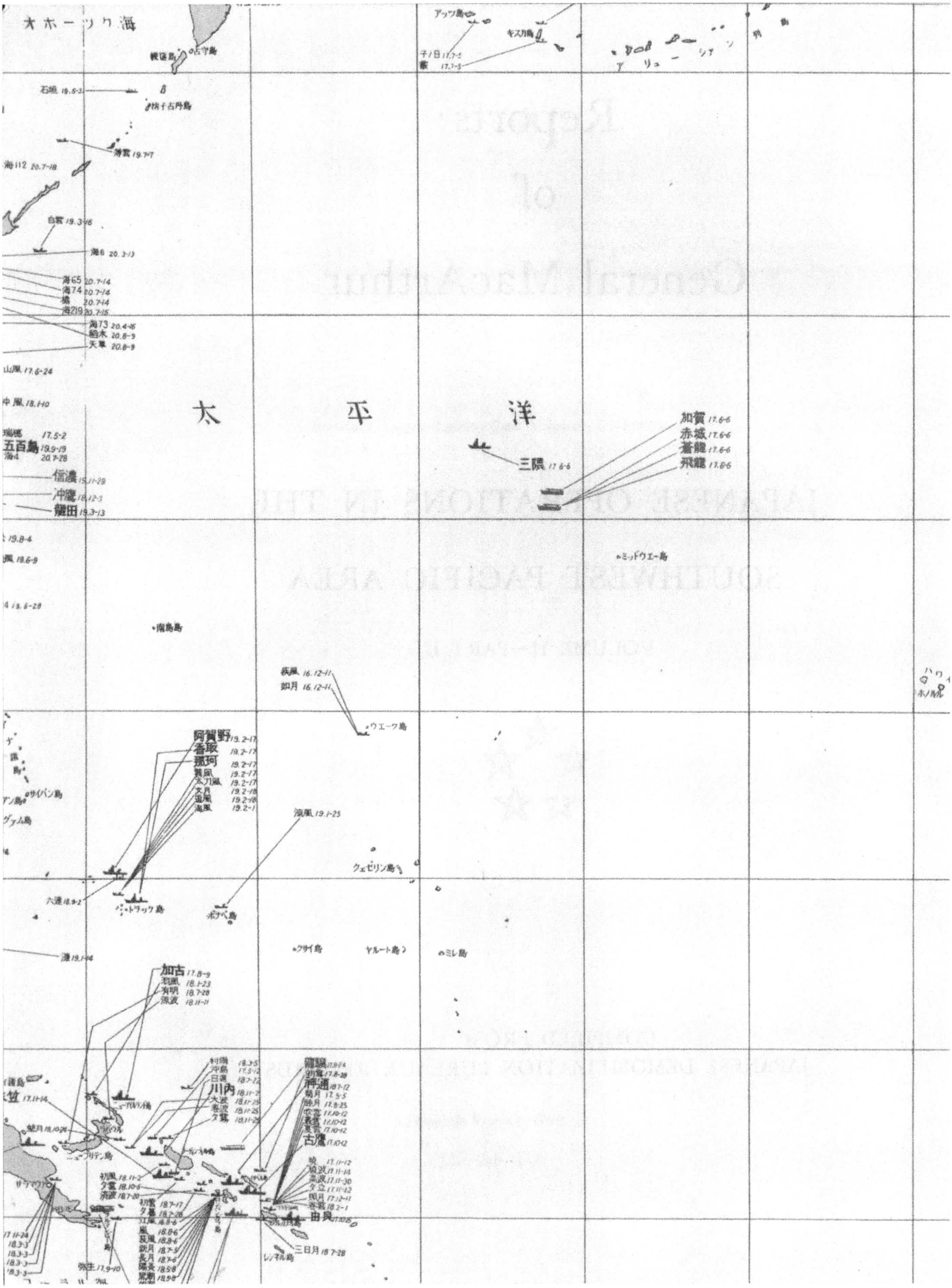

オホーツク海
太平洋
アッツ島
キスカ島
占守島
加賀 17.6-6
赤城 17.6-6
蒼龍 17.6-6
飛龍 17.6-6
三隈 17.6-6
ミッドウエー島
南鳥島
疾風 16.12-11
如月 16.12-11
ウエーク島
阿賀野 19.2-17
香取
那珂 19.2-17
サイパン島
グアム島
涼風 19.1-25
クェゼリン島
トラック島
ポナペ島
クサイ島
ヤルート島
ミレ島
信濃 19.11-29
五百島
加古 17.8-9
川内
神通
古鷹
由良
三日月 18.7-28
ニューブリテン島
ハワイ
ホノルル

FOREWORD

The *Reports of General MacArthur* include two volumes being published by the Department of the Army in four books reproduced exactly as they were printed by General MacArthur's Tokyo headquarters in 1950, except for the addition of this foreword and indexes. Since they were Government property, the general turned over to the Department in 1953 these volumes and related source materials. In Army and National Archives custody these materials have been available for research although they have not been easily accessible. While he lived, General MacArthur was unwilling to approve the reproduction and dissemination of the *Reports,* because he believed they needed further editing and correction of some inaccuracies. His passing permits publication but not the correction he deemed desirable. In publishing them, the Department of the Army must therefore disclaim any responsibility for their accuracy. But the Army also recognizes that these volumes have substantial and enduring value, and it believes the American people are entitled to have them made widely available through government publication.

The preliminary work for compiling the MacArthur volumes began in 1943 within the G-3 Section of his General Staff, and was carried forward after the war by members of the G-2 Section, headed by Maj. Gen. Charles A. Willoughby with Professor Gordon W. Prange, on leave from the University of Maryland, as his principal professional assistant. Volume II of the *Reports* represents the contributions of Japanese officers employed to tell their story of operations against MacArthur's forces. The very large number of individuals, American and Japanese, who participated in the compilation and editing of the *Reports* would make a complete listing of contributors relatively meaningless.

Volume I narrates the operations of forces under General MacArthur's command from the Japanese attack on Luzon in 1941 through the surrender in 1945. While service histories have covered much of the same ground in separate volumes, no single detailed narrative of General MacArthur's leadership as commander of the Southwest Pacific Area has yet appeared. Chapters dealing with the reconquest of Borneo, plans for the invasion of Japan, and the Japanese surrender make a distinctly new contribution. Volume I Supplement describes the military phase of the occupation through December 1948, reporting events not treated elsewhere in American publications. Volume II on Japanese operations brings together a mass of information on the enemy now only partially available in many separate works. Collectively, the *Reports* should be of wide interest and value to the American people generally, as well as to students of military affairs. They are an illuminating record of momentous events influenced in large measure by a distinguished American soldier.

Washington, D.C.
January 1966

HAROLD K. JOHNSON
General, United States Army
Chief of Staff

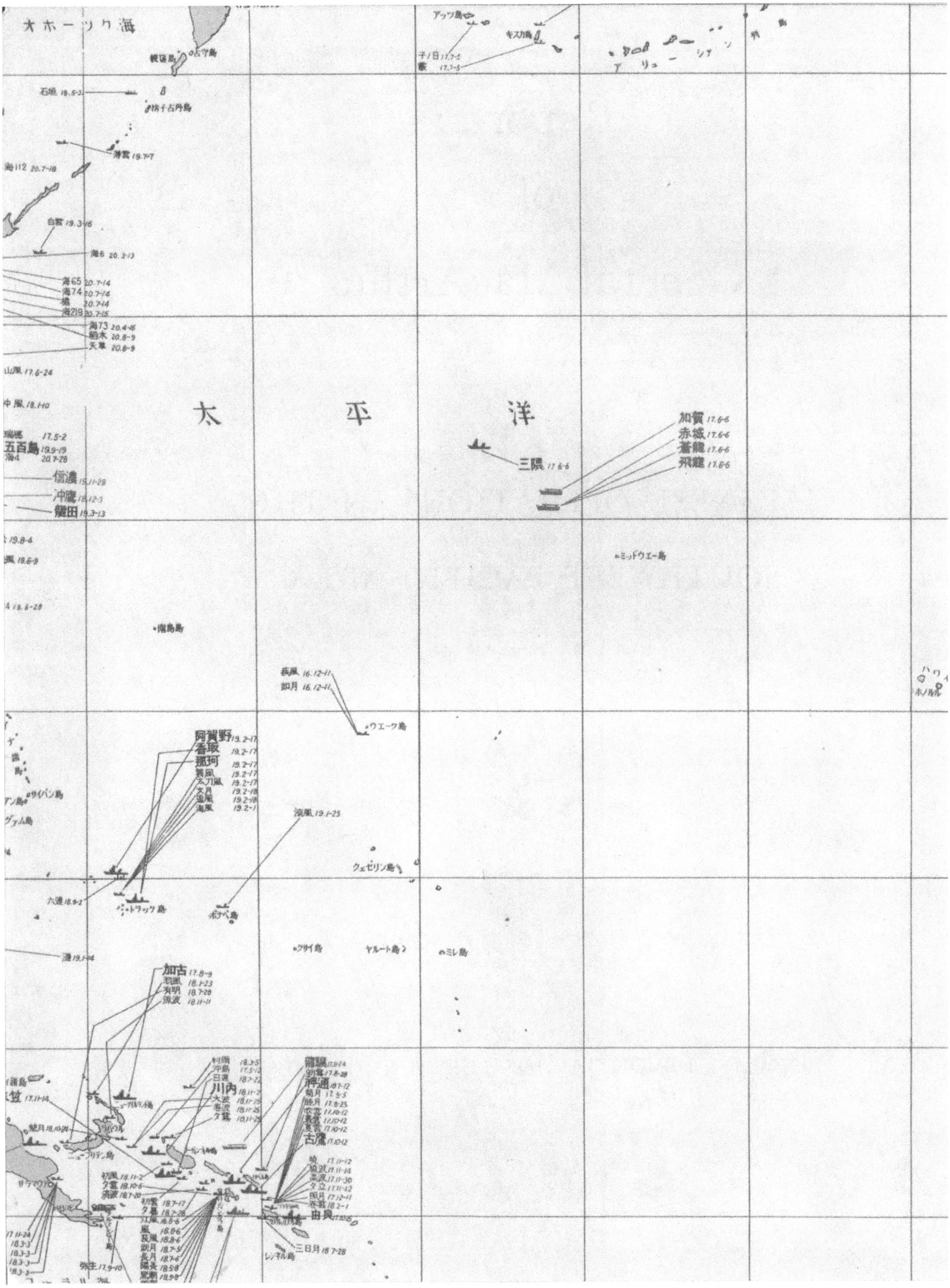

オホーツク海
占守島
幌筵島
アッツ島
キスカ島
アリューシャン列島
太平洋
三隈 17.6-6
加賀 17.6-6
赤城 17.6-6
蒼龍 17.6-6
飛龍 17.6-6
ミッドウエー島
南鳥島
如月 16.12-11
ウエーク島
阿賀野 19.2-17
香取 19.2-17
那珂 19.2-17
サイパン島
涼風 19.1-25
クェゼリン島
トラック島
ポナペ島
クサイ島
ヤルート島
ミレ島
加古 17.8-9
有明 18.7-28
川内
神通
古鷹
由良
ニューブリテン島
三日月 18.7-28
信濃 19.11-29
龍田 19.3-13
ホノルル

Reports
of
General MacArthur

JAPANESE OPERATIONS IN THE SOUTHWEST PACIFIC AREA

VOLUME II—PART II

COMPILED FROM
JAPANESE DEMOBILIZATION BUREAUX RECORDS

Library of Congress Catalog Card Number: 66–60007

Facsimile Reprint, 1994

CMH Pub 13–2

FOREWORD

The *Reports of General MacArthur* include two volumes being published by the Department of the Army in four books reproduced exactly as they were printed by General MacArthur's Tokyo headquarters in 1950, except for the addition of this foreword and indexes. Since they were Government property, the general turned over to the Department in 1953 these volumes and related source materials. In Army and National Archives custody these materials have been available for research although they have not been easily accessible. While he lived, General MacArthur was unwilling to approve the reproduction and dissemination of the *Reports,* because he believed they needed further editing and correction of some inaccuracies. His passing permits publication but not the correction he deemed desirable. In publishing them, the Department of the Army must therefore disclaim any responsibility for their accuracy. But the Army also recognizes that these volumes have substantial and enduring value, and it believes the American people are entitled to have them made widely available through government publication.

The preliminary work for compiling the MacArthur volumes began in 1943 within the G–3 Section of his General Staff, and was carried forward after the war by members of the G–2 Section, headed by Maj. Gen. Charles A. Willoughby with Professor Gordon W. Prange, on leave from the University of Maryland, as his principal professional assistant. Volume II of the *Reports* represents the contributions of Japanese officers employed to tell their story of operations against MacArthur's forces. The very large number of individuals, American and Japanese, who participated in the compilation and editing of the *Reports* would make a complete listing of contributors relatively meaningless.

Volume I narrates the operations of forces under General MacArthur's command from the Japanese attack on Luzon in 1941 through the surrender in 1945. While service histories have covered much of the same ground in separate volumes, no single detailed narrative of General MacArthur's leadership as commander of the Southwest Pacific Area has yet appeared. Chapters dealing with the reconquest of Borneo, plans for the invasion of Japan, and the Japanese surrender make a distinctly new contribution. Volume I Supplement describes the military phase of the occupation through December 1948, reporting events not treated elsewhere in American publications. Volume II on Japanese operations brings together a mass of information on the enemy now only partially available in many separate works. Collectively, the *Reports* should be of wide interest and value to the American people generally, as well as to students of military affairs. They are an illuminating record of momentous events influenced in large measure by a distinguished American soldier.

Washington, D.C.
January 1966

HAROLD K. JOHNSON
General, United States Army
Chief of Staff

FOREWORD TO THE 1994 EDITION

I determined for several reasons to republish General MacArthur's reports to commemorate the fiftieth anniversary of World War II. First, the *Reports of General MacArthur* still stand as a detailed account from MacArthur's perspective of his operations against the Japanese in the Southwest Pacific Area. Second, the *Reports* offer a unique Japanese version of their operations in the Southwest Pacific that remains one of the few English-language descriptions of Imperial Army campaigns during World War II. Third, excellent illustrations, many of them original artwork commissioned for the *Reports*, plus superb maps give these volumes an enduring value for military historians and the American public. Finally, while General MacArthur remains a towering figure in American historiography, the passage of fifty years has dimmed the contributions of the U.S. Army units that first checked the Japanese southward advance in Papua New Guinea, then spearheaded the counteroffensive along the north New Guinea coastline that enabled MacArthur to make good his promise to return to the Philippines. The veterans of these campaigns, both men and women, deserve to be remembered for their contributions to the Nation in its time of greatest peril. These are General MacArthur's *Reports*, but they are also his testament to the American soldiers who served under his command.

Washington, D.C.
31 January 1994

HAROLD W. NELSON
Brigadier General, USA
Chief of Military History

PREFACE

This volume parallels the record of Allied operations in the SWPA from the Defense of Luzon, 8 December 1941, to the Surrender Negotiations in Manila, 15 August 1945. It is the Japanese official record, contained in operational monographs furnished by the Japanese Demobilization Bureaux, the successors to the former War and Navy Ministries, developed by Officers of the Japanese Imperial Headquarters, Tokyo, and on the Staffs of major Japanese Commanders in the field. Like Volume I, the material is thus presented by eye witnesses to events, and is supported by official documentary evidence.

It is a record of bitter resistance and tenacious fighting by a first-class Army and Navy, led by Diplomats and Military Politicians through the holocaust of national destruction, an Army that was steeped in medieval cruelty, but fought with the most modern technical skill and savage valor, until superior skill and equal valor broke the spell of the Samurai and the legend of an invincible Empire.

DOUGLAS MACARTHUR

TABLE OF CONTENTS—PART II

ILLUSTRATIONS—PART II

Page

CHAPTER XIII

STRUGGLE FOR LEYTE

Activation of *Sho* Operation No. 1

In the growing light of dawn on 17 October, the Japanese naval radar station and lookout post on Suluan Island, lying athwart the entrance to Leyte Gulf, suddenly discovered the presence of an enemy task force standing inshore.[1] At 0719 an urgent warning was flashed to all Navy headquarters, reporting the approach of one enemy battleships, and six destroyers. This was followed an hour later by a second terse report: "Enemy elements have begun landing". Then the Suluan radio ceased transmitting.

Admiral Toyoda, Commander-in-Chief of the Combined Fleet, was still at Shinchiku, Formosa, when these reports were received.[2] Promptly interpreting the landing as preliminary to an invasion, he issued an order at 0809 alerting the entire Combined Fleet for *Sho* Operation No. 1.[3] This was followed at 0928 by an order directing Vice Adm. Kurita's First Striking Force to advance at once from Lingga anchorage to Brunei Bay, North Borneo, in preparation for a sortie against the enemy invasion fleet. A further order issued at noon directed the Task Force Main Body, under Vice Adm. Ozawa, to prepare for a coordinated sortie from home waters to facilitate the attack of the Kurita Force.[4]

Meanwhile, the Fourth Air Army had vainly been endeavoring to clarify the situation in Leyte Gulf. At 0800, the 2d Air Division was ordered to conduct search operations and prepare to attack any enemy shipping which might be discovered. However, bad weather hampered reconnaissance flights, and the situation remained obscure throughout the morning.[5] Lacking confirmation of the enemy landing

1 This chapter was originally prepared in Japanese by Maj. Toshiro Magari, Imperial Japanese Army. For duty assignments of this officer, cf. n. 1, Chapter XI. All source materials cited in this chapter are located in G-2 Historical Section Files, GHQ FEC.

2 Cf. Chapter XII, section on Formosa Air Battle. Admiral Toyoda left Formosa on 17 October to return to Tokyo, where he arrived on the 20th. The initial alert order to the Combined Fleet was sent out directly from Formosa, but subsequent orders were issued through Combined Fleet headquarters in Tokyo.

3 At 0946 on 17 October, Combined Fleet dispatched a message to all major subordinate commands transmitting the following estimate of enemy intentions: "Although not yet certain, the Allied landing on Suluan Island appears to be part of an invasion move against the south central Philippines. The enemy seems to be launching the invasion in accordance with previously prepared plans, despite the unsatisfactory progress of his operations in the Palau area and the substantial losses inflicted on his fleet (in the Formosa Air Battle)." *Hito Homen Kaigun Sakusen Sono Ni* 比島方面海軍作戰其二 (Philippine Area Naval Operations, Part II) 2d Demobilization Bureau, Oct 47, pp. 39–40.

4 The expected time of attack by the Kurita Force was provisionally set by this order at just before dawn on 22 October, assuming that the enemy main landing would be in the Leyte-Samar area. The Task Force Main Body was to sortie from the Inland Sea on 19 October and reach the area east of Luzon Strait by late evening of 21 October. *Dai Ichi Yugeki Butai Sento Shoho*: 第一遊擊部隊戰闘詳報 (First Striking Force Detailed Action Report) Second Fleet Headquarters, 1 Dec 44, p. 39.

5 A reconnaissance craft based on Leyte made a flight over the gulf area during 17 October, but no enemy ships were discovered inside the gulf. Visibility outside the gulf was so poor due to dense clouds that observation was impossible. The 16th Division sent a staff officer on this flight to make a first-hand study of the situation. *Hito Sakusen Kiroku Dai Sanki Dai Nikan Furoku Reite Sakusen Kiroku* 比島作戰記錄第三期第二卷附錄レイテ作戰記錄(Philippine Operations Record, Phase Three, Vol. II Supplement: Leyte Operations Record) 1st Demobilization Bureau, Oct 46, p. 34.

from Army sources, Southern Army headquarters at Manila regarded the Suluan reports with some suspicion in view of the earlier false landing scare at Davao.

In the early afternoon, however, corroborative evidence began to appear. At 1230 naval reconnaissance planes from Legaspi, favored by a brief break in the weather over the gulf, spotted the enemy task force off Suluan. The First Air Fleet immediately ordered an attack, but by the time the attack group of 13 planes reached the scene, the weather had closed in again, and the enemy ships were invisible. Enemy carrier aircraft had meanwhile launched attacks, beginning shortly after 0800, on air bases in the Manila area, at Legaspi and on Cebu, clearly indicating that hostile surface forces were operating off the Philippines.

These developments convinced Army headquarters by the afternoon of 17 October that the Suluan landing was indeed a fact, but there was still doubt as to whether or not it represented the beginning of a major invasion operation. The scattered enemy air attacks did not for the moment appear to point to imminent invasion,[6] and belief persisted that a major enemy venture was unlikely in view of the damage believed inflicted on the enemy's naval forces in the Formosa air battle. Both Southern Army and Fourteenth Area Army thus remained undecided.

By the night of 17 October, however, a more serious view began to prevail. Fourth Air Army headquarters concluded that the carrier air strikes, carried out in the face of extremely bad weather, indicated that some major enemy move was afoot. In addition, intercepts of enemy tactical radio messages gave evidence of unusual activity. Lt. Gen. Kyoji Tominaga, Fourth Air Army Commander, immediately relayed this information to Southern Army headquarters and recommended that steps be taken to activate *Sho* Operation No. 1.

Field Marshal Hisaichi Terauchi, Southern Army Commander-in-Chief, now agreed that the weight of evidence pointed to a probable invasion of the Philippines. He therefore sent an immediate recommendation to Imperial General Headquarters for the activation of *Sho* Operation No. 1 and ordered the Fourth Air Army to launch attacks on the enemy ships off Leyte without delay. Lt. Gen. Tominaga promptly assigned this mission to the 2d Air Division and also ordered the newly-organized 30th Fighter Group to assemble its strength with all possible haste at forward bases in the central Philippines.[8]

As day broke on 18 October, a violent storm with 65 mile-per-hour winds enveloped the entire Leyte Gulf area, preventing the 2d Air Division from executing its attack orders. During the morning, 16th Division lookout posts along the Leyte coast reproted sighting an undetermined number of enemy ships inside the gulf. The division radioed this information to Thirty-fifth Army headquarters on Cebu and also to Fourteenth Area Army headquarters at Manila. The report, however, noted the possibility that the enemy ships might have entered the gulf merely to take temporary refuge from the storm, a hypothesis which was supported by the fact that they had as yet

6 The total number of carrier planes which attacked the Manila area and Clark Field was calculated at 100, while about 50 attacked Legaspi. Philippine Area Naval Operations, Part II, op. cit., pp. 12–3.

7 (1) *Hito Koku Sakusen Kiroku Dai Niki* 比島航空作戰記錄第二期 (Philippine Air Operations Record, Phase Two) 1st Demobilization Bureau, Oct 46, p. 67. (2) Statement by Lt. Col, Katsuo Sato, Staff Officer (Operations) Fourth Air Army.

8 The 30th Fighter Group had been activated in Japan on 11 October. On 18 October, the group headquarters and the 16th Fighter Brigade were already in the Philippines, while the 12th Fighter Brigade and 200th Fighter Regiment had not yet arrived. (1) *Dairikumei Dai Senhyakugoju-go* 大陸命第千百五十號 (Imperial General Headquarters Army Order No, 1150) 11 Oct 44. (2) Philippine Air Operations Record, Phase Two, op. cit., pp. 69–71.

taken no hostile action.[9]

Despite the indecisive nature of the 16th Division report, the Thirty-fifth Army and Fourteenth Area Army commands shared Field Marshal Terauchi's estimate that the enemy was about to begin an invasion. It was still considered premature, however, to assume that the enemy's objective would necessarily be Leyte, though this appeared to be the greatest probability.

Meanwhile, Combined Fleet, in close consultation with the Navy High Command, had hastily drawn up a new battle plan for *Sho* Operation No. 1 to conform to changes in fleet disposition and carrier air strength resulting from the Formosa Air Battle. This plan, issued to the Fleet at 1110 on 18 October, was considered tentative, to apply only in the event of an immediate enemy invasion attempt at Leyte. Its essential points were as follows:[10]

1. The enemy's over-all landing plans have not yet been ascertained, but judging from the landing on Suluan Island, the sweeping of Surigao Strait, and the air attacks on Manila and Cebu, the possibility of a landing in the Tacloban area is considered great.

2. In the event such a landing attempt materializes, Combined Fleet surface and air forces will operate as follows:

a. The First Striking Force will advance through San Bernardino Strait and attack the enemy invasion forces.

b. In coordination with this attack, the Task Force Main Body will lure the enemy to the north and attack elements of the enemy forces at the most favorable opportunity.[11]

c. The Second Striking Force will come under the command of Southwest Area Fleet and will cooperate with the Army in executing counter-landings.[12]

d. The base air forces will concentrate in the Philippines for an all-out attack on the enemy carrier groups.

e. The Advance Submarine Force will attack damaged enemy vessels and amphibious convoys with all forces at its command.

f. The First Striking Force will execute its attack against the enemy invasion forces at the landing point on X-Day. The Task Force Main Body will advance to the area east of Luzon on X-1 or X-2.

g. X-Day will be announced by separate order. It is tentatively set at 24 October.

9 The 16th Division estimate suggested the possibility that the ships in the gulf might be a group of damaged vessels retiring from the Formosa battle, which the storm had forced to put in for shelter. *Hito Sakusen Kiroku Dai Sanki Dai Nikan: Reite Sakusen ni okeru Dai Juyon Homengun no Tosui* 比島作戰記錄第三期第二卷レイテ作戰に於ける第十四方面軍の統帥 (Philippine Operations Record, Phase Three, Vol. II: Fourteenth Area Army Command in the Leyte Operations) 1st Demobilization Bureau, Oct 46, p. 6.

10 (1) Philippine Area Naval Operations, Part II, op. cit., pp. 40–1. (2) First Striking Force Detailed Action Report, op. cit., pp. 40–1.

11 As a result of the transfer of the flying groups of the 3d and 4th Carrier Divisions to Second Air Fleet for the Formosa air battle, the Task Force Main Body had lost much of its planned striking power. The mission assigned to it was therefore less offensive than that assigned in the original *Sho-Go* battle plan. The original plan had called for a diversionary attack by the Ozawa Force against the enemy's main carrier task forces. (Cf. Chapter XI, p. 305) Under the more general terms of the 18 October plan, however, Vice Adm. Ozawa was left discretion to choose an attack target more commensurate with the strength of his force. (Statement by Capt. Toshikazu Ohmae, Staff Officer (Operations), First Mobile Fleet.

12 The Second Striking Force, for this newly-assigned mission, was to be composed only of the Fifth Fleet (two heavy cruisers, one light cruiser, seven destroyers) plus the 16th Cruiser Division (one heavy cruiser, one light cruiser, one destroyer) detached from the First Striking Force. Under this plan, the 16th Cruiser Division was ordered to proceed to Manila, while the Second Striking Force main body, then at Amami Oshima in the Ryukyus, was ordered to go first to Mako, refuel, and then proceed to the Philippines. While en route from Mako on 23 October, the main body received new orders from Southwest Area Fleet changing its mission to support of the First Striking Force in the attack on Leyte Gulf. Philippine Area Naval Operations, Part II, op. cit., pp. 63–4, 83.

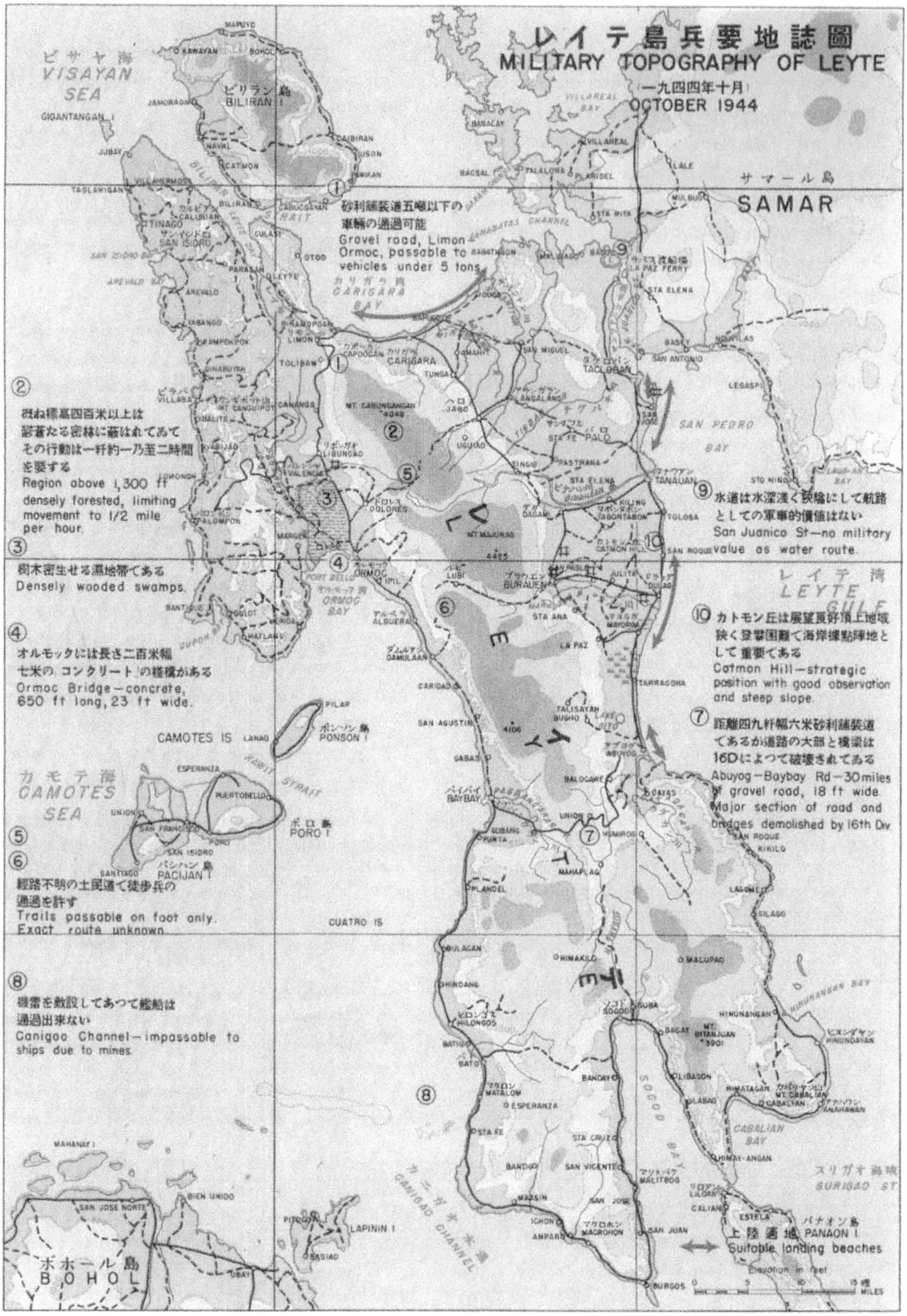

PLATE NO. 87

Military Topography of Leyte

In Leyte Gulf, events now began to move more rapidly. During the afternoon of 18 October, enemy naval units unleashed a heavy gunfire bombardment of the Dulag, Catmon Hill, and Tolosa sectors, while minesweepers boldly penetrated into the northern part of the gulf and began sweeping operations. Enemy carrier planes meanwhile flew a total of 400 sorties throughout the day against airfields on Luzon and in the Visayas. It was becoming clear that amphibious operations would be launched momentarily.

In Tokyo, the Army and Navy Sections of Imperial General Headquarters had been closely watching developments. Following receipt of Field Marshal Terauchi's message recommending activation of *Sho* Operation No. 1, the Army Section decided that the situation had become sufficiently clear to warrant such a step. Joint consultations had already been held with the Navy Section to secure coordinated action.

As a result of these consultations, the Navy Section, at 1701 on 18 October, issued a directive to Combined Fleet ordering the execution of *Sho* Operation No. 1.[13] This was followed immediately by an Army Section order activating the same operation for all Army forces concerned. The latter order stated:[14]

> *1. The Philippines are designated as the area of decisive battle for the Japanese armed forces.*
>
> *2. The Commander-in-Chief, Southern Army, in cooperation with the Navy, will conduct decisive battle operations against the main strength of the United States forces attacking the Philippines.*
>
> *3. The Commander-in-Chief, China Expeditionary Army, and the Commander, Tenth Area (Formosa) the Army, will take all necessary action to facilitate above operations.*

Simultaneously with its decision to activate the *Sho-Go* Operation, the Army Section re-examined the existing plan to withhold decisive ground action until the invasion of Luzon. The High Command was now convinced, on the basis of the results believed obtained in the Formosa Air Battle, that conditions had become highly favorable to the conduct of decisive ground operations in the central Philippines.

More than one-half of the American fleet's known carrier strength was still believed to have been destroyed or put out of action in the Formosa Battle. It consequently appeared that the enemy, counting upon the heavy damage he had inflicted on the Japanese air forces, was undertaking the invasion of the Leyte area with seriously reduced carrier strength. Land-based air support would have to come from the distant enemy bases on Palau and Morotai, neither of which was within effective fighter range.

Although Japanese air losses had admittedly been severe, the High Command estimated that these could be replaced with relatively greater speed than the enemy could replenish his carrier forces. The Japanese air forces therefore had good prospects of gaining at least temporary local superiority over the Philippines. This would make possible the safe movement of troop reinforcements and supplies to the Leyte invasion area, thus eliminating the major reason for the original decision to attempt only a delaying action in the central or southern Philippines.

Imperial General Headquarters therefore made the crucial decision to set aside the previous *Sho-Go* plan and throw all available ground strength into Leyte. The disadvantages inherent in a last-minute change of plan were believed greatly outweighed by the better

13 *Daikaishi Dai Yonhyakushichijuroku-go* 大海指第四百七十六號 (Imperial General Headquarters Navy Directive No. 476) 18 Oct 44.

14 *Daihonyei Rikugun Tosui Kiroku* 大本營陸軍統帥記錄 (Imperial General Headquarters Army High Command Record) 1st Demobilization Bureau, Nov 46, p. 269.

prospects of dealing a crushing defeat to the enemy through simultaneous and concerted commitment of air, sea and ground forces.[15]

To transmit this decision to the theater command, the Army Section of Imperial General Headquarters decided to dispatch a staff liaison group to Manila. This group, headed by Col. Ichiji Sugita of the Operations Section, left Tokyo by air on the morning of 19 October, stopping at Taihoku, Formosa, to effect necessary liaison with Tenth Area Army. Meanwhile, Field Marshal Terauchi and General Yamashita at Manila continued to shape their plans on the basis of the standing instructions to fight only a strategic delaying action in the central Philippines.[16]

Southern Army, early on 19 October, had issued an order to Fourteenth Area Army and Fourth Air Army, implementing the Imperial General Headquarters order of the 18th for activation of *Sho* Operation No. 1. This stated in substance:[17]

> 1. *Activation of Sho Operation No. 1 has been ordered.*
>
> 2. *The Southern Army, assembling all its fighting power, will seek decisive battle with the main strength of the enemy forces landing in the Philippines.*
>
> 3. *All forces must fulfill their assigned missions by every means at their disposal and must fight this battle to a successful conclusion.*

Pursuant to this order, Fourteenth Area Army on the same day directed Lt. Gen. Suzuki, Thirty-fifth Army Commander, to commit maximum strength to Leyte for the execution of the Army's assigned mission of strategic delay in that sector.[18] Lt. Gen. Suzuki simultaneously ordered activation of the *Suzu* No. 2 Operation (Cf. Chapter XI) and directed the 16th Division to repel enemy landing forces and secure the Leyte airfields.

Enemy activity in Leyte Gulf meanwhile was accelerating rapidly. Beaches along the Leyte east coast were aggressively reconnoitered by small landing parties.[19] Shore installations were shelled and bombed. Various types of shipping continued to enter the gulf in ever increasing numbers. The main enemy landing effort was obviously imminent.

Despite the optimism of the Army High Command with regard to the prospects of gaining eventual air superiority in the battle area, it was already evident that the air phase of the *Sho-Go* Operation plans was not working out as intended. Those plans had envisaged mass air attacks against the invading enemy naval forces and troop convoys beginning prior to their arrival at the landing point. Thus, the enemy would suffer severe losses before his troops could hit the invasion beaches.[20]

The enemy, however, had forestalled these

---

15 Data regarding the reasoning which resulted in the Imperial General Headquarters decision to shift the decisive ground battle area to Leyte furnished by Col. Takushiro Hattori, Chief, Operations Section, Imperial General Headquarters, Army Section.

16 Statements by Lt. Gen. Jo Iimura, Chief of Staff, Southern Army, and Maj. Gen. Toshio Nishimura, Deputy-Chief of Staff, Fourteenth Area Army.

17 *Reite-To ni okeru Sakusen Keika Gaiyo narabini Kyokun* レイテ島に於ける作戰經過概要竝に教訓 (Outline of Operations and Lessons of the Leyte Campaign) Imperial General Headquarters, Army Section, Feb. 45, p. 23.

18 Philippine Operations Record, Phase III, op. cit. Vol. II, pp. 6–7.

19 The 16th Division reported that its troops had fired upon and repulsed enemy forces attempting to land on 19 October. Philippine Operations Record, Phase III, op. cit. Vol. II Suppl., pp. 39–40. (American Editor's Note: The forces engaged by the 16th Division on 19 October were not, as the Japanese assumed, the initial waves of the main assault forces, but merely amphibious reconnaissance patrols and underwater demolition teams dispatched to prepare the way for the main landings.)

20 Cf. Chapter XI, pp. 304, 308.

plans by striking at Leyte before the planned concentration of Japanese air strength in the Philippines had been completed, and before the Navy's base air forces had time to recover from losses sustained in the Formosa Air Battle and earlier enemy carrier strikes on the Philippines. The First Air Fleet, already in the Philippines, had been reduced to an operational strength of less than 50 aircraft.[21] The Second Air Fleet, which had lost half its strength in the Formosa Battle, had not yet begun its redeployment from Formosa to the Philippines.

Although the Fourth Air Army had sustained relatively lighter losses, its strength was widely dispersed. Before it could operate effectively in the Leyte area, it had to concentrate at forward bases in the central Philippines,[22] an operation rendered both difficult and dangerous by enemy action, bad weather, and the virtually useless condition of many of the forward fields due to continuous rains. Under such unfavorable conditions, the concentration required a minimum of several days, and in the meanwhile the enemy was able to operate in Leyte Gulf against extremely light air opposition.

On 19 October, when weather conditions finally permitted an air attack against the enemy invasion fleet, no more than five naval and three Army aircraft could be mustered against the steadily increasing concentration of enemy shipping in Leyte Gulf.[23]

Defensive Dispositions

Air reconnaissance during 19 October gave the first clear indication of the size of the invasion force which the enemy was throwing against Leyte. At 0810 only 18 transports were reported in the gulf, together with some 34 naval ships, including six carriers. By 1800 of the same day, the number of transports reported in or near the gulf had increased to 100, and the naval units to 46, including six carriers and ten battleships. At the same time, two covering naval task groups, one with four carriers and the other with two, were reported maneuvering off the east coast of Luzon.[24] In the Admiralties and at Hollandia, further large concentrations of invasion shipping were reported in readiness for departure.[25]

It was evident from these reports that the invading forces would be formidable. The only major combat unit available to counter the first wave of the assault pending the concentration of forces from other sectors of the Philippines was Lt. Gen. Shiro Makino's 16th Division, veterans of the initial Japanese conquest of the Philippines and strongest of the Thirty-fifth Army's field divisions.

Estimating that enemy landings would most probably occur in the Dulag-Tarragona-Abuyog sector, Lt. Gen. Makino had disposed the bulk

21 Philippine Area Naval Operations, Part II, op. cit., p. 10.

22 The Fourth Air Army had only a handful of planes already based in the central and southern Philippines, and some of its flying units were scattered into North Borneo, Malaya and the Celebes. The main strength of the 2d Air Division and part of the 30th Fighter Group, aggregating about 70 planes, were on Luzon. Remaining strength of the 30th Fighter Group had not yet begun deploying from the Homeland. Philippine Air Operations Record, Phase Two, op. cit., pp. 68–9, Chart No. 9.

23 (1) Philippine Area Naval Operations, Part II, op. cit., p. 13. (2) Outline of Operations and Lessons of the Leyte Campaign, op. cit., p. 28, Appended Chart No. 3.

24 *Senkyo Shubo* 戦況手簿 (Daily Record of the War Situation) Operations Section, Imperial General Headquarters, Army Section, 11 Oct 44—Jul 45, pp. 1–2.

25 The concentration in the Admiralties was reported to consist of a naval force with six carriers and a transport group of 30 ships. The group at Hollandia was reported to number about 150 ships, of which about 80 were believed to be transports. Philippine Area Naval Operations, Part II, op. cit., p. 13.

of his troops in prepared coastal positions between Abuyog on the south and Palo on the north, with the heaviest concentration of strength in the vicinity of Dulag. Construction of a secondary defense line through Tabontabon and Santa Ana and of the projected main line along an axis running through Dagami and Burauen, was still incomplete due to terrain difficulties and guerrilla interference.[26] Division headquarters and rear installations remained in the Tacloban sector, to the north of the main defensive dispositions.[27] (Plate No. 88)

Except for minor elements left on Luzon and Samar, Lt. Gen. Makino had at his immediate disposal the full combat strength of the 16th Division, in addition to 4th Air Division ground units in the area, and part of the service forces of Thirty-fifth Army. Elements of the 36th Naval Garrison Unit at Ormoc and Tacloban were to come under his command immediately upon an enemy invasion.[28]

Tactically the main combat forces were organized in two groups, the first responsible for defense of the right flank sector south of and including Dulag and Burauen, and the second charged with operations in the sector extending from Catmon Hill (inclusive) to Palo (exclusive). The main strength of the division reserve (33d Infantry, less 2d Battalion) was stationed in the Palo sector, with elements at Dagami (1st Battalion, 20th Infantry), Tagungtong (2d Battalion, 33d Infantry), and north of Burauen (7th Independent Tank Company). Artillery

---

26 (1) Statement by Maj. Gen. Yoshiharu Tomochika, Chief of Staff, Thirty-fifth Army. (2) 20th Infantry Regiment Operations Order No. 219, 20 Sep 44. XXIV Corps ADVATIS Translation No. 1, 3 Nov 44.

27 Both Fourteenth Area Army and Thirty-fifth Army felt that the retention of 16th Division headquarters at Tacloban, on the flank of the main coastal defense positions, was unwise. In August, they instructed Lt. Gen. Makino to relocate his headquarters and command facilities in the rear of the main defenses, near Dagami or Burauen, but since this required the prior establishment of communication, transport, supply and other necessary facilities, as well as the pacification of guerrilla forces active in the mountainous region west of Burauen, the transfer had not yet been effected when the enemy invasion of Leyte began. (1) Philippine Operations Record, Phase Three, op. cit. Vol. II, p. 4. (2) Statement by Maj. Gen. Tomochika, previously cited.

28 Order of battle of the Japanese forces on Leyte at the time of the Allied landing (company level and above) was as follows:

Army Troops:

16th Division
- 9th Infantry (less three companies on Samar)
- 20th Infantry
- 33d Infantry (less one company on Samar)
- 22d Artillery Regt. (less 3d Bn. on Luzon)
- 16th Engineer Regt. (less 2d Co.)
- 2d Co., 16th Transport Regt.
- 7th Independent Tank Co.
- 16th Division Special Troops
- Thirty-fifth Army depot units
- Elms, 63d Indep. Motor Transport Bn.
- 316th Indep. Motor Transport Co.
- 317th Indep. Motor Transport Co.

34th Air Sector Command
- 98th Airfield Bn.
- 114th Airfield Bn.
- 54th Airfield Co.
- 2d Airfield Construction Unit
- 11th Airfield Construction Unit
- Misc. service elms.

Naval Land Forces:
- Elms, 36th Naval Garrison Unit
- 311th Naval Construction Unit
- Misc. service elms.

(1) Philippine Operations Record, Phase Three, op. cit. Vol. II Suppl., Attached Chart II. (2) Philippine Area Naval Operations, Part II, op. cit., p. 78. (3) *Koku Butai Haichi Yozu* 航空部隊配置要圖 (Air Unit Disposition Chart) Home Depot Division, 21 Jul 47. Attached Chart VI. (4) Situation Map, 16th Division. ATIS Bulletin No. 1624, 17 Dec 44. (5) Situation Report, 16th Division, 8 Oct 44. ATIS Bulletin No. 1678, 4 Jan 45. (6) Chart of Air Force Ground Units in the Tacloban Sector, 20 Aug 44. ADVATIS Bulletin No. 75, 15 Dec 44.

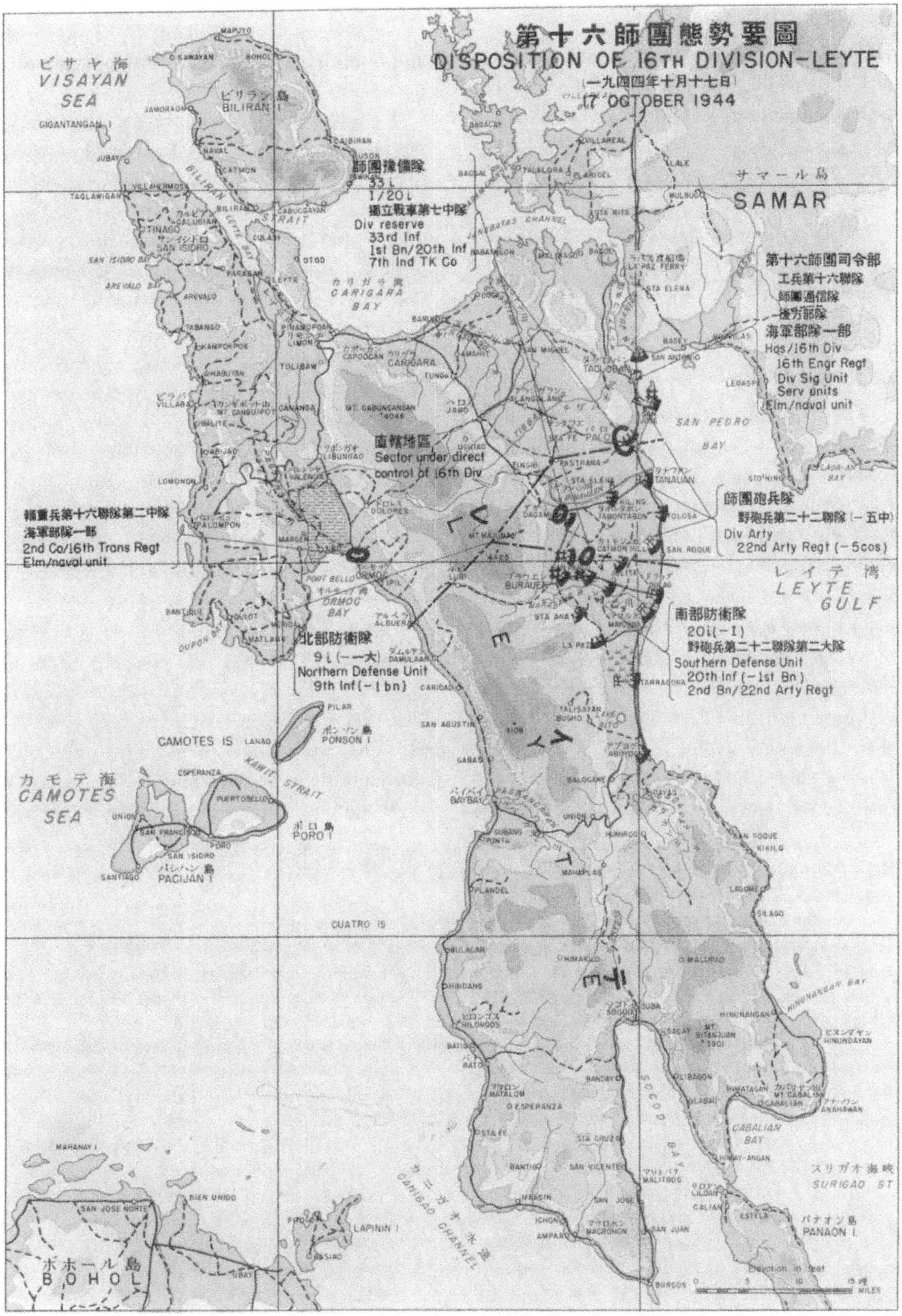

PLATE NO. 88
Disposition of 16th Division—Leyte, 17 October 1944

retained under division control was emplaced at Catmon Hill, San Jose, and Tolosa. The tactical allocation of units was as follows:[29]

Southern Defense Unit:	*20th Inf. (less 1st Bn.)*
	2d Bn., 22d Arty. Regt. (less 6th Co.)
Northern Defense Unit:	*9th Inf. (less 2d, 5th, and 7th Cos.)*
Division Reserve:	*33d Inf.*
	1st Bn., 20th Inf.
	7th Ind. Tank Co.
Division Artillery:	*22d Arty. Regt. (less 5 cos.)*

These dispositions remained in effect until 18 October, when the enemy's pre-landing operations in Leyte Gulf afforded Lt. Gen. Makino his first definite clue as to where the invading forces would go ashore. These operations appeared to confirm the previous estimate that the major landing effort would be made in the Dulag area. However, the presence of a large number of enemy ships, as well as minesweeping activity in the northern section of Leyte Gulf also gave ominous indications that the enemy might execute a secondary landing closer to Tacloban, outflanking the main concentration of 16th Division strength and immediately threatening the division headquarters itself.

To meet the completely unexpected threat on the north,[30] Lt. Gen. Makino swiftly ordered the 33d Infantry to move its main strength, supported by two batteries of division artillery, into positions in the Palo-San Jose coastal sector.[31] Simultaneously, the 2d Battalion of the regiment, which was standing by at Tagungtong in the southern sector, was released from the reserve and assigned to reinforce the Southern Defense Unit. This battalion moved up to Julita to give closer support to the southern flank. The 1st Battalion, 20th Infantry, stationed in reserve at Dagami, was ordered to prepare for rapid motorized movement to any sector where further reinforcements might be required.[32] Execution of these orders got under way late on 18 October.

Despite these last-minute moves, Lt. Gen. Makino's forces were ill prepared to meet the overwhelming assault which impended. Defensive works, where they had been completed, were weak and imperfectly disposed.[33] Many positions were demolished by the heavy naval gunfire preparation, which continued

29 (1) Outline of Operations and Lessons of the Leyte Campaign, op. cit., Attached Chart I. (2) 22d Artillery Regt. Operations Order, 16 Jul 44. XXIV Corps ADVATIS Translation No. 1, 3 Nov 44. (3) Situation Report, 16th Division, 8 Oct 44. ATIS Bulletin 1678, 4 Jan 45.

30 In all defensive planning for Leyte prior to the actual enemy invasion, neither Thirty-fifth Army nor the 16th Division command had given serious consideration to the possibility of an enemy landing in the Tacloban area. The principal reason for this was the belief that the enemy would not risk sending large numbers of assault craft into the narrow confines of San Pedro Bay, where they would be more vulnerable to attack. Landings in the Dulag area, farther south, would not involve this risk. (Statement by Maj. Gen. Tomochika, previously cited.)

31 In compliance with this order, the 33d Infantry dispatched its 1st Battalion to the San Jose sector, while the 3d Battalion was ordered to take up positions in the vicinity of Palo. The reinforcing batteries of division artillery moved into position at San Jose. Regimental headquarters was set up on the heights north of Palo. (1) 33d Infantry Regiment Operations Order No. 433, 18 Oct 44. X Corps ADVATIS Translation No. 5, 28 Oct 44. (2) *Reite To ni okeru Dai Juroku Shidan Sakusen Keika no Gaiyo* レイテ島に於ける第十六師團作戰經過の概要 (Summary of Operations of the 16th Division on Leyte) Summarized by Survivors of the Leyte Campaign, 23 July 49, p. 12.

32 (1) 33d Infantry Regiment Operations Order No. 433, 18 Oct 44. X Corps ADVATIS Translation No. 5, 28 Oct 44. (2) 16th Division Staff Dispatch No. 231, 18 Oct 44. ATIS Bulletin No. 1616, 8 Dec 44.

33 The field positions were old-type long connecting trenches prepared in three echelons. These were difficult to defend and easily discovered from the air since camouflage was lacking. The only effective positions were the cave emplacements for artillery, which had been constructed on Catmon Hill. (Interrogation of Col. Shujiro Kobayashi, Chief, Operations Section, Fourteenth Area Army.)

throughout 19 October. The bulk of the 22d Artillery Regiment's guns, with the exception of the battery emplaced on Catmon Hill, were also put out of action by the rain of shells. Effective resistance was further handicapped by a general breakdown in division communications as a combined result of the enemy bombardment and damage wrought by the typhoon of 17–18 October.[34]

Launching of the Invasion

On the morning of 20 October, the enemy finally began his main landing operations. While naval units subjected the shore defenses to a final preparatory bombardment, the enemy's vast transport convoy maneuvered into take-off positions and put more than 200 assault craft into the water. The first waves of the invasion force, now estimated at about three divisions,[35] began hitting the beaches at 1130 Japan time (1030 Philippine time).[36]

The assault was delivered in two main thrusts, one aimed at the Dulag sector, and the other at San Jose, just five miles southeast of Tacloban. (Plate No. 89) As the enemy troops moved ashore, they were subjected to scattered fire from artillery emplacements not destroyed by the naval gunfire and air preparation.[37] In the Dulag sector, Col. Keijiro Hokoda's 20th Infantry Regiment put up what initial resistance it could from its shore positions but was unable to prevent the establishment of a small but firm enemy beachhead.

The simultaneous landing near Tacloban was far more critical. The two battalions of the 33d Infantry, which Lt. Gen. Makino had ordered to the Palo-San Jose coastal sector on 18 October, had barely enough time to move into position before the enemy assault began, and the absence of adequate prepared defenses in the area made effective resistance impossible. These elements and the small number of naval troops already in the San Jose sector were forced to fall rapidly back from the beaches. The 1st and 3d Battalions, 33d Infantry, took up strong positions on the heights north and northwest of Palo. The naval troops withdrew northward to Tacloban. By nightfall of 20 October the enemy had overrun Tacloban airfield and penetrated into the outskirts of the town itself.[38]

Already on the day prior to the enemy landings, Lt. Gen. Makino had decided to establish his command post at Dagami in order to be in a more central location for directing subsequent operations. The division commander and his staff moved out of Tacloban during the night of 19–20 October. Most of the division headquarters and special troops were ordered to

34 Signal communications were impossible. Roads were washed out and impassable. Bridges were down. . . . For about a week from the time the storm first hit the island, elements of the division were scattered and out of contact with one another. While trying to assemble its forces for operations, the division was hit by enemy. . . .attack. Ibid.

35 Three divisions was the strength initially estimated by Japanese intelligence. On 26 October, the presence of a fourth division, the 24th, was ascertained. Philippine Operations Record, Phase Three, op. cit. Vol II Suppl., Appended Map IV.

36 *Gunreibu Socho no Sojosho* 軍令部總長の奏上書 (Reports to the Throne by Chief of Navy General Staff) 21 Oct 44.

37 A majority of the gun emplacement on the beach were destroyed, but defiladed positions and cave defenses remained intact, with only light casualties. Maj. Gen. Yoshiharu Tomochika, *Gun Sambocho no Shuki: Hito Haisen no Shinso* 軍參謀長の手記比島敗戰の眞相 (The Truth of the Philippine Defeat: Notes of an Army Chief of Staff) Tokyo, 46, p. 19. (Maj. Gen. Tomochika was Chief of Staff of Thirty-fifth Army and was present on Leyte through the entire campaign after 30 October.)

38 These enemy forces also seized a base which was secretly under construction on Cancabato Bay for the use of naval special-attack forces.

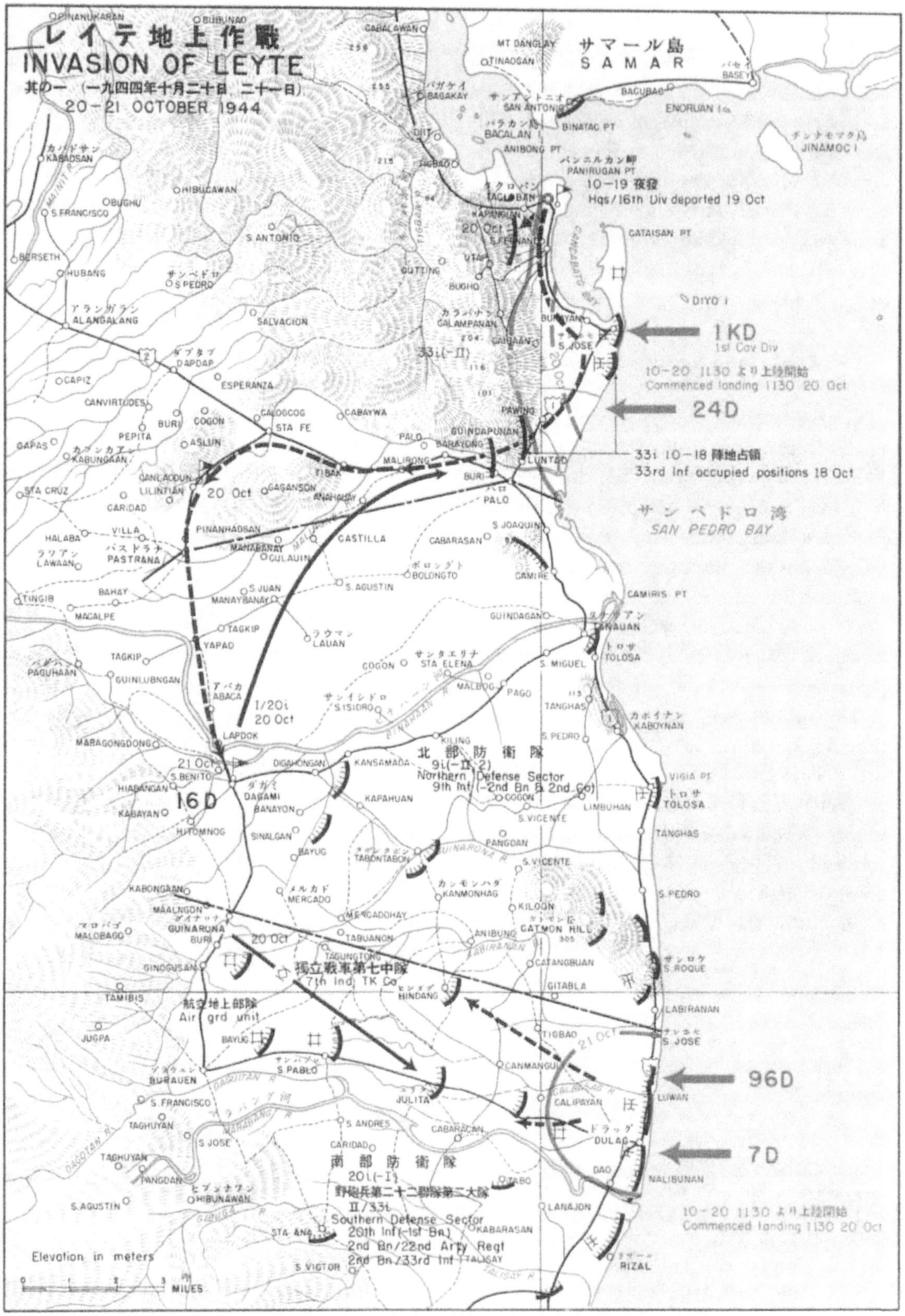

PLATE NO. 89
Invasion of Leyte, 20–21 October 1944

follow as soon as possible. The command group was proceeding along the highway west of Palo when reports of the first enemy landings reached them about midday on the 20th.

Immediately perceiving the serious danger to his northern flank as a result of the San Jose landing, Lt. Gen. Makino ordered the division mobile reserve (1st Battalion, 20th Infantry) at Dagami to proceed at once to the Palo-San Jose sector to back up the 33d Infantry. Meanwhile, to bolster the equally hard-pressed forces in the south, the 7th Independent Tank Company was released from division control and ordered to move up from Buri to the Dulag front. Lt. Gen. Makino and his staff continued on to Dagami, arriving early on 21 October.[39]

The evacuation of Tacloban by the division rear echelon, which began early on 20 October, necessitated the abandonment of permanent wireless installations and resulted in complete severance for forty-eight hours of all contact between the 16th Division and higher headquarters at Cebu and Manila.[40] During this critical period, Fourteenth Area Army and Thirty-fifth Army were completely without knowledge of developments on Leyte.

Swiftly exploiting their success on the northern flank, the enemy pushed ahead to occupy Tacloban completely during the night of 20–21 October. The naval garrison elements and 16th Division rear echelon personnel retreated into the hills west of the town. These small forces were all that remained on the extreme north flank, and they now found themselves cut off from the main Japanese forces to the south.

On the southern flank of the San Jose beachhead, the enemy also pressed forward vigorously against the 33d Infantry positions on the heights northwest of Palo. Although enemy elements infiltrated past this strongpoint, the regiment held its positions throughout the 21st and even launched minor counterattacks against the enemy left flank.

In the Dulag area to the south, the enemy forces, after completely occupying Dulag itself, drove on to take the airfield west of the town. They then continued to exert heavy pressure in the direction of Burauen, opposed by the Southern Defense Unit and airfield garrisons of the 34th Air Sector Command. Artillery emplaced on Catmon Hill, which had not yet come under ground assault, shelled the enemy beachhead area, while 9th Infantry elements engaged advancing enemy forces along the northern perimeter of the beachhead above Dulag. Enemy naval bombardment slackened on 21 October, but carrier planes immediately took over the task of supporting the ground forces.

Despite a gradual stiffening of 16th Division resistance, the enemy, by noon of 21 October, had succeeded in extending his northern beachhead from Tacloban south to the mouth of the Palo River, while his southern foothold had been expanded to a roughly semi-circular beachhead about two miles in depth, centered on Dulag.[41]

Air opposition to the enemy invasion remained on a minor scale as Fourth Air Army and the naval air forces continued the gradual

39 (1) Statements by Lt. Col. Toshii Watanabe, Staff Officer (Intelligence later Operations), Thirty-fifth Army; and Maj. Eizo Hori, Staff Officer (Intelligence), Fourteenth Area Army. (2) Summary of Operations of the 16th Division on Leyte, op. cit., pp. 11–2. (3) 16th Division Operation Order No. 837, 22 Oct 44. XXIV Corps ADVATIS Translation No. 13, 20 Nov 44.

40 This break in communications was complete from 2400 on 20 October until 2200 on 22 October, by which time field stations were put into operation at the division command post at Dagami. Philippine Operations Record, Phase Three, op. cit. Vol. II Suppl., p. 45.

41 Ibid., p. 41.

concentration of their strength in the Philippines in preparation for the all-out air offensive tentatively set to begin on 24 October.[42] On 20 October, 37 Army and Navy planes attacked invasion shipping off Leyte, claiming to have sunk or damaged several enemy craft. The number of attacking planes dropped to 21 on 21 October, and on 22 and 23 October no attacks were carried out.[43]

Both the 2d Air Division, containing the major combat flying elements of the Fourth Air Army, and the First Air Fleet had undergone command changes just as the enemy invasion of Leyte got under way. Lt. Gen. Isamu Kinoshita took over command of the 2d Air Division from Lt. Gen. Masao Yamase after the latter was wounded in the enemy air strike on Clark Field on 19 October. On the 20th, Vice Adm. Takijiro Onishi relieved Vice Adm. Teraoka as First Air Fleet Commander.[44]

The assembly of Fourth Air Army strength for the big offensive was proceeding less rapidly than anticipated. Not only was it proving extremely difficult to move already available fighter strength to forward bases, but it appeared unlikely that the transfer from the Homeland and China of the 30th Fighter Group's most powerful units, the 12th Fighter Brigade and 200th Fighter Regiment, could be completed before 22 or 23 October. Second Air Fleet, however, had completed its transfer to Clark and Nichols Fields by the 22d.

Action by Army Headquarters

Urgent 16th Division dispatches reporting the launching of the enemy invasion reached Thirty-fifth Army headquarters at Cebu and Fourteenth Area Army and Southern Army headquarters at Manila around noon on 20 October. These headquarters tensely waited for further information, but before any detailed reports could come through, the black-out of 16th Division communications drew a cloak of obscurity over the situation on Leyte.

Though handicapped by inadequate information, Thirty-fifth Army took immediate action to move reinforcements to Leyte in accordance with the *Suzu* No. 2 Operation plan, which had already been activated on 19 October. Orders were issued during the 20th directing the following units to advance immediately to Leyte, where they were to come under 16th Division command:[45]

1. 41st Infantry Regiment (less one battalion) of the 30th Division (Army reserve) from Mindanao.

2. 169th Infantry Battalion of the 102d Division from the Visayas sector.

3. One infantry battalion of the 57th Independent Mixed Brigade, from Cebu.[46]

Since Imperial General Headquarters' decision to fight a decisive ground action on Leyte had not yet been transmitted to Southern Army, all Army headquarters in the Philippines were

42 On 20 October the Fourth Air Army ordered the 7th Air Division in the Celebes area to dispatch immediately to the Philippines the bulk of its strength, consisting of the 3d and 9th Composite Air Brigades.

43 (1) Outline of Operations and Lessons of the Leyte Campaign, op. cit., Appended Chart III. (2) Philippine Area Naval Operations, Part II, op. cit., pp. 13–4, 101.

44 Vice Adm. Onishi reached Manila on 17 October from Formosa, where he had been conferring with Admiral Toyoda, Commander-in-Chief of the Combined Fleet, regarding the missions of the First Air Fleet in the *Sho-Go* operation. Although his new command did not become effective until 20 October, Vice Adm. Onishi began on the night of the 19th to organize the first *Kamikaze* attack units within the First Air Fleet. Cf. Chapter XVII.

45 Outline of Operations and Lessons of the Leyte Campaign, op. cit., p. 30.

46 This unit, designated as the Tempei Battalion, was a provisional organization consisting of filler troops destined for the 57th Independent Mixed Brigade, currently being organized in the Celebes by Second Area Army. The battalion was awaiting transshipment at Cebu when the Leyte invasion began, and Thirty-fifth Army, on its own authority, decided to assume command of the unit and use it to reinforce Leyte. Philippine Operations Record, Phase Three, op. cit. Vol. II Suppl., p. 54.

still operating on the assumption that only a strategic delaying action would be conducted in accordance with the original *Sho-Go* plans. The staff mission dispatched by Imperial General Headquarters to transmit this vital decision reached Manila at about 2000 on 20 October, immediately launching into a series of night conferences.[47] Developments on Leyte subsequent to the enemy landing remained entirely unknown at this time.[48]

The High Command directive evoked diametrically opposite reactions from Southern Army and Fourteenth Area Army. Field Marshal Terauchi was entirely in agreement with the decision to wage a decisive ground action on Leyte, since Southern Army headquarters had become increasingly skeptical of the practicability of the original *Sho-Go* plans.[49] Field Marshal Terauchi and his staff held that not only the sea and air forces, but the ground forces as well, must be fully committed in defense of Leyte, since the establishment of an enemy foothold in the central Philippines, involving the acquisition of air bases, would render decisive ground action on Luzon an impossibility.

The reaction of General Yamashita and the Fourteenth Area Army staff, on the other hand, was one of shocked surprise.[50] The principal reasons for this doubt on the part of the Area Army command were the following:[51]

1. Judging from the strength of enemy air attacks in support of the Leyte invasion,[52] it appeared highly questionable that the enemy's carrier air strength had been so seriously depleted in the Formosa Air Battle as to assure Japanese ability to gain air superiority. Without such superiority, the movement of large-scale reinforcements to Leyte would be subject to grave risk.

2. The transport of the requisite troop strength to Leyte, and its logistical support, would require shipping far in excess of the amount available to Fourteenth Area Army, and the assembly of shipping from other areas would so delay troop movement as to weigh heavily against the chances of success on Leyte.

3. It was still difficult to determine with certainty whether the invasion of Leyte was the main enemy effort, or whether it would prove to be a limited objective operation preliminary to a quickly following, major attack on Luzon.

47 The Chief of Staff, Deputy-Chief of Staff and other staff officers of Southern Army took part in these conferences. Immediately thereafter, Maj. Gen. Nishimura, Deputy-Chief of Staff of Fourteenth Area Army, was called to Southern Army headquarters and informed. (Statements by Col. Ichiji Sugita, Staff Officer (Operations), Imperial General Headquarters, Army Section; and Maj. Gen. Nishimura, previously cited.)

48 Outside of the fact that the enemy had landed, Fourteenth Area Army had absolutely no information on the progress of the ground battle on Leyte until 25 October. (Statement by Lt. Col. Shigeharu Asaeda, Staff Officer (Operations), Fourteenth Area Army.)

49 Statement by Lt. Gen. Iimura, previously cited.

50 These orders were a bolt from the blue to the Fourteenth Area Army. . . .The Chief and Deputy-Chief of the Army General Staff had outlined this plan [to fight the decisive ground battle on Luzon] to General Yamashita when he stopped in Tokyo to confer with Imperial General Headquarters on his way to the Philippines. . . .General Yamashita had confirmed his own understanding of the plan by further questioning. Maj. Gen. Nishimura, Fourteenth Area Army Deputy-Chief of Staff, and others who had been with General Yamashita at the time were now dumbfounded and highly indignant. Unpublished Memoirs of the Philippines Campaign written by Lt. Gen. Akira Muto, Chief of Staff, Fourteenth Area Army, *Hitosen no Jisso* 比島戰の實相 (The Truth of the Philippines Campaign), 15 Jun 47, pp. 1–8.

51 (1) Philippine Operations Record, Phase Three, op. cit. Vol. II, pp. 9–10. (2) Statement of Maj. Gen. Nishimura, previously cited. (3) The Truth of the Philippines Campaign, op. cit., pp. 7–8.

52 The total number of reported sorties by enemy aircraft over the Philippines amounted to approximately 350 on 17 October, 970 on 18 October, and 600 on 19 October. Outline of Operations and Lessons of the Leyte Campaign, op. cit., Appended Chart III.

4. Luzon was eventually certain to be invaded by reason of its strategic and political importance to the enemy. Therefore, if Luzon were stripped of troops to reinforce Leyte and, despite this, Leyte were lost, a decisive ground defense of the more strategically important area of Luzon would be seriously prejudiced.

Despite these views, General Yamashita recognized that much depended upon the actual development of the situation on Leyte, and even more upon the outcome of the impending sea and air offensive. He therefore proffered no formal opposition at this time and directed the Area Army staff to begin working out plans to implement the revised operational policy.

Pending the elaboration of these plans, Fourteenth Area Army took steps to obtain Navy cooperation in transporting the 30th Division elements already ordered to Leyte by Thirty-fifth Army. Southwest Area Fleet promptly assented to a request that the 16th Cruiser Division and available naval transports be dispatched to Cagayan, northern Mindanao, by the evening of 24 October to pick up these troops and ferry them to Ormoc, on Leyte.[53] Fourteenth Area Army also confirmed Thirty-fifth Army's decision to use the Tempei Battalion and decided, in addition, to make available the 20th Antitank Battalion from Luzon. An Area Army order implementing these arrangements was issued at 1700 on 21 October.[54]

The following morning Field Marshal Terauchi summoned the Fourteenth Area Army and Fourth Air Army commanders to Southern Army headquarters and transmitted to them a formal order directing compliance with the decision to fight the decisive ground battle on Leyte. This order stated:[55]

1. The opportunity to annihilate enemy is at hand.

2. The Fourteenth Area Army, in cooperation with the Navy and the Air forces, will muster all strength possible and destroy the enemy on Leyte.

In accordance with this order, General Yamashita promptly ordered the 26th Division, presently on Luzon, to prepare for early shipment to Leyte.[56] He also decided that the 1st Division and 68th Brigade, scheduled to reach the Philippines shortly, would be allocated to Thirty-fifth Army as additional reinforcements.[57] An urgent dispatch was immediately sent to the Thirty-fifth Army Commander

53 The 16th Cruiser Division, already under orders to proceed from Brunei to Manila, left Brunei at 1800 on 21 October. The division was not re-ordered to proceed direct to Cagayan under the reinforcement plan because Southwest Area Fleet wished to keep it outside the range of enemy carrier strikes until the scheduled time for the reinforcement operation, and also desired to brief the division commander at Manila prior to the start of the operation. (Statement by Vice Adm. Gunichi Mikawa, Commander, Southwest Area Fleet.)

54 This order formally placed the Tempei Battalion (n. 46) under Thirty-fifth Army command, effective immediately. The 20th Antitank Battalion was to embark for Leyte from the Manila area on or about 23 October. Outline of Operations and Lessons of the Leyte Campaign, op. cit., p. 24.

55 The Truth of the Philippines Campaign, op. cit., p. 7.

56 The 26th Division was in no condition to be moved at this time. Part of the division personnel and all of its equipment was deployed in the Tarlac and Bongabon areas. A detachment of 6000 men was on special duty in Manila as a port labor force, and preparations for a tactical move to a combat area would take a great deal of time. On 22 October, Gen. Yamashita ordered the division to prepare selected units for quick shipment to Leyte. The division began to prepare the 12th Independent Infantry Regiment for this mission. On the 24th, the Area Army commander further ordered that the main strength of the division would eventually be shipped to Leyte.

57 Orders to that effect were issued on 27 October. The 1st Division had been ordered to move to Luzon from the Shanghai area on 22 September, and was currently en route. It reached Manila on 27 October. The 68th Brigade was still on Formosa, although scheduled under the *Sho-Go* plans to move to Luzon and come under Fourteenth Area Army command. (Cf. Chapter XI, p. 302 and Chapter XII, p. 329.)

notifying him of the vital change in the Army's mission and the reinforcement plans.

Upon receipt of this dispatch, Lt. Gen. Suzuki and his staff began formulating a new operational plan covering the deployment of forces on Leyte. This plan, completed within the next few days, was essentially as follows:[58]

1. Operational policy:

a. The Army will act immediately in cooperation with the decisive operations of the naval and air forces.

b. Reinforcements will be concentrated on the plain near Carigara.

c. Enemy troops which have landed near Tacloban and in the Dulag area will be destroyed.

d. The direction of the initial main effort will be against the enemy in the Dulag area.

e. The general attack will begin on or about 10 November.

2. Allocation of missions:

a. The 16th Div. will hold the Dulag area, Catmon Hill, and the heights west of Tacloban in order to cover the concentration of the main forces of the Thirty-fifth Army.

b. The following units, after landing at the ports indicated, will concentrate on the Carigara plain:

1st Div.—Carigara

26th Div.—Carigara

102d Div. (Hq. and three battalions)—Ormoc

c. After the concentration of the Army's main forces on the Carigara plain and adjacent areas to the southeast, operations will begin with the objective of destroying the enemy in the Dulag and then the Tacloban area.

d. In conjunction with the Army counteroffensive from the Carigara plain, the following units will perform the indicated operations:

(1) 30th Div. (Hq. and three battalions) will land at Albuera or Baybay, advance to the Dulag area, and operate against the enemy's left rear.

(2) If circumstances permit, the 68th Brig. will stage a counterlanding south of Dulag. Failing this, the brigade will land at Ormoc, advance to the Dagami area, and join in the main assault.

3. The headquarters of the Thirty-fifth Army will advance to Leyte.

Owing to the obscurity surrounding the situation on Leyte, the Thirty-fifth Army plan of operations was formulated on the basis of serious miscalculations. First, the enemy invasion force was estimated from the initial intelligence reports at not more than three divisions. Second, it was estimated that the enemy, as in previous instances, would not attempt to advance inland until his Tacloban and Dulag beachheads were joined, thus allowing adequate time for the planned concentration of reinforcements in the Carigara area. Third, even if such an attempt were made, Thirty-fifth Army overconfidently believed that the combat-tested 16th Division would be able to contain the enemy in the eastern part of Leyte long enough to cover the assembly of reinforcements.[59]

Preparations for Sea-Air Attack

While plans were being made for committing maximum ground strength in the defense of Leyte, the sea and air forces were rushing to complete preparations for the concerted assault on the enemy invasion fleet, which constituted the first phase of the *Sho-Go* Operations.

At 0813 on 20 October, a few hours prior to the Leyte landings, Admiral Toyoda issued a Combined Fleet order confirming, with minor changes, the tentative battle plan issued two days earlier. The essentials of this order were

58 *Dai Sanjugo Gun Hatsuchaku Bunsho Utsushi* 第三十五軍發着文書寫 (Document Files, Thirty-fifth Army headquarters) Oct—Dec 44, pp. 21–2, 25.

59 Philippine Operations Record, Phase Three, op. cit. Vol. II Suppl., pp. 47, 49–51.

as follows:[60]

. . . .

2. *The Combined Fleet, in cooperation with the Army, will annihilate the enemy invading the Central Philippines in accordance with the following operational plan:*

a. *The First Striking Force will penetrate to the Tacloban area at dawn on 25 October (X-Day) and destroy the enemy transport group and its covering naval escort forces.*

b. *In coordination with the attack of the First Striking Force, the Task Force Main Body will maneuver in the area east of Luzon in order to lure the enemy to the north. It will also attack the enemy at the most favorable opportunity.*

c. *The Southwest Area Force (Southwest Area Fleet) will command all naval air units assembled in the Philippines and will utilize them to destroy the enemy carrier and invasion forces in conjunction with the penetration of the First Striking Force. It will also effect immediate amphibious counterlandings on Leyte in cooperation with the Army.*

d. *The main strength of the 6th Base Air Force [Second Air Fleet] will advance to the Philippines and come under the command of Southwest Area Force. It will execute a general attack against enemy task forces on 24 October (Y-Day).*

e. *The Advance Submarine Force will continue its previously assigned missions, as will other forces participating in the Sho-Go Operation.*

While this order was being disseminated, the surface forces scheduled to participate in the attack were already maneuvering into position. Vice Adm. Kurita's First Striking Force reached Brunei Bay, North Borneo, at noon on 20 October, and began making final preparations. Far to the north, Vice Adm. Ozawa's Task Force Main Body sortied from the Bungo Channel, at the southern entrance to the Inland Sea, on the afternoon of the same day, immediately after receiving the Combined Fleet battle order.

The mission of the Ozawa force was of vital importance to the success of the over-all plan. It was to act as a decoy to draw off the main strength of the enemy naval forces covering the invasion operations in Leyte Gulf, thus allowing the First Striking Force to penetrate to the landing point and smash the enemy's troop and supply ships. To heighten its effectiveness as a lure, the Ozawa Force sortied with all of the 3d Carrier Division, made up of the regular carrier *Zuikaku* and the light carriers *Zuiho*, *Chitose*, and *Chiyoda*. The total number of aircraft available to put aboard these ships, however, was only 108. These belonged to the poorly trained air groups of the 1st Carrier Division and represented about half the normal complement. In addition to the half-empty carriers, the force comprised two battleships (*Ise*, *Hyuga*),[61] three light cruisers(*Oyodo*, *Tama*, *Isuzu*) and eight destroyers (31st Destroyer Squadron).[62] Vice Adm. Ozawa fully anticipated that his fleet would be completely wiped out, but this sacrifice was deemed essential to achieve the primary objective—destruction of the entire enemy invasion force.[63]

---

60 First Striking Force Detailed Action Report, op. cit., pp. 8–9.

61 *Ise* and *Hyuga* (4th Carrier Division) had been converted for use as semi-carriers by removing their after gun turrets and installing aircraft lifts and launching catapults. However, neither ship had ever been used as a carrier in combat. They were included in the decoy force mainly because of their firepower, each ship having eight 14-inch guns.

62 Philippine Area Naval Operations, Part II, op. cit., pp. 71–3.

63 "The chief concern was to lure your [the American] forces. . . .north; we expected complete destruction. I thought that, if Kurita's Fleet succeeded in attacking your landing forces, I would be satisfied; if they destroyed the transports there in Leyte Gulf, I would have been satisfied." (Interrogation of Vice Adm. Jisaburo Ozawa, Commander, Task Force Main Body. United States Strategic Bombing Survey (Pacific), Naval Analysis Division, Report, *Interrogations of Japanese Officials*. Vol. I, p. 223.)

レイテ戰に於ける日本艦隊の兵力部署

TACTICAL GROUPING OF JAPANESE NAVAL FORCES FOR THE LEYTE BATTLE[1]

COMBINED FLEET 聯合艦隊 (Adm. Toyoda)

- TASK FORCE MAIN BODY 機動部隊本隊 (Vice Adm. Ozawa) — **Ozawa Force 小澤部隊**
  - 3rd Carrier Division (CV Zuikaku & 3 CVL Chitose, Chiyoda & Zuiho)
  - 4th Carrier Division (2 BB Hyuga & Ise)
  - 31st Destroyer Squadron
    - CL Isuzu
    - 41st Destroyer Division (DD Shimotsuki)
    - 43rd Destroyer Division (4 DD Maki, Kuwa, Sugi & Kiri)
    - 61st Destroyer Division (3 DD Akitsuki, Hatsutsuki & Wakatsuki)
  - 2 CL Oyodo & Tama
- FIRST STRIKING FORCE[2] 第一遊擊部隊 (Vice Adm. Kurita)
  - First Task Group and Second Task Group — **Kurita Force 栗田部隊**
  - First Task Group
    - 4th Cruiser Division (4 CA Atago, Takao, Maya & Chokai)
    - 1st Battleship Division (3 BB Yamato, Musashi & Nagato)
    - 5th Cruiser Division (2 CA Myoko & Haguro)
    - 2nd Destroyer Squadron
      - CL Noshiro & DD Shimakaze
      - 2nd Destroyer Division (2 DD Hayashimo & Akishimo)
      - 31st Destroyer Division (4 DD Kishinami, Okinami, Asashimo & Naganami)
      - 32nd Destroyer Division (2 DD Hamanami & Fujinami)
  - Second Task Group
    - 3rd Battleship Division (2 BB Kongo & Haruna)
    - 7th Cruiser Division (4 CA Kumano, Suzuya, Tone & Chikuma)
    - 10th Destroyer Squadron
      - CL Yahagi & 2 DD Nowaki & Kiyoshimo
      - 17th Destroyer Division (4 DD Urakaze, Isokaze, Yukikaze & Hamakaze)
  - Third Task Group — **Nishimura Force 西村部隊**
    - 2nd Battleship Division (2 BB Yamashiro & Fuso)
    - CA Mogami & DD Shigure
    - 4th Destroyer Division (3 DD Mitsushio, Asagumo & Yamagumo)
- SOUTHWEST AREA FORCE 南西方面部隊 (Vice Adm. Mikawa)
  - Second Striking Force 第二遊擊部隊 (Vice Adm. Shima) — **Shima Force 志摩部隊**
    - 21st Cruiser Division (2 CA Nachi & Ashigara)
    - 1st Destroyer Squadron
      - CL Abukuma
      - 7th Destroyer Division (2 DD Akebono & Ushio)
      - 18th Destroyer Division (2 DD Shiranuhi & Kasumi)
      - 21st Destroyer Division (3 DD Wakaba, Hatsuharu & Hatsushimo)
    - 16th Cruiser Division[3] (CA Aoba, CL Kinu & DD Uranami)
  - Fifth Base Air Force (1st Air Fleet) approx. 35[4][5] (Vice Adm. Onishi)
  - Sixth Base Air Force (2nd Air Fleet) approx. 200[4][6] (Vice Adm. Fukudome)
  - Philippine Force 比島部隊
- ADVANCE SUBMARINE FORCE 先遣部隊 (Vice Adm. Miwa)
  - Approx. 33 submarines of which about 19 were available for Sho Operation No. 1.

(1) Includes only those forces which participated directly in the naval battle for Leyte Gulf.
(2) Combined Fleet assumed direct command of First Striking Force 20 Oct.
(3) After activation of Sho Operation No. 1, 16th Cruiser Division was assigned the mission of transporting troops from Mindanao to Leyte and operated independently during the Leyte battle.
(4) Fifth and Sixth Base Air Forces were grouped under a unified command designated First Combined Base Air Force 25 Oct.
(5) Number of planes in operational condition as of 18 Oct.
(6) Number of planes which had advanced to Philippines by 23 Oct.

PLATE NO. 90

Tactical Grouping of Japanese Naval Forces for the Leyte Battle

Vice Adm. Kurita's First Striking Force, by the night of 21 October, had almost completed final preparations for its sortie from Brunei. By order of Admiral Toyoda, it was to operate henceforth under direct Combined Fleet command instead of under Vice Adm. Ozawa, who thus became responsible only for directing the Task Force Main Body.[64] Late on 21 October, Vice Adm. Kurita issued to his subordinate commanders a final battle plan calling for twin, coordinated thrusts at Leyte Gulf from north and south. This plan was as follows:[65]

1. Mission: In accordance with Combined Fleet dispatch order No. 363, this Force, in cooperation with action of the base air forces and the Task Force Main Body, will sortie to the Tacloban area at daybreak on 25 October.[66] *After destroying enemy naval forces in the vicinity, it will annihilate enemy transports and landing forces.*

2. Dispositions and Operational Procedure:

a. I will personally lead the 1st and 2d Task Groups, starting from Brunei at 0800 22 October. After penetrating through San Bernardino Strait at sunset on 24 October, I will destroy the enemy surface forces in night battle east of Samar and then proceed to the Tacloban area at daybreak on 25 October to destroy the enemy transport convoy and landing forces.

b. The 3d Task Group under command of Vice Adm. Nishimura will leave Brunei at a time fixed by the commander and move up separately. At daybreak on 25 October, this force will sortie to the Tacloban area through Surigao Strait and cooperate with the main body in annihilating the enemy transports and landing forces.

While the Kurita Force was preparing to weigh anchor, the Second Striking Force under Vice Adm. Shima was enroute from the Pescadores to Manila, where it was to operate with the 16th Cruiser Division in moving troops to Leyte. Since Fourteenth Area Army had indicated that no troops were ready to embark, however, Combined Fleet suggested to Southwest Area Force, now commanding the Second Striking Force, that Vice Adm. Shima's ships be utilized temporarily to support Vice Adm. Kurita's attack. Southwest Area Force concurred and radioed Vice Adm. Shima to take his force to Coron Bay and refuel there from fleet tankers in preparation for a possible sortie through Surigao Strait.[67]

At 0800 on 22 October, the main body of

64 The change of command became effective at 1802 on 20 October upon transmission of the Combined Fleet dispatch. First Striking Force Detailed Action Report, op. cit., p. 9.

65 Ibid., pp. 4-5.

66 The attack date of 25 October set by Combined Fleet was actually too early for Vice Adm. Kurita's force to meet without difficulty. In order to reach the target area on scheduled time, he could not send his main force via the safest route skirting west of the Dangerous Ground, in the South China Sea, but had to navigate up the Palawan Passage, which was known to be infested with enemy submarines. The shortness of time was also a factor in Vice Adm. Kurita's decision to send Vice Adm. Nishimura's 3d Task Group through the still more dangerous Sulu Sea and Surigao Strait. The 3d Task Group included old, slow-speed battleships, which necessitated taking a shorter route in order to reach the Leyte Gulf area in time. (1) Interrogation of Vice Adm. Takeo Kurita, Commander, First Striking Force USSBS, *Interrogations of Japanese Officials*, op. cit. Vol. I, p. 36. (2) Statement by Vice Adm. Kurita, previously cited. (3) First Striking Force Detailed Action Report, op. cit., p. 10.

67 During the evening of 21 October, Southwest Area Force dispatched a recommendation to Combined Fleet headquarters that Vice Adm. Shima's force, since it was to operate in the same waters as the First Striking Force, be placed under Vice Adm. Kurita to assure unified command. Combined Fleet, however, did not act on this recommendation since Vice Adm. Kurita was on the verge of departure from Brunei with his battle plan already decided. To revise the plan would have required considerable long-range radio communication, possibly compromising the secrecy of the attack plan. Also, Combined Fleet felt that Vice Adm. Shima's force should remain under Southwest Area Force command at Manila to preserve flexibility in case it became possible to start reinforcement transfers to Leyte sooner than anticipated. (Statement by Rear Adm. Toshitane Takata, Vice Chief of Staff, Combined Fleet.)

the First Striking Force sortied from Brunei Bay as planned, and seven hours later Vice Adm. Nishimura's 3d Task Group cleared the anchorage en route for Surigao. Far to the north, Vice Adm. Ozawa's imposing but actually weak decoy force was due east of Okinawa, heading south into the Philippine Sea. The surface phase of the *Sho-Go* Operation plan had swung into action.

Meanwhile, the Army and Navy air forces were speedily concentrating their strength in the Philippines for the start of the coordinated air offensive on 24 October, one day in advance of the fleet attack. (Plate No. 91) This was essential to the success of the entire plan, both to cover the penetration of Vice Adm. Kurita's attack groups to the landing point and to destroy a portion of the opposing naval strength, particularly carriers, in advance of Kurita's attack.

The Second Air Fleet was assembled in the Manila area (principally Clark Field) with all units expected to reach full operational readiness at their new bases by 23 October. Fourth Air Army, however, was experiencing more difficulty in assembling its scattered forces and barely expected to get them deployed at their battle stations in time. Lt. Gen. Tominaga, Fourth Air Army Commander, nevertheless had issued the following order on 21 October directing final dispositions for the attack :[68]

1. The enemy is continuing to land on Leyte. The main strength of our Navy has started operations with the objective of destroying the enemy in the waters east of the Philippines.

2. Fourth Air Army will employ its entire strength in an effort to annihilate enemy ships engaged in landing troops in the Leyte Gulf area.

3. Pending concentration of the 30th Fighter Group in the central Philippines, the 2d Air Division will continue attacking the enemy on the present basis. Upon arrival of the 30th Fighter Group, the division will assume command of this unit and launch attacks in force as soon as possible. These attacks will be continued in full strength thereafter.

4. The 30th Fighter Group will deploy in the central Philippines by the evening of 23 October and come under command of the 2d Air Division.

5. Fourth Air Army will establish an advance command post at Bacolod by the evening of 23 October.

The deployment plan of the 2d Air Division, which was to command all Army combat flying elements assembling in the Philippines, envisaged retaining heavy and light bomber units at Clark Field and Lipa, on Luzon. Fighter, fighter-bomber, and reconnaissance units, because of their shorter range, were to operate from bases around Bacolod, on Negros Island. As the deployment of the latter progressed, the taxi strips at the Negros bases became so soft and muddy under prolonged rainfall that the useability of the fields was seriously reduced. However, alternate bases were lacking, and time was so short that plans could not be changed.[69]

In spite of many difficulties, Fourth Air Army succeeded, by late on 23 October, in gathering together the bulk of its widely scattered units and deploying them for the decisive battle. This redeployment operation was as follows :[70]

68 Philippine Air Operations Record, Phase Two, op. cit., p. 72.

69 Statement by Lt. Col. Shiro Nonogaki, Staff Officer (Operations), 2d Air Division.

70 A number of units arrived too late to participate in the general air offensive, but early enough to engage in the first stages of the Leyte campaign. These were the 200th Fighter Regiment which arrived from the Homeland between 23–26 October, the 38th Air Reconnaissance Regiment from the Homeland on the 24th, the 33d Fighter Regiment from Malaya on the 29th, and the remaining one-half of the 12th Fighter Brigade from the Homeland and China on 24 October. Philippine Air Operations Record, Phase Two, op. cit., pp. 24, 45–7, 68–9, and 76.

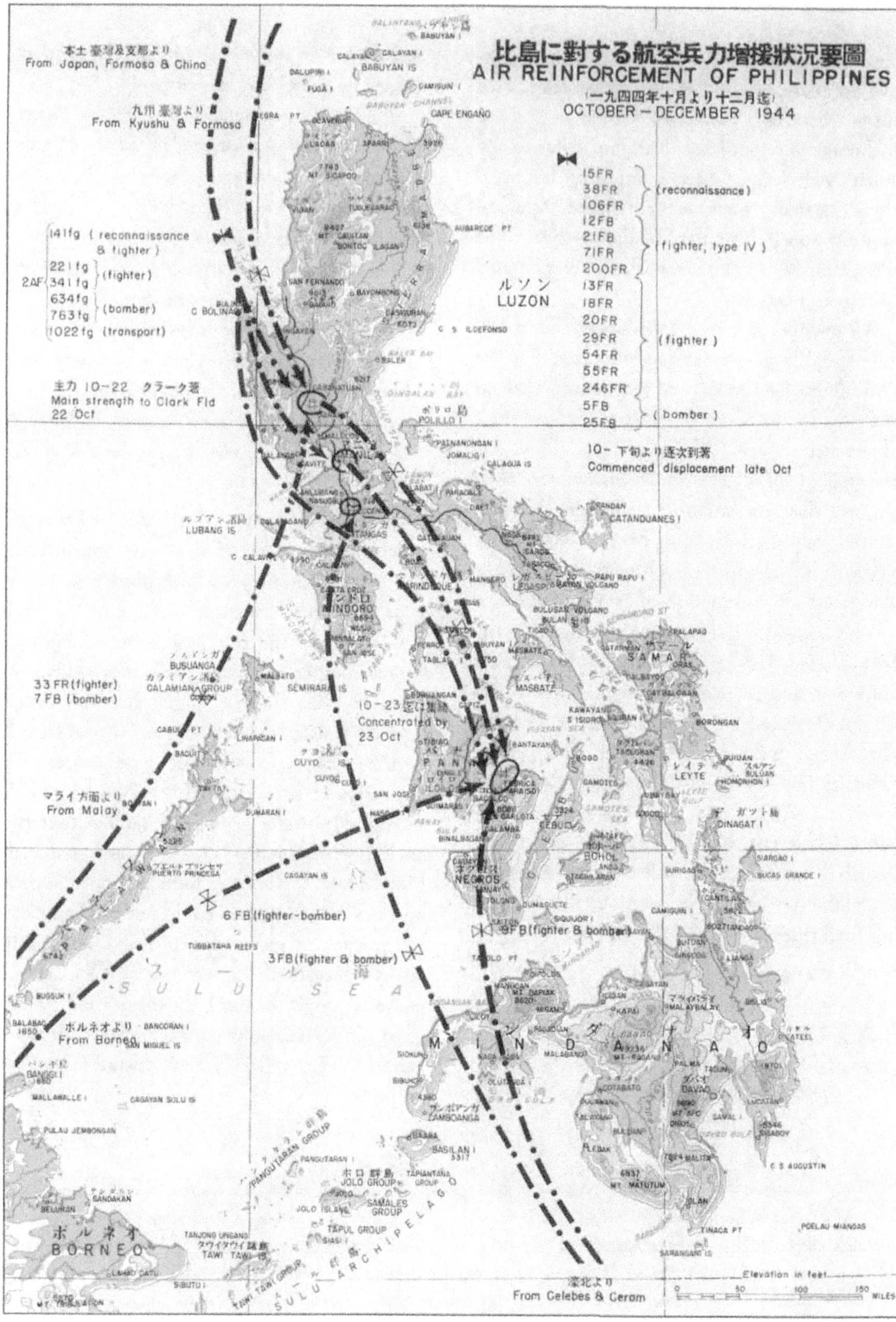

PLATE NO. 91

Air Reinforcement of the Philippines, October–December 1944

To Bacolod:

16th Fighter Brigade from Clark Field

22d Fighter Brigade from Manila

Fighter complement of the 9th Composite Air Brigade from Celebes

6th Fighter-Bomber Brigade from North Borneo

2d Air Reconnaissance Regt. from Clark Field

20th Fighter Regt. From Formosa

To Manila:

One-half of the 12th Fighter Brigade (30th Fighter Group) from the Homeland and China

To Lipa:

Bomber complement of the 3d Composite Air Brigade from Celebes

3d Light Bomber Regt. (25th Bomber Brigade) from Formosa

To Clark Field:

7th Heavy Bomber Brigade from Malaya

14th Heavy Bomber Regt. (25th Bomber Brigade) from Formosa

On the evening of 23 October, Lt. Gen. Tominaga proceeded from Manila to Bacolod and established the advanced command post of Fourth Air Army. Nearly 400 Army and Navy craft were now poised at Philippine bases in readiness for the decisive air and sea battle.[71]

Leyte Sea Battle: First Phase

At dawn on 23 October, the main body of Vice Adm. Kurita's powerful surface attack force, speeding northward from Brunei, sustained the first of series of telling blows which ultimately reduced the force of 32 ships to less than half its original strength before it had reached the objective. This initial blow was struck by enemy submarines. (Plate No. 92)

At 0634, as the Force passed up the western side of Palawan, a spread of torpedoes suddenly entered the cruising disposition. Four of these squarely hit the flagship, the heavy cruiser *Atago*, which sank inside of 19 minutes, the destroyer *Kishinami* taking off Vice Adm. Kurita and his staff. Two more torpedoes struck the heavy cruiser *Takao*, inflicting such severe damage that she had to be ordered back to Brunei under destroyer escort.

Twenty-two minutes after the first attack, another spread of torpedoes found its mark on the heavy cruiser *Maya*, which blew up and sank in four minutes. Within the space of less than half an hour, the Kurita Force had thus lost three major combat units, not counting the destroyers *Naganami* and *Asashimo* detached to accompany the crippled *Takao* back to port. Far more serious, the submarine contact had alerted the enemy two full days in advance of the scheduled assault on Leyte Gulf.

Vice Adm. Kurita's undamaged ships sailed on at high speed after the submarine attacks, not slowing down until 1623, when it was considered safe to effect the transfer of the commander and his staff from *Kishinami* to the super-battleship *Yamato*, which he had decided to make his flagship. The force then continued northward, heading into Mindoro Strait during the night.

During the 23d, Vice Adm. Nishimura's southern force (3d Task Group) had meanwhile negotiated Balabac Strait without incident and begun crossing the Sulu Sea. Vice Adm. Shima's Second Striking Force, now under definite orders to penetrate Surigao Strait on the night of 24 October in support of the

71 Over-all strength of the forces deployed for the air offensive by night of 23 October was as follows: Navy—First Air Fleet, about 50 aircraft of all types; Second Air Fleet, 196 aircraft. Army—Fourth Air Army, 150 aircraft, with approximately 80 additional planes to reach Philippine bases on 24 October. (1) Philippine Area Naval Operations, Part II, op. cit., p. 14. (2) *Dai Yon Kokugun oyobi Dai Juyon Homengun Shorui Tsuzuri* 第四航空軍及第十四方面軍書類綴 (Documents of Fourth Air Army and Fourteenth Area Army) Plan for General Annihilation Attack on Leyte Gulf, 22 Oct 44.

Nishimura group,[72] had reached Coron Bay on schedule and was making final preparations to sortie.[73] To the north, Vice Adm. Ozawa's decoy force, after refueling its destroyers from larger units about 800 miles east of Formosa, had set a southwesterly course, at the same time opening up a powerful long-wave radio transmitter aboard the flagship in order to attract enemy attention.

At dawn on 24 October, the critical air phase of the battle plan swung into action. At 0630, the full strength of the First and Second Air Fleets—199 planes—took off from Clark Field to sweep the waters east of Luzon.[74] This force, after nearly two and a half hours' flying, finally spotted an enemy group with a nucleus of six carriers 160 miles east of Manila and immediately attacked. Action reports claimed one battleship set afire, one carrier and one cruiser damaged, and 32 enemy planes shot down. Japanese losses were heavy, 67 planes failing to return to base.[75]

While this attack was in progress, a search plane discovered a second enemy group with two carriers about 40 miles to the north, and at 0940 a third group including three carriers was spotted due east of San Bernardino Strait. The importance of knocking out this third group to facilitate the penetration of Kurita's force was fully realized, but the all-out effort of the morning made it impossible to mount another attack until early evening. An attack group of 24 planes then sortied but, unable to locate the enemy force off San Bernardino due to poor visibility, turned north to attack the enemy groups east of Luzon instead. The results of these attacks were not ascertained.[76]

Army air units had meanwhile launched full-scale attacks against enemy invasion shipping in Leyte Gulf itself. Beginning at 0800, the 2d Air Division mounted three major attacks during the 24th, using 80 planes in the first attack, 38 in the second, and 29 in the final attack at dusk. Results reported, however, were extremely meager. In all three attacks, only one seaplane tender, one cruiser and one landing craft were believed sunk, with five transports and two cruisers damaged.[77]

The first day of the land-based air offensive had thus failed to achieve one of its most vital objectives, the neutralization of the enemy carrier forces threatening Vice Adm. Kurita's two-pronged thrust toward Leyte Gulf. At the same time, Vice Adm. Ozawa's parallel effort to lure these forces to the north, where they could not interfere with Kurita's advance, had produced no apparent results by late afternoon of the 24th, a failure that was not, however, attributable to lack of persistence on the part of the Task Force Main Body. Indeed Vice

72 These orders were received by Vice Adm. Shima at 1645 on 23 October, while the Second Striking Force was still en route to Coron Bay. The 21st Destroyer Division, which had been on detached service transporting ground elements of Second Air Fleet to Manila, was ordered to join Vice Adm. Shima's force immediately.

73 When the Second Striking Force put into Coron Bay at dusk on 23 October, the fleet tankers from which it was to refuel had not yet arrived. Since the time schedule made it impossible to wait, destroyers were immediately refueled from the cruisers. The force sortied at 0400 on 24 October, still without the 21st Destroyer Division which followed about four hours behind.

74 An earlier attack force had sortied before dawn but returned to base without having located any enemy carrier groups. Philippine Area Naval Operations, Part II, op. cit., p. 15.

75 Ibid. (Access to Allied records after the war revealed to the Japanese that the U. S. Fleet carrier *Princeton* had actually been sunk in this attack.)

76 Ibid., p. 15–6.

77 (1) Philippine Air Operations Record, Phase Two, op. cit., pp. 75–7. (2) Outline of Operations and Lessons of the Leyte Campaign, op. cit., Appended Chart III. (3) 2d Air Division Telegram, No. 943, 25 Oct 44. ADVATIS Bulletin No. 170, 22 Jan 45.

Adm. Ozawa, convinced that the luring operation could only succeed if the enemy were given a physical demonstration of the proximity of his force, had actually closed to within 150 miles of the northernmost enemy task group and sent all his operable aircraft, totalling 56 of all types, to carry out an attack.[78] When this attack still drew no enemy retaliation, Vice Adm. Ozawa in mid-afternoon ordered his advance guard (*Ise*, *Hyuga* and four destroyers) under Rear Adm. Chiaki Matsuda to break away from the carriers, proceed south, and forcibly divert the enemy by a night attack.[79]

The combined ineffectiveness of the air offensive and of Vice Adm. Ozawa's decoy operation had meanwhile brought further disaster to the Kurita Force as it headed across the Sibuyan Sea toward San Bernardino Strait. At 0810 on the 24th, enemy scout planes spotted the Force in Tablas Strait, and at 1025 a group of about 30 carrier aircraft swept in for the first attack. The super-battleship *Musashi* and heavy cruiser *Myoko* both sustained aerial torpedo hits, and in addition *Musashi* was hit by a heavy bomb. Damage to the strongly-armored *Musashi* was negligible, but *Myoko*, her speed cut to 15 knots, dropped out of formation and was ordered back to Brunei.[80]

Almost simultaneously with this first attack on the Kurita Force, Vice Adm. Nishimura's 3d Task Group diving across the Sulu Sea was discovered and attacked by 22 enemy carrier planes just south of the Cagayan Islands. The battleship *Fuso* and destroyer *Shigure* received bomb hits which caused only superficial damage, and the force raced on toward Surigao Strait,[81] followed closely by Vice Adm. Shima's Second Striking Force, which had sortied from

---

78 At 0910 Vice Adm. Ozawa received reports of the Second Air Fleet's attacks on an enemy carrier group east of Luzon. He immediately flew off a search mission, which reported at 1105 that it had spotted an enemy group 180 miles southwest of the decoy force. After closing to about 150 miles, Vice Adm. Ozawa launched his carrier aircraft at noon for an attack. Due to bad weather conditions and the fact that most of the fliers were insufficiently trained in operating from carriers, the attack groups were instructed to proceed to bases on Luzon if, after carrying out the attack, they were unable to return to the carriers. Only a few aircraft actually returned, and these had failed to locate the enemy. However, 26 planes from his flagship, *Zuikaku*, had succeeded in making an attack on the enemy group, those planes not shot down continuing on to shore bases. The rest of the attacking groups were engaged by enemy interceptors and failed to find their targets. Philippine Area Naval Operations, Part II, op. cit., pp. 73–4.

79 Account of operations of the Ozawa Force is based on the following sources: (1) Ibid., pp. 73–8. (2) Interrogations of Rear Adm. Chiaki Matsuda, Commander, 4th Carrier Division, Vice Adm. Ozawa and Capt. Ohmae. USSBS, *Interrogations of Japanese Officials*, op. cit. Vol. I, pp. 219–27, 153–60, 277–83. (3) *Gunkan Senji Nisshi* 軍艦戦時日誌 (Battle and Flag Logs of IJMS) Vol. I, *Zuikaku Zuiho* and *Chitose*, Oct 44.

80 Account of operations of the Kurita Force is based on the following sources: (1) Philippine Area Naval Operations, Part II, op. cit., pp. 36–57. (2) First Striking Force Detailed Action Report, op. cit. (3) Battle and Flag Logs IJMS, op. cit. Vol. I, 7th Cruiser Division; Vol. II, *Yamato, Nagato,* and *Musashi*. (4) Interrogations of Rear Adm. Tomiji Koyanagi, Chief of Staff, First Striking Force; Capt. Kenkichi Kato, Executive Officer, IJMS *Musashi*; Comdr. Tonosuke Otani, Staff Officer (Operations), First Striking Force; Vice Adm. Kurita. USSBS, *Interrogations of Japanese Officials*, op. cit. Vol. I, pp. 32–55, 147–52, 171–5, 171; Vol. II, p. 362.

81 Account of operations of the Nishimura Force is based on the following sources: (1) Philippine Area Naval Operations, Part II, op. cit., pp. 57–62. (2) Interrogation of Comdr. Shigeru Nishino, Commanding Officer, IJMS *Shigure*. USSBS, *Interrogations of Japanese Officials*, op. cit. Vol. II, p. 341. (3) Statements by Comdr. Ichio Shibayama, Commanding Officer, IJMS *Asagumo*, and Lt. Comdr. Tomoo Tanaka, Commanding Officer, IJMS *Michishio*.

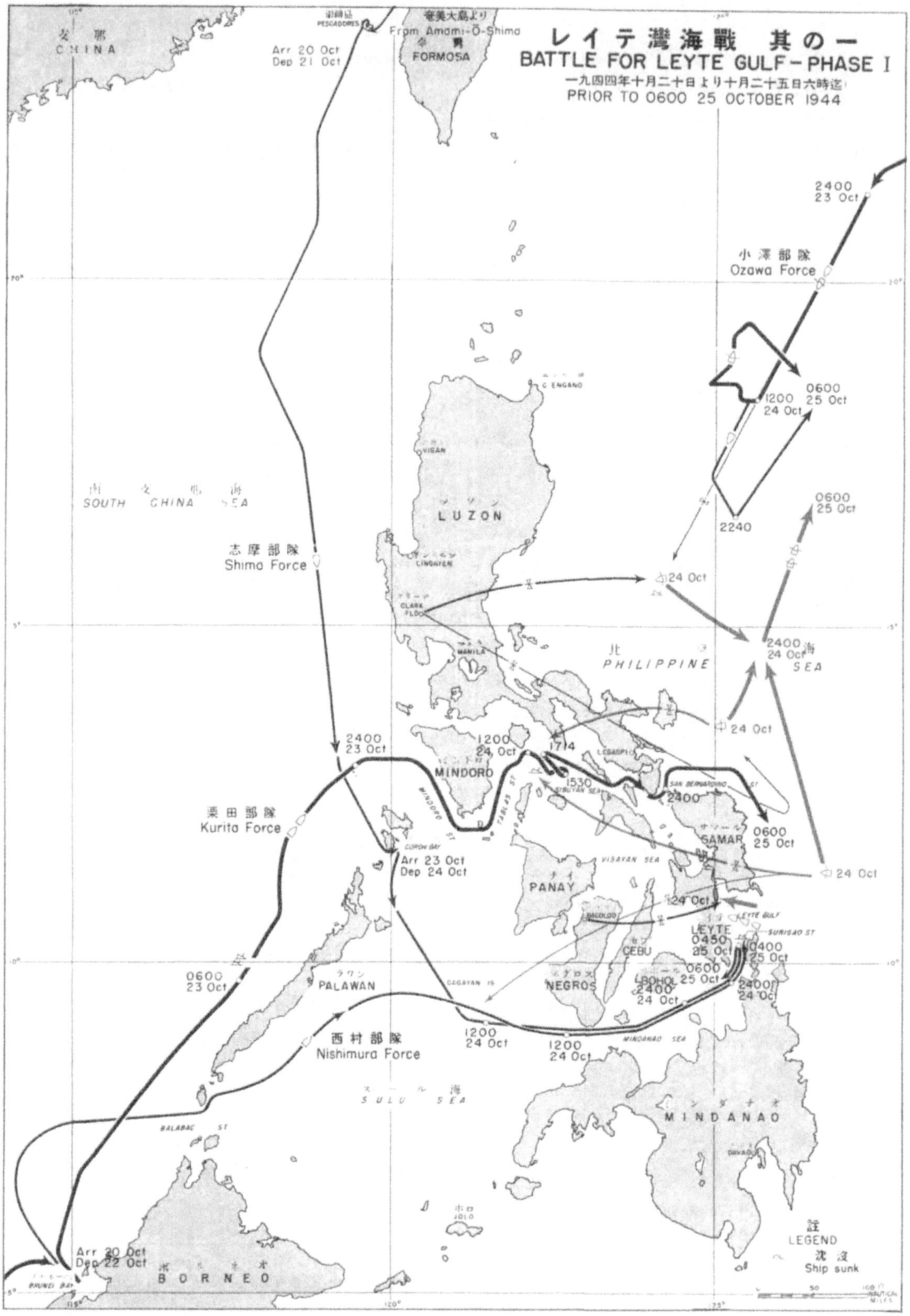

PLATE NO. 92

Battle for Leyte Gulf, Phase I, Prior to 0600 25 October 1944

Coron Bay at 0400.

Enemy air attacks on the Kurita Force were renewed with deadly intensity on the afternoon of the 24th. At 1207 the second wave of about 30 carrier aircraft attacked, concentrating on *Musashi*, which took three more torpedoes and two large bombs. The speed of the 64,000-ton battleship was reduced to 22 knots, and the disposition had to slow down to keep her in formation. At 1315, hoping to spur the air forces and the Ozawa Force into more effective action to ease enemy pressure on his group, Vice Adm. Kurita dispatched the following radio to Vice Adm. Ozawa and Southwest Area Force:[82]

> *I am receiving repeated torpedo and bomb attacks by enemy carrier aircraft. Request urgent information concerning situation as regards attacks on, or contacts with, the enemy by your forces.*

Immediately after the dispatch of this message, a third enemy attack wave swept in to inflict new damage. Five more aerial torpedoes struck *Musashi*, while four bomb hits shattered her superstructure. Down badly at the head, the huge ship lost further speed and fell slowly astern of the disposition. Vice Adm. Kurita's flagship *Yamato* also received a torpedo hit, but damage was superficial. A fourth attack one hour later resulted in further slight damage to *Yamato* and to the battleship *Nagato*.

With *Musashi* too badly damaged to continue, Vice Adm. Kurita at 1452 ordered the destroyer *Kiyoshimo* to escort her back to Mako in the Pescadores. Just as the retirement was getting under way, however, the fifth and heaviest air attack of the day began, enemy carrier aircraft completing the destruction of *Musashi* with 11 more torpedoes and ten large bombs.[83] A bomb also hit the escorting *Kiyoshimo*, reducing her speed to 20 knots.

Vice Adm. Kurita was now convinced that to continue would risk the destruction of his entire force before it could reach the objective.[84] He therefore decided to execute a temporary retirement and, at 1530, ordered his force to reverse its course to the northwest. At 1600 he originated a dispatch to Combined Fleet reporting this action and indicating that he deemed it best "to withdraw temporarily outside the range of enemy planes while awaiting the results of our other operations and standing by to cooperate as developments warrant"[85].

For an hour and a half the force continued to retire until approaching nightfall brought safety from further air attacks. Vice Adm. Kurita was now expecting fresh instructions from Combined Fleet in response to his message of 1600, but since the retirement had put his force dangerously behind schedule in case Combined Fleet ordered execution of the attack as planned, he ordered his ships at 1714 to put about once more and head for San Bernardino.

---

82 First Striking Force Detailed Action Report, op. cit., p. 17.

83 *Musashi* had now taken a total of 20 aerial torpedoes and 17 large bombs, the most terrific punishment ever sustained by any battleship. Even then, she remained afloat for some four hours after the last attack, finally capsizing and sinking at 1935 with the loss of approximately half of her 2,400-man complement. The damage which finally caused her to sink came from three torpedoes, one of which reached a magazine, while two others struck the side successively in the same spot, thus penetrating the inner armor. Cumulative bomb damage hindered efforts to keep the ship afloat by pumping. Survivors of the sinking of *Musashi* were returned to Coron Bay in the destroyers *Hamakaze* and *Kiyoshimo*.

84 Thus far Vice Adm. Kurita had lost one battleship and two heavy cruisers sunk, and two heavy cruisers and three destroyers which had to be sent back to base. This reduced the number of ships by one-fourth, but the loss in gunnery strength was considerably greater since the casualties were mainly in heavy units.

85 First Striking Force Detailed Action Report, op. cit., p. 18.

The force was already one hour along on its new heading when, at 1815, the following message, addressed to all naval units engaged in the *Sho-Go* Operation, was received from the Commander-in-Chief, Combined Fleet:[86]

All forces will dash to the attack, trusting in divine assistance.[87]

Now committed to carry out the attack, Vice Adm. Kurita studied the situation in the light of meager reports received from the Ozawa and Nishimura Forces. A dispatch from Vice Adm. Ozawa, which had been received at 1603, confirmed the conclusion which Vice Adm. Kurita had drawn from the fierce air attacks on his force, namely that the decoy operation had not yet taken effect, and disclosed Ozawa's plan to divert enemy attention forcibly by a night attack.[88] Actually, less than three hours after this dispatch was sent, a search plane from the enemy's northernmost task group had finally spotted Ozawa's carriers, starting a chain of reactions that soon brought the decoy plan into belated but effective operation.

From the Nishimura Force, now well into the Mindanao Sea, Vice Adm. Kurita received a dispatch at 2020 indicating that the force expected to penetrate through Surigao Strait into Leyte Gulf by 0400 on 25 October. This was much earlier than his own force, delayed by its retirement, could possibly reach the objective, and a revision of the coordinated timetable was clearly necessary. Vice Adm. Kurita therefore sent off a dispatch at 2145 modifying the attack plan as follows:[89]

1. The main body of the First Striking Force will pass through San Bernardino Strait at 0100, 25 October, and will then proceed southward down the east coast of Samar, reaching Leyte Gulf at about 1100.

2. The 3d Task Group will penetrate into Leyte Gulf as scheduled and will then join the main body at a rendezvous point ten miles northeast of Suluan Island 0900, 25 October.

Less than three hours later, the Kurita Force sped through San Bernardino Strait slightly ahead of schedule, debouching into the Philippine Sea at approximately 0035 on 25 October. Vice Adm. Kurita now headed around the northeastern tip of Samar, expecting to encounter enemy surface opposition at any moment. In fact, far less was in opposition than he knew, for the bulk of the enemy naval forces covering the northern approaches to Leyte Gulf was already speeding to attack the Ozawa group.[90] The lure had begun to work.

Now, however, sudden disaster enveloped the southern prong of the attack. The Nishimura Force, ignored by enemy aircraft since the morning of the 24th, entered Surigao Strait at 0130 on the 25th, immediately becoming the target of continuous torpedo attacks by enemy PT boats. His ships maneuvering sharply to

---

86 Actually this message was sent out by Combined Fleet headquarters at 1813 prior to receipt of Vice Adm. Kurita's dispatch. The latter did not reach Combined Fleet until 1959, at which time Admiral Toyoda sent off a further message ordering the First Striking Force to carry out the attack. (1) Statement by Rear Adm. Takata, previously cited (2) First Striking Force Detailed Action Report, op. cit., p. 19.

87 Ibid.

88 This message had been dispatched by Vice Adm. Ozawa at 1439.

89 First Striking Force Detailed Action Report, op. cit., p. 20. (There is no record clarifying whether this dispatch reached Vice Adm. Nishimura.)

90 The Ozawa Force was at this time still operating in two groups. From shortly after 2000 until about midnight on the 24th, the carrier main group ran northward to readjust its schedule in view of Vice Adm. Kurita's temporary retirement, then turned southeast to rendezvous with Rear Adm. Matsuda's advance guard. The latter had meanwhile failed to make contact with the enemy and was heading north again for the rendezvous point.

avoid being hit, Vice Adm. Nishimura pressed on, making brief contacts at 0300 and 0315 with small enemy destroyer groups, which retired as the Japanese force opened fire.

At about 0320, a heavy torpedo attack struck the formation from both flanks, setting fire to the flagship *Yamashiro*, badly damaging the heavy cruiser *Mogami*, and sinking the destroyers *Yamagumo* and *Michishio* outright. Another destroyer, *Asagumo*, received heavy damage and dropped out of formation, sinking a few hours later.[91] *Yamashiro* and *Mogami*, though damaged, moved on with the formation.

With the destroyer *Shigure* in the van, *Yamashiro*, *Fuso* and *Mogami* pressed on through the strait, dodging spread after spread of enemy torpedoes. At 0350 shells from heavy enemy surface units at the northern end of the strait began crashing into the formation. In five minutes, *Mogami*, heavily hit and afire, was forced to withdraw. The flagship *Yamashiro* sank a few minutes later under a hail of shells, and *Fuso* followed her to the bottom at about 0410. *Shigure*, well out in front, had sped on toward the enemy until 0403, when she turned back to find that the rest of the force had vanished. Her commander decided to retire, and the destroyer ran south to become the sole survivor of the Nishimura Force.[92]

Vice Adm. Shima's Second Striking Force had meanwhile entered Surigao Strait at 0300, approximately two hours behind the 3d Task Group. Aboard his flagship, the heavy cruiser *Nachi*, Vice Adm. Shima overheard Vice Adm. Nishimura giving orders by radio-telephone for evasive action against torpedo attack, and before long enemy PT boats were attacking his own formation. At 0321 the light cruiser *Abukuma* was suddenly hit and fell out of formation. The rest of the force plunged on through heavy smoke drifting down from the battle ahead, and at 0420 the burning *Mogami* was sighted off the starboard bow.

At this juncture enemy targets were identified by radar off the port. *Nachi* and the heavy cruiser *Ashigara* immediately swung to starboard to fire torpedoes, and in executing this maneuver *Nachi* collided with the crippled *Mogami*, receiving underwater damage which reduced her speed.[93] From *Mogami*, Vice Adm. Shima learned that both *Yamashiro* and *Fuso* had been sunk.[94] Rather than sail into the evident enemy trap, he decided to execute a temporary withdrawal from the strait in order to regroup his strength. Laying a smoke screen, the force put about at 0450 and ran south.

Actually, the Surigao battle was now at an end. Vice Adm. Shima's Force, after safely emerging into the Mindanao Sea,[95] underwent

91 At 0330 Vice Adm. Nishimura dispatched a message to Vice Adm. Kurita reporting the damage sustained by his force in this attack. The message, however, did not reach the First Striking Force commander until 1425. First Striking Force Detailed Action Report, op. cit., p. 30.

92 Preceding account of the 3d Task Group action in Surigao Strait is based on the following: (1) Interrogation of Comdr. Nishino, USSBS, *Interrogation of Japanese Officials*, op. cit. Vol. II, p. 341. (2) Statements by Comdr. Shibayama and Lt. Comdr. Tanaka, previously cited (3) First Striking Force Detailed Action Report, op. cit., pp. 19–20, 29–30.

93 The bridge watch on *Nachi*, due to poor visibility, thought that *Mogami* was dead in the water and consequently misjudged the turn. Actually, *Mogami* was making about 18 knots on a converging course.

94 At 0449 Vice Adm. Shima sent a dispatch to Vice Adm. Kurita reporting the "complete destruction" of the 2d Battleship Division (*Yamashiro* and *Fuso*) and the fact that *Mogami* was heavily damaged and afire. This message was received by the First Striking Force at 0532, and was the only information on the outcome of the Surigao battle to reach Vice Adm. Kurita prior to his own final thrust toward Leyte Gulf. First Striking Force Detailed Action Report, op. cit., p. 30.

95 At 0630, as the force reached the southern entrance of Surigao Strait, it underwent a torpedo attack by enemy PT boats but escaped without damage. (Statement by Comdr. Kokichi Mori, Staff Officer (Torpedo), Second Striking Force.)

two enemy air attacks, which indicated that a renewed penetration attempt in daylight would almost certainly risk destruction.[96] All ships were running short of fuel, and the situation of Vice Adm. Kurita's main striking force remained unknown. At 0900, therefore, Vice Adm. Shima finally abandoned all thought of penetrating Surigao.[97]

The task of destroying the enemy invasion fleet in Leyte Gulf now rested solely on the shoulders of the Kurita Force, already heavily engaged off the coast of Samar.

Leyte Sea Battle : Finale

Daybreak on 25 October found Vice Adm. Kurita's battle formation speeding south toward Leyte Gulf, now less than 100 miles away. At 0623 *Yamato*'s radar screen suddenly showed the presence of enemy aircraft, and Vice Adm. Kurita ordered the force to shift to antiaircraft cruising disposition. The signal for this maneuver was being sent out from the flagship when, at 0644, lookouts in her crow's nest spotted the masts of enemy ships on the horizon to the southeast. One minute later the contact was identified as a group of about six carriers in the act of launching their planes.

Since the enemy ships were erroneously identified as Fleet carriers, Vice Adm. Kurita believed that he had surprised an element of Admiral Halsey's fast carrier Task Force and was thus presented with an opportunity to strike a telling blow at the enemy's main carrier strength.[98] He immediately decided to attack and ordered his ships to deploy for action. (Plate No. 93)

With the wind from the northeast, Vice Adm. Kurita set the mean course of deployment at 110 degrees so as to bring his force around to the north of the enemy group. This was to permit the attack to be delivered downwind, forcing the enemy carriers to run in that direction and preventing them from launching more of their aircraft. The maneuver, however, had the effect of partially nullifying the advantage in speed which Kurita's ships possessed and lengthening the time required to close.

The order to engage was given at 0658. *Yamato* opened fire with her 18-inch batteries, marking the first and only time in naval history that guns of this caliber were used in surface action. The enemy carriers, with their screening ships, began fleeing eastward, engaging in evasive action and laying down heavy smoke.[99] The smoke and scattered rain squalls gave effective concealment, and Kurita's ships had to cease firing temporarily for want of targets.

---

96 The first air attack was carried out at about 0730 by more than ten enemy carrier planes, and, the second at 0900 by 30 planes. In the second attack a direct hit was made on *Mogami*, retiring with the Shima Force, and the ship finally had to be sunk by friendly torpedoes later the same morning. Ibid.

97 Ibid.

98 Vice Adm. Kurita's belief that he was engaging a portion of Admiral Halsey's main carrier forces persisted throughout the action off Samar. Actually, the group sighted and engaged by his force was Rear Adm. C. A. F. Sprague's Task Unit 77.4.3, composed of six escort carriers, three destroyers and four destroyer escorts. Two other groups with a total of 12 escort carriers were disposed to the south and southeast of Rear Adm. Sprague's group. Both latter groups sent their aircraft to attack Kurita but were not engaged in the surface action, although the group to the southeast was spotted and briefly pursued by one of Kurita's flanking battleships. Cf. n. 101.

99 At 0702 a dispatch was sent to Vice Adm. Nishimura's 3d Task Group reporting that the Kurita Force was engaging an enemy force with six carriers off Samar, and ordering the 3d Task Group to join up as speedily as possible. This dispatch was sent because Vice Adm. Kurita, although informed of the loss of Nishimura's two battleships and damage to *Mogami* in the Surigao battle, thought that *Mogami* and the remaining elements of the 3d Task Group might have succeeded in getting through Surigao Strait. First Striking Force Detailed Action Report, op. cit., p. 23.

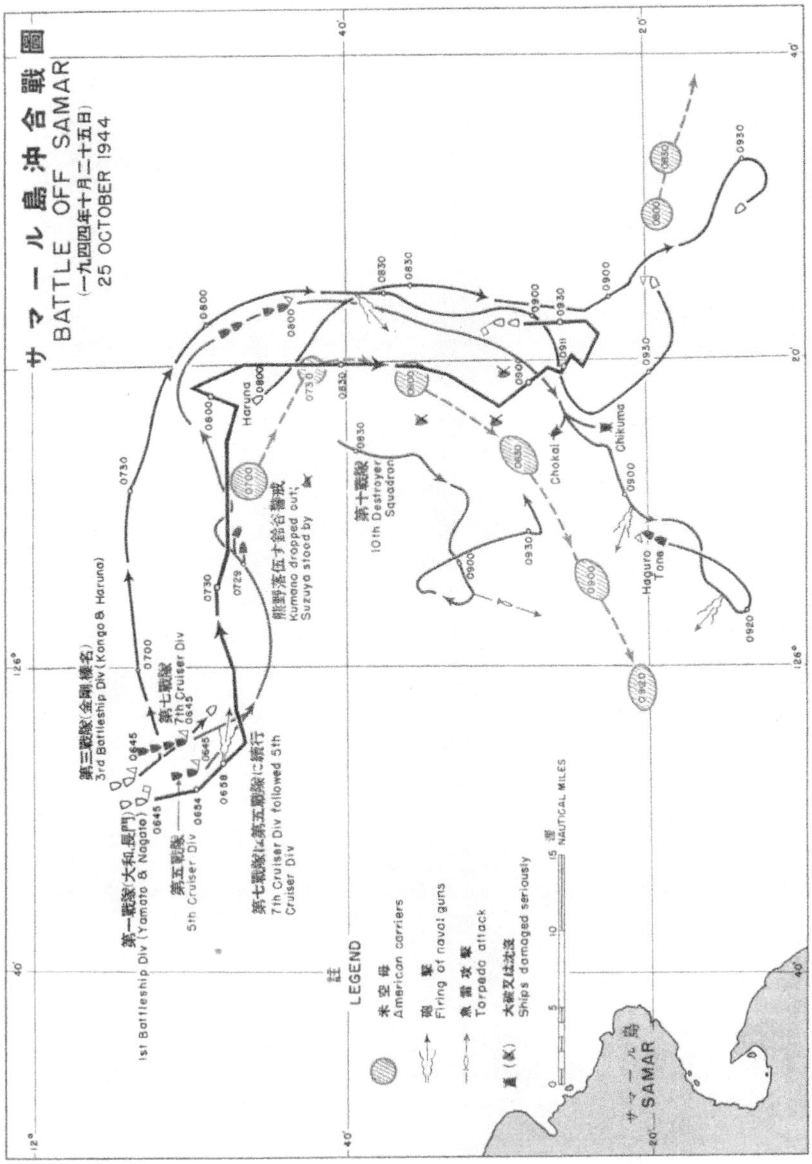

PLATE NO. 93
Battle off Samar, 25 October 1944

At 0710 planes from the enemy carriers began scattered attacks, causing superficial damage to some of the Japanese ships. Enemy escort vessels simultaneously dashed through the smoke and rain making torpedo runs. At 0725 *Yamato* drew first blood when she brought her main battery to bear on an enemy ship believed to be a cruiser, sinking her in two minutes. Almost immediately, however, an enemy torpedo caught the heavy cruiser *Kumano*, causing her to lose speed and fall behind the formation. At 0729 Vice Adm. Kurita released his cruisers, ordering all ships to pursue the enemy at full speed.

Still receiving intermittent air and torpedo attacks, the Kurita force gradually closed on the fleeing enemy ships, the direction of the pursuit swinging from east to south. At 0754 *Yamato* and *Nagato* were forced to maneuver to evade a spread of six torpedoes and fell several miles astern of the lead ships—the fast battleships *Haruna* and *Kongo* and four heavy cruisers: *Chikuma*, *Chokai*, *Tone* and *Haguro*. The latter drove on at top speed, narrowing the range to about ten miles and gradually maneuvering the enemy into position for a torpedo attack by the 10th Destroyer Squadron.

At 0826 *Kongo* reported that she had knocked out one of the enemy carriers, and fifteen minutes later *Yamato* sank another ship believed to be a cruiser. From about 0830 enemy planes again began attacking in increasing numbers, concentrating on the cruisers which were leading the Japanese pursuit. *Chikuma* and *Chokai* were both hit and put out of action by 0900. At 0902 *Kongo* reported that she had sunk an enemy destroyer attempting a torpedo run on the outside of the formation.

Leaving the crippled *Chikuma* and *Chokai* astern, *Tone* and *Haguro* pressed on, closing the range to about six and a half miles by 0910.

The enemy ships were now running southwest, with *Tone* and *Haguro* taking them under fire from off the port beam and the 10th Destroyer Squadron attacking with torpedoes from the starboard.[100] The complete destruction of the enemy appeared imminent.

Yamato's evasive maneuvers under torpedo attack had meanwhile put her still farther astern of the lead cruisers and battleships. Between the flagship and the scene of action lay rain squalls and several curtains of smoke, which made direct observation of the progress of the battle impossible. Due to overloaded battle circuits, radio reports from the ships engaged were fragmentary, and attempts at air observation by means of spot planes launched from *Yamato* netted little additional information due to enemy fighter interference.[101] Vice Adm. Kurita consequently did not know that, beyond

100 The 10th Destroyer Squadron had begun firing its torpedoes at 0905 while still at extreme range of about eight miles, completing firing its first salvoes at 0915. It continued on a course parallel to the enemy while reloading for another attack, but at 0930 was obliged to break off the attack in response to an order to rendezvous with the main body to the north. At 0925 the squadron reported by radio to *Yamato* that it had sunk a fleet carrier and set another afire. Battle and Flag Logs of IJMS, op. cit. Vol. I, 10th Destroyer Squadron. (American Editor's Note: The 10th Destroyer Squadron's claims regarding the results of its torpedo attack were entirely inaccurate. Actually, the torpedoes failed to reach the American carrier group, presumably because they had been fired at more than maximum range. Damage received by the carriers at this stage of the battle came from the guns of the Japanese cruisers and battleships attacking from port.)

101 The second spot plane launched at 0851 reported another group of four enemy carriers, distinct from the group engaged by the Kurita Force, running south under a smoke screen laid by four destroyers. A short while earlier (0810), the battleship *Haruna* had reported sighting two enemy carriers, screened by four destroyers, to the southeast of the scene of action. *Haruna* started in pursuit but broke off at 0930 without having succeeded in getting within effective firing range.

the smoke and rain, his cruisers and destroyers had barely begun the destruction of a virtually helpless enemy.

Although ignorant of the situation in the van, Vice Adm. Kurita was forced both by events and by consideration of the basic mission to make a rapid decision concerning his next move. At this point enemy air attacks, steadily mounting in frequency, had inflicted considerable damage to the First Striking Force. In addition—and even more important—there was a danger that further continuation of the fight would exhaust the fuel needed for execution of the primary mission—a penetration into Leyte Gulf. Vice Adm. Kurita still labored under the misapprehension that he was pitted against a group of Admiral Halsey's fast carriers and that it would be impossible to close the enemy with his main battle strength. All these considerations impelled the First Striking Force commander to seek an early opportunity to disengage.

The moment now seemed propitious. Knowing that the 10th Destroyer Squadron had already launched its heavy torpedo attack on the fleeing carriers and estimating that the cruisers out of sight in the van had, after a two hour action, inflicted devastating damage on the hapless enemy, Vice Adm. Kurita decided to reassemble his force for the final penetration attempt. At 0911 the flagship began sending out the order to break off action and reassemble on a northerly course. By 0930 the order had reached all units, scattered over 25 miles of ocean, and the cruisers and destroyers hammering at the fleeing enemy carrier group immediately put about to the north. The reassembly, carried out under sporadic air attacks, was not completed until 1030.[102]

Vice Adm. Kurita now had an opportunity to evaluate the results of the battle on the basis of initial reports made by the returning units. Three or four enemy fleet carriers were believed to have been sunk, in addition to two heavy cruisers and a destroyer also sunk and several other units heavily damaged.[103] These optimistic figures totaled up to a substantial victory, but the cost had not been light.

The heavy cruiser *Kumano* and destroyer *Hayashimo*, both heavily damaged, were retiring independently. *Chikuma* and *Chokai*, mortally wounded by enemy aerial torpedoes and bombs, remained on the scene of action with two destroyers standing by.[104] At 1114 the heavy cruiser *Suzuya* was added to the casualty list when an enemy bomb hit started uncontrollable fires.[105] Thus, for the final thrust into Leyte Gulf, Vice Adm. Kurita could muster but 15 of his original 32 ships—four battleships, two heavy cruisers, two light cruisers and seven destroyers.

The regrouping of his units completed, Vice Adm. Kurita at 1120 gave the order to change course to the southwest for the crucial run into the gulf. Some members of his staff, however, were now of the opinion that the penetration attempt should be abandoned, and discussion continued even as the force headed toward its final objective. Vice Adm. Kurita now had to make the most fateful and dramatic decision of the entire Leyte sea battle.

The background against which this decision was finally taken was as follows: Dispatches from Vice Adm. Ozawa reporting that the

102 First Striking Force Detailed Action Report, op. cit., p. 24.

103 Ibid., p. 51.

104 *Chokai*'s crew was later taken off by the destroyer *Fujinami*, which then sank the crippled cruiser by torpedo fire. *Chikuma* and the destroyer *Nowaki* were never heard from again and presumed sunk.

105 *Suzuya* was left behind with the destroyer *Okinami* standing by. Fire aboard the cruiser finally reached her torpedo locker, setting off violent explosions. The *Suzuya* sank at 1300, 25 Oct and the *Okinami* took aboard survivors.

decoy force was under heavy attack by enemy carrier planes had not reached Vice Adm. Kurita, who therefore had no certainty that the luring operation had been effective. On the other hand, enough was now known of the events which had occurred in Surigao Strait to make it plain that the southern prong of the attack had failed, and that any penetration into Leyte Gulf would have to be made by Kurita's depleted force single-handed.

The question of time was of even more vital importance. The engagement off Samar had, in conjunction with the time consumed in regrouping his scattered force, so retarded Vice Adm. Kurita's advance toward the gulf that it would be mid-afternoon before he could reach the objective. With ample warning of his approach, it seemed highly probable that the enemy would have evacuated his transports and supply ships from the gulf.

There were also indications that the enemy was mustering strong forces for the defense of the gulf. Radio intercepts revealed not only that enemy carrier aircraft were concentrating on Tacloban airfield, but that powerful enemy surface reinforcements would reach the Leyte area within about two hours. Without any visible support by friendly air forces,[106] it seemed likely that a thrust into the gulf would therefore be not only fruitless but suicidal.

Reports had meanwhile been received from Southwest Area Force to the effect that an enemy carrier task force had been spotted northeast of Samar. Vice Adm. Kurita and the majority of his staff, increasingly dubious of the prospects of success in a thrust into Leyte Gulf, estimated that a more effective blow could be struck by running north to deliver a surprise attack on the newly-reported enemy carrier force.[107]

At 1236, with the entrance to Leyte Gulf only 45 miles away, Vice Adm. Kurita made his definitive decision. Ordering his force to put about to the northwest, he radioed Combined Fleet and all other naval commands concerned that the penetration of Leyte Gulf had been abandoned, and that he was proceeding north to seek battle with an enemy task force, after which he would retire through San Bernardino Strait.[108]

Enemy carrier planes continued to harass the formation during the afternoon but inflicted no serious damage. Just before sunset, Vice Adm. Kurita's ships reached the area in which the enemy task force had been reported, only to find that the force had vanished.[109] With nothing to attack and his destroyers beginning to run short of fuel, Vice Adm. Kurita ordered his force to turn west and run for San Bernardino.

The message communicating Kurita's last-minute decision to abandon the penetration of Leyte Gulf caused disappointment and indignation in Combined Fleet headquarters. To Admiral Toyoda and his staff, Kurita's action, especially after the success reported in the battle off Samar, seemed tantamount to throwing away an excellent opportunity. With the First Striking Force rapidly steaming north, it was recognized that a fresh order to execute the attack

106 Soon after ordering his ships to set course again toward Leyte Gulf, Vice Adm. Kurita had sent off a dispatch to Southwest Area Fleet urgently requesting air support.

107 Sources covering the reasons for Vice Adm. Kurita's decision to abandon the penetration of Leyte Gulf are as follows: (1) First Striking Force Detailed Action Report, op. cit., pp. 26–7. (2) Philippine Area Naval Operations, Part II, op. cit., pp. 54–6. (3) Interrogations of Vice Adm. Kurita and Rear Adm. Koyanagi, USSBS, *Inierrogation of Japanese Officials*, op. cit. Vol. I, pp. 32. (4) Statement of Comdr. Otani, previously cited. (5) Reports to the Throne by Chief of Navy General Staff, op. cit., 26 Oct 44.

108 (1) First Striking Force Detailed Action Report, op. cit., p. 28. (2) In the Detailed Action Report of the Flagship *Yamato*, it is recorded that the telegrams from Ozawa were received at 1215 and 1430, but were not brought to the attention of Adm. Kurita until evening. Battle and Flag Logs of IJMS, op. cit. Vol. II, *Yamato*.

109 Actually, the original contact report was erroneous, and no enemy carrier force was in the area at the time.

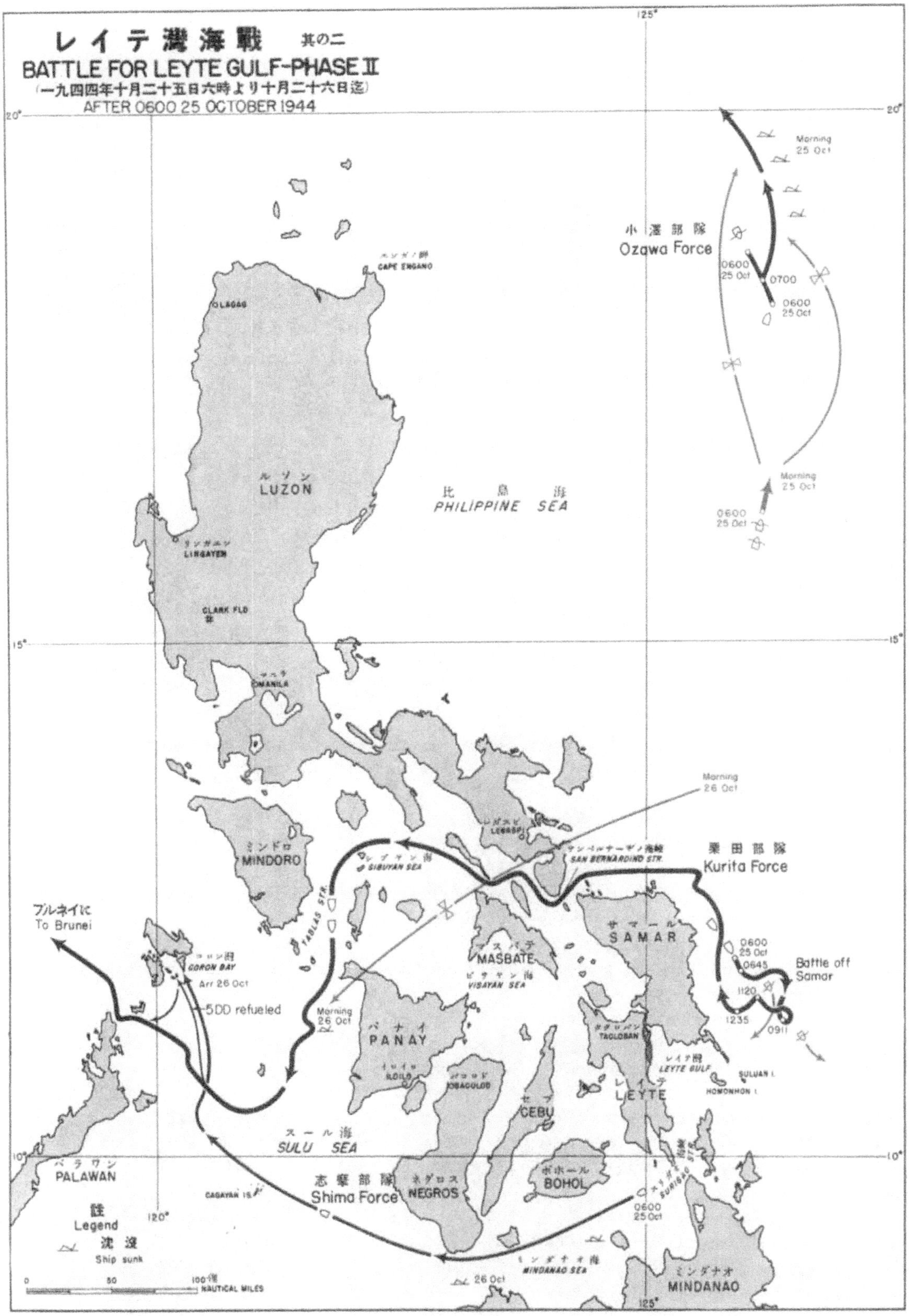

PLATE NO. 94

Battle for Leyte Gulf, Phase II, after 0600 25 October 1944

as planned could probably not be carried out. Nevertheless, hoping to spur Vice Adm. Kurita into a more positive course of action, Admiral Toyoda at 1647 originated the following dispatch order:[110]

1. If the opportunity arises, the First Striking Force will tonight attack and destroy remnants.[111] Other forces will cooperate with this attack.

2. If no opportunity presents itself, the Task Force Main Body and the First Striking Force will return to their supply bases, such movement to be carried out as ordered by the respective commanders.

By the time this message was received by *Yamato* at 1925, the First Striking Force was about to begin its run through San Bernardino Strait. Vice Adm. Kurita, since he had failed to make contact with the reported, but actually non-existent, enemy task force northeast of Samar, now saw no hope of carrying out a night attack, and the fuel situation of his ships was becoming steadily more critical. He therefore decided to continue on to Brunei.

Vice Adm. Kurita steamed away from the scene of battle unaware that the naval air forces,[112] resuming their offensive on the morning of the 25th, had actually scored results which would have facilitated his penetration into Leyte Gulf. *Kamikaze* units of the First Air Fleet, in the first real demonstration of their effectiveness, had heavily hit one enemy carrier group to the east of Surigao Strait,[113] following this up with a damaging attack on the group with which the Kurita Force had just broken off action east of Samar.[114] Second Air Fleet units based at Legaspi also flew repeated regular sorties throughout the day, but these achieved little or no success.[115]

Vice Adm. Kurita also did not know that the Ozawa Force had so effectively fulfilled its sacrifice mission that it had lost a total of seven ships, including all four of its carriers, during 25 October. At 0630 of the fateful day, Vice Adm. Ozawa, after rendezvousing with Rear Adm. Masuda's advance guard, turned his force northward and ran away from the Philippines at top speed in order to lure the enemy after him. At 0815, just at the height of the critical battle off Samar, planes from Admiral Halsey's fast carriers, which had sped northward during the night, began attacking Vice Adm. Ozawa's

110 First Striking Force Detailed Action Report, op. cit., p. 31.

111 The attack mission was couched in these vague terms to avoid committing Vice Adm. Kurita to a specific course of action which might be impossible under the circumstances prevailing at the time he received the order. However, the phraseology implied that Kurita was to take offensive action of some kind, preferably against remaining enemy forces in the Leyte Gulf area. (Statement by Rear Adm. Takata, previously cited.)

112 The First and Second Air Fleets, effective 25 October, were combined as the First Combined Base Air Force, Vice Adm. Shigeru Fukudome, Second Air Fleet Commander, assuming overall command. Philippine Area Naval Operations, Part II, op. cit., p. 18.

113 This attack was made by a *Kamikaze* group, which took off from Davao at 0630. One regular enemy carrier was reported sunk. (Actually, American sources indicate that two escort carriers, USS *Santee* and USS *Suwanee*, were seriously damaged, and a third escort carrier, USS *Sangamon*, lightly damaged.)

114 This attack was executed by the *Shikishima* Unit, based at Clark Field. Two planes were reported to have crashed into and sunk one enemy medium carrier, and another plane hitting but not sinking a second carrier. (According to American sources, the escort carrier USS *Saint Lo* was sunk, and three other escort carriers sustained varying degrees of damage in this attack. Only one of six suicide dives made by the attacking aircraft failed to hit a target.)

115 During the morning of 25 October, an attack force of over 50 bombers and fighters took off for a strike against enemy carrier groups off the east coast of Samar. The force failed to make contact, however, and returned to base. An aggregate of more than 70 planes sortied in the afternoon, between 1215 and 1630. These groups likewise were unsuccessful in executing attacks. Reports to the Throne by Chief of Navy General Staff, op. cit., 26 Oct 44.

ships in the last dramatic act of the battle for Leyte Gulf.

Between 0815 and 1700 wave after wave of enemy planes drove in to pound the empty Japanese carriers and their protecting units. In the first attack the carrier *Chitose* and destroyer *Akitsuki* were sunk, the light cruiser *Tama* seriously damaged,[116] and minor damage received by other ships, including Vice Adm. Ozawa's flagship, the regular carrier *Zuikaku*. During this action, Vice Adm. Ozawa radioed to Vice Adm. Kurita that the decoy force was under attack by huge numbers of enemy carrier planes, but this message was never received.

By noon the light carrier *Chiyoda* had received such heavy damage that she had to be abandoned.[117] Meanwhile, communications on the *Zuikaku* had broken down due to bomb damage, and at 1100 Vice Adm. Ozawa transferred his flag to the light cruiser *Oyodo*. *Zuikaku*, the last surviving carrier which had taken part in the raid on Pearl Harbor, went down under heavy enemy attack at 1414. *Zuiho* followed her to the bottom at 1527.

Having knocked out all of Vice Adm. Ozawa's carriers, enemy planes now began concentrating their attacks on the battleships *Ise* and *Hyuga*. Both ships were rocked by scores of near misses but escaped with only minor damage. Darkness finally brought respite, and the battered remnants of the decoy force, leaving three destroyers behind to pick up survivors, withdrew northward.[118]

The main action in the historic battle for Leyte Gulf was now ended, but still further losses were sustained by the retiring forces as they made their way back to base under continuing enemy air attacks.[119] These losses brought the total destruction suffered by the Fleet to four carriers, three battleships, six heavy cruisers, three light cruisers and nine destroyers, representing almost half the participating tonnage and 34.4 percent of Japan's total naval surface tonnage at the start of the Leyte battle.[120]

Exaggerated estimates of the losses inflicted on the enemy served, temporarily at least, to obscure the disastrous nature of the setback suffered in the Leyte battle.[121] It nevertheless

116 *Tama*, her speed cut to 14 knots, was ordered to retire alone to Okinawa. She was never heard from again.

117 Vice Adm. Ozawa ordered the light cruiser *Isuzu* and destroyer *Maki* to take off *Chiyoda's* crew and sink her. The carrier was still afloat, however, on the night of 25 October, when she was finally sent to the bottom by gunfire from American cruisers.

118 Two of the three destroyers left behind rejoined the main body late on 25 October. The third, *Hatsuzuki*, was believed sunk by enemy surface action while covering the retirement. Vice Adm. Ozawa thus returned with two light cruisers, two battleships, and six destroyers out of his original force of 17 ships.

119 The Kurita Force lost the light cruiser *Noshiro* and the destroyers *Hayashimo* and *Fujinami*, all by enemy carrier plane attack. The Shima Force lost the light cruiser *Abukuma*, when she was attacked by 30 enemy land bombers while limping back to Coron Bay and the destroyer *Wakaba* of the 21st Destroyer Division which was sunk by enemy carrier planes south of Mindoro on the morning of 24 October as the division was trying to catch up with Shima's main body.

120 Tonnage lost in the Leyte Battle was 271,000, out of a total participating tonnage of 556,800. Total surface tonnage of the fleet prior to the battle was 787,550. Losses in the Leyte battle were 26.1 percent of the aggregate losses suffered by the Japanese Navy for the entire period of the war.

121 Final estimates of the losses inflicted on the enemy by the surface and naval air forces in the Leyte battle (24–25 October) were as follows: By the First Striking Force: three or four carriers (including one of the *Enterprise* class), two heavy cruisers, and one destroyer sunk. By the First Combined Base Air Force: two carriers, one transport, and one light cruiser sunk: four carriers, two battleships, and one cruiser damaged. Reports to the Throne by Chief of Navy General Staff, op. cit., 25–26 Oct 44. (According to American naval sources, actual losses on 24–25 October were one light carrier and one escort carrier sunk by air attack (escort carrier *Saint Lo* by *Kamikaze*): one escort carrier, two destroyers and one destroyer escort sunk by surface action.)

was soon apparent that the sea-air phase of the *Sho-Go* plan to crush the enemy invasion of the Philippines had failed of its objective.

Deterioration on Eastern Leyte

On Leyte itself, the ground situation had deteriorated rapidly during the period of the vital sea and air operations. In the northern beachhead, although the 33d Infantry continued to hold its key hill positions west and northwest of Palo, blocking an enemy thrust inland over the Palo-Carigara highway, the enemy on 22 October began driving in the makeshift Japanese defenses west of Tacloban and at the same time pushed south from Palo to widen the beachhead to the Binahaan River. In the south, a powerful enemy thrust from Dulag toward the Burauen airfields was simultaneously gaining momentum, the spearhead of the attack reaching a point three miles west of Dulag by 22 October. (Plate No. 95)

Watching these developments from his command post at Dagami, Lt. Gen. Makino decided on the night of the 22nd that an immediate realignment of the southern sector forces was imperative. Since commitment of the main strength of these forces in frontal resistance to the enemy drive along the Dulag-Burauen road would leave the flank of the 9th Infantry in the Catmon Hill-Tanauan coastal sector dangerously exposed, he ordered the 20th Infantry to disengage along the road and shift its main strength northward to the Hindang area, where it was to occupy prepared positions and make ready for a possible counterattack on the enemy flank. One element of the regiment was to be left at San Pablo to join ground personnel of the 34th Air Sector command in defending the Burauen group of airfields.[122]

Swift developments, however, soon rendered this plan ineffective. Weakened by the shift of the 20th Infantry's main strength, the right-flank positions on the Dulag-Burauen road gave way under heavy enemy assault, and by 24 October the easternmost of the three Burauen airfields, near San Pablo, had been overrun. At the same time, the main body of the 20th Infantry, moving into its new positions in the Hindang sector, was suddenly hit and driven into retreat by enemy elements which had infiltrated inland, by-passing the 9th Infantry strongpoint on Catmon Hill.[123]

Lt. Gen. Makino now recognized that the loss of the eastern coastal plain could not be averted and laid plans to pull back the division main strength to rear-line positions in the mountains west of Dagami and Burauen, with the 33d Infantry to continue covering the division left flank north of the Binahaan River.[124] Orders for the execution of the withdrawal were transmitted to the 9th and 20th Infantry Re-

---

122 This order also directed the 9th Infantry to secure its Catmon Hill-Tanauan positions, screening the reassembly of the 20th in the Hindang area, and the 33d Infantry to continue resistance north and west of Palo in order to deny the enemy access to the vital Palo-Carigara highway, which must be used by Japanese reinforcements advancing to the front from western Leyte. 16th Division Operations Order No. 837, 22 Oct 44. XXIV Corps ADVATIS Translation No. 13, 20 Nov 44.

123 Col. Keijiro Hokoda, 20th Infantry Regiment commander, had already been killed in action on the Dulag front on 23 October. Philippine Operations Record, Phase Three, op. cit. Vol. II Suppl., p. 41.

124 Under these plans, the 9th Infantry and attached units were to take up positions to the northwest of Dagami, being redesignated as the Left Sector Unit; the 20th Infantry and attached units were to become the Central Sector Unit with positions northwest of Guinarona; and the airfield troops of the 34th Air Sector command were to become the Right Sector Unit with positions to the west of Burauen. Central Sector Unit (20th Infantry Regiment) Operations Orders No. 2, 27 Oct 44; No. 6, 29 Oct 44; and No. 13, 1 Nov 44. XXIV Corps ADVATIS Translation No. 1, 3 Nov 44.

PLATE NO. 95
Leyte Campaign, 21 October—1 November, 1944

giments—the latter already retreating from Hindang—on 26 October. The 9th Infantry was directed to start evacuating its Catmon Hill positions after the 27th.[125]

In the meantime the situation throughout the southern sector had deteriorated still further. By 27 October, enemy forces pushing on from San Pablo had taken both North and South Burauen airfields, as well as Burauen itself, placing all five major airfields on eastern Leyte in enemy hands.[126] Other enemy elements were hammering at the 9th Infantry's secondary positions at Tabontabon, three miles northwest of Hindang, while on the coast a juncture of enemy forces in the vicinity of Tanauan had linked the northern and southern beachheads.

Although rendered hazardous by the speed of the enemy's advances, the withdrawal of the southern forces proceeded with reasonable success. The remnants of the 20th Infantry reached their assigned positions northwest of Guinarona by 27 October, and at the same time the airfield troops driven out of Burauen took up positions on the right flank. Except for a small rear-guard, the 9th Infantry troops on Catmon Hill pulled out on the night of 29 October and carried out a dangerous but successful infiltration past Tabontabon and Dagami, both of which had fallen to the enemy in the interim. What remained of the regiment closed into its new positions northwest of Dagami by 31 October, completing the withdrawal. Approximately 3,500 troops were now assembled in the heights west of the Burauen-Dagami road.[127]

Meanwhile, resistance in the north had crumpled disastrously after 25 October. West of Tacloban, the enemy had crossed the Tigbaw River against scattered resistance and were now in a position to launch a flanking drive up the west side of San Juanico Strait. Simultaneously the 33d Infantry was forced to yield its positions west and northwest of Palo,[128] and the enemy immediately drove through the breach to begin a swift advance toward Jaro along Highway 2. Resistance in the northern sector was now so disorganized that 16th Division headquarters had no exact knowledge of the extent of enemy penetrations inland.[129]

During the final week of October, the Fourth Air Army and First Combined Base Air Force (Navy) kept up a sustained effort in support of the ground operations, gradually shifting the main weight of their attacks from amphibious shipping and carriers to the enemy-occupied airfields at Tacloban and Dulag.[130]

---

125 Outline of Operations and Lessons of the Leyte Campaign, op. cit., p. 35.

126 Ibid. (According to Japanese sources, both the Burauen North and Burauen South airfields were evacuated on 26 October. This differs from American records, which state that the Bayug (Burauen South) airfield was taken on 24 October, and the Buri (Burauen North) airfield on 27 October, the latter against strong Japanese resistance.)

127 Ibid., p. 38. These troops included the remaining strength of the 2d Battalion, 33d Infantry, which had fought under 20th Infantry command in the southern sector, as well as a small number of 33d Infantry survivors who had made their way back to the division main body from the Palo front.

128 Prior to the loss of these positions, the regimental commander, Col. Tatsunosuke Suzuki, had been killed in action while leading a raid into Palo on 23 October. Philippine Operations Record, Phase Three, op. cit. Vol. II Suppl., p. 42.

129 The 16th Division command was entirely unaware that an element of the enemy's northern forces had moved amphibiously to the western end of San Juanico Strait and was gradually infiltrating into the area directly northeast of Carigara. It also had no information of the landing on southwestern Samar by elements of the enemy 1st Cavalry Division. (Statement by Lt. Col. Watanabe, previously cited.)

130 By agreement between Southern Army and Southwest Area Force, all the Army Air units and certain regular Navy Air units joined in the attack on ground installations (including airfields). The balance of the regular Navy units maintained shipping interdiction, while the *Kamikaze* concentrated on carriers.

This air support bolstered 16th Division morale but did not perceptibly slow the enemy's ground offensive. Moreover, the scale of the effort gradually declined as losses mounted[131] and as increasing air force strength had to be diverted to the protection of convoys carrying troop reinforcements to Leyte.[132]

In conjunction with the main air effort against enemy supply shipping and airfields, Navy *Kamikaze* units continued to carry out suicide attacks against enemy carrier task forces operating off the east coast of the Philippines. These attacks were carried out by small numbers of aircraft but continued to show a relatively high degree of effectiveness.

Initial Reinforcement

Back at Cebu, Thirty-fifth Army headquarters had received virtually no information regarding the critical developments on eastern Leyte and had deferred the advance of the headquarters to Ormoc pending the arrival of a liaison officer from Fourteenth Area Army bringing essential data with regard to general operational plans and the scheduled movement of troop reinforcements from Luzon.[133] In the meantime, the initial reinforcements ordered to Leyte by Thirty-fifth Army and Fourteenth Area Army had already begun arriving at Ormoc.

Beginning with the Tempei Battalion, which arrived from Cebu on 23 October, a total of approximately 6,150 troops mainly of the 30th and 102d Divisions disembarked at Ormoc from naval transports, warships, powered sailing vessels and small craft over a period of eight days ending on 30 October.[134] Despite spo-

---

131 Replacements available to Fourth Air Army units during the latter part of October averaged only about 52 percent of the losses sustained in combat. As of 31 October, the total operational strength of the Fourth Air Army had decreased to 148 planes of all types, while the First Combined Base Air Force was down to a total of 149 aircraft. (1) Philippine Air Operations Record, Phase Two, op. cit., Attached Chart 13. (2) Outline of Operations and Lessons of the Leyte Campaign, op. cit., Attached Chart 3. (3) Philippine Area Naval Operations, Part II, op. cit., p. 19.

132 Total daily sorties flown by Army and Navy air units in attack operations during the period 25–31 October were as follows (figures in parentheses represent *Kamikaze* aircraft):

	Fourth Air Army	First Combined Base Air Force
25 October	162	195 (18)
26 October	161	70 (5)
27 October	70	52 (14)
28 October	98	51
29 October	43	31 (14)
30 October	2	6 (6)
31 October	51	28

(1) Philippine Area Naval Operations, Part II, op. cit., p. 97. (2) Outline of Operations and Lessons of the Leyte Campaign, op. cit., Attached Chart 3.

133 This officer reached Cebu on 29 October, and the following day the Thirty-fifth Army Chief of Staff, Maj. Gen. Yoshiharu Tomochika, left for Ormoc with an advance party. The Area Army liaison officer informed Thirty-fifth Army that the 1st Division and a detachment from the 26th Division (Imabori Detachment) would land at Ormoc on 1 November, with the rest of the 26th Division and the 68th Brigade to follow to Ormoc or Carigara at dates to be communicated later. Philippine Operations Record, Phase Three, op. cit. Vol. II Suppl., pp 52–3, 58.

134 The units disembarked were: 41st Infantry Regiment (30th Division); 169th and 171st Independent Infantry Battalions (102d Division); Tempei Battalion (57th Independent Mixed Brigade); 20th Antitank Battalion; and an element of the 364th Independent Infantry Battalion (55th Independent Mixed Brigade). Shortage of shipping made it impossible to transport in the initial movement all the reinforcements which Thirty-fifth Army planned to dispatch from within the Army area. The headquarters, artillery and engineer units of the 102d Division did not move to Leyte until mid-November, and the 172d Independent Infantry Battalion of the same division was never sent due to a change in plan. Elements of the 77th Infantry, 30th Division reached Ormoc sector in late November and early December, but the division headquarters and special troops never moved from Mindanao. Ibid., pp. 53–6, 76–7.

radic enemy air attacks, no serious troop losses were sustained during these movements. Returning to Manila, however, the 16th Cruiser Division, with two naval transports, was severely attacked by enemy carrier planes in Jintololo Channel on 26 October, losing the light cruiser *Kinu*, the destroyer *Uranami*, and one transport.[135]

At Ormoc, the reinforcement units found themselves completely unable to establish communications contact with 16th Division headquarters, under whose command they were now supposed to operate, and began advancing independently via Highway 2 with the intention of joining the division in the Dagami-Burauen area. The 41st Infantry (Tempei Battalion attached), starting soon after its disembarkation on 25 October, led the van and by the 28th was passing through Carigara. Immediately behind it came the 169th and 171st Independent Infantry Battalions.[136]

The 41st Infantry had proceeded only a short distance east of Carigara when it suddenly came into contact with a small enemy force, which withdrew northward after a brief skirmish. Despite this indication that the enemy had already infiltrated into the coastal area northeast of Carigara,[137] the 41st Infantry commander decided to continue the advance along Highway 2, leaving the Tempei Battalion to secure the east flank.

On 30 October, as the regiment moved down the road toward Jaro, it again ran into the enemy, this time in considerable strength. Taken badly by surprise, the regiment was forced to fall back through the hills toward Carigara. The enemy continued up the road to attack the 169th Independent Infantry Battalion, which had advanced behind the 41st to the vicinity of Tunga, and the battalion was likewise driven into retreat through the hills to the southwest. The Tempei Battalion, at this time, had just moved into Carigara, toward which the enemy pressed on unopposed.

The situation was in this precarious state when the advance party of Thirty-fifth Army headquarters, led by Maj. Gen. Yoshiharu Tomochika, finally reached Ormoc on the night of 30 October. Nothing was yet known at Ormoc, however, regarding the developments near Jaro and Tunga; nor was Maj. Gen. Tomochika able to obtain any clearer picture of the situation from an Army liaison officer who had left 16th Division headquarters for Ormoc on 27 October. The report of this officer indicated an enemy break-through on the division's northern flank but gave no clue as to the extent of penetration.

Consequently unaware that the enemy was already preparing to strike at Carigara from two sides, Maj. Gen. Tomochika believed that the Thirty-fifth Army plan to use the Carigara

135 The sinking of the *Kinu* and *Uranami* virtually wiped out the 16th Cruiser Division, since its only other unit, the heavy cruiser *Aoba*, had been heavily damaged by submarine attack on 23 October off the west coast of Luzon and was laid up at Manila. Philippine Area Naval Operations, Part II, op. cit., pp. 83–5.

136 The 364th Independent Infantry Battalion remained at Ormoc to secure the area and assist in future debarkation operations, while the 20th Antitank Battalion, arriving on the night of 30 October, was not ready to begin moving toward Carigara until 2 November.

137 This was the first knowledge the Japanese gained of the presence of enemy troops in the Barugo area. It was ascertained from local informants that they had landed by boat. (1) Philippine Operations Record, Phase Three, op. cit. Vol. II Suppl., p. 68. (2) Field Diary of a Unit of the 30th Division. ADVATIS Bulletin No. 116, 4 Jan 45.

"A completely unexpected move was the American advance through San Juanico Strait. . . . Our naval intelligence had estimated that the channel could not be navigated except by small native craft. . . . This surprise maneuver upset the plans for the recapture of Tacloban." Interrogation of Maj. Chuji Kaneko, Staff Officer (Intelligence and Operations) 102d Division.

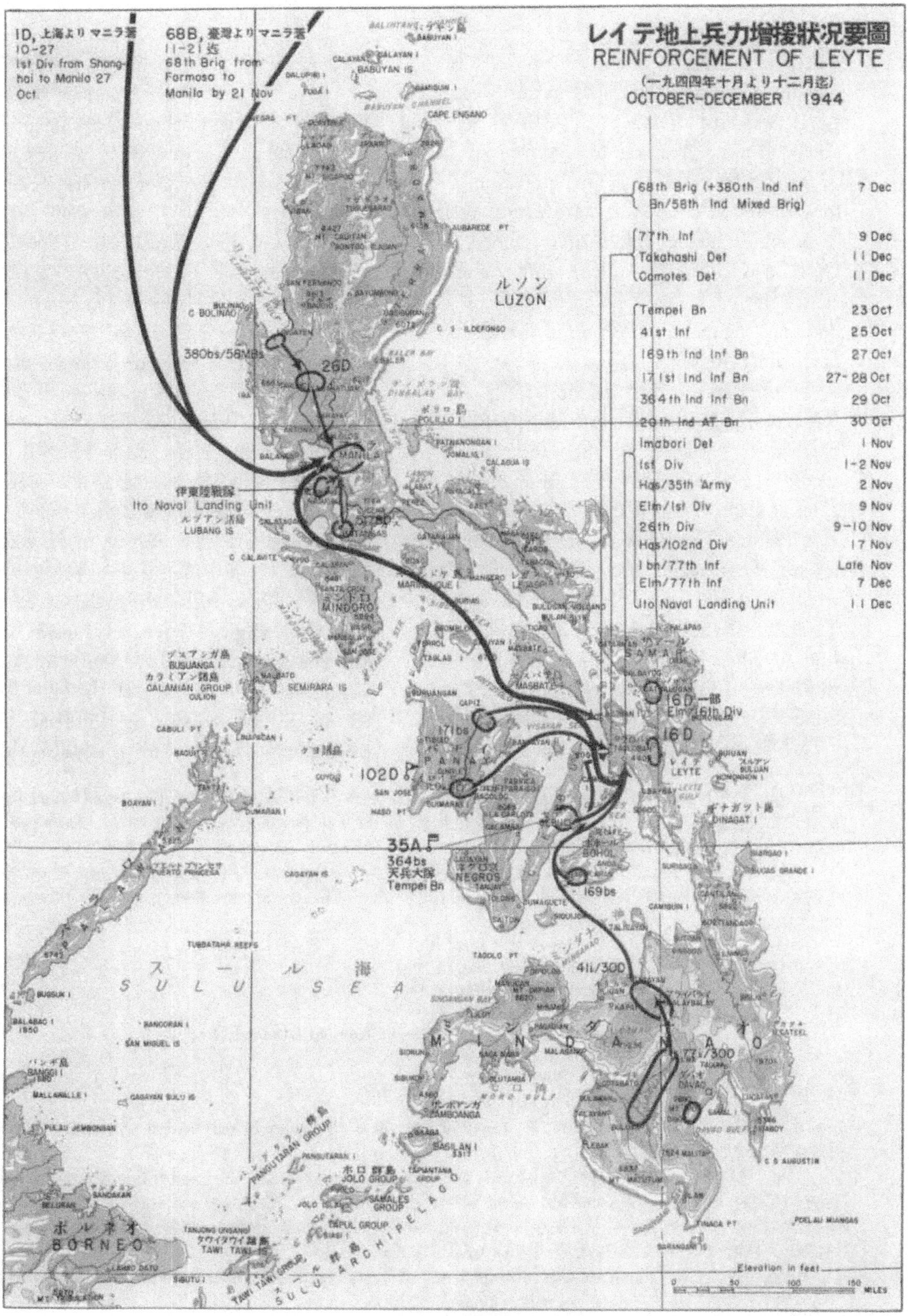

PLATE NO. 96

Reinforcement of Leyte, October—December 1944

area as the massing point for the main forces scheduled to undertake the recapture of eastern Leyte was still feasible. However, it was obviously necessary to set up a unified command over the reinforcement groups already moving forward and to secure the Carigara area against a possible enemy penetration. Since 16th Division could not exercise this command effectively, Maj. Gen. Tomochika decided to undertake temporary direction of the reinforcement units himself.

To implement this decision, Maj. Chuji Kaneko, 102d Division staff officer, was dispatched to Carigara on the morning of 31 October carrying instructions from the Army chief of staff. These instructions called for the 41st Infantry, if possible, to continue its advance toward the Dagami area, and for the 169th and 171st Independent Infantry Battalions, the Tempei Battalion and 20th Antitank Battalion to take up positions on a river line three miles east of Carigara for the purpose of covering the assembly of the main Army forces.

While Maj. Kaneko was on this mission, three naval transports entered Ormoc harbor in mid-afternoon of 1 November and swiftly disembarked the Imabori Detachment of the 26th Division,[138] representing the first element of the major troop reinforcements allocated by Fourteenth Area Army. The same night, the main body of the convoy, with four large transports, reached Ormoc and began unloading the veteran 1st Division from Manchuria.[139] By noon of 2 November, a total of 13,000 troops had safely landed, concluding the most successful major reinforcement operation of the Leyte campaign.[140] (Plate No. 96)

Meanwhile, on the basis of staff studies completed between 22 and 25 October, Fourteenth Area Army had estimated that five additional divisions were needed to fill long-term operational requirements in the Philippines, three for Thirty-fifth Army use on Leyte and two as strategic reserve on Luzon.[141] This estimate was submitted to Southern Army on 28 October, and on the 30th the latter forwarded an initial request to Imperial General Headquarters for the earliest possible shipment

138 The Imabori Detachment was composed of the 12th Independent Infantry Regiment (less 2d Battalion), plus reinforcing elements. It had been ordered to Leyte by the 26th Division commander on 22 October in compliance with Gen. Yamashita's orders. (Cf. p. 352)

139 As it landed at Ormoc, the division was short one battalion from each infantry regiment, these battalions having been constituted as a second echelon for later shipment. Component units disembarked on 1 November were:

Headquarters
1st Infantry (less 2d Bn.)
49th Infantry (less 3d Bn.)
57th Infantry (less 1st Bn.)
1st Reconnaissance Regt.
Two light tank cos. (attached from 2d Armored Div.)
1st Field Artillery Regt.
1st Engineer Regt.
1st Division Special Troops

Dai Ichi Shidan Sakusen Kodo no Gaiyo 第一師團作戰行動の概要 (Summary of 1st Division Operations) Home Depot Division, 1 Mar 47, pp. 1–4, 9.

140 The convoy had moved from Luzon under strong air and surface escort and had undergone no serious attacks en route despite being spotted by enemy planes on 1 November. One ship was sunk in Ormoc harbor on 2 November, but the enemy air attack came after all troops and the bulk of the equipment and supplies were already ashore. Philippine Area Naval Operations, Part II, op. cit., pp. 85–7.

141 The three-division estimate for employment on Leyte excluded the 68th Brigade, which had not yet arrived from Formosa but was already earmarked for Thirty-fifth Army. It also excluded the 26th Division, only one element of which had so far been transported from Luzon to Leyte. (Statement by Maj. Gen. Nishimura, previously cited.)

of one to two more divisions.[142] Shortly thereafter Southern Army was informed that the transfer of the 23d Division from Manchuria was tentatively scheduled.

Plans for the continuing reinforcement of Leyte underlined the necessity of bolstering air strength, since the diversion of planes to convoy escort duty was seriously reducing the ability of Fourth Air Army and the First Combined Base Air Force to carry out primary operational missions. An Army-Navy Central Agreement was therefore concluded on 28 October, earmarking the bulk of new aircraft and flying units for the Philippines as they became available.[143] Under this agreement, the Army planned to dispatch a number of additional air regiments and earmarked all available replacement pilots and aircraft for the Fourth Air Army. The Navy planned to deploy no additional units but to replenish the existing base air forces through the allocation of maximum replacements.

Limon Battle: Phase I

On the morning of 1 November, Maj. Kaneko reached the Carigara area to discover that the situation was actually so critical as to render Maj. Gen. Tomochika's instructions to the reinforcement groups, as well as the general plan to hold the Carigara plain as the assembly area for the Army main strength, impossible of execution.

The enemy's drive along Highway 2 to within a few miles of Carigara not only blocked the 41st Infantry's route of advance to the Dagami area but outflanked the proposed river line east of Carigara even before it had been established. Troop strength immediately available to defend this line was inadequate. The 41st Infantry and the Tempei Battalion were in the general vicinity of Carigara, with the 171st Independent Infantry Battalion about a day's march to the west, but the 169th Independent Infantry Battalion was far off in the mountains southwest of Tunga, and the 20th Antitank Battalion was still at Ormoc. (Plate No. 97)

Since immediate action was necessary, Maj. Kaneko decided to abandon Maj. Gen. Tomochika's instructions and effect new dispositions. Contacting the commanders of the reinforcements units in the Carigara area, he advised them to occupy the heights southwest of Capoocan instead of the river line east of Carigara. By the night of 2 November, the 41st Infantry and the Tempei Battalion had pulled back from Carigara, joining the 171st Independent Infantry Battalion in the new positions.

Maj. Kaneko had dispatched a courier to Ormoc on 1 November with a report of the situation and the emergency dispositions he planned to effect. The courier, however, did not reach Army headquarters until the morning of 3 November. In the meantime, on 2 November, the Army Commander, Lt. Gen. Suzuki, had arrived at Ormoc from Cebu and, on the basis of an outdated briefing by the chief of staff, had drawn up a plan for an early counteroffensive from the Carigara area. Aware of the fact that the heaviest enemy pressure was being brought against the northern flank

142 This was in the nature of an emergency interim requisition made pending further examination by Southern Army of the Area Army estimate. Southern Army further requested the shipment of 800 replacement aircraft, to arrive not later than 6 November.

143 The agreement further specified the missions of the Army and Navy air forces as follows: *Army air forces*—(1) to neutralize enemy land air bases; (2) to secure air control over the enemy anchorage area; (3) to attack transports and interdict unloading operations; (4) to render close support to ground troops. *Navy air forces*—(1) to intercept enemy supply and reinforcement from the sea, including attack on enemy transports in anchorage; (2) to attack enemy carriers and gunfire support ships. Imperial General Headquarters Army High Command Record, op. cit., pp. 272–3.

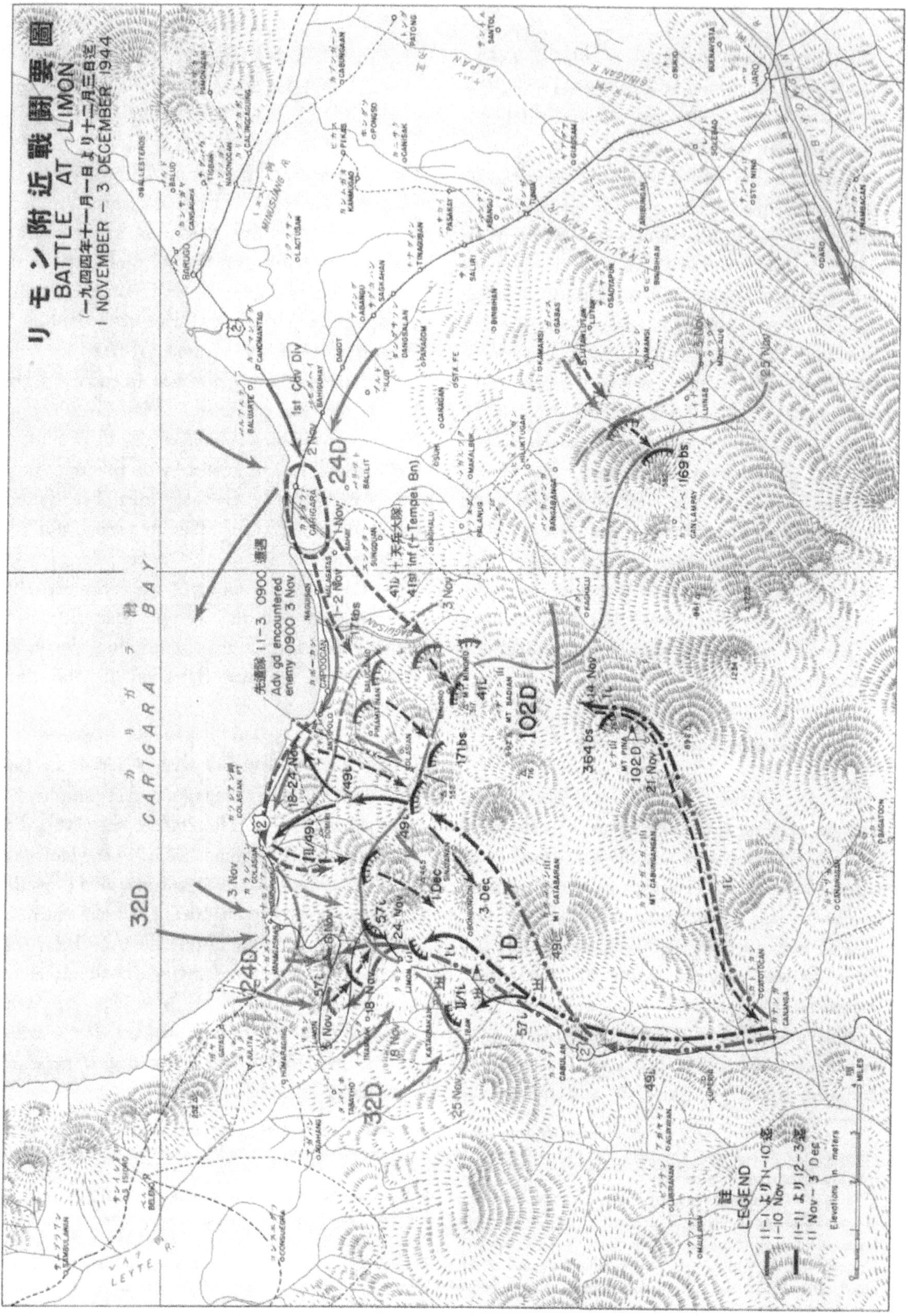

PLATE NO. 97
Battle at Limon, 1 November—3 December 1944

of the 16th Division, the Army Commander decided to shift the main weight of the counteroffensive toward Tacloban instead of Dulag as originally planned. This plan provided:[144]

1. Upon completion of the assembly of the 1st Division in the Carigara area, the Army will begin an offensive in the direction of Tacloban and will destroy the enemy.

2. The 1st Division will make the main effort.

3. The Imabori Detachment will occupy the northeastern foot of the heights southwest of Jaro and cover the assembly of the main body of the 26th Division.

4. After landing, the 26th Division will assemble in the vicinity of Dolores. It will then advance to the Jaro area and launch an attack toward Tacloban.

The 1st Division was meanwhile beginning its march northward to Carigara. The advance guard, consisting of the 1st Reconnaissance Regiment (reinf.), moved out of Ormoc on Highway 2 at 1430 on 2 November. The division commander, Lt. Gen. Tadatsu Kataoka, and the first serial of the division main body, composed of the 57th Infantry Regiment (reinf.), followed about nine hours later.

When Maj. Kaneko's report finally arrived, the full seriousness of the situation in the Carigara area became clear. Lt. Gen. Suzuki immediately approved the emergency dispositions already taken and issued orders temporarily attaching the various reinforcement units to the 1st Division. It was impossible, however, to notify Lt. Gen. Kataoka immediately of the changed situation, and the division advance guard moved into the vicinity west of Capoocan at 0900 on 3 November to find itself unexpectedly engaged by the enemy.[145]

Badly surprised, the advance guard fell back toward Colasian with the enemy in pursuit. Meanwhile, at 1000, Lt. Gen. Kataoka reached the high ground south of Managasnas. Finding his advance guard engaged, he immediately ordered the 1st Reconnaissance Regiment and the 20th Antitank Battalion[146] to check the enemy advance. At the same time he dispatched an order to the division main body to close up at forced march.[147]

While the 1st Reconnaissance Regiment fought in Colasian during the afternoon to check the enemy push from Capoocan, an enemy amphibious force estimated at one battalion, with more than four tanks, suddenly landed in the regiment's rear at Pinamopoan, just east of Managasnas. Lt. Gen. Kataoka thought that this small force could easily be pinched off, and he ordered the 57th Infantry to move up from the Limon area and attack, with the remainder of the division to complete its concentration in the vicinity of Cananga as rapidly as possible. Meanwhile, however, the enemy, after a heavy artillery bombardment, had moved into Managasnas. The 1st Reconnaissance Regiment fell back to the southwest and occupied the east end of a long ridge overlooking the coast.

At 1800 on 4 November, the 57th Infantry began moving up into attack positions on the left of the 1st Reconnaissance Regiment. During the 5th and 6th, attacks were launched toward Managasnas, but enemy artillery superiority caused heavy casualties, and the 57th

144 Philippine Operations Record, Phase Three, op. cit. Vol. II Suppl., pp. 67–8.

145 The enemy had entered Carigara on 2 November to find that the Japanese had already evacuated.

146 The 20th Antitank Battalion had left Ormoc shortly before the advance guard of the 1st Division, but had been overtaken by the latter's mechanized reconnaissance elements. The battalion henceforth operated as part of the division.

147 Narrative of the operations in the Limon area is based on the following sources: (1) Summary of 1st Division Operations, op. cit. (2) *Dai Hyakuni Shidan Sakusen Keika no Gaiyo* 第百二師團作戰經過ノ概要 (Summary of 102d Division Operations) Home Depot Division, Apr 46. (3) Outline of Operations and Lessons of the Leyte Campaign, op. cit., pp. 41–84. (4) Philippine Operations Record, Phase Three, op. cit. Vol. II Suppl., pp. 56–152.

Infantry was obliged to yield part of the coastal ridge.

The 1st Division main body had by this time almost completed its concentration in the Cananga area. On 6 November Lt. Gen. Kataoka issued a new operational plan envisaging commitment of the division's full strength, plus the reinforcement units under division control, in a wide four-pronged attack. The essentials of this plan were as follows:[148]

1. The 1st Division will attack the enemy in the Carigara area, the 41st Infantry on the right, the 1st Infantry on the right center, the 49th Infantry on the left center, and the 57th Infantry on the left. The line of departure for the attack and unit fronts will be as indicated below.

2. The 41st Infantry, reinforced by the Tempei Battalion, will secure the division right flank by occupying the heights west of Santo Nino to Hill 582.

3. The 169th Independent Infantry Battalion will advance to the upper reaches of the Naguison River and, upon arrival, will be attached to the 1st Infantry.

4. The 171st Independent Infantry Battalion will assemble near Hill 517 and will be attached to the 49th Infantry.

5. The 1st Infantry (reinf., less 2d Bn.) will advance to the high ground astride the Naguison River, maintaining liaison with the 41st Infantry on the right and the 49th Infantry on the left.

6. The 49th Infantry (reinf., less 3d Bn.) will advance to a line extending from the northern foot of Hill 517 to the northern foot of Hill 465 and prepare to attack in the direction of Capoocan.

7. The 57th Infantry (reinf., less 1st Bn.) will continue its present mission in the sector south of Managasnas.

8. The 1st Artillery Regiment will support the attack from positions around Mt. Catabaran.

Two factors, however, rendered Lt. Gen. Kataoka's planned scheme of maneuver difficult to execute. First, the 1st and 49th Infantry Regiments had to deploy over the mountainous terrain to the east of the Ormoc-Limon Highway to reach their assigned line of departure; and second, the reinforcement units were widely scattered and, in some cases, far from their designated sectors. Progress of the deployment was extremely slow, and the 57th Infantry, with the 1st Reconnaissance Regiment and 20th Antitank Battalion, meanwhile had to carry the attack alone. These units fought a savage two weeks' action but were barely able to hold their own in the strategic heights south of Managasnas.[149]

While the Limon battle thus entered a temporary stalemate, over-all developments in the battle for Leyte were impelling higher Army headquarters in Manila to undertake a serious review of both tactical plans and strategic policy.

Command Decisions

Two main factors led Fourteenth Area Army to decide early in November on a revision of the tactical plan being followed by the Thirty-fifth Army on Leyte. The first of these factors was the enemy's unexpectedly rapid sweep across northern Leyte to the Carigara area. The second was the steadily increasing use by enemy land-based air power of the captured airfields in the Tacloban and Dulag-Burauen sectors.[150]

148 1st Division Operations Order No. 41, 6 Nov 44. ATIS Bulletin No. 1656, 28 Dec 44.

149 On top of the failure of the 1st and 49th Infantry Regiments to reach their attack positions, heavy rains and enemy air attack hindered the deployment of the 1st Artillery Regiment, with the result that supporting fire to neutralize enemy artillery in the Carigara sector was not opened until 8 November. Another severe rainstorm on 9 November hampered operations by logistic difficulties.

150 On 5 November air intelligence reports indicated that about 150 medium bombers and fighters were based on the Tacloban airfield, and approximately 100 aircraft of various types on the fields in the Burauen area. Philippine Operations Record, Phase Three, op. cit. Vol. II Suppl., p. 90.

Thirty-fifth Army had formulated its initial plan of operations on the basis of the estimate that the enemy could be contained on eastern Leyte until the main Army forces could be concentrated in the Carigara area. The concentration completed, these forces were to move up to the Jaro sector, on the left flank of the 16th Division, and from this line of departure launch the main counteroffensive effort toward Dulag.

The enemy's penetration of the Carigara area even before the 1st Division had time to move up from Ormoc, changed the tactical picture drastically. If Thirty-fifth Army now adhered to its plan of advancing via Carigara and Jaro, it would face enemy resistance from the start. It would also be seriously handicapped by the necessity of conducting operations on two widely separated fronts, with the main forces engaged around Carigara and the 16th Division fighting an isolated action to hold its positions west of Dagami and Burauen.

The Area Army command estimated that the combined effect of these disadvantages would be to prolong the operations for recapture of eastern Leyte to such an extent that their ultimate success would be jeopardized. Speed was of the utmost importance since the enemy, if left free to expand the use of the eastern Leyte airfields, would gain a steadily increasing air advantage fatal to the reinforcement and supply of the Thirty-fifth Army. It therefore seemed imperative to modify the operational plan with the primary objective of recapturing the airfields as quickly as possible.

Since the 16th Division in the Burauen-Dagami area still held advantageous mountain positions within striking distance of three of the five major airfields occupied by the enemy, Fourteenth Area Army decided that the most effective plan was to restrict operations in the Carigara area to a holding action and throw the main weight of the Thirty-fifth Army's forces into an offensive on the 16th Division front. About 5 November, therefore, an order was dispatched to Lt. Gen. Suzuki directing him to deploy the Army main strength toward Burauen and launch offensive operations in that sector as speedily as possible.[151]

At the same time that this modification of the tactical plan was ordered, General Yamashita and his staff were becoming increasingly convinced that the overall situation afforded scant hope of victory on Leyte and that the whole strategic policy of fighting decisive ground action there should be abandoned, involving the cancellation of further reinforcements to Thirty-fifth Army. On 7 November, one day prior to the scheduled embarkation for Ormoc of the 26th Division, this view was verbally transmitted to Lt. Gen. Iimura, Southern Army Chief of Staff, by Lt. Gen. Muto, Chief of Staff of Fourteenth Area Army. The major reasons advanced in support of the Area Army view were as follows:[152]

> *1. The evident failure of the decisive operations carried out by the sea and air forces to bring about any significant weakening of enemy forces on Leyte or of enemy carrier air strength supporting the invasion.*
>
> *2. Deterioration of the overall air situation due to attrition of Japanese air strength as against a steady build-up of enemy air forces operating over the Philippines.*[153]

---

151 (1) Philippine Operations Record, Phase Three, op. cit. Vol. II, pp. 15–16. (2) Statement by Maj. Gen. Nishimura, previously cited. (3) Interrogation of Col. Kobayashi, previously cited.

152 Statements by Lt. Gen. Iimura and Maj. Gen. Nishimura, both previously cited.

153 In addition to an intensification of enemy air operation from land bases on eastern Leyte and on Morotai, carrier forces continued intermittent but heavy attacks on Luzon. On 5 and 6 November, enemy carrier planes attacked Manila, Clark Field, and Legaspi in a total of 1,170 sorties. Philippine Area Naval Operations, Part II, op. cit., pp. 22–3.

輸送部隊霧の中を進む　栗原　信　昭和二十三年三月十七日

Original Painting by Shin Kurihara

PLATE NO. 98
Transport Division Proceeding Through Fog

3. The extremely unfavorable development of the tactical situation on Leyte.

4. The increasing difficulty of transporting troop reinforcements due to enemy air activity, coupled with inability to check the steady reinforcement of enemy forces on Leyte.

5. The danger of stripping the defenses of Luzon without reasonable prospects of securing a decisive victory on Leyte.[154]

Despite these representations, the Southern Army Chief of Staff declined to consider any departure from the policy fixed by Imperial General Headquarters, and the embarkation of the 26th Division proceeded on schedule. The convoy carrying the division main strength, together with the remaining three infantry battalions of the 1st Division, sailed from Manila on the morning of 8 November under strong air and surface escort.[155]

The first section of the convoy, carrying the 1st Division elements, reached Ormoc on 9 November, unloaded, and left the harbor without incident. The 26th Division transports, however, underwent air attack as they neared Ormoc, resulting in damage to one ship. The transports entered the harbor at 1830 and began disembarking troops, severely hampered by heavy seas and a shortage of landing craft.[156]

By morning of 10 November, all troops were ashore, but no machine-guns, heavy weapons, supplies or equipment had been unloaded. Despite this, the convoy commander, in order to escape the enemy's regular morning air attack. ordered the transports to suspend

---

154 The Fourteenth Area Army estimate of overall additional troop requirements for the Philippine theater (cf. p. 380) had been pared down by Southern Army from five to four divisions. On 4 November, Southern Army requested Imperial General Headquarters to effect the transfer of the 23d and one other division as speedily as possible, and to assign two additional divisions for shipment to the Philippines prior to 31 December. On the same date, Southern Army received notification of the official transfer of the 23d Division to Fourteenth Area Army, but until 10 November there was no indication of Imperial General Headquarters' reaction to the troop request as a whole. This increased Fourteenth Area Army's skepticism with regard to continuation of decisive ground action on Leyte. (1) *Nampo Gun Sakusen Kiroku* 南方軍作戰記錄 (Southern Army Operations Record) 1st Demobilization Bureau, Jul 46, pp. 155–6. (2) Statements by Lt. Gen. Iimura and Maj. Gen. Nishimura, both previously cited.

155 Units embarked on this convoy were the following:

- 26th Division
  - Headquarters
  - 13th Independent Infantry Regt.
  - 2d Bn., 12th Independent Infantry Regt.
  - 3d Bn., 11th Independent Infantry Regt.
  - 26th Division Reconnaissance Regt.
  - 11th Ind. Field Arty Regt. (less one bn.)
  - 26th Engineer Regt.
  - 26th Division Special Troops
  - 25th Antitank Bn.
  - 21st Mortar Unit
  - 65th Independent Engr. Bn.
- 1st Division
  - 1st Bn., 57th Infantry
  - 2d Bn., 1st Infantry
  - 3d Bn., 49th Infantry

(1) Philippine Operations Record, Phase Three, op. cit. Vol. II Suppl., pp. 82–3. (2) Summary of 1st Division Operations, op. cit., p. 9. (3) Outline of Operations and Lessons of the Leyte Campaign, op. cit., p. 36.

156 Insufficient landing craft were available at Ormoc because Thirty-fifth Army had been informed that no large ships would henceforth be employed in transport operations to Leyte. The Army was therefore employing most of its landing craft elsewhere. Philippine Operations Record, Phase Three, op. cit. Vol. II Suppl., pp. 81–2.

unloading and sail for Manila. This precaution proved fruitless, for enemy planes caught the convoy shortly after it left the harbor, sinking two transports and one escort vessel. The surviving ships continued on to Manila.

On 11 November, a follow-up convoy of five transports reached Ormoc with ammunition cargoes and elements of 26th Division special troops. Just as unloading began, 300 enemy planes unleashed an attack which sank all the transports and five of seven escort vessels. The reinforcement operation, at an exceedingly heavy cost in ships, had ended in partial failure.[157]

At Manila, Fourteenth Area Army meanwhile pressed its recommendation for abandonment of decisive ground operations on Leyte. Following a direct interview on 9 November between General Yamashita and Lt. Gen. Iimura, Southern Army Chief of Staff, Field Marshal Terauchi summoned a conference of the Southern Army staff on 10 November to discuss the issue. On the basis of the decision reached at this conference, Field Marshal Terauchi personally summoned General Yamashita to Southern Army headquarters on 11 November and informed him that the following basic decision had been reached:[158]

1. The Leyte decisive operation will be continued to the best of the Army's ability, and, at the very least, the enemy will be denied the use of the Leyte airfields.

2. The defenses of Luzon will be strengthened but not so as to compromise the successful continuation of the Leyte operation.

In compliance with this decision, Fourteenth Area Army proceeded to draw up long-range plans for the continuation of the campaign, taking into account a notification to Southern Army on 10 November that Imperial General Headquarters planned to assign the 10th and possibly one other division to the Area Army in the near future.[159] On 15 November, a general directive incorporating these plans was issued:[160]

1. The Thirty-fifth Army will endeavor to accomplish the destruction of the enemy on Leyte, setting as its minimum objective the disruption of the enemy's use of air bases.

2. Forces:

a. (Troop units already on Leyte enumerated)

b. Projected reinforcements:

68th Brigade (main part to arrive Manila 17–18 Nov.)

23d Division (main part to arrive Manila 22–23 Nov; remainder from mid to late Dec.)

10th Division (arrive Manila early Dec.)

3. Dependent upon the availability of sufficient air cover and surface escort, these reinforcements will be transported to Leyte as required by the local situation. Under unfavorable conditions, the transport of troops may be delayed or temporarily suspended, but every effort will be made to maintain supply shipments adequate for the existing forces.

4. In the event that further troop shipments cannot be accomplished, Luzon will become the main theater of future operations in the Philippines.

Expansion of air force strength in the Philippines was meanwhile progressing under the

157 Philippine Area Naval Operations, Part II, op. cit., p. 89.

158 Statement by Lt. Gen. Iimura, previously cited.

159 The 10th and 19th Divisions, in Formosa and Korea respectively, were officially assigned to Fourteenth Area Army on 20 November. With this action, Imperial General Headquarters had assigned three of the four additional divisions requested by Southern Army. Due to the shipping situation, it was deemed impossible to move a fourth division to the Philippines before 31 December. (Statement by Col. Hattori, previously cited.)

160 Documents of Fourth Air Army and Fourteenth Area Army, op. cit. Outline of Operation Plans for the Leyte Decisive Battle, 15 Nov 44.

歩兵機関銃隊の前進　宮本三郎　昭和二十三年三月十七日

Original Painting by Saburo Miyamoto

PLATE NO. 99
Advancing Infantry Machine Gun Unit

Army-Navy Central Agreement of 28 October. Of eleven fresh Army air regiments scheduled for deployment during November and December, five (29th, 54th and 246th Fighter Regts.; 74th and 95th Heavy Bomber Regts.) had reached the Fourth Air Army by mid-November, and shipments of replacement aircraft were increasing.[161] On 23 November, the Air Army reorganized its subordinate units, making the 4th Air Division, hitherto charged only with base activities and anti-submarine patrol, fully operational, with responsibility for the Luzon and Mindoro sectors, the 2d Air Division remaining responsible for operations in the central and southern Philippines.[162]

Replenishment of naval air strength was also proceeding satisfactorily, replacement aircraft to the First Combined Base Air Force exceeding losses by 26 per cent during November.[163] On 15 November Combined Fleet decided to discontinue the training of carrier air groups in favor of concentrating on the expansion of the base air forces,[164] and the Third Air Fleet, stationed in the homeland, was assigned the primary mission of training replacement units for deployment to the Philippines. At the same time, the "T" Attack Force was ordered to advance from Formosan to Philippine bases, and its missions were expanded to include attacks on transports and enemy airfields.[165]

With plans completed for continuing the decisive battle on Leyte, Field Marshal Terauchi and his staff departed Manila for Saigon on 17 November in order to establish the Southern Army command at a more central location within the Army zone of responsibility. The transfer had been authorized by Imperial General Headquarters prior to the Leyte invasion but had then been deferred so that Southern Army could more effectively direct the preparations and planning for the decisive battle.[166]

---

161 Four additional air regiments (18th, 55th, and 71st Fighter Regts. and 106th Reconnaissance) arrived during the last half of November, and two (72d and 73d Fighter Regts.) during the early part of December. In addition, the 15th Air Reconnaissance Regt., which had earlier returned to the Homeland for rest and replenishment, and the 13th Fighter Regt., which was in the Celebes, were redeployed to the Philippines in early December. The ratio of replacements to aircraft losses increased from 52 percent for October to 79 percent for the first three weeks in November. (1) Philippine Air Operations Record, Phase Two, op. cit. Chart XIV. (2) Daily Record of the War Situation, op. cit., pp. 14–32.

162 Four fighter regiments (the 29th, 246th, 33d, and 20th) were assigned to the 4th Air Division, the 2d Air Division retaining control of eleven fighter regiments (the 17th, 18th, 22d, 24th, 27th, 31st, 45th, 54th, 55th, 71st, and 200th), one fighter-bomber regiment (the 67th), and two reconnaissance regiments (the 2d and 38th). The 30th Fighter Group, temporarily released to 2d Air Division command on 23 October, returned to direct command of Fourth Air Army on 23 November, when it was ordered back from Bacolod to Clark Field for rest and replenishment. The remainder of the air units in the Philippines, including the entire heavy bomber force, was under direct Fourth Air Army command. Documents of Fourth Air Army and Fourteenth Area Army, op. cit. Outline of Air Operations of the Fourth Air Army in the Leytet Decisive Battle, 15 Nov 44; and Fourth Air Army Operations Order No. 593, 23 Nov 44.

163 Philippine Area Naval Operations, Part II, op. cit., pp. 23, 98.

164 As a result of this decision, the Third Fleet, then engaged in training carrier air units, was broken up. The 634th Air Group was assigned to Second Air Fleet, and most of the pilots and aircraft of the 601st Air Group were used as replacements to the First Combined Base Air Force. Headquarters, Third Fleet and the 653d Air Group were deactivated.

165 Philippine Area Naval Operations, Part II, op. cit., p. 27.

166 Marshal Terauchi had consistently maintained that Southern Army headquarters should be centrally located on the continent to assure more effective strategic control of operations throughout the southern area. In August, Imperial General Headquarters agreed in principle to a transfer of Southern Army headquarters to Saigon but left the time to be decided later. Transfer was finally authorized in October following a staff visit to Manila by Maj. Gen. Joichiro Sanada, Chief, 1st Bureau (Operations), Imperial General Headquarters, Army Section. (Statement by Lt. Gen. Iimura, previously cited.)

Execution of the transfer involved no change in the chain of command, Southern Army retaining control of both Fourteenth Area Army and Fourth Air Army. Subject to this control, the Area Army and Fourth Air Army remained equal theater commands for ground and air operations respectively, and were to effect local coordination between themselves and with Southwest Area Fleet.[167]

Limon Battle: Phase II

The debarkation of the 26th Division and the second echelon of the 1st Division at Ormoc on 9–10 November found the situation on the Limon front still stalemated. The 1st Division elements set out immediately for the Limon area, while the 26th Division main strength, despite its lack of weapons, was ordered to assemble at Dolores, five miles north of Ormoc, in preparation to move up and join the Imabori Detachment engaging the enemy southwest of Jaro.[168]

The 1st Division elements were well on the way to Limon when the Fourteenth Area Army order of 5 November, delayed in transmission, reached Thirty-fifth Army on 10 November. To comply with the directed shift of the Army's main front toward Burauen, Lt. Gen. Suzuki abandoned his plan to commit the 26th Division on the Jaro front and ordered Lt. Gen. Tsuyuo Yamagata, the division commander, to move his troops as rapidly as possible to Albuera, nine miles southeast of Ormoc, and prepare for operations eastward. To replace the 26th Division in future operations on the Jaro front the Army Commander planned to use the 30th Division. The division, which had not yet left Mindanao, was accordingly sent orders to land at Ipil, concentrate north of Ormoc, and prepare to reinforce the Imabori Detachment.

By 13 November the newly-arrived battalions of the 1st Division closed into the Limon area, and Lt. Gen. Kataoka immediately decided to regroup his forces for a new offensive. The scheme of maneuver embodied in the 6 November battle plan was discarded, since the 1st Infantry had still not reached its scheduled attack positions astride the Naguison River. Lt. Gen. Kataoka decided to launch the offensive on a narrower front, shifting the 1st Infantry to the left flank to support the 57th Infantry, under heavy enemy pressure to the south of Managasnas. On 14 November the following dispositions were ordered:[169]

1. Disposition of second echelon units: The 1st Battalion, 57th Infantry, will revert to 57th Infantry control. The 3d Battalion, 49th Infantry, will be attached to the 57th Infantry. The 2d Battalion, 1st Infantry, will be the division reserve.

2. The 57th Infantry (reinf.) will attack along the main road toward Managasnas.

3. The 49th Infantry (less 3d Bn.), with the 171st Independent Infantry Battalion attached, will envelop the enemy left flank in the direction of

167 At the time of the transfer, a proposal was advanced in favor of placing the Fourth Air Army under command of Fourteenth Area Army to assure more effective coordination of ground and air operations. Southern Army rejected the proposal, however, on the principal ground that it was necessary to maintain a strategic mobility of air forces between all sectors in the Southern Army zone of responsibility. (Statement by Col. Yozo Miyama, Chief, Operations Section, Southern Army and Col. Hottori, previously cited.)

168 The Imabori Detachment had moved up from Ormoc to the vicinity of Daro, about three miles southwest of Jaro, by 4 November. Since that date it had been engaged by the American 96th Division.

169 (1) Philippine Operations Record, Phase Three, op. cit. Vol. II Suppl., pp. 103–4. (2) Summary of 1st Division Operations, op. cit., p. 9–11. (3) Field Message, CO 49th Infantry Regt to CO 2d Bn, undated. X Corps ADVATIS Translation No. 23, 25 Nov 44.

Colasian.

4. The 1st Infantry (less 2d Bn.) will concentrate in the sector south of Limon.[170]

5. The 1st Field Artillery Regiment will support the advance of the 57th and 49th Infantry from its present positions, priority of fire to the 57th Infantry.

6. The 41st Infantry (reinf.) will cover the division right flank by operations from the heights southwest of Capoocan.

7. The 2d Battalion, 1st Infantry, will be the division reserve.

In execution of these orders, the 57th Infantry launched an attack on 14 November, pushing forward to the summit of the ridge overlooking Managasnas but suffering heavy casualties in an enemy counterattack with flame-throwing tanks. The 49th Infantry, which did not attack until 16 November, made even slower progress in the densely-forested foothills to the southwest.

While the battle continued, the commander and headquarters of the 102d Division, together with a battalion of field artillery and part of the divisional engineers, landed at Ormoc from Iloilo on the night of 17 November.[171] Thirty-fifth Army immediately ordered the reversion to 102d Division control of the 169th and 171st Independent Infantry Battalions and also directed the division to take command of the 41st Infantry Regiment (reinf.) and the 364th Independent Infantry Battalion.[172] With these forces, Lt. Gen. Shimpei Fukue, 102d Division commander, was assigned the mission of securing the heights south and southwest of Capoocan and harrassing enemy artillery and rear installations by raiding tactics.[173]

The force of its attack spent, the 57th Infantry was now losing ground south of Managasnas and fighting desperately to prevent an enemy breakthrough into the Pagsangahan Valley corridor leading south to Ormoc. On the right flank, however, the 49th Infantry succeeded on 18 November in infiltrating the enemy left to launch a two-pronged thrust toward Colasian and Capoocan. Using raiding tactics, the 1st Battalion reached the sea west of Capoocan, while the 2d Battalion penetrated to the vicinity of Colasian.

This pressure on the enemy's rear line of communications brought no let-up in his attack on the 57th Infantry front. A fresh enemy division (32d) was thrown into battle on 18

170 The 169th Independent Infantry Battalion, previously ordered attached to the 1st Infantry, was not included in the 14 November plan. The battalion was still separated from the right flank elements of the 1st Division by a six-mile gap, and communication contact was so poor that the battalion commander did not know that he was under 1st Division command. (1) Field Message, CO 169th Indep Inf Bn to CG 102d Div, 7 Nov 44. ADVATIS Bulletin No. 63, 13 Dec 44. (2) Summary of 102d Division Operations, op. cit., p. 15.

171 With the arrival of these elements, the total strength of the 102d Division on Leyte, less units attached after arrival, was approximately 3,600. The 172d Independent Infantry Battalion was left behind on Negros to guard the Bacolod airfields against guerrillas, and the remaining five battalions were dispersed on Negros, Panay, Masbate, Bohol and other islands of the Visayas. *Reite Tonyu Heiryoku Hyo* レイテ投入兵力表 (Chart of Forces Committed on Leyte) 1st Demobilization Bureau, Dec 48.

172 The 364th Independent Infantry Battalion was now up to its full complement of 1,000 troops as a result of the movement of its remaining elements from Cebu. Upon attachment to the 102d Division, the battalion was relieved of its guard mission in the Ormoc area. (1) Ibid. (2) Philippine Operations Record, Phase Three, op. cit. Vol. II Suppl., pp. 54, 112.

173 The division command post was established in the sector southwest of Mt. Pina on 21 November, but communication difficulties made it impossible to establish liaison with scattered subordinate units until about 25 November. By Army order, eight guns were released by the divisional artillery unit to the 26th Division. (1) Philippine Operations Record, Phase Three, op. cit. Vol. II Suppl., pp. 111–112. (2) Field Message, Chief of Staff, 102d Division, 22 Nov 44. ADVATIS Bulletin No. 94, 24 Dec 44.

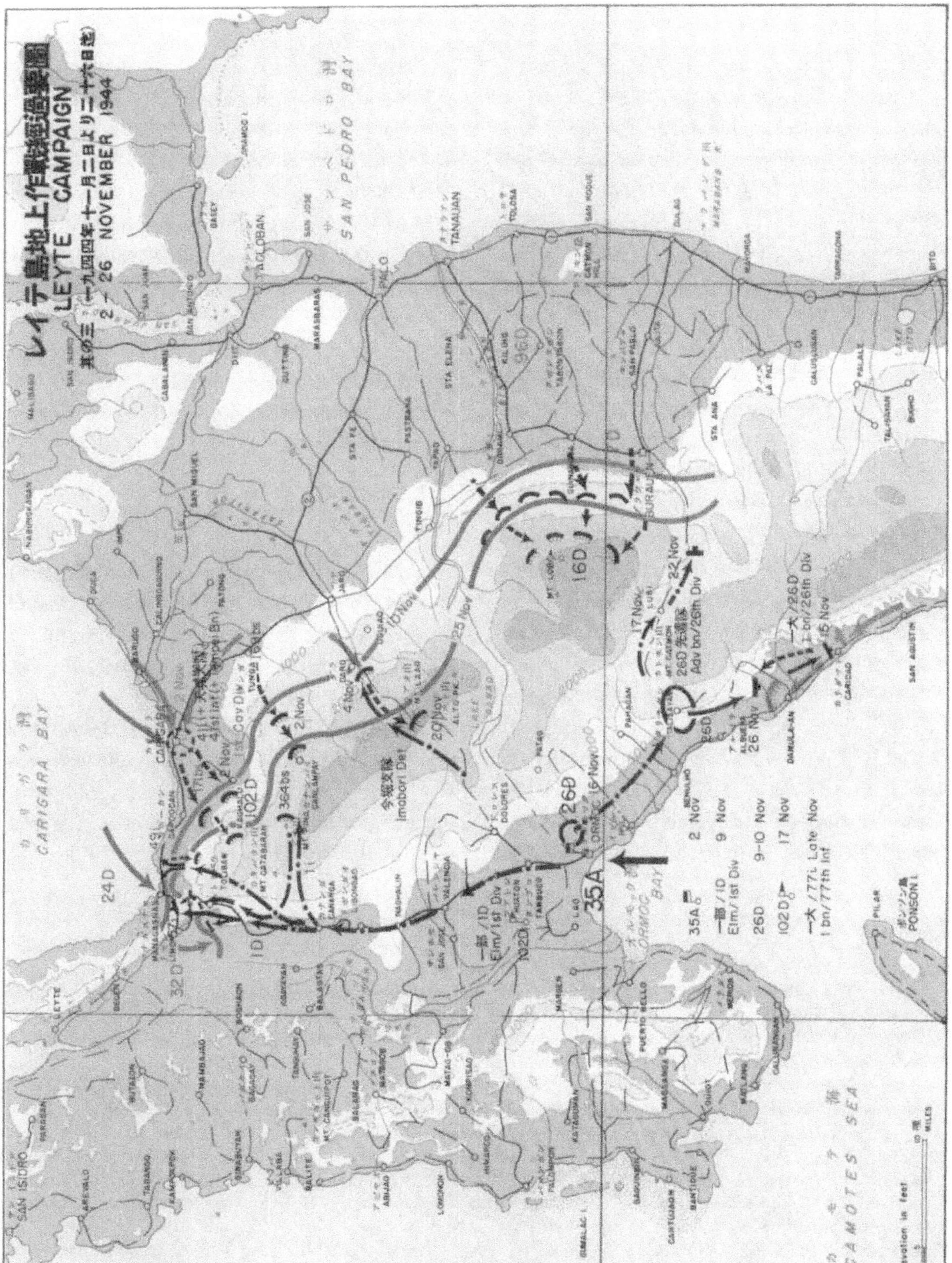

PLATE NO. 100

Leyte Campaign, 2–26 November 1944

November, the left flank of the 1st Division was turned, and the strategic heights north of Limon were lost. Lt. Gen. Kataoka now decided to commit the 1st Infantry around the left of the 57th Infantry to restore the line. Before this could be done, however, enemy elements drove between the 57th Infantry and the Leyte River to capture Limon on 24 November, and a second enemy spearhead crossed the Leyte River west of Limon, threatening the positions of the 1st Artillery Regiment.[174] The 1st Infantry, reaching its scheduled attack positions on the left rear of the 57th, found itself already outflanked and forced to go on the defensive.

The 1st Division was now in a tenuous situation. With no reserves available for commitment, Lt. Gen. Kataoka had to suspend the 49th Infantry's operations in the Colasian-Capoocan sector and make these troops available to support the division left flank along the Limon-Ormoc highway.[175] The withdrawal of the 49th opened a wide gap between the 1st and 102d Divisions, and the enemy swiftly advanced through the breach to attack rear echelon elements of the 1st Division along the highway south of Limon.

On 5 December the Thirty-fifth Army Commander ordered the Imabori Detachment to disengage its main forces leaving only a security detachment on the front southwest of Jaro, pull back to Dolores, and then hasten northward to bolster the disintegrating 1st Division. While the detachment was engaged in this maneuver, the situation on the north grew steadily worse. The 1st Division was now enveloped on both flanks, and the offensive to retake Carigara had become a desperate struggle to hold the enemy away from Ormoc.

Battle of Burauen

On the 16th Division front west of Burauen and Dagami, November witnessed determined Japanese efforts to break the enemy's grip on the Burauen airfields, but these efforts, too, proved unavailing.[176]

The beginning of the month found the division, reduced to less than 3,500 effectives and with almost no heavy weapons, endeavoring to consolidate its positions in the hills west of the Burauen-Dagami road.[177] On the

174 To relieve the threat to the 1st Artillery Regiment, Lt. Gen. Kataoka committed the division reserve (2d Battalion, 1st Infantry) to an attack along the west side of the Limon-Ormoc highway in the vicinity of Tabayho. This attack did not succeed in driving the enemy back across the Leyte River but did save the artillery positions from direct attack. Summary of 1st Division Operations, op. cit., pp. 12–3

175 The 49th Infantry units operating in the Colasian-Capoocan sector were subjected to strong enemy counterattacks on 24 November and began pulling back the same day to positions south of Colasian. On 1 December, the 2d Battalion was ordered to evacuate these positions to reinforce the division left flank, and two days later the 1st Battalion was likewise ordered to the left flank sector. Ibid. pp. 11, 13.

176 Narrative of the battle of Burauen is based on the following: (1) Philippine Operations Record, Phase Three, op. cit. Vol. II, pp. 20–6; Vol. II Suppl. op. cit., pp. 131–44. (2) Philippine Air Operations Record, Phase Two, op. cit., pp. 94–7. (3) Outline of Operations and Lessons of the Leyte Campaign, op. cit., pp. 71–4, 81–7, 91–2. (4) Philippine Area Naval Operations, Part II, op. cit., pp. 29–32. (5) *Gohoku Sakusen Kiroku Furoku Dai Ni Dai Sanjuni Shidan (Kaede) Morotai To Sento Gaishi* 濠北作戰記錄附錄第二第三十二師團楓モロタイ島戰鬪概史 (North of Australia Operations Record, Supplement II, General Outline of the 32d Division (Kaede) Operations on Morotai) 1st Demobilization Bureau, Jul 46. pp. 25–6. (6) Statement by Lt. Col. Sato, previously cited. (7) 16th Division Report on the *Wa* Operation, 15 Dec 44. ADVATIS Bulletin No. 152, 15 Jan 45. (8) Airborne Raiding Unit, 2d Parachute Group. Operations Order No. 11, 28 Nov 44. ADVATIS Bulletin No. 146, 12 Jan 45. (9) Situation on Leyte, 19–20 Nov 44. X Corps ADVATIS Translation No. 39, 14 Dec 44.

177 By 10 November total strength, including 800 attached naval base personnel, was down to 3,350. The division staff was almost intact, but most of the company commanders and all battalion commanders and above in the infantry units, half the battalion and company commanders in the artillery, and all engineer officers had been killed. Only three infantry guns and 20 machine guns remained in the division.

night of 6 November, the enemy launched a savage attack, supported by armor, which pushed Lt. Gen. Makino's troops back to the foot of Mt. Lobi and a point about three miles southwest of Burauen. The supply situation was critical, but the division continued small-scale night raids on enemy positions and ambush attacks on enemy patrols. In order to aid the 16th Division in these efforts, Thirty-fifth Army hastily organized and dispatched to the Burauen front a provisional infantry company.

The 26th Division was meanwhile preparing to move up from Albuera toward the Burauen front. As far as Lubi, a road unusable by vehicles was the only route of advance, and beyond Lubi even foot-trails were non-existent. Lt. Gen. Yamagata ordered the 3d Battalion of the 13th Independent Infantry Regiment ahead as an advance guard, with the 26th Engineer Regiment to follow and repair the road to Lubi. On 18 November the advance guard reached Lubi and moved on into uncharted terrain. Behind it, the engineers made slow progress, hampered by bad weather, lack of equipment, and constant enemy air interdiction.

Meanwhile, a new threat was developing to the south. On 10 November, an enemy force had been reported moving up the western coastal road north of Baybay, and Lt. Gen. Yamagata had ordered the 2d Battalion of the 12th Independent Infantry Regiment to move down the road from Albuera to verify the report and cover the division right flank. On 15 November, the battalion was attacked by guerrillas and shortly thereafter made contact with elements of an enemy force which was apparently concentrating in the vicinity of Caridad.[178] Notified of this development, Thirty-fifth Army ordered the 26th Division to contain any eventual enemy thrust north toward Albuera while proceeding with the movement of its main strength toward Burauen.

It was apparent, however, that the division would remain only effective unless its lack of heavy weapons, ammunition and equipment was remedied. To meet this need, Fourteenth Area Army planned to run two supply convoys into Ormoc sometime after 23 November and requested Fourth Air Army to provide adequate cover against enemy air attack.

In view of the increasingly stringent enemy air blockade around Leyte, Fourth Air Army recognized that the usual escort measures would probably be insufficient. Lt. Gen. Tominaga therefore planned to supplement these by a large-scale offensive against enemy air bases and shipping, coupled with a neutralization attack on the Burauen airfields by a small force of airborne troops. The latter mission was assigned to the Kaoru Airborne Raiding Detachment, with a strength of 80 men.[179] On 22 November this unit was alerted, and the following operational instructions were issued:

> *1. The Fourth Air Army will cover the movement of transport convoys to Leyte during the period 23–27 November.*
>
> *2. Throughout this period enemy airfields will be subjected to concentrated neutralization attacks, and attacks will also be carried out against enemy vessels*

178 Since no amphibious movements of enemy troops had been detected around the southern coast of Leyte, it was presumed that the force in the Caridad area had advanced from eastern Leyte over the Abuyog-Baybay road. Thirty-fifth Army had discounted the possibility of an enemy advance over this road in view of 16th Division reports to the effect that the road had been rendered impassable by road blocks and demolitions.

179 The Kaoru Detachment was organized from part of the strength of the *Gi-Go* (義號) Airborne Raiding Unit, a specially-trained and equipped infantry force of about company size, attached to Fourth Air Army. The *Gi-Go* Unit originally belonged to the 1st Raiding Unit (Cf. n. 25, Chapter XII), the bulk of which had been sent to western New Guinea and Morotai. One company remained in the Philippines and was trained for airborne operations, Fourth Air Army intending to use it on Morotai. This company was redesignated as the *Gi-Go* Airborne Raiding Unit.

in anchorage.

3. In the evening of 26 November the Kaoru Raiding Detachment will be transported by air to the Burauen North and South Airfields. The transports will crash land on the airfields with landing gear retracted and the detachment will then demolish airfield installations. After landing, the detachment will come under the command of the Thirty-fifth Army.

The airborne attack on the Burauen airfields was thus conceived as a limited hit-and-run operation. Fourteenth Area Army and Fourth Air Army immediately proceeded, however, to draw up a more ambitious plan for a coordinated airborne and ground offensive to effect the permanent recapture of the Burauen airfields. On 23 November Fourteenth Area Army issued a series of orders embodying the plan for this offensive, henceforth designated as the *Wa* Operation. The essential points of the plan were as follows:

1. General outline: As much air power as possible will be assembled in order to gain temporary air superiority. Airborne forces will be utilized to seize the Burauen and San Pablo airfields in a surprise attack. To further insure capture of the above areas, the 16th and 26th Divisions will launch a general land offensive.

2. X-Day to be announced. It will fall in the period 5–10 December.

3. Preliminary Attacks

a. 1–3 December—General air offensive aimed at airfields.

b. 4–7 December—Naval air support of ground operations in the Limon sector to create a diversion.

4. Main Attacks

a. The 2d Parachute Group[180] *will secure an airborne lodgement at Burauen Army Airfields (North and South) and at San Pablo Airfield on the night of X-1.*

b. Thirty-fifth Army will execute strong patrolling and infiltration operations on the Dagami-Burauen front on X-1.

c. On X-Day, the 16th and 26th Divisions will execute a coordinated attack. The 16th Division objective will be Burauen North Airfield and the 26th Division objectives will be Burauen South and San Pablo Airfields.

On 24 November the air offensive to cover the moment of the supply convoys got under way, with sixty 2d Air Division planes striking at the Leyte airfields and 30 Navy aircraft attacking enemy shipping in Leyte Gulf. Daily attacks were mounted over the Leyte area during the next four days, while the 7th Air Division carried out coordinated strikes against enemy bases on Morotai. On the evening of 26 November, transports carrying the Kaoru Detachment took off from Lipa Airfield, on Luzon, reporting by radio two hours later that they were over the target area. No further reports were received, but since no enemy planes appeared over Ormoc Bay on 27 November, it seemed that the operation had been at least partially successful.[181]

Despite these measures, the supply operation proved a failure. The first section of the first convoy, consisting of three transports with one escort, was caught by enemy carrier aircraft on 24 November off Masbate, and all four ships were sunk. The second section, identically composed, was attacked the following day

180 The 2d Parachute Group was composed of the 3d and 4th Parachute Regiments. These regiments had been ordered to the Philippines by Imperial General Headquarters on 25 October for employment on Leyte. On 6 November, Imperial General Headquarters activated the 2d Parachute Group to command the two regiments and an air regiment assigned to provide airlift. The Group was assigned to Fourth Air Army. (Statement by Col. Hattori, previously cited.)

181 Post-war access to American documentary sources establishes that the Kaoru Detachment raid actually failed. Three of the transports overshot their targets and crash-landed in the surf at Bincay, about five miles south of Dulag, on the coast of Leyte Gulf. A fourth transport tried to land at Burauen North (Buri) Airfield but was turned back by antiaircraft fire. All detachment personnel were killed or captured.

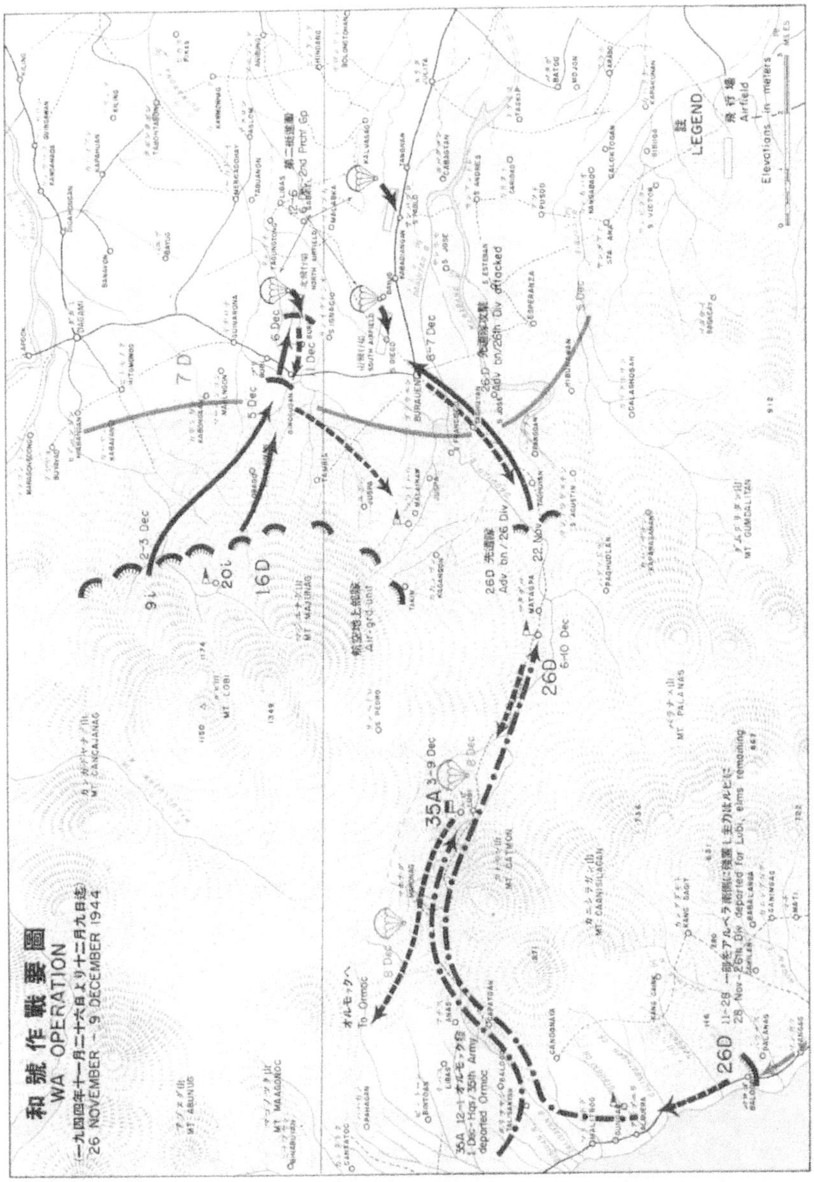

PLATE NO. 101
Wa Operations, 26 November—9 December 1944

northeast of Mindoro, losing two ships while the remaining two put back to Manila. The second convoy, with two transports, two subchasers and a patrol boat, reached Ormoc safely on 27 November, only to be attacked the same night by enemy PT boats operating in Ormoc Bay. Enemy land-based planes followed up with a dawn strike, and every ship in the convoy was sunk.

Thirty-fifth Army now faced almost insuperable obstacles to the execution of its mission under the *Wa* Operation plan. Of the two divisions earmarked for the ground assault, the 16th Division was down to 2,000 men by the end of November and was losing an average of about 75 men per day. The main body of the 26th Division was not yet moving toward the attack area and still lacked weapons and supplies. The battalion guarding the division flank against the enemy threat from the south had been pushed back, forcing Lt. Gen. Yamagata to move his main strength from Albuera toward Caridad to relieve the pressure. At the end of November, the 3d Battalion of the 13th Infantry in the Lubi area was the only element of the 26th Division within a week's march of Burauen.

Under these circumstances, Thirty-fifth Army felt that no coordinated attack on the Burauen front could be mounted before 7 December. Fourteenth Area Army nevertheless designated 5 December as X-Day, and Lt. Gen. Suzuki was forced to try to meet the deadline. Instructions were sent to the 16th Division to launch an attack toward the Burauen North Airfield on the night of 5 December, and, the 26th Division, newly reinforced by the 3d Battalion of the 77th Infantry, 30th Division, which had just landed in the vicinity of Ipil, between Ormoc and Albuera, was ordered to move immediately to the Lubi area and prepare for an attack on Burauen South Airfield. Three battalions were to remain at Albuera to contain the enemy moving north on the coast highway. The 68th Brigade, scheduled to arrive about 6 December, was to land at Ipil and assume the role of Army flank guard against the threat from the Caridad sector, thus relieving the balance of the 26th Division for operations on the Burauen front.

Pursuant to instructions, the 9th and 20th Infantry Regiments, 16th Division, hastily drew their remaining effectives together into provisional composite assault units. The 9th Infantry, 400 strong, and the 20th Infantry, with 800, assembled west of Dagami. Meanwhile, the main strength of the 26th Division disengaged in the Caridad sector and began advancing toward Burauen. So slow was its progress that by 3 December it had only reached Lubi, and it was apparent that the movement could not be completed in time for the attack. Despite this, the advance battalion of the division assembled in a mountain defile west of Burauen and prepared to attack on 5 December as planned.

On 1 December, the Thirty-fifth Army Commander left Ormoc to establish an advance command post at Lubi. Two days later, advice was received from Manila that the airborne assault had been postponed until the night of 6 December due to predictions of adverse weather. Thirty-fifth Army immediately took steps to notify the 16th and 26th Divisions, but a communications breakdown made it impossible to get through to General Makino's headquarters.

On the night of 5 December, the 16th Division composite assault units, unaware of the change in schedule, set out from their assembly point west of Buri, infiltrated the enemy lines, and attacked Burauen North Airfield during the morning of the 6th. Initial enemy reaction was confused, and the Japanese had successfully occupied about half the airfield by the time the first wave of airborne troops parachuted down on the target area at 1920. The combined forces renewed the attack, completing the occupation of the airfield during the night of 6–7 December.

This initial success was shortlived. Planes carrying the second wave of airborne troops from Lipa were forced back to base by bad weather, and the third wave, also waiting at Lipa, never got off the ground. On 8 December the enemy launched a strong counterattack on Burauen North Airfield, supported by artillery, mortars and tanks. The small Japanese force resisted stubbornly for three days but, by 11 December, was forced to withdraw into the mountains to the west after demolishing a number of enemy aircraft, guns and vehicles.

Simultaneously with the attack on Burauen North Airfield, other parachute units had jumped on Burauen South and San Pablo Airfields, but these units were never heard from and were presumed to have been wiped out.[182] A coordinated ground attack toward Burauen by the 3d Battalion, 13th Independent Infantry Regiment of the 26th Division, on 6–7 December had likewise ended in failure due to stubborn enemy resistance. The bid to regain possession of the eastern Leyte airfields was at an end.

Ormoc—Final Collapse

While the abortive *Wa* operation was in progress, the already critical situation on northwestern Leyte was rendered still more desperate by a sudden enemy amphibious landing directly on the eastern shore of Ormoc Bay.[183] Thirty-fifth Army, though aware of increasing enemy patrol boat activity in the Camotes Sea, had had neither time nor adequate forces to remedy the almost complete lack of defenses in the bay area.[184]

To the authorities in Manila, the latent enemy threat to the Camotes Sea area was a source of great concern. Southwest Area Force headquarters had for some time been studying possible countermoves to an enemy penetration of the area, and, on 4 December, Fourteenth Area Army issued an order establishing a Camotes Islands Garrison Detachment, the strength being fixed at two provisionally organized infantry companies, a battery of artillery, and an engineer platoon. This

182 Post-war access to American documentary sources reveals that these units landed on their assigned targets but were immediately engaged by elements of the enemy 11th Airborne Division and forced to withdraw from both Burauen South and San Pablo Airfields by noon of 7 December. In addition to the major airborne landings on the Burauen group of airfields, small parachute detachments were dispatched to attack the airfields at Dulag and Tacloban. The transports carrying these units, however, were unable to reach the targets due to anti-aircraft fire and never returned to base.

183 The narrative of the Ormoc battle is based on: (1) Philippine Operations Record, Phase Three, op. cit. Vol. II, pp. 26–33. (2) Philippine Operations Record, Phase Three op. cit. Vol. II Suppl., pp. 58, 62, 139–194. (3) Philippine Air Operations Record, Phase Two, op. cit., pp. 95–8, 100–01, Chart XVII. (4) Outline of Operations and Lessons of the Leyte Campaign, op. cit., pp. 92–111. (5) Daily Record of the War Situation, op. cit., pp. 25–33. (6) Document Files, Thirty-fifth Army, op. cit. (7) Reports to the Throne by Chief of the Navy General Staff, op. cit. (8) Philippine Area Naval Operations, Part II, op. cit., pp. 32–6, 95–5. (9) Statements of Maj. Gen. Tomochika, Lt. Col. Asaeda, and Lt. Col. Watanabe, all previously cited. (10) Field Message, Fourteenth Area Army to CG, 26th Division, 19 Nov 44. Eighth Army ADVATIS Translation No. 11, 12 Jan 45.

184 When the initial plans were formulated for the defense of Leyte, the possibility of an amphibious attack on the west coast was considered negligible since an enemy convoy would have to negotiate the treacherous and heavily mined Canigao Channel and sail close to the naval and air bases on Cebu. In mid-November, enemy PT boat activity in the Ormoc Bay area drew attention to the need of stronger shore defenses, and plans were laid for the construction of field fortifications in the foothills of the coastal range. However, since combat troops could not be spared and shipping unit personnel in Ormoc were fully engaged, little construction was ever accomplished.

detachment was to leave Manila for the Camotes Islands as soon as possible.

At 0305 on the morning of 7 December, naval patrol bombers operating over the Camotes Sea detected an 80 vessel convoy of cruisers, destroyers, and transports about ten miles southwest of Baybay on a northerly course. The alarm was flashed to Manila and to Army and Navy units throughout the central Philippines. At 0800, the gunnery ships of the enemy formation began a violent bombardment of Ipil and Albuera. Between 0920 and 1100 enemy amphibious troops estimated at more than a regiment swarmed ashore at Albuera.

Meanwhile, the wheels of Japanese counteraction had been set in motion. Southwest Area Force ordered First Combined Base Air Force to hit the enemy amphibious force immediately and in full strength. At 0940 Fourth Air Army ordered the 2d Air Division to throw its full strength against the enemy landing force as soon as its planes could be put into the air. The 5th Air Brigade, comprising the main heavy bomber strength of the Air Army, and special attack units of the 4th Air Division were likewise ordered to begin attack operations immediately. Fifty planes of the 2d Air Division, 12 heavy bombers of the 5th Air Brigade, and 20 suicide aircraft promptly sortied against the enemy landing point. In addition, 56 naval aircraft of the First Combined Base Air Force from Clark Field joined the attack.[185]

Fourteenth Area Army meanwhile dispatched orders to the Thirty-fifth Army commander to suspend the *Wa* operation at once and throw the main strength of his forces against the new enemy landing. The 68th Brigade, which was already enroute to Leyte under orders to land at Ormoc, was ordered to go ashore instead at Palompon, 15 miles west of Ormoc, from which a first-class metalled road gave quick access to Highway 2 and the Ormoc corridor.

After landing, the 68th Brigade was to rush north and shore up the disintegrating lines of the 1st Division, thus making it unnecessary for Lt. Gen. Suzuki to weaken his forces in the south for that purpose. Fourth Air Army meanwhile agreed to land the remaining strength of the 2d Parachute Group at Valencia airdrome, eight miles north of Ormoc, for use against the enemy beachhead. Fourteenth Area Army also began considering the advisability of shipping additional ground units from Luzon to northwestern Leyte.

The situation on the Ormoc front was desperate in the extreme. The only troops within close striking distance of the enemy's Albuera beachhead were the Mitsui Shipping Unit reinforced by newly-arrived minor elements of the 77th Infantry, 30th Division, the 26th Division flank guard detachment already engaging elements of the U. S. 7th Division south of Albuera, and some 26th Division and Thirty-fifth Army rear echelon personnel. Beside these fragments, the only regular line unit within reasonable distance was one battalion of the Imabori Detachment at Dolores. The remainder of the detachment was strung out in march column along the mountain trails to the east, still engaged in its withdrawal from the Jaro front.

At Thirty-fifth Army headquarters in Ormoc, the deputy chief of staff, Maj. Gen. Tomochika, in the absence of the Army Commander

185 Results claimed in these attacks were one battleship, one cruiser, two destroyers, five transports, and more than ten landing barges sunk, and eight transports damaged. (According to American sources, one destroyer, one large transport and a number of amphibious craft were damaged. The damage to amphibious shipping hampered the logistic support of the 77th Division.)

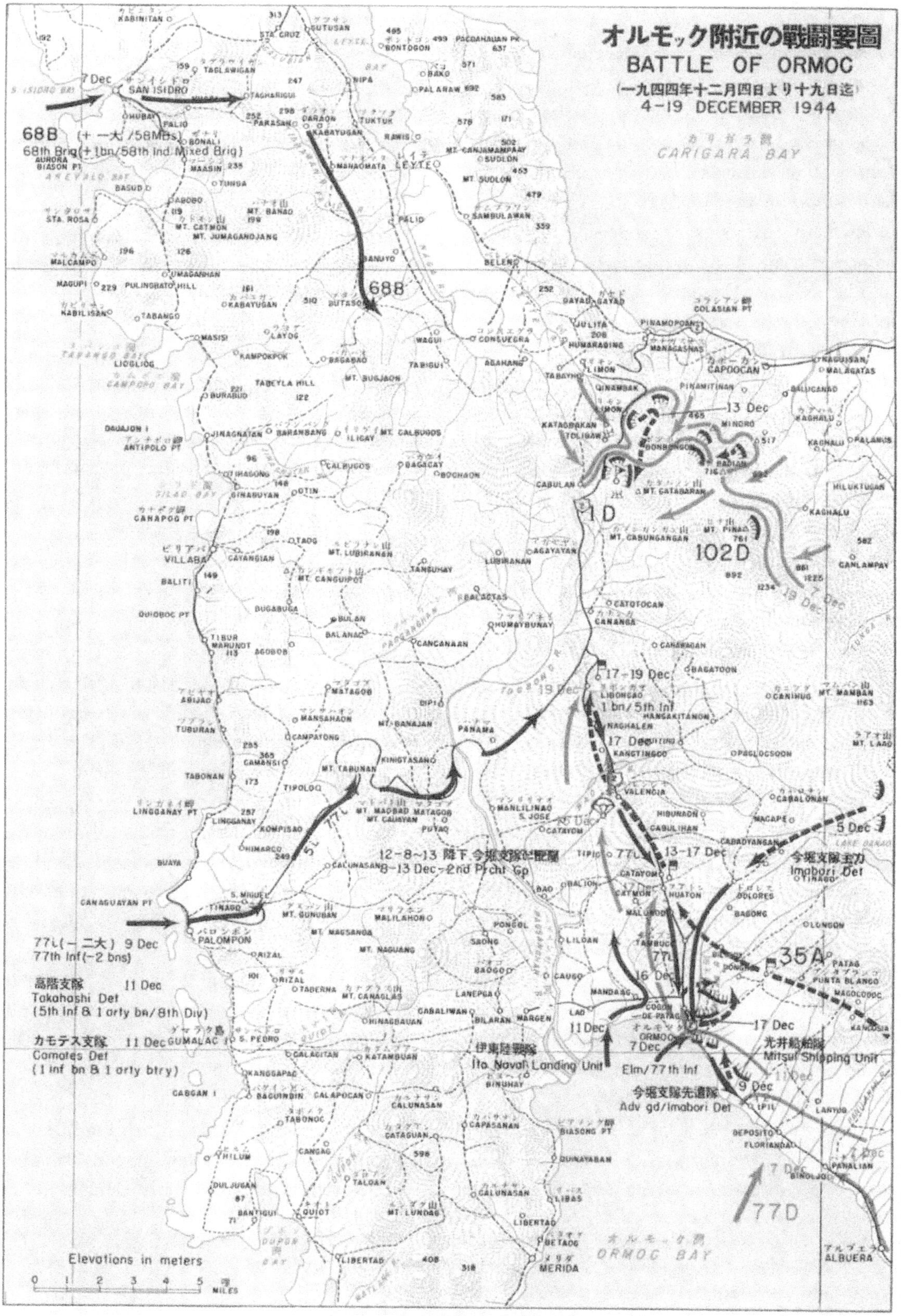

PLATE NO. 102
Battle of Ormoc, 4–19 December 1944

on the Burauen front, took immediate steps to meet the situation.[186] The Mitsui Shipping Unit was instructed to occupy defense positions on Red Roof Hill, a commanding terrain feature about one and a half miles southeast of Ormoc, while the Imabori Detachment was ordered to rush south through Ormoc and engage the enemy advancing from Albuera.

The Mitsui Shipping Unit immediately moving into its assigned positions, while the commander of the Imabori Detachment hurried to Ormoc, arriving there late on the 7th. Pending the arrival of the detachment main body, the battalion already in the Dolores sector was ordered to move south toward Ipil as an advance guard. In the Albuera sector, the enemy landing was being opposed only by small elements of the 26th Division flank guard detachment and by division and Army rear echelon troops. Brushing aside this weak opposition, the enemy advanced the beachhead north to take Ipil late on 7 December.

On the evening of 8 December, Lt. Gen. Suzuki received word from rear headquarters at Ormoc that Fourteenth Area Army had ordered suspension of the *Wa* operation, and he immediately directed the 16th and 26th Divisions to initiate a general withdrawal from the Burauen-Dagami front. The 26th Division was to move back through Lubi and attack the enemy on the Albuera front, while the 16th Division was to follow the 26th over the Lubi trail to be available for possible use in the Ormoc sector. Early on 9 December, the Army Commander left the advance command post at Lubi to return to Ormoc.

Meanwhile, the battle for Ormoc had already begun. On the night of 8 December, the advance guard battalion of the Imabori Detachment arrived in the vicinity of Ipil and began to dig in. On the morning of the 9th, before it had been able to complete organization of the position, the battalion was subjected to a counterattack by elements of the enemy 77th Division. Heavily outnumbered, the battalion was driven back to the Mitsui Shipping Unit positions on Red Roof Hill. While this action was being fought, the balance of the Imabori Detachment arrived in the Ormoc sector and was joined by elements of the 2d Parachute Brigade, which had begun to land at Valencia Airfield on the 8th. In view of the precarious situation south of Ormoc, Col. Imabori decided to station the detachment main strength on the heights northeast of the town and ordered the hasty preparation of defense positions.

As the opening round of the battle for Ormoc was being fought, Fourteenth Area Army and Southwest Area Force completed plans for counteraction. On 8 December, the Camotes Islands Garrison Detachment finally left Manila. The next day, the Area Army further designated the Takahashi Detachment (5th Regimental Combat Team, 8th Division) as an emergency reinforcement to leave for Leyte immediately. Southwest Area Force ordered the Ito Naval Landing Unit[187] to accompany the Takahashi Detachment, and both units embarked for Palompon late on 9 December.

Fourteenth Area Army regarded the dispatch of these units only as an emergency measure and proceeded to plan further counteraction. On 10 December, when the 39th Infantry Regiment, 10th Division arrived on Luzon, General Yamashita promptly ordered the regiment to prepare for a counterlanding in the Carigara Bay area on the night of 16–17 December. This

186 Under a reorganization and expansion of Thirty-fifth Army headquarters, effected on 14 November, Lt. Gen. Takaji Wachi, Deputy Chief of Staff of Southern Army, replaced Maj. Gen. Tomochika as Chief of Staff of Thirty-fifth Army, Maj. Gen. Tomochika becoming Deputy Chief of Staff.

187 The Ito Naval Landing Unit had a strength of 400, with 5 amphibious tanks and 16 light howitzers.

was to be known as the *Ketsu* (決) Operation.

Enemy air attacks were meanwhile disrupting troop movements to Leyte. On 7 December, the 68th Brigade,[188] scheduled to land at Palompon, was forced ashore by air attack at San Isidro, on the northwest tip of Leyte, where it was unable to establish communication with Thirty-fifth Army. Headquarters and one battalion of the 77th Infantry, 30th Division landed successfully at Palompon on 9 December, but the convoy carrying the Takahashi Detachment from Luzon underwent a heavy air attack which sank one ship with the loss of an entire battalion of the 5th Infantry Regiment. The remainder of the detachment landed at Palompon on 11 December, together with the Camotes Islands garrison unit, which was unable to carry out its planned mission due to strong enemy forces in Ormoc Bay. The Ito Naval Landing Unit moved further south and landed near the mouth of the Pagsangahan River, west of Ormoc.

The situation by this time was virtually hopeless. On 11 December, the Mitsui Shipping Unit and advance guard battalion of the Imabori Detachment, after delaying the enemy for two days, were forced to evacuate Red Roof Hill under heavy pressure. While these units assembled astride the road north of Ormoc, the enemy moved into the deserted ruins of Ormoc on the same day. The Thirty-fifth Army had now lost its main supply port, and the retirement route of the 16th and 26th Divisions was completely severed.

Lt. Gen. Suzuki reached the Army command post at Huaton, five miles north of Ormoc, at 2020 on 13 December. By this time, the 1st and 102d Divisions to the north were engaged in desperate, last-ditch fighting east and west of Mt. Catabaran. To the south, only miscellaneous elements, including the Imabori Detachment and Mitsui Shipping Unit, stood between the command post and the enemy.[189]

Despite this hopeless situation, Lt. Gen. Suzuki, counting upon the arrival of additional 77th Infantry reinforcements from Palompon, decided on a final offensive effort. On 14 December he issued the following order to activate the operation, fixing 17 December as the attack date :

1. The Imabori Detachment (reinforced by elements of the 2d Parachute Group and the Ito Landing Unit) will, on 15 December, gradually push attack positions forward. At dusk on 17 December, the detachment will assault Red Roof Hill, southeast of Ormoc, and thence will penetrate into the outskirts of the town.

2. The 77th Infantry (less two battalions) will relieve the Imabori Detachment on the main road north of Ormoc on 15 December, and, on 17 December, will attack Ormoc from the north.

3. The 1st Division will dispatch a battery of field artillery to the south of Huaton and will support the Imabori Detachment's assault on Red Roof Hill.

4. The Mitsui Shipping Unit will hold its present position northeast of Ormoc and facilitate the Imabori Detachment's attack.

Although there was considerable doubt at this time as to the situation of the 26th Division south of Ormoc, Lt. Gen. Suzuki felt that it could play a part in the forthcoming operation by harrassing the enemy's rear. Accord-

188 The 68th Brigade, commanded by Maj. Gen. Takeo Kurisu, was roughly equivalent to a regimental combat team in the American Army. The nuclear regiment was the 126th Infantry, supported by brigade artillery, engineer, signal, and medical units, aggregating (less attachments) about 5,000 men. Attached for this operation was the 380th Independent Infantry Battalion of the 58th Independent Mixed Brigade, raising the total strength to about 6,200 men. In landing at San Isidro, four ships of the 68th Brigade convoy were heavily damaged by enemy air attack and had to be grounded. The brigade got ashore but without most of its equipment, supplies and heavy guns.

189 The Ito Naval Landing Unit had been ordered under the command of the Imabori Detachment but was unable to reach the detachment positions north of Ormoc due to enemy capture of intervening territory.

ingly, at 1700 on 14 December, he dispatched the following order:

1. The 26th Division will endeavor to secure the heights around Talisayan with its main strength and interrupt the enemy's northward advance.

2. An element will be stationed on the heights east of Albuera to cover the rear of the 16th Division.

As for the 16th Division, marooned in the mountain fastness of central Leyte, its plight seemed utterly hopeless. The Thirty-fifth Army Commander, however, hoped to extricate a portion of it for further operations, and, on the afternoon of the 15th, sent the following order to Lt. Gen. Makino, the division commander:[190]

The main strength of the 16th Division will move to the sector north of Talisayan via Lubi and aid the 26th Division in interrupting the enemy's northward advance.

Before preparations for the 17 December offensive against Ormoc could be completed, the enemy again seized the initiative with a series of surprise thrusts. Advancing through the swampy lowlands of the Pagsangahan River valley on the west side of Highway 2, an enemy force attacked Valencia airfield on 15 December. The next day the Imabori Detachment found itself beating off a strong enemy assault on its right flank at the juncture of Highway 2 and the Antilao River. The 77th Infantry, which had just come up to relieve the right flank elements of the detachment, was also hit by the enemy attack, suffering heavy casualties. On the morning of 17 December, an enemy combat patrol raided the Army command post at Huaton and almost captured the Army commander.

Although these developments entirely disrupted Lt. Gen. Suzuki's attack plan, the Imabori Detachment nevertheless assaulted Red Roof Hill on the evening of 17 December and succeeded in carrying the summit. The detachment subsequently encountered such severe resistance, however, that it was unable to approach Ormoc and finally had to yield Red Roof Hill and retire to the 77th Infantry and Mitsui Shipping Unit positions in the hills northeast of Ormoc.

In the meantime, the enemy had completely overrun Highway 2 as far north as Valencia, and Lt. Gen. Suzuki, at his new command post at Libongao, was cut off from all units to the south. On the night of 17 December the Army commander committed one battalion of the Takahashi Detachment southward along the highway in an abortive attempt to restore Communications. The same day, a discouraging message was received from Fourteenth Area Army that the *Ketsu* Operation had been cancelled. No reason was given.

On 18 December, whatever fleeting hopes were still entertained that the Leyte decisive battle could be continued were dealt a final, crushing blow when the news arrived from Fourteenth Area Army that the enemy had landed a force of about one division on Mindoro, at the very threshold of Luzon. It was plain that no further reinforcement could possibly be sent to Leyte. This message was quickly followed by another containing a policy guide from Gen. Yamashita with the following general provisions:

1. It is contemplated that the mission of Thirty-fifth Army will be changed to one of strategic delay.

2. In preparation for official notification of such a change, the Army should promptly plan for self-sufficient, sustained resistance.

Lt. Gen. Suzuki and his staff therefore, drew

190 Enemy airborne landings at points along the Talisayan-Lubi trail, carried out simultaneously with the amphibious landing on Ormoc Bay, rendered the withdrawal of the 16th and 26th Divisions doubly difficult. Remnants of the 26th Division did not reach the mountains west of Ormoc until the end of February, followed in mid-March by a handful of survivors of the 16th Division.

up a tentative plan for self-sufficient action in the Palompon area and issued it to the troop units at noon on 19 December. The 1st and 102d Divisions, which were still holding off the enemy in the north, were instructed to await further orders before initiating a withdrawal toward Palompon, but in case such orders were never received, they were to act on their own initiative.

At 1400 on the 19th, the Thirty-fifth Army command post at Libongao was raided by the enemy. As a consequence, the headquarters was displaced to Kompisao, a small barrio about five miles northeast of Palompon, closing into the new location on 20 December. Meanwhile, the Takahashi Detachment arrived at Libongao on the evening of the 19th.

Reports received during 20–21 December at the new Thirty-fifth Army command post were extremely fragmentary, but they all continued to reflect the hopelessness of the situation and the imminence of total collapse. Although no official order confirming Gen. Yamashita's policy guide of the 18th had been received, the circumstances were such that Lt. Gen. Suzuki saw no further reason to delay the planned general retreat. Orders were therefore dispatched on 21 December to the Takahashi Detachment to cover the withdrawal of all Army forces into the Palompon area.

The huge and costly effort to turn the tide of battle in the Pacific by a defeat of the enemy on Leyte had ended in complete failure. Japanese air, sea and ground force strength had been consumed in vast quantities, and the enemy now possessed a forward base in the heart of the Philippines, from which to continue operations northward and expand his range of air control over the slim remaining lifeline between the homeland and the southern area.

Already, by his advance to Mindoro, the aggressive enemy served notice that the invasion of Luzon was rapidly approaching.

CHAPTER XIV

PRELUDE TO THE DEFENSE OF LUZON

Estimates and Plans, November 1944

By the first part of November 1944, the Fourteenth Area Army command was already convinced that the loss of Leyte could not be averted.[1] Closely watching the unfavorable development of the struggle for this strategic island, General Yamashita and his staff concluded that enemy control of Leyte would be firmly established before mid-December, and that General MacArthur's forces would then be ready to launch the next move in the planned conquest of the Philippines.

The question of what this move would be was given serious study. Although a direct jump from Leyte to Luzon was not ruled out,[2] the Area Army command considered it more probable that General MacArthur would preface his assault against Luzon by an intermediate operation in the central Philippines. The objective of this operation, it was surmised, would be to acquire advance air bases covering the most direct route of approach to the west coast of Luzon via the Sulu Sea.[3]

In the opinion of General Yamashita's staff, the enemy would probably seek such bases in the western Visayas. The Iloilo sector of Panay was considered a highly probable target of invasion because its coastline was suited for amphibious landing operations and because well-developed airfields already existed there. Cebu and Negros were rated secondary possibilities. Mindoro, although situated in greater proximity to the anticipated landing areas on central Luzon, received almost no consideration since it possessed relatively few favorable sites for the construction of operational airfields.[4]

---

1 This chapter was originally prepared in Japanese by Maj. Hiroshi Komatsu, Imperial Japanese Army. Duty assignments of this officer were as follows: Student, Army War College, Nov 41–Nov 42: Staff Officer (Training), 51st Air Training Division, Dec 42–May 44; Staff Officer (Intelligence and Supply), Fifth Air Army, May 44–Mar 45; Staff Officer (Operations), 13th Air Division, Mar 45–Aug 45. All source materials cited in this Chapter are located in G-2 Historical Section Files, GHQ FEC.

2 If such a move materialized, it was expected to be a limited operation employing three to five divisions and directed at the east coast of Luzon. The purpose of this attack, it was estimated, would be to draw Japanese strength eastward preparatory to later landings by the main invasion forces on the west coast of Luzon. Baler, Dingalan and Lamon Bays were considered the most probable landing points along the east coast, with possible secondary landings at Legaspi and/or Aparri. *Dai Yon Kokugun oyobi Dai Juyon Homengun Shorui Tsuzuri* 第四航空軍及第十四方面軍書類綴 (Documents of Fourth Air Army and Fourteenth Area Army): Operational Policy Outline for Luzon, 14 Nov 44.

3 Following the invasion of Leyte, the consensus of opinion of the Area Army staff was that the main effort of the enemy's eventual assault on Luzon would be directed at the west rather than the east coast of the island. Greater suitability of the west coast terrain for the deployment of large forces and proximity of the landing sites to the key strategic area of Manila were the main reasons for this estimate. (Statement by Maj. Eizo Hori, Staff Officer (Intelligence), Fourteenth Area Army.)

4 Fourth Air Army had surveyed Mindoro for air base sites which could be developed for operational use, but had found the terrain generally unfavorable. The only project undertaken by the Air Army was the improvement of an already existing strip near San Jose, on the southwest coast of the island, for auxiliary and emergency use. (Statement by Col. Misoo Matsumae, Staff Officer (Operations), Fourth Air Army.)

Already short of troops for the defense of Luzon itself, Fourteenth Area Army could not contemplate serious resistance to a move into the western Visayas. It was therefore expected that the enemy would rapidly achieve his preliminary objectives and at the same time complete the build-up of his forces for the final blow against Luzon. This assault, it was estimated, would come around 7–10 January and would involve a total enemy strength of more than ten divisions. The main landing was expected either at Lingayen Gulf or in the Batangas area, with possible secondary landings at one or more of several points—Aparri on the north, Lamon Bay on the east, and Legaspi on the south. Airborne landings at key positions on the central Luzon plain were also considered probable.[5]

In the light of this assessment of enemy plans and capabilities, General Yamashita felt that it was imperative to stop diverting troops to Leyte immediately in order to reserve full strength, including all reinforcements either already allocated or promised to Fourteenth Area Army, for the defense of Luzon. This view was strongly pressed upon Southern Army between 7 and 10 November. On 11 November, however, Field Marshal Terauchi bluntly informed General Yamashita that the decisive battle on Leyte had to be continued at all costs, and that defensive preparations on Luzon might be expanded only insofar as they would not interfere with the execution of this policy.[6]

To comply with Field Marshal Terauchi's decision, Fourteenth Area Army promptly drew up a new reinforcement schedule. This called for the progressive commitment to Leyte of the 68th Brigade, the first elements of which had recently reached Luzon from Formosa,[7] and the 23d and 10th Divisions, just allocated to the Area Army by Imperial General Headquarters. Since the Tokyo High Command currently was considering the allocation of only one additional division, the new schedule meant that the Area Army would divert the bulk of the reinforcements still to reach the theater to the Leyte battlefront.

General Yamashita, though keenly aware that execution of the reinforcement schedule would bar any attempt to revitalize the weakened defenses of Luzon, was prepared to carry it out as far as future developments allowed. At the same time, however, he could not ignore the fact that the rapid consolidation of the enemy's hold on Leyte and his ever-tightening control of the central Philippine sea lanes made it highly improbable that any reinforcements reaching Luzon later than the 68th Brigade could actually be moved to the battle zone. There was a strong possibility, therefore, that the 23d and 10th Divisions and any other forces later allocated would be available for the defense of Luzon. Again, however, this was contingent upon whether these forces could be successfully moved to Luzon from rear areas before the enemy invasion of Luzon was launched.

The existence of so many uncertain factors made it impossible at this stage to draw up any clear-cut tactical plan for the deployment of forces on Luzon. General Yamashita's staff nevertheless drafted a tentative outline of operational policies to meet the major situations which might arise. This outline, dated 14

5 The preceding summary of Fourteenth Area Army's estimate of enemy plans and capabilities as of 7 November is based on the following sources: (1) Statement by Maj. Hori, previously cited. (2) Documents of Fourth Air Army and Fourteenth Area Army, op. cit: Operational Policy Outline for Luzon, 14 Nov 44.

6 The staff discussions here referred to are dealt with more fully in Chapter XIII, pp. 384–7 and n. 158, p. 387.

7 The 68th Brigade had already been earmarked for Leyte in October and was scheduled to move to Manila early in November for transshipment. When the new Leyte reinforcement schedule was drawn up, only the first elements of the brigade had reached Luzon, and its movement was not actually completed until about 21 November.

November, was the first indication that the Area Army command was beginning to think in terms of a delaying defense of Luzon rather than a decisive engagement of forces. A summary of the essential points of the outline follows :[8]

1. In the event of an early enemy invasion of Luzon (prior to about mid-December), the Area Army will conduct operations as indicated below:

a. In general, the Area Army will conduct a strategic delaying action, making use of existing defense positions, but local offensives will be carried out as permitted by the tactical situation. Particularly, should the enemy effect a landing only in the Lamon Bay area, the 8th Division and 2d Armored Division will be diverted to that front for offensive action.

b. If, prior to or accompanying a landing on central Luzon, the enemy carries out landings on northern Luzon or in the Legaspi area, the 103d and 105th Divisions will prevent enemy seizure of air bases in their respective areas of responsibility. If, however, the enemy lands directly on central Luzon without a prior or accompanying operation in the Bicol area, the 105th Division will be withdrawn to central Luzon to reinforce the force on that front.

c. Units other than those mentioned above will generally conduct delaying actions designed to weaken enemy strength. However, in certain circumstances, the Manila Defense Force may be reinforced by the 2d Armored Division for the purpose of conducting a major defense of the Manila area.

2. In the event of a full-scale enemy invasion of Luzon after mid-December, the Area Army will conduct operations as indicated below:

a. If reinforcements have not yet reached Luzon, strategic delaying operations will be carried out by existing forces with the objective of consuming maximum enemy strength.

b. If reinforcements have already arrived, operational plans will be decided on the basis of the strength of such reinforcements and the prevailing situation. However, in case adequate forces have been deployed in central Luzon prior to the enemy invasion, decisive battle operations will be conducted as outlined in the Area Army operational plan of 11 October.[9] Either the Lingayen Gulf or Batangas area will be the decisive battle zone.

3. To prepare for the worst eventuality, the Area Army will organize positions for final sustained resistance in the following key areas:

a. Northern Luzon—mountains around Baguio and north of San Jose.

b. Western Luzon—mountains west of Tarlac.

c. Central Luzon—mountains east of Manila.

4. In anticipation that the enemy will enjoy sea and air control during operations on Luzon, special combat methods will be fully exploited. Particular importance will be placed on the use of special-attack boat units, which will prepare to operate in the Batangas, Lingayen Gulf, Manila Bay and Lamon Bay areas.

5. Minor troop elements will be stationed on islands adjacent to Luzon to act as outpost guards and hinder the use of these islands by the enemy. Troops may be dispatched by small craft to reinforce these elements, depending upon the situation.

In conformity with this rough outline of operational policies, General Yamashita immediately dispatched staff officers from Area

8 Documents of Fourth Air Army and Fourteenth Area Army, op. cit: Operational Policy Outline for Luzon, 14 Nov 44.

9 This plan, drawn up immediately after the arrival of General Yamashita in Manila to assume command of Fourteenth Area Army, was based on the original *Sho-Go* Operation concept of fighting the major ground battle in defense of the Philippines on Luzon. The essence of the plan was the concentration of three divisions, designated as "decisive battle forces", on the central Luzon plain ready for speedy deployment against enemy forces landing at one or more of three anticipated major invasion points, i. e. Lamon Bay, Lingayen Gulf and Batangas. If possible, decisive battle was to be fought on or near the beaches. Documents of Fourth Air Army and Fourteenth Area Army, op. cit: Operational Policy Outline for Decisive Battle in the Luzon Area, 11 Oct 44.

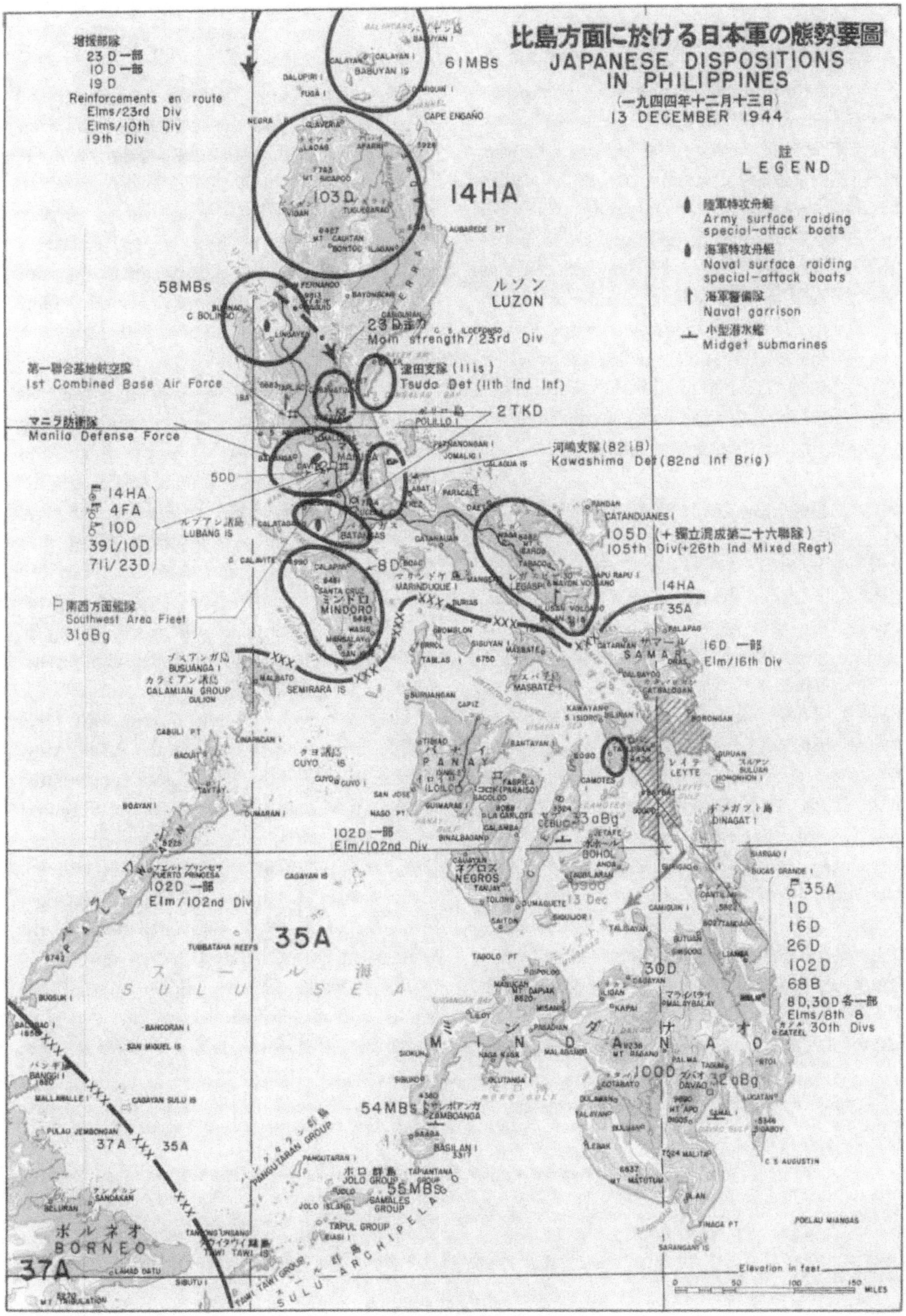

PLATE NO. 103

Japanese Dispositions in Philippines, 13 December 1944

Army headquarters to carry out a preliminary survey of the three mountain districts designated as areas of final resistance. Late in November the transfer of essential base installations and reserve stores of ammunition and supplies to these areas began.[10] To guard transport movements against increasingly troublesome attacks by guerrillas, the 2d Armored Division was ordered to augment its security measures protecting the central highway network leading out of the Manila area. Meanwhile, Area Army forces elsewhere on Luzon continued to strengthen their defense positions in preparation for the expected enemy assault.

Situation on 13 December

Despite these preparatory steps for the defense of Luzon, the main effort of Fourteenth Area Army still remained centered on maintaining the flow of troops and supplies to Leyte in compliance with Field Marshal Terauchi's orders. To provide adequate reinforcements to the Thirty-fifth Army before its resistance collapsed entirely, General Yamashita was obliged to weaken still further the forces reserved for the defense of Luzon. Early in December, the 380th Independent Infantry Battalion of the 58th Independent Mixed Brigade was released to reinforce the 68th Brigade upon its final embarkation for Leyte. On 9 December the 5th Infantry Regiment of the 8th Division was shipped to northwestern Leyte in a last effort to prevent enemy capture of the Ormoc plain.

Of the five full divisions and two mixed brigades which had been present on Luzon and adjoining islands at the beginning of the Leyte operation, only two inferior divisions, the 103d and 105th, and one mixed brigade still remained unweakened. The Area Army had diverted to Leyte almost all of the 26th Division and varying portions of the 2d Armored Division, 8th Division, and 58th Independent Mixed Brigade.

The remainder of these original forces was thinly distributed over Luzon and adjoining islands. (Plate No. 103) On the Batan and Babuyan Islands to the north of Luzon was the 61st Independent Mixed Brigade. The 103d Division was spread over northern Luzon, with its main strength around Aparri. The 58th Independent Mixed Brigade, reinforced by the 12th Surface Raiding Regiment with about 80 special-attack boats, guarded the vital coastal sector of Lingayen Gulf.[11]

The 2d Armored Division, the only large mobile force under Fourteenth Area Army control, had the bulk of its forces concentrated in the San Miguel-Cabanatuan sector to protect the highway system of north-central Luzon. One tank regiment of the division was stationed south of Laguna de Bay, in Batangas Province, to act as a spearhead force in the event of an enemy landing in that area.

The Manila Defense Force covered the Manila area and Bataan Peninsula.[12] As of 13 December, this force had a combat strength

10 The Area Army attributed particular importance to the build-up of the three key defense areas but at the same time made every effort to maintain secrecy regarding the extent of the program in order to avoid a possible adverse effect on the morale of the forces on Luzon. (Statement by Maj. Gen. Toshio Nishimura, Deputy-Chief of Staff, Fourteenth Area Army.)

11 The 12th Surface Raiding Regiment formed part of the 3d Surface Raiding Force, assigned to the Lingayen area. Additional elements of the force were at Manila awaiting transshipment north. When the enemy landed on Mindoro on 15 December, however, Fourteenth Area Army decided to keep these elements in Manila Bay to strengthen the bay defenses. The 3d Surface Raiding Force was transferred the same day to direct Area Army command and assumed control of all Army surface raiding units in the Manila Bay area. Organization and tactics of Army and Navy surface raiding units are treated in Chapter XVII.

12 The headquarters of the Manila Defense Force was activated at Manila on 1 November, under orders from Tokyo, to command the miscellaneous troops units in the Manila area.

of four provisional infantry battalions and five infantry companies. One provisional infantry battalion was stationed on Bataan Peninsula, one company each at Olongapo, Bagac, and Mariveles.

The 8th Division, less elements transferred to Leyte, continued to keep its main strength disposed at key points in Batangas Province. The division also had under its command the 2d Surface Raiding Force with about 220 special-attack boats, based in Batangas and Balayan Bays. Mindoro Island was included in the divisional area of responsibility, but because of inadequate troop strength and the belief that an enemy landing there was unlikely, only a small security force of two provisional infantry companies under command of the 17th Infantry Regiment was stationed on the island.[13]

Defense of the Baler and Dingalan Bay sectors, on the east central coast of Luzon, was charged to the Tsuda Detachment, composed of the 11th Independent Infantry Regiment (reinf.) of the 26th Division. The 82d Infantry Brigade of the 105th Division, organized as the Kawashima Detachment and operating under direct Area Army command, occupied positions in the Lamon Bay area, farther to the south, where it was reinforced by the 1st Surface Raiding Force with about 180 special attack craft. The 105th Division, reinforced by the 26th Independent Mixed Regiment (less 1st Battalion on northwest Samar), was deployed on southern Luzon, principally in the Legaspi area and at Naga.[14]

It was not until December that elements of the additional divisions allocated by Imperial General Headquarters began to arrive on Luzon. The first echelon of the 23d Division debarked at San Fernando, on northwestern Luzon, on 2 December and began a period of regrouping and re-equipment necessitated by severe losses inflicted by enemy submarines while the convoy was en route from Formosa.[15] On 10 December the first echelon of the 10th Division arrived at Manila, followed on the next day by the second echelon of the 23d Division.[16] No elements of the 19th Division, newly assigned to Fourteenth Area Army on 20 November, had yet reached Luzon.

Immediately upon the arrival of its first echelon, the 10th Division was ordered by Fourteenth Area Army to organize a force to execute a counterlanding on northern Leyte under the *Ketsu* Operation plan.[17] This force, designated the Nagayoshi Detachment, was

13 *Hito Sakusen Kiroku Dai Sanki Dai Yonkan Mindoro To Sento Gaishi* 比島作戰記錄第三期第四卷ミンドロ島戰闘概史 (Philippine Operations Record, Phase Three, Vol. IV: General Outline of Mindoro Operations) 1st Demobilization Bureau, Nov 49, pp. 1–2.

14 General sources covering the disposition of Fourteenth Area Army forces in the Luzon area as of 13 December are as follows: (1) *Hito Sakusen Kiroku Dai Sanki Dai Sankan Ruson To ni okeru Sakusen* 比島作戰記錄第三期第三卷呂宋島に於ける作戰 (Philippine Operations Record, Phase Three, Vol. III: Operations on Luzon) 1st Demobilization Bureau, Oct 46, pp. 49–51 and Attached Map No. 1-A. (2) Fourteenth Area Army Operations Order No. A-217, 8 Dec 44. ADVATIS Bulletin No. 414, 13 Mar 45, pp. 1–4.

15 The first echelon included 23d Division headquarters and the 64th and 72d Infantry Regiments. Submarine attacks sank several troop transports in the convoy, with heavy losses to both regiments as well as to the division staff. (Statement by Lt. Col. Masaichi Takahashi, Staff Officer (Operations), 23d Division.)

16 The first echelon of the 10th Division comprised the major portion of the 39th Infantry Regiment and some division troops. The division headquarters moved to Manila by air about the same time as the arrival of the first echelon convoy. The second echelon of the 23d Division included the 71st Infantry Regiment, the division engineer regiment, and an independent artillery battalion. (1) *Dai Ju Shidan (Tetsu Heidan) Shijitsu Shiryo* 第十師團(鐵兵團)史實資料 [Historical Data, 10th Division (*Tetsu* Force)] Home Depot Division, 15 Feb 47. (2) Statement by Lt. Col. Takahashi, previously cited.

17 Cf. Chapter XIII, pp. 399, 401.

organized from the 39th Infantry and supporting troops. Although the detachment immediately began preparing to re-embark, tightening enemy control of the central Philippine sea routes made it virtually certain that the rest of the 10th Division and all of the 23d Division would never move to Leyte as scheduled in mid-November. The Area Army therefore began fitting these units into its plans for the defense of Luzon. The first step was an order to the 23d Division on 8 December, directing it to move to the San Jose and Umingan sectors and organize defense positions. One infantry battalion was to be released by the division to reinforce the 58th Independent Mixed Brigade in the Lingayen sector.[18]

In addition to the Army forces on Luzon, approximately 25,000 naval ground combat troops were distributed in the Manila Bay area, at Clark Field, and at Legaspi.[19] These troops, with the exception of the naval garrison unit stationed at Clark Field, were under command of the 31st Special Base Force headquarters in Manila, which in turn was controlled by Southwest Area Fleet, the top theater naval command. The Clark Field unit was commanded by the First Combined Base Air Force.

The numerical weakness of the ground combat forces was aggravated by a dangerously low level of ammunition and food reserves. A large proportion of the stores built up on Luzon for the *Sho-Go* Operation had been drained off to supply the Thirty-fifth Army on Leyte.[20] Enemy submarine activity against Japanese supply lines leading to Luzon made it increasingly difficult to replace the amounts expended.[21] As of mid-December, each division on Luzon had on hand 1,600/2,000 tons of ammunition, in addition to an Area Army reserve of not more than 1,000 tons per division, mostly stored in the Manila area.[22] With the arrival of additional troops, the shortage of available reserves became even more accentuated.

The situation in regard to air strength was equally ominous. Despite an increased flow of air replacements from the Homeland during November, constant enemy attacks on Japanese air bases and heavy losses incurred in supporting the Leyte operation had gradually whittled down the army and navy air forces in the theater. On the eve of the Mindoro invasion, they could muster a combined operational strength of only about 230 aircraft of all types. The Fourth Air Army, on 9 December, had a total of 133 operational planes, divided between bases on Luzon and in the Bacolod area of Negros.[23] On 14 December, the Navy's First Combined Base Air Force, including the First and Second Air Fleets, was down to about 100

18 Fourteenth Area Army Operations Order No. A-217, ADVATIS Bulletin No. 414, 13 Mar 45.

19 This figure included the crews from sunken vessels who had been absorbed by the naval ground combat forces. *Senkyuhyakuyonjuyon oyobi Yonjugo Nen Zai Hito Nihon Gun Heiryoku* 一九四四及四五年在比島日本軍兵力 (Strength of Japanese Forces in the Philippines, 1944–45) 1st and 2d Demobilization Bureaus, Nov-Dec 49.

20 During November over 23,000 tons of ammunition and rations were shipped from Luzon to Leyte, more than half the amount being sunk in transit. *Hito Sakusen Kiroku Dai Sanki Dai Nikan Furoku Reite Sakusen Kiroku* 比島作戰記錄第三期第二卷附錄レイテ作戰記錄 (Philippine Operations Record, Phase Three, Vol. II Supplement: Leyte Operations) 1st Demobilization Bureau, Oct 46, pp. 149–50.

21 About 80 per cent of food supplies shipped to the Philippines from Thailand and French Indo-China, which were the major sources, were lost in transit due to ship sinkings. Unpublished memoirs of Lt. Gen. Akira Muto, Chief of Staff, Fourteenth Area Army, *Hitosen no Jisso* 比島戰の實相 (The Truth of the Philippines Campaign) 15 Jun 47, p. 17.

22 Philippine Operations Record, Phase Three, op. cit. Vol. III, p. 46.

23 *Senkyo Shubo* 戰況手簿 (Daily Record of the War Situation) Operations Section, Army General Staff. 11 Oct 44—Jul 45.

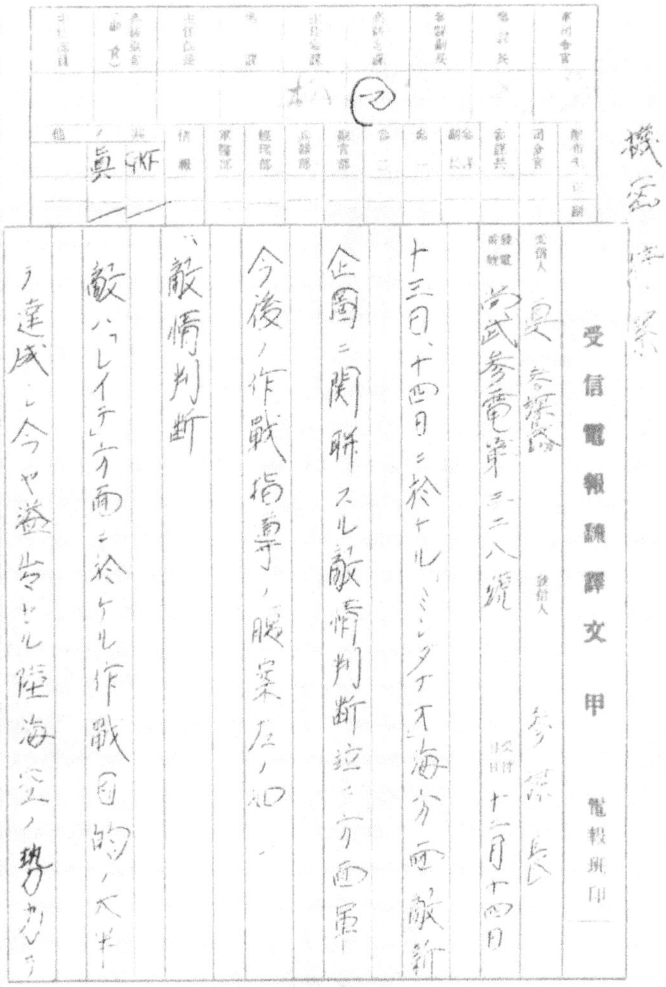

受信電報飜譯文 甲

電報班印

受信人 四 参謀長

發信人 参謀長

發電番號 尚武参電第三二八號

發信日時 十一月十四日

十三日、十四日ニ於ケル「ミンダナオ」海方面敵新
企圖ニ関聯スル敵情判断竝ニ方面軍
今後ノ作戰指導ノ腹案左ノ如シ
一、敵情判断
敵ハ「レイテ」方面ニ於ケル作戰目的ノ大半
ヲ達成シ今ヤ益々盛ナル陸海空ノ勢力ヲ

機密至急

PLATE NO. 104

Facsimile of Original *Shobu* (Fourteenth Area Army) Staff Radio No. 328 (Secret Urgent) to the Chief of Staff, Fourth Air Army

aircraft, based mainly at Clark Field on Luzon.[24]

The condition of the naval surface forces also precluded strong fleet intervention against an enemy move into the northern Philippines. Vice Adm. Shima's Second Striking Force, which had retired from Brunei to Lingga in November after the Battle for Leyte Gulf, was the only surface force of any consequence remaining in the southern area. Its operational strength on 13 December comprised only the two hybrid battleship-carriers *Ise* and *Hyuga*, three cruisers and three destroyers.[25] All other major fleet units were in the shelter of Japanese home waters. Of these, only three battleships, one cruiser and five first-line destroyers were fit for action. All six existing carriers were immobilized by lack of air complements.[26]

Naval units actually present in Philippine waters were limited to small craft. Five escort destroyers still remained in Manila Bay on 13 December, but the initiation of heavy enemy air attacks on the Manila area led Southwest Area Fleet on the 14th to order them to Camranh Bay in French Indo-China, and to Singapore. The withdrawal of these units left in the Philippines a total of two submarine chasers, 19 PT boats, ten midget submarines, and about 180 naval special-attack craft. The last were assembled mainly in Manila Bay.[27]

Such was the over-all situation of the Japanese ground, air, and sea forces available for the defense of the northern Philippines when the approaching end of Thirty-fifth Army resistance on Leyte foreshadowed the launching of General MacArthur's next assault. The Fourteenth Area Army command had estimated correctly the time of this assault, but the enemy again struck at a point where invasion was least expected.

Enemy Advance to Mindoro

First warning that an enemy amphibious force was on the move came at 0900 on 13 December. An Army reconnaissance plane radioed back to its base that a formation of about 80 ships was moving westward through the Mindanao Sea 45 miles north of Cagayan. Further air searches confirmed this sighting, and at 1710 three naval *Kamikaze* aircraft with fighter escort took off from Cebu to attack the formation as it passed south of Negros Island. Two cruisers were reported set afire by suicide crashes.[28] Army special-attack planes launched additional strikes during the day and claimed hits on one naval unit and one transport.[29]

---

24 In addition to these forces, 150 special-attack planes were currently on Formosa. These were ordered on 14 December to proceed to the Philippines as reinforcements for the First Combined Base Air Force. *Dai Niji Sekai Taisen Ryakureki Otsu* 第二次世界大戰略歷乙 (Abridged Chronicle of World War II, B) 2d Demobilization Bureau, Mar 46, Part III, p. 25.

25 *Gunreibu Socho no Sojosho* 軍令部總長の奏上書 (Report to the Throne by Chief of Navy General Staff) 14 Dec 44. (After surviving the Battle for Leyte Gulf, in which they operated with Vice Adm. Jisaburo Ozawa's northern decoy force, *Ise* and *Hyuga* were ordered to Lingga in November to strengthen the Second Striking Force.)

26 Operational strength in home waters as of mid-December compiled by Capt. Toshikazu Ohmae, Chief, Planning Sub-Section, Imperial General Headquarters, Navy Section. (The Battle for Leyte Gulf had reduced fleet carrier strength to five ships. Two carriers were subsequently commissioned, of which one, the 62,000-ton supercarrier *Shinano*, was sunk by enemy submarine attack on 29 November during her shakedown run.)

27 (1) Report to the Throne by the Chief of Navy General Staff, op. cit., 14 Dec 44. (2) Statements by Comdr. Tadao Kusumi and Comdr. Shigeichi Yamamoto, both Staff Officers (Operations), Southwest Area Fleet.

28 *Hito Homen Kaigun Sakusen Sono San* 比島方面海軍作戰其の三 (Philippine Area Naval Operations, Part III) 2d Demobilization Bureau, Nov 47, pp. 3–5. (American naval records covering the Mindoro operation indicate that the flagship of the invasion task force, the U.S.S. Nashville, was severely damaged in this attack, forcing the transfer of the command staff to a destroyer.)

29 *Hito Koku Sakusen Kiroku Dai Niki* 比島航空作戰記錄第二期 (Philippine Air Operations Record, Phase Two) 1st Demobilization Bureau, Oct 46, Appended Chart No. 17.

Although the size and composition of the enemy force left no doubt that an amphibious landing was intended, the target remained uncertain. The position of the convoy by nightfall on the 13th had eliminated Cebu as a possible objective, but it was still expected that the landing would be made in the western Visayas, either on Negros or on Panay. Ground forces throughout the threatened area were alerted. On Luzon, the Fourth Air Army and First Combined Base Air Force commands prepared to throw all available strength into an aerial assault on the invasion convoy at dawn of the following day.[30] Southwest Area Fleet simultaneously ordered the Second Striking Force to move up from Lingga to Camranh Bay in preparation for a possible sortie.

To pinpoint the convoy for the planned air attack, search planes took off before daybreak on 14 December and combed the coastal waters off Panay and Negros, expecting to find the enemy already starting landing operations. Actually, the invasion force was steaming on northward through the Sulu Sea, and the search failed to re-establish contact. The main attack groups poised at Clark Field nevertheless took off at 0715 to make a sweep of the general target area. Flying south over Batangas, they ran headlong into a formation of enemy carrier-borne fighters, and only part of the force was able to elude pursuit and continue the mission. Most of these planes failed to locate the enemy convoy, and results were not clear.

Further attack operations from Luzon bases were interdicted throughout the day by heavy and sustained enemy carrier-plane raids involving an estimated total of 560 sorties. In the meantime, however, Army aircraft based at Bacolod had picked up the invasion fleet off the southwestern tip of Panay at 0745. Intermittent strikes were flown by small formations of planes from Negros bases during the morning and afternoon, with undetermined results.[31] At 1330 the enemy force was reported still heading northwest through the Sulu Sea at a speed of 15 knots.

By late afternoon of the 14th the location of the convoy in the southern waters of Mindoro Strait made it clear that the objective lay beyond the western Visayas. For the first time Fourteenth Area Army estimated that the landing would take place on Mindoro and ordered the 8th Division to alert its outposts there.[32] It was also a serious possibility, however, that the enemy might strike directly at Luzon. General Yamashita therefore issued a warning to Area Army forces throughout central Luzon. As a further precaution, the Nagayoshi Detachment, which was awaiting shipment to Leyte, and the 71st Infantry Regiment of the 23d Division, which had just reached Manila, were ordered to deploy immediately to Bataan Peninsula and Batangas, respectively, to meet potential enemy landings in those sectors.[33]

---

30 Philippine Area Naval Operations, Part III, op. cit., pp. 5–6.

31 The total number of sorties flown against the invasion convoy during 14 December was 69, broken down as follows: By Fourth Air Army—29, of which 16 were special-attack planes. By First Combined Base Air Force—40, of which 33 were *Kamikaze* planes. Twenty-one Army and 18 Navy planes failed to return. (1) Philippine Air Operations Record, Phase Two, op. cit., Appended Chart 17. (2) *Hito Homen Kaigun Sakusen Sono Ni* 比島方面海軍作戦其の二 (Philippine Area Naval Operations, Part II) 2d Demobilization Bureau, Oct 47, p. 103.

32 (1) Statement by Major Hori, previously cited. (2) 17th Infantry Regiment (8th Division) Operations Order, 14 Dec 44. Eighth Army ADVATIS Translation No. 11, 12 Jan 45.

33 The Nagayoshi Detachment was relieved by this order of its mission, assigned four days earlier, to prepare for the *Ketsu* Operation on Leyte, and was placed under command of the Manila Defense Force. Fourteenth Area Army assumed direct command of the 71st Infantry Regiment, 23d Division, which had arrived in Manila on 11 December.

The Navy and air commands in Manila saw much greater probability of a landing on Luzon than on Mindoro. They therefore directed all subordinate units and installations to prepare for action against both enemy amphibious forces and possible airborne attack groups.[34]

Developments early on 15 December proved the Area Army's eleventh-hour estimate correct. At 0530 a navy plane reported that the invasion force was standing at anchor off San Jose, on the southwest coast of Mindoro, apparently starting to put troops ashore from 30 transports. An element of the 1st Provisional Infantry Company, 359th Independent Infantry Battalion, and a small number of naval seaplane base personnel were the only Japanese forces present in the area. Incapable of offering any serious resistance to the enemy landing, which was estimated to involve an entire division, these forces hastily withdrew inland along the trail toward Bulalacao, on the southeast coast of the island.[35] (Plate No. 105)

Air opposition to the landing during 15 December was restricted by a renewal of heavy enemy carrier-plane raids on Luzon bases, as well as by cumulative operational losses sustained during the two preceding days. Before the enemy raids began, however, a group of 13 navy suicide planes took off from Clark Field and penetrated to the landing area without fighter escort to execute a determined attack. Only two planes returned, reporting three transports sunk and several other ships damaged. A further sortie against the landing point was attempted in the afternoon by Cebu-based naval aircraft but was unsuccessful.[36]

The enemy's new invasion move had already set in motion a vital exchange of communications between General Yamashita and Field Marshal Terauchi at Saigon. Early on 14 December, less than twenty-four hours after the enemy convoy was first sighted, the Fourteenth Area Army Commander had radioed to Field Marshal Terauchi recommendations for a basic modification of operational policy for the Philippines. These recommendations were premised explicitly on the assumption that the enemy was about to invade Negros or Panay, in the western Visayas. They provided that further operations in the Thirty-fifth Army zone, including both Leyte and the Visayas, should be limited to delaying actions by the forces

---

34 (1) Southwest Area Fleet estimated that the enemy invasion force would penetrate through Mindoro or Tablas Strait during the night of the 14th and attempt a dawn landing in Tayabas Bay or the vicinity of Batangas. Landings in Manila or Subic Bays during the 15th were considered possible alternatives. Philippine Area Naval Operations, Part III, op. cit., pp. 7–8. (2) Statement by Col. Matsumae, previously cited.

35 Available Japanese sources covering ground operations on Mindoro indicate that the small elements present in the San Jose area attempted virtually no hostile action against the enemy landing force prior to withdrawing toward Bulalacao. This is confirmed by American operational accounts, which mention minor skirmishing on Caminawit Point, location of the navy seaplane base, as the only action which took place. The day following the landing, the 17th Infantry Regiment at Batangas ordered the Bulalacao garrison unit, consisting of another small element of the 1st Provisional Infantry Company, to go to the assistance of the San Jose force. On 17 December, this unit took up positions in the hills northwest of Bulalacao and was there joined on the 18th by the elements retreating from San Jose. The combined forces continued to occupy positions in this sector until 24 January when they were attacked by an enemy force and compelled to disperse inland. *Mindoro To Sento ni kansuru Hokoku* ミンドロ島戦闘に關する報告 (Mindoro Operations Special Report) Home Depot Division, undated.

36 Actual results of the first attack, according to American records, were two LST's sunk and one destroyer damaged. Army air action against the landing was limited to an attack by one suicide plane at dusk on the 15th. In addition to the attacks on the landing point, 17 sorties were flown by naval aircraft based on Cebu and Mindanao against enemy invasion shipping in the Sulu Sea. (1) Philippine Air Operations Record, Phase Two, op. cit., Appended Chart 17. (2) Philippine Area Naval Operations, Part II, op. cit., p 103.

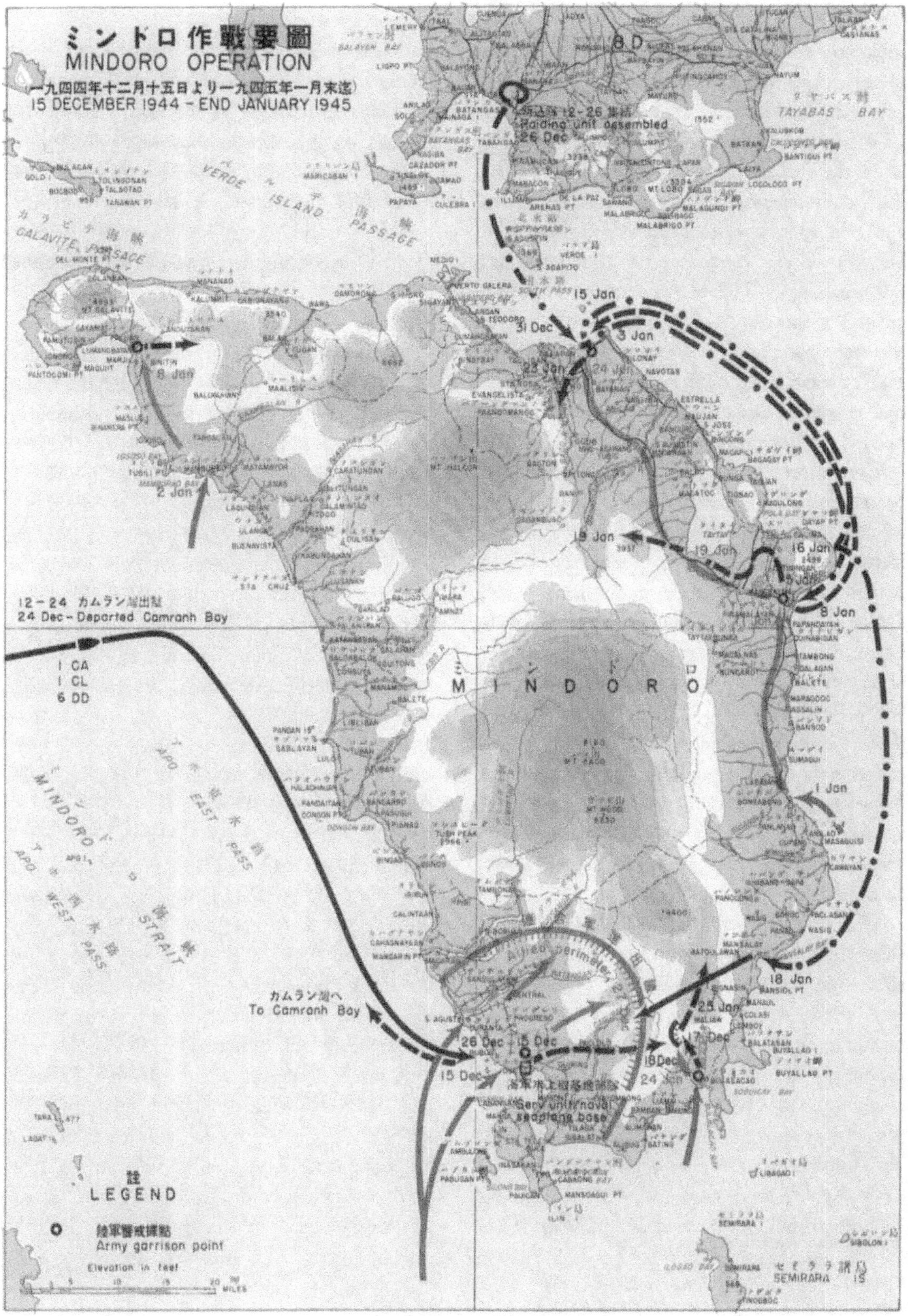

PLATE NO. 105

Mindoro Operation, 15 December 1944—End January 1945

already present, and that the Area Army should immediately concentrate all effort on preparing for the final defense of Luzon.[37]

When the enemy landed at Mindoro instead of in the western Visayas, General Yamashita did not withdraw his recommendations of the 14th. Mindoro was not under Thirty-fifth Army but within the Area Army's zone of direct responsibility. The enemy landing therefore created a situation not specifically covered in the recommendations. Despite this fact, General Yamashita felt that the broad policies he had urged in his message to Southern Army needed no basic revision.

Even more clearly than if the Visayas had been invaded, the establishment of an enemy foothold on southwestern Mindoro meant the forced termination of all assistance from Luzon to Thirty-fifth Army forces in the central and southern Philippines. Mindoro was scarcely 150 miles from Manila Bay, the nerve-center of supply and reinforcement operations for the entire Philippine area. Enemy air power based there would completely block all southward movement. Just as clearly, the enemy move forewarned that the final advance on Luzon would be launched at an early date. This made it imperative to complete defense preparations with all possible speed.

General Yamashita, however, still had to decide the immediate question of what to do about the situation on Mindoro. This issue was hastily studied on 15 December, immediately upon confirmation of the enemy landing. Although the invaded area was dangerously close to Luzon, the Area Army staff decided that any attempt to bolster the meager forces on Mindoro was not feasible. Enemy control of the air and sea, the impossibility of amassing enough shipping to move an effective force, and the delay which a diversion of troop strength from Luzon would impose upon preparations for defense of the main island ruled out any attempt at sending reinforcements.[38]

At Saigon, the policy recommendations transmitted by General Yamashita on 14 December were being studied by Field Marshal Terauchi when another dispatch from Manila, dated the 15th, reported the enemy landing on Mindoro. It was clear from this dispatch that the Area Army had no intention of undertaking aggressive action. Field Marshal Terauchi, however, took an opposite view. In his opinion, the acquisition of advanced bases on Mindoro would so clinch enemy air domination to the north that the later defense of Luzon would become impossible. He therefore dispatched a radio message to Fourteenth Area Army on the same day, strongly urging a counterlanding. On 16 December Lt. Gen. Jo Iimura, Southern Army Chief of Staff, departed for Manila to consult with the Area Army on all questions of future strategy.

Reaching Manila on 17 December, Lt. Gen. Iimura immediately conferred with Lt. Gen. Akira Muto, Area Army Chief of Staff. The latter explained in detail General Yamashita's strategic plans, and Lt. Gen. Iimura, convinced that these plans were sound, sent a radio to Field Marshal Terauchi on 18 December advising blanket approval of the Area Army Commander's proposals of the 14th. He further urged that full discretion be left to the theater command in tactical matters. Before this message reached Saigon, Field Marshal Terauchi had again radioed to Manila reiterating the need of speedy action to counter the enemy landing on Mindoro. On 19 December, however, a further communication from Saigon transmitted concurrence in Lt. Gen. Iimura's

37 Documents of Fourth Air Army and Fourteenth Area Army, op. cit: Chief of Staff, Fourteenth Area Army Radio No. 328 (Secret Urgent) to Chief of Staff, Southern Army, 14 Dec 44. (Plate No. 104 is the facsimile of the same message as received by Fourth Air Army)

38 Statement by Maj. Gen. Nishimura, previously cited.

recommendations.

Despite Southern Army's final agreement to give Fourteenth Area Army full latitude in tactical decisions, the question of a counterlanding on Mindoro continued to be debated in Manila under pressure from other quarters. Southwest Area Fleet was already making independent plans for a hit-and-run surface force attack on enemy invasion ships off San Jose. On 20 December, just as final orders were sent to Vice Adm. Shima's Second Striking Force at Camranh Bay to execute this attack, the Area Fleet received instructions from Combined Fleet to discuss the possibilities of a full-scale counterlanding with Fourteenth Area Army.[39] Both Fourth Air Army and the First Combined Base Air Force strongly supported a counterlanding because the advance of enemy land-based air power to Mindoro would virtually terminate operations from major Japanese bases on central Luzon.

Imperial General Headquarters in Tokyo had also studied the strategic implications of the Mindoro invasion and decided upon a general revision of the *Sho* No. 1 Operation plan to meet the new situation in the Philippines. This revision was embodied in two draft outlines dated 18 December, the first jointly concurred in by the Army and Navy Sections, and the second drawn up by the Army Section operations staff. Although intended only as a basis for discussion with the theater command, both these documents contained specific provisos for action to prevent or at least impede enemy utilization of air bases on Mindoro.[40]

Immediately after the formulation of these tentative policy directives, the Army Section of Imperial General Headquarters dispatched a staff mission from Tokyo headed by Lt. Gen. Shuichi Miyazaki, Chief of First Bureau (Operations), to confer with Lt. Gen. Iimura and General Yamashita's command regarding future strategy in the Philippines. Reaching Manila on 21 December, Lt. Gen. Miyazaki began a series of staff consultations in which the question of bolstering the defense of Mindoro again arose. No attempt was made to force any definite course of action upon the Area Army, but Lt. Gen. Miyazaki indicated that Imperial General Headquarters shared Southern Army's opinion that all possible steps should be taken to hinder enemy utilization of air bases on Mindoro. The Area Army therefore agreed on 23 December to re-examine the possibility of

39 On 16 December, the day following the enemy landing at San Jose, Southwest Area Fleet radioed to the 43d Destroyer Division, which had left Manila on the 14th en route to Indo-China, to turn back and execute a hit-and-run night attack on the enemy anchorage. The commander of the destroyer division, however, failed to obey this order because of insufficient fuel, and his ships continued on to Indo-China. On 20 December, therefore, Southwest Area Fleet ordered Vice Adm. Shima, Second Striking Force commander, to organize and dispatch a task group to execute the attack. This plan was not changed upon receipt of the Combined Fleet instructions to press for a counterlanding, because Southwest Area Fleet was anxious to take advantage of a momentary laxity in enemy precautions off the San Jose beachhead and also doubted that a counterlanding operation would materialize. (Statement by Comdr. Kusumi, previously cited.)

40 The joint policy outline drawn up by the Army and Navy Sections stated: "Extreme effort will be made to obstruct the utilization by the enemy of bases on Mindoro." The Army Section's supplementary draft, bearing on the same point, stated: "The defenses of air bases in the Bacolod (Negros) area, on Palawan and Mindoro will be strengthened; and even in the worst circumstances troop strength will be reinforced as necessary with the objective of obstructing enemy use of these bases." Documents of Fourth Air Army and Fourteenth Area Army, op. cit: Draft Outline of *Sho* No. 1 Operational Policy to Meet Current Conditions; and Draft Basis for Imperial General Headquarters Study of Policy to Meet Current Conditions in the *Sho* No. 1 Area, 18 Dec 44.

To Fourth Air Army Military Top Secret
No. 18 of 20 Copies
18 December 1944
Imperial General Headquarters,
Army Section
Draft Basis for Imperial General
Headquarters Study of Policy
to Meet Current Conditions
in the *Sho* No. 1 Area

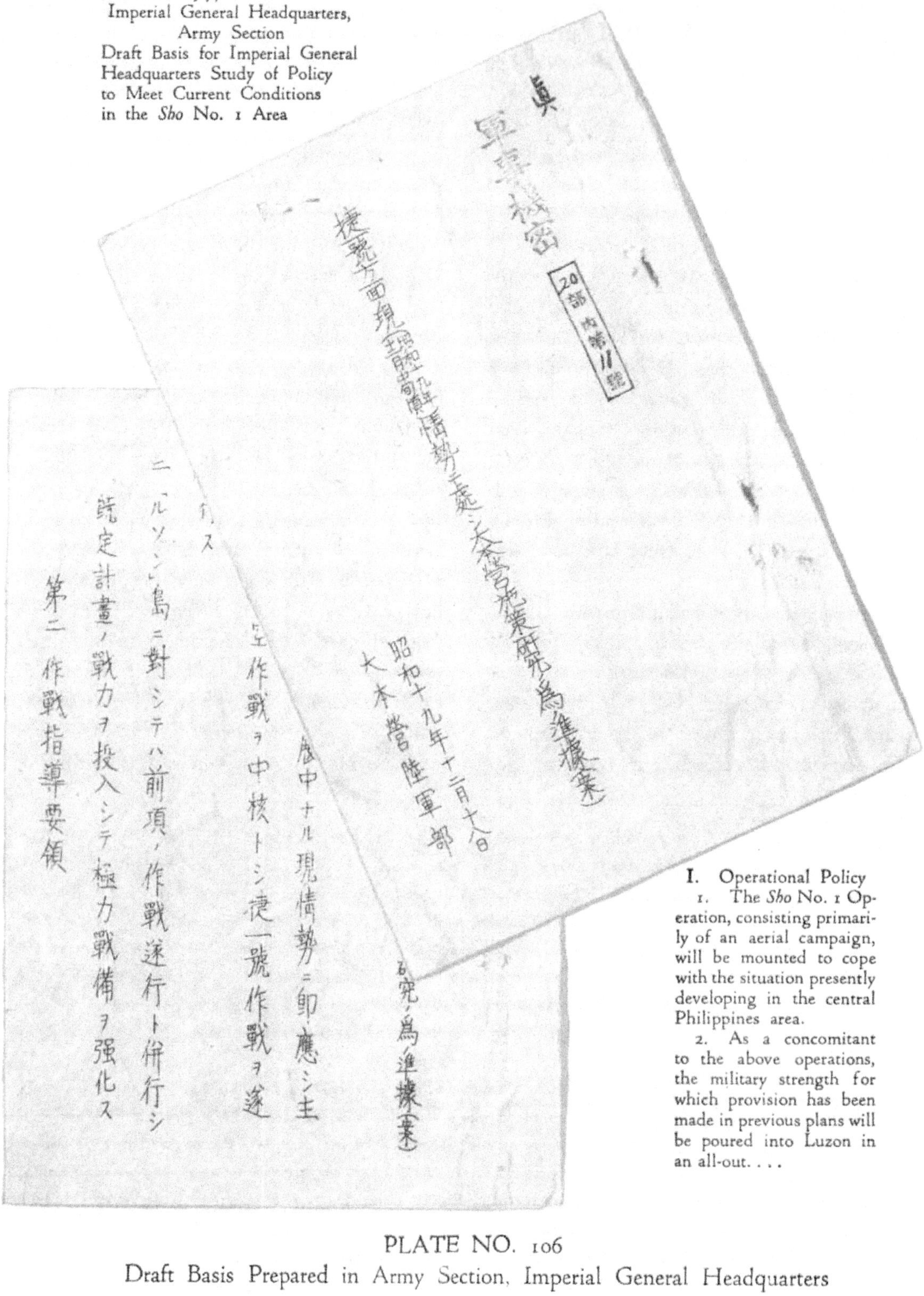

I. Operational Policy

1. The *Sho* No. 1 Operation, consisting primarily of an aerial campaign, will be mounted to cope with the situation presently developing in the central Philippines area.

2. As a concomitant to the above operations, the military strength for which provision has been made in previous plans will be poured into Luzon in an all-out. . . .

PLATE NO. 106
Draft Basis Prepared in Army Section, Imperial General Headquarters

dispatching minor troop reinforcements.[41]

The result of this re-examination was a half-hearted compromise. The Area Army staff adamantly maintained that it would be impossible to move any sizeable forces to Mindoro but agreed to dispatch a small raiding unit to hamper enemy development of airfields in the San Jose area. On or about 24 December, an order was issued to the 8th Division to organize a task unit for this mission as soon as possible.

While the counterlanding issue was being thrashed out in Manila, Vice Adm. Shima's Second Striking Force at Camranh Bay had completed preparations for the planned surface thrust at enemy invasion shipping in the San Jose area. On 24 December, a task group composed of the heavy cruiser *Ashigara*, one light cruiser, and six destroyers set out from Camranh Bay to execute the attack.[42]

As the task group neared Mindoro on the evening of the 26th, it was spotted by enemy aircraft. A severe air attack shortly thereafter sank one destroyer and inflicted minor damage on the *Ashigara* and two other ships. The force nevertheless continued toward the objective and at 2300 broke into the anchorage area. There it briefly shelled shore installations and launched torpedo attacks against enemy ships, four of which were claimed sunk. The task group withdrew at midnight without encountering surface opposition and headed back to Camranh Bay.[43] This marked the last sortie by Japanese fleet units into Philippine waters.

On the same day that the naval attack was executed, a small ground raiding detachment organized by the 8th Division assembled at Batangas and prepared to embark for Mindoro.[44] On 31 December the detachment finally moved by landing craft across the Verde Island Passage to Calapan, on northeastern Mindoro, and thence to Pinamalayan, where it arrived on 5 January. Three days later, just as the invasion of Luzon was beginning, the detachment encountered an enemy force advancing up Mindoro's east coast and was forced to sail back to Calapan. Although it later succeeded in moving to Mansalay, on the southeast coast of Mindoro, and then infiltrated overland toward San Jose, all attempts to raid enemy airfields were abortive.[45]

During the latter part of December and early January, the debilitated Army and Navy air forces based on Luzon and Negros also endeavored to impede the development of enemy bases on Mindoro by attacking resupply convoys en route from Leyte and raiding airfields in the San Jose area.[46] The air effort, however, was equally ineffectual in retarding the enemy's swift accomplishment of his objectives. By late December reconnaissance

41 (1) Data concerning the policy discussions which took place in Manila during the period 17–23December are based on a joint statement by the following high-level participants: Lt. Gen. Shuichi Miyazaki, Chief, First Bureau (Operations), Imperial General Headquarters, Army Section; Lt. Gen. Jo Iimura, Chief of Staff, Southern Army; Col. Ichiji Sugita, Staff Officer (Operations), Imperial General Headquarters, Army Section. (2) Statement by Maj. Gen. Haruo Konuma, Deputy-Chief of Staff, Fourteenth Area Army.

42 Philippine Area Naval Operations, Part III, op. cit., p. 15.

43 Ibid., pp. 15–6.

44 The raiding detachment was a specially organized provisional unit which included a small number that originally belonged to the *Gi-Go* Airborne Raiding Unit. (Cf. Chapter XIII, p. 393, n. 179) (1) Philippine Operations Record, Phase Three, op. cit. Vol. IV, p. 11. (2) Statement by Sgt. Naoji Sato, member of the Mindoro raiding detachment.

45 Philippine Operations Record, Phase Three, op. cit. Vol. IV, pp. 11–2, and Attached Map No. 2.

46 During the period 16 December to 5 January, Army aircraft flew a total of 213 sorties, of which 53 were special-attack planes, against enemy resupply convoys en route to Mindoro. Naval planes flew a total of 110 sorties against enemy shipping and 86 sorties against airfields on Mindoro. (1) Philippine Air Operations Record, Phase Two, op. cit., Appended Chart 17. (2) Philippine Area Naval Operations, Part III, op. cit., pp. 12–4.

reports indicated that at least two air bases had been completed and put into full operational use.

Final Luzon Defense Plans

During the short breathing spell afforded by the Mindoro operation, Fourteenth Area Army poured its full effort into speeding final plans and preparations to meet the impending enemy onslaught against Luzon. The Area Army command now anticipated that the invasion would come between 10 and 20 January, with the main thrust directed either at the Batangas area or at Lingayen Gulf.[47]

Before formulating detailed plans, General Yamashita first had to decide the basic tactics to be employed in the defense of Luzon. When the tentative plans of mid-November were formulated, uncertainty as to the troop strength which would be available had made it impossible to foresee whether the Area Army would be capable of waging another decisive battle as on Leyte or whether it would be obliged to fall back upon defensive tactics calculated to delay the enemy and wear down his strength. At that period, however, General Yamashita had hoped that he would have adequate forces at his command to challenge the enemy in decisive battle.

By the time Mindoro was invaded, no such optimism remained. Although the Area Army could count upon employing all incoming reinforcements for the defense of Luzon,[48] shipping hazards and delays made it improbable that these forces would be assembled on Luzon in time to deploy for battle by the expected date of invasion. Moreover, in view of the steady decimation of the air forces, it was certain that operations would have to be conducted under a smothering blanket of enemy air power.

Under these circumstances, General Yamashita feared that an all-out offensive effort in the initial phase of the battle for Luzon might result in the early destruction of his own forces. Coordinated delaying action, including local counter-offensives under favorable circumstances, seemed to assure better prospects of prolonged resistance. Such tactics would consume maximum enemy strength and gain precious time for the reinforcement of Japanese defenses on Formosa and the Ryukyus. The Area Army operations staff was therefore instructed to draw up final plans based on the concept of a protracted delaying action.[49]

---

47 Immediately after the invasion of Mindoro, the Batangas area was rated the most probable landing point of the main enemy forces. In this event, a secondary landing was considered likely in the Lingayen area. Conversely, if the main landing took place at Lingayen, a secondary landing in Batangas was expected. In either case, it was considered probable that the enemy would also make an additional landing at Aparri and effect airborne landings on central Luzon, especially in the vicinity of Manila. (Statement by Maj. Hori, previously cited.)

48 With the invasion of Mindoro, Fourteenth Area Army scrapped the November reinforcement plan for Leyte, under which the 23d and 10th Divisions were to have been shipped to that island. This action was sanctioned by the Imperial General Headquarters draft outline of future operational policy for the Philippines, dated 18 December, which provided that the 23d, 10th and 19th Divisions should be assembled on Luzon as speedily as possible to hasten preparations for the defense of the main island. Documents of Fourth Air Army and Fourteenth Area Army, op. cit: Draft Basis for Imperial General Headqurters Study of Policy to Meet Current Conditions in the *Sho* No. 1 Area, 18 Dec 44.

49 The conclusions reached by the Area Army command generally accorded with the views of the Army Section of Imperial General Headquarters respecting Luzon defense. In the event of a full scale enemy landing on Luzon, the maximum strength permitted by the over-all situation, particularly air and sea sure-destructive [special-attack] forces, will be committed to that area in order to destroy the invading forces. Even in the most unfavorable situation, the main strength [of the forces on Luzon] will employ key areas in central and northern Luzon as bases and conduct long-term operations with the objective of destroying or consuming enemy combat strength, thereby obstructing his plans with respect to other areas (Ryukyus, Formosa, China coast, etc.) and facilitating over-all operations of the armed forces. Ibid.

On 19 December a general outline of operation was completed. Detailed appendices were issued the next day specifying the disposition, missions and command status of forces in two broad operational zones, one covering northern Luzon and the other, central and southern Luzon. These plans envisaged an initial effort by the forces defending coastal areas to inflict maximum losses upon the enemy at the time of landing. Delaying actions would then follow with the purpose of retarding capture of key inland communication points and airfields. The last phase would be a prolonged last-ditch stand in the three mountain regions previously designated as areas of final resistance. Commitment of the main Area Army forces in a large-scale offensive was not entirely excluded. Such action, however, would be undertaken only in an exceptionally favorable tactical situation. The general plan of operations was outlined as follows:[50]

I General Policy

The Area Army will secure the vital northern Luzon area with its main strength, and the mountainous regions east of Manila and west of Clark Field with elements of its forces. The forces in each of these areas will coordinate their operations with the objective of containing the main body of the American forces on Luzon and destroying its fighting strength, and at the same time will prepare for protracted resistance on an independent, self-sufficient basis.

II Outline of Operations

1. To facilitate long-term resistance, the Area Army will speedily establish a major base of operations in the northern Luzon area and secondary bases in the mountainous districts east of Manila and west of Clark Field. Other sectors will be utilized as forward areas, in which elements of Area Army strength will seek to obstruct the enemy's advances and consume his strength.

2. In the event of attempted enemy landings in the central or southern Luzon sectors, forces occupying coastal positions and surface raiding units will first endeavor to repel the landings. Should the enemy (succeed in landing and) attempt a deep penetration inland, elements will delay his advance as long as possible by securing key communications points and simultaneously carrying out the complete destruction of communications facilities and essential installations. Every effort will be made to hold airfields on central Luzon, but if this becomes impossible, the airfields will be destroyed.

3. In the event of landings by enemy airborne troops on the central Luzon plain, mobile forces will first contain and then destroy these troops, thereby preventing their seizure of airfields.

4. Against an enemy force attempting to land in the northern Luzon sector, every effort will be made to crush the landing in its initial phase. If this is impossible, defending forces will check any deep enemy penetration inland by holding positions directly behind the coast. In the worst eventuality, the main defensive position in the Baguio-Mankayan-Bambang area will be secured. Air bases in the vicinity of Aparri, Echague and Tuguegarao will be kept operational as long as possible.

5. After the enemy's main strength has penetrated into central Luzon, Area Army forces in the (three key) defense areas will carry out closely-coordinated raiding attacks employing substantial strength. The objective of these operations will be to contain the enemy's main forces, consume his fighting strength and especially to hinder his utilization of air and sea bases. Under favorable cirumstances the Area Army may commit its main forces in offensive operations but will guard against a premature exhaustion of its fighting strength.

6. In the event the enemy should effect a deep penetration inland prior to the completion of the Area Army's new defensive dispositions, operations

50 Documents of Fourth Air Army and Fourteenth Area Army, op. cit: Operational Policy Outline for Luzon, 19 Dec 44. (Portions textually quoted are Parts I and II. Part III outlined the general missions of forces, more specifically covered in the separate appendices issued 20 December.)

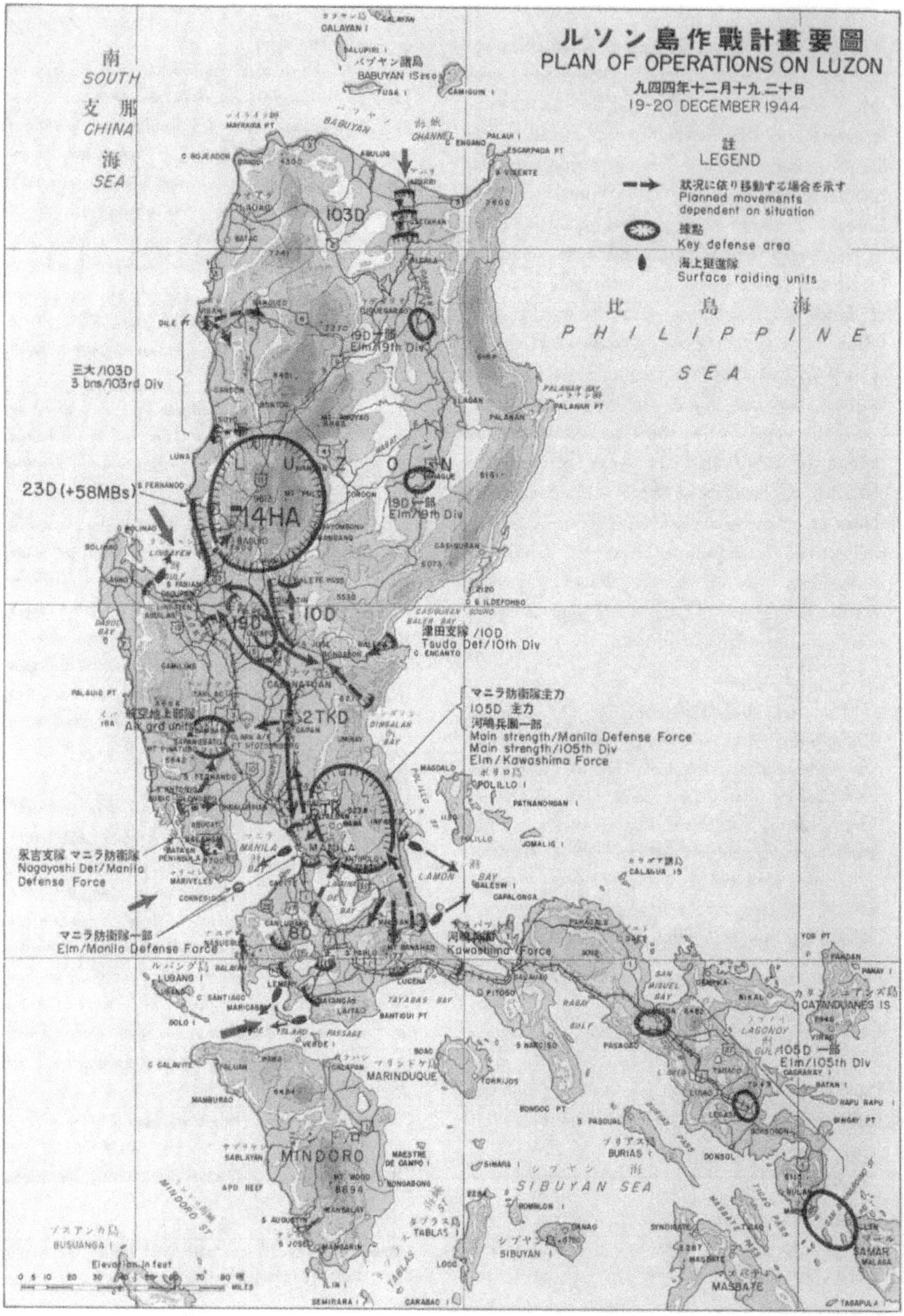

PLATE NO. 107

Plan of Operations on Luzon, 19–20 December 1944

will be decided on the basis of the existing situation, but in general available forces, in particular the armored division, will execute vigorous short-term counterattacks.

7. Operational plans for the northern Luzon sector and the southern and central Luzon sector will be further outlined in separate appendices.

8. The Area Army will count upon the air and naval forces to carry out the following support missions:

a. Air force—to execute all-out attacks on enemy ships at the time invasion operations begin, thereby reducing the combat strength of the invading forces; and thereafter, to facilitate the operations of the Area Army main forces by conducting raids from bases in the Echague and Tuguegarao areas.

b. Naval surface forces—primarily to protect harbors on northern Luzon.

c. Air and naval ground units—to participate in ground combat, maintaining appropriate command relationships with local army forces.

The greater part of the Area Army was to be committed to the defense of the northern sector for two reasons: first, because of its strategic importance for the purpose of delaying subsequent enemy operations against Formosa and the Ryukyus; secondly, because its mountain fastnesses and few routes of entry from the central Luzon plain provided the most favorable tactical conditions for extended resistance. The plans of 19–20 December therefore shifted the center of gravity of Area Army troop dispositions to the north. The heaviest concentration of forces was to be disposed along the southern perimeter of the great northern defense sector, running from the eastern shore of Lingayen Gulf to the vicinity of Baler Bay. Southern Luzon was to be virtually stripped of troops in order to bolster the second largest concentration of forces in the mountains east of Manila. Planned dispositions and mission assignments of Area Army forces were as follows:[51] (Plate No. 107)

Northern Zone

103d Division:

1. Division main strength to be deployed in the Aparri coastal sector. This force will, if possible, repel any enemy landing attempt in the Aparri area by a firm defense of its coastal positions. If a landing is made, it will prevent an enemy penetration of the Cagayan Valley by defending successive positions in the vicinity of Lallo and north of Alcala.

2. An element of the division (about three infantry battalions, reinf.) to occupy strategic points on the north-west coast. This force will come under direct Area Army command at the appropriate time. In case of an enemy landing in this sector, it will endeavor to inflict maximum losses at the time of landing and will then check enemy advances inland via the Abra River valley and the Suyo-Sabangan or Lagangilang-Lubuagan roads.

23d Division (58th Ind. Mixed Brig. attached):

1. Main strength to occupy positions near the coast along the eastern shore of Lingayen Gulf.[52] *The division will endeavor to smash an enemy landing attempt from these positions and, depending upon development of the situation, may be made the pivot of a counteroffensive by other Area Army forces. If the situation becomes untenable, the main strength will hold key positions around Baguio and strive to reduce enemy fighting strength; and elements will defend mountain positions east of San Fernando.*

51 Documents of Fourth Air Army and Fourteenth Area Army, op. cit: Operational Policy Outline for Luzon, 20 Dec 44.

52 An Area Army order issued on 17 December had already directed the 23d Division to assume command of the 58th Independent Mixed Brigade and take over the defense of the entire Lingayen area below the southern boundary of the 103d Division near Suyo. When it received this order, the division had just completed moving its main strength from San Fernando, where it had disembarked on 2 December, to Umingan and San Jose under a previous Area Army order to organize defensive positions in those sectors. Cf. pp. 409–10.

2. An element of the division to be deployed on the western side of Lingayen Gulf. This element will endeavor to inflict losses on an enemy landing force by raiding tactics and will strongly defend the coastal defile landing from the western peninsula to Lingayen.

3. Other elements to occupy positions guarding the lower Agno River crossing points and endeavor to prevent an enemy breakthrough onto the central plain.

10th Division (Tsuda Detachment attached):

1. Main strength to be deployed to the San Jose, Umingan and Natividad sectors.[53] *Positions will be organized in these sectors, and at the same time the division will prepare to support a possible counteroffensive on the 23d Division front or to counterattack east of Bongabong against an enemy force advancing inland after a landing in the Baler-Dingalan Bay area. In case of an enemy penetration onto the central Luzon plain, the division will hold its prepared positions against attack and, depending on the situation, may be ordered to counterattack toward Munoz or act as the pivot of a counteroffensive by other Army forces. If the situation becomes untenable, it will occupy mountain positions and check an enemy advance toward the Bambang area.*

2. An element of the division (Tsuda Detachment) to secure coastal positions on Baler and Dingalan Bays. In case of an enemy landing, this force will endeavor to inflict maximum losses at the time of landing and will then check an enemy advance inland by defending mountain position in depth.

2d Armored Division:

1. Main strength to remain deployed in its present location (Cabanatuan—San Miguel area). The division will destroy enemy airborne forces landing on the central Luzon plain and, depending on the situation, may be ordered to launch a counteroffensive on the Lingayen or Manila front, or to move north to Tuguegarao and defend that area against an enemy penetration of the Cagayan Valley.

2. The 6th Tank Regiment presently in Manila to remain in the Manila area and prepare to operate against enemy airborne troops landing in the vicinity.[54] *Thereafter, if circumstances permit, it will rejoin the division in the Cabanatuan sector.*

19th Division:

1. As it arrives on Luzon, the division to assemble its main strength to the south of San Leon. This force is designated Area Army second-line reserve, its mission to be assigned later as the situation warrants.

2. Elements of the division will secure the vital areas around Tuguegarao and Echague, and will also prepare for action against possible enemy airborne landings.

Southern and Central Zone

Manila Defense Force:

1. Main strength to occupy positions forming one wing of the key defenses in the mountains east of Manila. It will secure these positions and also prepare to sortie at any time for a counterattack.

2. A strong element to occupy positions in and around Manila. In case of an enemy attack, this force will inflict maximum losses by a determined defense of its positions and endeavor to create a favorable tactical opportunity for a counterattack by the main strength.

3. Garrison forces in the Corregidor Channel area (in conjunction with naval garrison units) will defend this area to the last and cooperate closely with the naval surface raiding force in smashing any attempted penetration into Manila Bay by enemy naval forces.

4. Garrison forces in the Bataan area (Naga-

53 The Area Army, on 18 December, had already ordered the headquarters and first echelon elements of the 10th Division, less the 39th Infantry Regiment, reinf., which had already been deployed to Bataan under command of the Manila Defense Force, to move to the San Jose area and construct defensive positions in the zone vacated by the 23d Division.

54 After the enemy landing on Mindoro, the 6th Tank Regiment had moved back from Los Banos, in Batangas, and was in the vicinity of Manila when the new operational plans were issued on 19–20 December.

yoshi Detachment[55]) *will occupy established positions and endeavor to repel any enemy landing attempt in this area. If a landing is made, it will hinder an advance from the beachhead and endeavor to consume enemy strength. Depending on the situation, the Nagayoshi Detachment may be ordered to the Angeles sector to reinforce the airfield units defending the mountain positions west of Clark Field.*

8th Division:

1. Main strength to occupy positions guarding key communication points to the east and west of Lake Taal. In the event of an enemy landing in the Batangas area, it will secure these positions and check an enemy advance inland, covering the redeployment of 105th Division and other forces to the key positions east of Manila. As this redeployment progresses or as the situation on other fronts may require, it will withdraw upon Area Army order to the sector north of Laguna de Bay and reinforce the key positions east of Manila. The division will at this time take command of all Army and Navy forces in the area.

2. Elements of the division, composed mainly of heavy gun units and surface raiding units, to be deployed at key coastal points in Batangas. These elements, cooperating with naval surface raiding units, will resist to the last any enemy landing attempt and seek to inflict maximum losses.

105th Division:

1. Main strength to redeploy immediately from southern Luzon to the key area east of Manila. Upon assembling in this area, it will organize defense positions.

2. An element of the division to remain deployed in the Bicol area. This force will secure the San Bernardino Strait area, and in the event of an enemy landing will endeavor to obstruct enemy seizure and utilization of the airfields at Legaspi and Naga.

Kawashima Force (82d Inf. Brig., reinf./105th Div.):

1. As the redeployment of the 105th Division progresses, the Kawashima Force will gradually transfer troop strength from the southern part of its operational sector, particularly around Atimonan, to the area east of Manila. Such troops will revert to the command of the 105th Division.

2. The remainder of the force will continue to occupy coast positions in the Lamon Bay area. In the event of an enemy landing, these elements will endeavor to inflict maximum losses at the time of landing, and thereafter will impede the enemy advance by resisting on successive positions in depth in the mountains. Depending on the general situation, the Area Army may order these elements to withdraw to positions in the key area east of Manila or to revert to 105th Division control and participate in a counterattack on the Tayabas or Lipa front.

Under the foregoing Area Army plans, airfield troops and base personnel of the Fourth Air Army and First Combined Base Air Force were to shoulder the responsibility of defending the Clark Field sector and the projected mountain strongpoint to the west. These forces were not under Fourteenth Area Army control. A draft agreement appended to the detailed operational plan for southern and central Luzon provided that these troops should come under the unified command of the ranking air commander in the Clark Field sector. However, Fourth Air Army, with Area Army assistance, was to plan and direct the construction of defense positions.[56]

The Area Army plans further envisaged immediate preparations for the transfer of General Yamashita's headquarters from Manila

55 Upon its attachment to the Manila Defense Force on 14 December for the reinforcement of Bataan, the Nagayoshi Detachment was ordered to assume command of the minor Manila Defense Force elements already stationed in the Bataan area.

56 The draft agreement further provided that the 31st Special Naval Base Force commander should exercise unified command of Army and Navy units assigned to defense of the Manila Bay mouth. After the start of land fighting, however, Fourteenth Area Army was to control the base force in all matters connected with ground operations. Documents of Fourth Air Army and Fourteenth Area Army, op. cit: Operational Policy Outline for Luzon.

to Baguio and the establishment of an Area Army command post at Ipo, northeast of Manila, to facilitate control of operations in the central and southern zone.

Battle Preparations

As the margin of time before General MacArthur's anticipated assault narrowed steadily, the Japanese forces on Luzon plunged into intense activity to implement the newly formulated battle plans. The Area Army accelerated the displacement of base installations and of munition and supplies not required in the first phase of operations into the three key defense areas. At the same time, the first troop movements required by the northern sector plans were begun.

Between 19 and 22 December, the main body of the 23d Division in the Umingan-San Jose area moved back to the division's newly-assigned sector of responsibility along Lingayen Gulf. Headquarters was established at Sison, five miles inland from the coast. Meanwhile, the 71st Infantry, which had been restored to division control on 18 December, moved north from the Batangas area toward Lingayen.[57] On 23 December the division third echelon from Formosa, made up of supporting elements, disembarked at San Fernando, in the northern part of the Lingayen section.[58]

The headquarters and first echelon elements of the 10th Division, less the Nagayoshi Detachment, had already moved north from Manila to occupy the San Jose sector vacated by the 23d Division. By 20 December, these elements were assembled at San Jose. Three days later the division second echelon reached Luzon from Formosa. The bulk of the troops disembarked at San Fernando and immediately prepared to move inland to join the division headquarters.[59] As a result of a last-minute Area Army order, however, the major portion of the 10th Infantry Regiment was diverted to Aparri and attached to the 103d Division to reinforce the defenses of the Cagayan Valley.[60]

The 19th Division was now the only major reinforcement unit allocated to Fourteenth Area Army which had not yet reached Luzon. On 27 December, however, the main strength of the division finally arrived from Formosa, without enemy interference and disembarked at

57 The headquarters and 1st Battalion reached Binalonan on 24 December, followed by the 2d Battalion on 31 December—1 January. The 3d Battalion remained in the Manila area under temporary command of the Manila Defense Force until 28 December and only rejoined the regiment in the Sison sector on 13 January, four days after the enemy landing on Lingayen Gulf. (1) Statement by Lt. Col. Takahashi, previously cited. (2) Manila Defense Force Operations Order No. 115, 28 Dec 44. XIV Corps ADVATIS Translation XIVCAET 0079, 27 Feb 45.

58 Arrival of the third echelon completed the transshipment of the division from Formosa to Luzon. Losses sustained in transit by all three echelons aggregated 30 per cent of division strength, in addition to large quantities of equipment and supplies. Philippine Operations Record, Phase Three, op. cit. Vol. I, Appended Chart.

59 Shipment of the 10th Division was carried out in only two echelons, the second including all division strength which had not been transported to Manila in the first part of December. (Cf. n. 16.) Main components of the second echelon were the 10th and 63d Infantry Regiments, the remaining strength of the 39th Infantry Regiment, and elements of division troops. Submarine attacks on the San Fernando-bound section of the convoy shortly before it reached port resulted in the loss of three infantry battalions and most of the division transport regiment. Historical Data, 10th Division (Tetsu Force), op. cit.

60 Under the 19–20 December plans, this mission was assigned to elements of the 19th Division in order to permit the concentration of maximum 10th Division strength in the San Jose-Umingan-Natividad area. The change in plan was caused by fear that it would be dangerous to postpone bolstering the defenses of the Cagayan Valley until arrival of the 19th Division, which might be delayed by shipping difficulties. (Statement by Maj. Gen. Konuma, previously cited.)

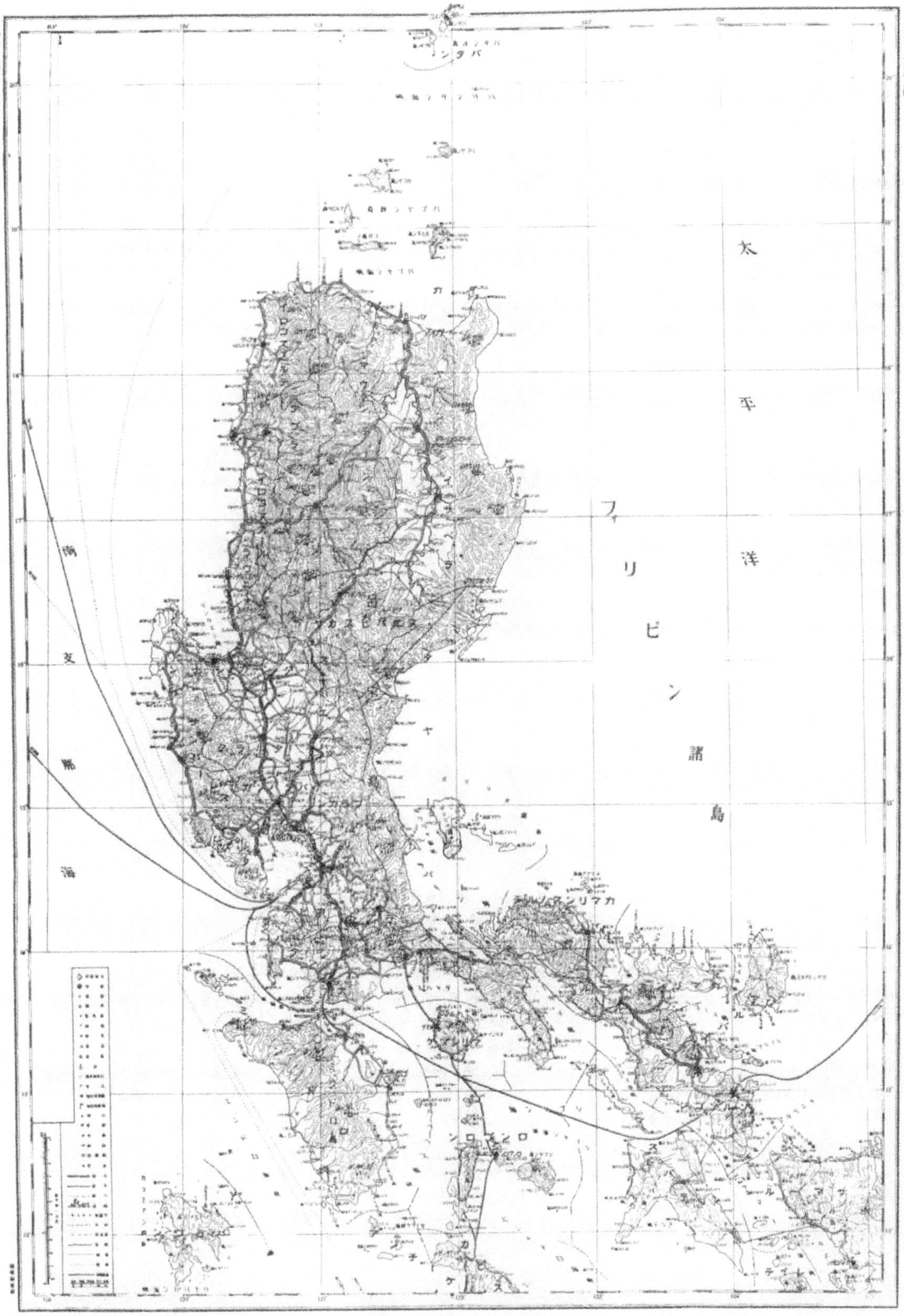

PLATE NO. 108
Map of Luzon Printed in 1941 by Land Survey Department, Imperial General Headquarters

San Fernando.[61] The division immediately began hauling its equipment and supplies to an inland assembly area at Naguilian, southeast of San Fernando, preparatory to moving to its assigned positions in the San Leon sector on the south rear flank of the 23d Division.

Meanwhile, Fourteenth Area Army had reconsidered the problem of assisting the Fourth Air Army and First Combined Base Air Force units in the preparation of defenses in the Clark Field sector. It was originally intended to limit such assistance to the dispatch of staff officers as technical advisors. As a result of an on-the-spot survey, however, the Area Army decided that this would be insufficient and ordered the 2d Armored Division to send the main strength of the 2d Mobile Infantry Regiment, with a tank company attached, to the Clark Field area. This force arrived about 24 December and immediately began organizing defensive positions. Shortly thereafter, the 2d Glider Infantry Regiment and miscellaneous smaller units of the 1st Airborne Raiding Group, assigned to Fourth Air Army, landed at San Fernando and were ordered to proceed at once to the Clark Field sector.[62] These were the last seaborne reinforcements to reach the Philippines.

In the central and southern zone, troop shifts to effect the new dispositions envisaged in the 19–20 December plans were not started until the last week in December. On the 23d, Fourteenth Area Army issued the first implementing order, which directed the 105th Division to begin transferring its main strength from the Bicol area to the sector just north of Laguna de Bay. Here it was to organize the southern flank of the key defense area east of Manila.[63] Division elements remaining in the San Bernardino Strait and Legaspi sectors were formed into the Noguchi Detachment and placed under direct Area Army control.[64]

---

61 The first and second echelons of the division, comprising the 73d Infantry Regiment, the 75th Infantry Regiment less 3d Battalion, the 76th Infantry Regiment less 2d and 3d Battalions, and the major portion of supporting division troops, arrived by the same convoy. The third echelon, including all remaining elements, never left Formosa due to the complete interdiction of Luzon-bound shipping just prior to and following the Lingayen landing. (Statement by Lt. Gen. Yoshiharu Ozaki, commander, 19th Division.)

62 Formation of the 1st Airborne Raiding Group, which was to absorb the 2d Parachute Group and Troop Carrier Air Regiment already in the Philippines, was ordered by the Inspectorate General of Army Aviation on 26 November. The Group headquarters was organized in Kyushu but was ordered to remain in Japan until additional component units had moved to the Philippines. The additional airlift units flew to Clark Field prior to the waterborne movement of the 1st and 2d Glider Infantry Regiments and supporting machine-gun, engineer and signal units. Transports carrying the 1st Glider Infantry Regiment and the engineer unit were sunk en route, and these elements never reached the Philippines. The headquarters was ordered on 27 December to proceed by air to Clark Field, where it arrived on 8 January. (1) Inspectorate General of Army Aviation Operations Order No. A-29, 26 Nov 44. Sixth Army ADVATIS Translation 6AAET 0361, 26 Feb 45. (2) Statements by Maj. Koji Tanaka, Staff Officer (Air), Imperial General Headquarters, Army Section, and Col. Yasuji Okada, Chief of Staff, 1st Airborne Raiding Group. (3) Philippine Operations Record, Phase Three, op. cit. Vol. III, Suppl. 2 (Kembu Group Operations in the Clark Sector), pp. 2–3.

63 The division was to assume command of miscellaneous elements already in the area and organize its main positions from the heights west of Antipolo to the vicinity of Morong, on the north shore of Laguna de Bay. The left flank was to be guarded by secondary positions north and east of Pililla. Documents of Fourth Air Army and Fourteenth Area Army, op. cit: Fourteenth Area Army Operations Order No. A-268, 23 Dec 44.

64 (1) Ibid. (2) The Noguchi Detachment, organized under this order, was made up of the 81st Infantry Brigade headquarters, the 182d and 185th Independent Infantry Battalions, and the 26th Independent Mixed Regiment less 3d Battalion. (Statement by Lt. Gen. Yoshitake Tsuda, commander 105th Division.)

While the 105th Division was preparing to execute this order, the Area Army decided upon the significant modification of the 19–20 December plans. These plans had originally envisaged a gradual withdrawal of forces into the key mountain positions east of Manila, possibly continuing after the start of the enemy invasion. However, the serious dislocation of troop and supply transport by increasingly persistent enemy air attacks[65] and guerrilla destruction created a danger that such movement might become impossible if delayed too long. General Yamashita therefore decided to concentrate maximum forces in the area east of Manila with all possible speed.

To implement this decision, the Area Army on 27 December ordered the 8th Division to transfer its main strength, previously scheduled to occupy positions east and west of Lake Taal, to the area east of Manila. Effective 1 January, the division commander was to assume command of all Area Army forces on the southern half of Luzon below a line running roughly from the Manila area on the west to the Lamon Bay area on the east. These forces were designated the Shimbu Group. The Noguchi Detachment was released from its previously assigned mission of securing the San Bernardino Strait area, and the Shimbu Group commander was authorized to move both this detachment and the Kawashima Force (Lamon Bay sector) toward the main concentration of forces in the area east of Manila.[66]

In accordance with the Area Army order, Lt. Gen. Shizuo Yokoyama, 8th Division commander, immediately ordered the main body of the division to concentrate at Ipo, Wawa and Antipolo. The 17th Infantry Regiment (less 3d Battalion, but with 3d Battalion, 31st Infantry, attached) was simultaneously designated the Fuji Force and ordered to remain in the Batangas area to defend key points. On 1 January Lt. Gen. Yokoyama moved to Manila with his headquarters, there assuming command of the Shimbu Group. An order issued the next day to the forces newly placed under his control called for the following dispositions:[67]

1. The 105th Division main body will defend positions near Antipolo.

2. The Manila Defense Force, less the Nagayoshi Detachment,[68] will occupy positions in the Wawa sector with its main strength. Elements will secure the Manila area.

3. The Kawashima Force will occupy positions in the Ipo sector.

4. The Kogure Detachment, to be organized by the headquarters of the 1st Surface Raiding Base

---

65 From 17 December until shortly before the enemy landing at Lingayen, carrier-based air attacks on Luzon were inconsequential. However, land-based planes operating from Leyte and later from Mindoro conducted almost daily bombing raids throughout this period, forcing movements over the railway and highway network to be limited largely to night time. For the period 17–31 December, an average of sixty-five planes a day were reported attacking Luzon targets. *Senkun Tokuho Dai Yonjuyon-go Ruson To Sakusen ni okeru Kyokun narabini Keika no Gaiyo* 戰訓特報第四十四號「ルソン」島作戰に於ける教訓竝に經過の概要 (Battle Lessons Special Report No. 44: Summary of Operations and Battle Lessons, Luzon Campaign) Imperial General Headquarters, Army Section, 29 Mar 45, Appended Chart No. 2.

66 *Shobu Sakumei Ko Dai Nihyakukyujuni-go* 尚武作命甲第二九二號 (Fourteenth Area Army Operations Order No. A-292) 27 Dec 44.

67 Statement by Col. Shujiro Kobayashi, Staff Officer (Operations), Fourteenth Area Army. (Upon formation of the Shimbu Group, this officer was temporarily assigned to duty on the staff of the Group headquarters.)

68 On 28 December, following the organization of the Shimbu Group, the Bataan area, with the exception of the southern tip of the peninsula, was transferred from the Manila Defense Force sector of responsibility to that of the 2d Armored Division. Effective the same date, the Nagayoshi Detachment came under 2d Armored Division command.

Force, will defend the Lamon Bay area.

5. The Noguchi Detachment, less the 26th Independent Mixed Regiment, will continue to secure the Bicol Peninsula. The 26th Independent Mixed Regiment will move to the area east of Manila and come under command of the Shimbu Group.

When this order was issued, the assembly of the 105th Division in the Antipolo sector was already well under way, and the first elements of the 8th Division were arriving at their new positions at Ipo, Wawa and Antipolo. By 1 January also, the Kawashima Force and the major portion of the Manila Defense Force had completed necessary preparation and were ready to move to the Ipo and Wawa sectors, respectively.

General Yamashita at this time was directing battle preparations from the Area Army's southern command post at Ipo, where he and a skeleton operational staff had arrived from Manila on 26 December. Shortly after his arrival, he was notified by Southern Army that all Fourth Air Army forces in the Philippines would come under Area Army control beginning 1 January 1945. In preparation for this change, the Area Army on 29 December issued an order from Ipo specifying the missions of the Fourth Air Army. General objectives of the Air Army were stated in this order as follows:[69]

The Fourth Air Army will make its main objective the destruction of enemy forces seeking to land on Luzon. Prior to the start of such operations, the Air Army will utilize favorable opportunities to destroy enemy shipping, neutralize opposing air strength and discover enemy moves to invade Luzon.

As soon as its control over the Army air forces became effective, the Area Army took further action to hasten the organization of ground defenses in the Clark Field sector. Lt. Gen. Yoshiharu Iwanaka, 2d Armored Division commander, was ordered to assume full responsibility for the execution of this mission and to coordinate the activities of the 2d Armored Division elements already in the Clark Field sector with those of the local Army and Navy airfield units. The dispatch of additional ground combat troops to the area, however, was not contemplated.

During the latter part of December, battle preparations proceeded with discouraging slowness. Overburdened transport facilities, enemy strafing and bombing attacks, guerrilla interference and an acute shortage of automotive fuel impeded progress in every direction. On the other hand, there were numerous indications that General MacArthur was virtually ready to strike. In the Batangas area, enemy air reconnaissance was conspicuously frequent, while the dropping of dummy parachutists and the activity of small surface craft along the coast also caused grave alarm in the 8th Division. Other reports indicated that guerrilla forces were beginning to assemble in the mountains east of Manila, and that enemy submarines were delivering arms to guerrillas in the Lamon Bay area.

The concentration of these activities in the south-central area of Luzon exerted a marked influence on the Area Army's estimate of the location of the enemy's main attack. After the first shock of the unexpected landing on Mindoro had passed, belief that Batangas would be the principal invasion point gradually diminished. Instead, there was an increasing tendency to believe that basic considerations of terrain and strategy would lead General MacArthur to land his main forces at Lingayen. The pronounced enemy activity in the Batangas and Manila areas, however, countered this tendency and made it impossible for the Area Army to arrive at any defi-

69 *Shobu Sakumei Ko Dai Sambyakuichi-go* 尙武作命甲第三〇一號 (Fourteenth Area Army Operations Order No. A-301) 29 Dec 44.

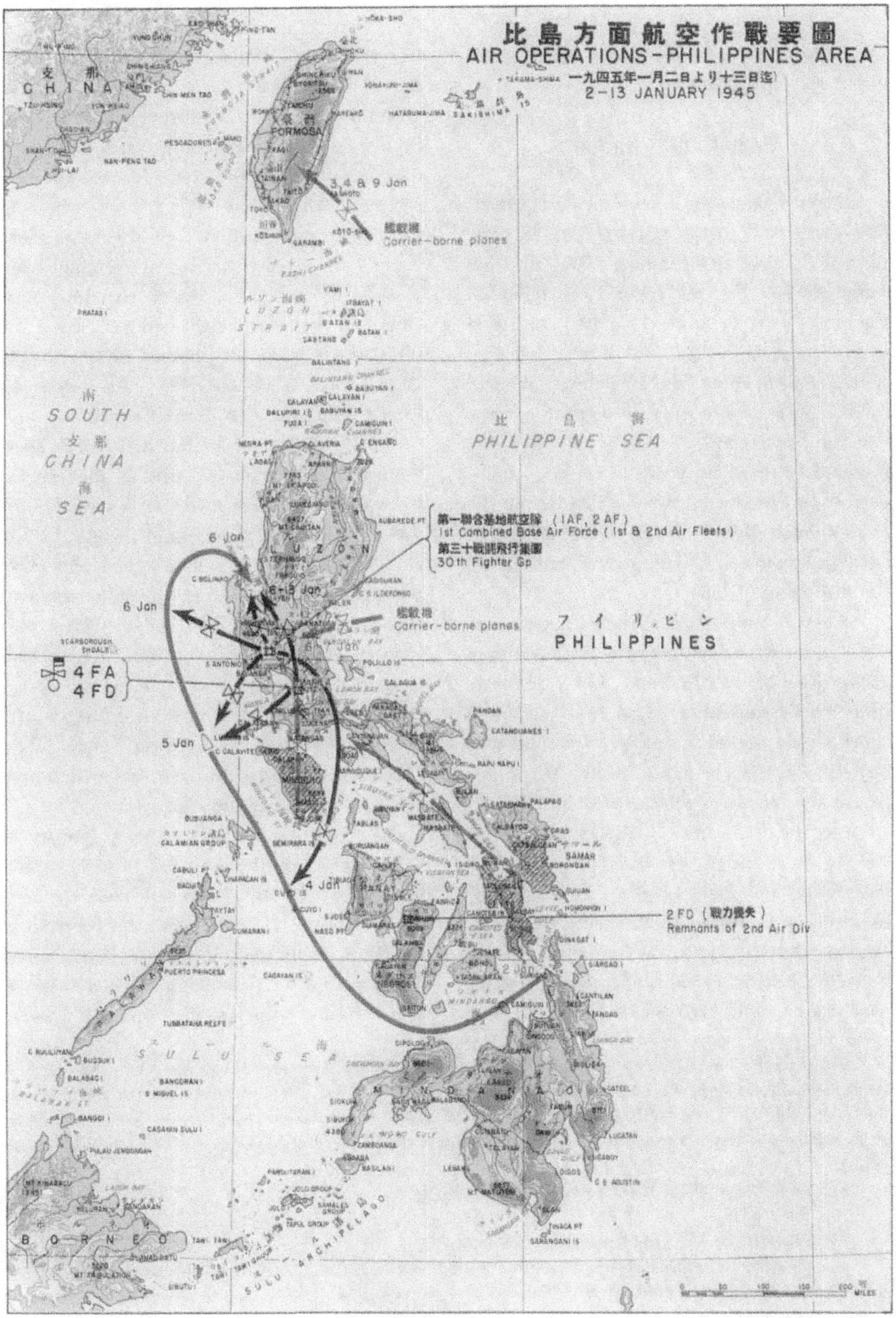

PLATE NO. 109

Air Operations—Philippines Area, 2–13 January 1945

nite conclusion. As December ended, this uncertainty still prevailed.[70]

Prelude to Invasion

General Yamashita's tense and expectant forces were still racing against time to complete their battle preparations when the first warnings came that the enemy was again on the move. At 1322 on 2 January, a naval lookout post on Surigao Strait reported that a long train of about 80 enemy ships, screened by destroyers, was moving through the strait into the Mindanao Sea. Other reports which followed immediately warned that large numbers of invasion craft were gathered in the Leyte, Palau and Marianas areas, and that enemy submarines had appeared in force off the west coast of Luzon.[71]

At first it was thought that the enemy convoy was merely a reinforcement and resupply group destined for Mindoro. On 3 January, however, developments began to point more conclusively toward a full-scale amphibious operation against Luzon. In the afternoon, just as the initially sighted enemy group passed into the Sulu Sea, a large enemy naval task force, including 12 escort carriers, four battleships and eight cruisers, was reported following in its wake through Surigao Strait. Another delayed report stated that an enemy transport convoy of 90 ships, under strong naval escort, had been spotted the previous day northwest of Palau, heading in the direction of Leyte Gulf.[72]

With such powerful enemy forces in motion, Southwest Area Fleet headquarters in Manila now considered an invasion of Luzon virtually certain. Area Army staff officers remaining in the Philippine capital took the same view. Their estimate of the situation was immediately radioed to Baguio, where General Yamashita and part of his staff had proceeded the same day to effect the planned transfer of the Area Army headquarters.[73] All forces on Luzon were promptly alerted for action.

Small numbers of aircraft had struck spasmodically at the enemy ships as they moved through the Mindanao Sea on 3 January. It was not until late afternoon of the 4th, however, that the first really determined and effective attacks were carried out by fighter-escorted suicide planes of the Fourth Air Army's 30th Fighter Group. Sortieing from Clark Field, these planes struck leading elements of the invasion force off southwestern Mindoro and between the Cuyo Islands and Panay. Five planes were reported to have executed suicide crash attacks.[74] (Plate No. 109)

Air action was intensified on 5 January as the vanguard of the invasion armada continued on past Mindoro and sailed up the west coast of Luzon. Suicide and other aircraft of the First Combined Base Air Force joined army planes in launching a succession of fierce attacks in the late afternoon as the enemy

---

70 It was not until 4 January, two days before the enemy invasion fleet entered Lingayen Gulf, that the Area Army concluded definitely that Lingayen would be the main point of attack. The main reason for this conclusion was the sudden shift of enemy bombing attacks from airfields to key bridges between Lingayen and Manila. The Area Army continued to regard a secondary landing in the Batangas area as probable. (Statement by Maj. Hori, previously cited.)

71 Daily Record of the War Situation, op. cit., 2 Jan 45.

72 Ibid., 3 Jan 45.

73 Statement by Maj. Gen. Konuma, previously cited.

74 Philippine Air Operations Record, Phase Two, op. cit., Appended Chart 17. (American naval records show that only one ship, the escort carrier *Ommaney Bay*, was hit in these attacks. Damage was so severe that the carrier had to be sunk by an accompanying destroyer.)

ships passed to the west of Manila Bay. Although extremely strong fighter opposition was encountered, returning aircraft reported a number of successful suicide hits on enemy vessels.[75]

By this time General Yamashita and his staff were firmly convinced that the enemy's ultimate destination was Lingayen Gulf. The forces in that area, caught with their dispositions for battle still incomplete, feverishly prepared to meet the impending onslaught. Lt. Gen. Muto, Area Army Chief of Staff, hastily inspected the 23d Division sector on the afternoon of the 5th and discovered that the division was organizing its main line of resistance so that it veered away from the coast at a point northeast of San Fabian and skirted the mountains along the northern edge of the central plain. From the point where the main line swerved eastward, a forward line of positions branched off southward to the Cabaruan Hills. This forward line was manned by relatively minor forces and was to be held only until enemy pressure became too severe. The defending elements were then to fall back on the main mountain positions. (Plate No. 110)

Lt. Gen. Muto considered these dispositions weak and inadequate. When he returned to Baguio late the same day, he urged the Area Army Commander to order the 23d Division to strengthen the forward positions on the south flank and make them part of the main line of resistance. Such an order was immediately issued although it was recognized that any major shift of troops strength could probably not be effected in time. It was hoped, nevertheless, that the order would spur the division to make a maximum effort in defense of the forward positions.[76] As a further step toward strengthening the sector, General Yamashita ordered the 2d Armored Division to send a detachment to Urdaneta, just northeast of the Cabaruan Hills.[77]

Developments on the morning of 6 January left no doubt about the enemy's intentions. Dawn of that day found leading elements of the invasion force entering Lingayen Gulf, and shortly thereafter naval units began a preliminary bombardment of the shore defenses, concentrating primarily on targets in the San Fernando area. Simultaneously, enemy land-based and carrier aircraft launched attacks on the Manila and Clark Field areas in an effort to neutralize air opposition to the landing operation.[78]

Despite these air assaults, Fourth Air Army and the First Combined Base Air Force threw every available plane into a daylong series of strikes both at the invasion ships in Lingayen Gulf and at follow-up elements along the route to the invasion area. In three major attacks on the Lingayen Gulf and San Fernando areas by *Kamikaze* of the First Combined Base Air Force, over fifteen suicide hits were reported,

75 Available Japanese documentary sources covering air operations during this period give figures only for the number of suicide aircraft and escorting fighters employed in special-attack missions and contain no data regarding ordinary attack missions. The total number of planes sent out on special-attack missions on 5 January was 30, of which only five were escort fighters. More than 18 suicide planes failed to return, but only seven were definitely reported to have crashed into enemy ships. (1) Philippine Area Naval Operations, Part II, op. cit., p. 103 (2) Philippine Air Operations Record, Phase Two, op. cit., Appended Chart 17. (According to American official sources, nine ships were actually damaged on 5 January by suicide plane crashes or near misses; no ships were sunk.)

76 (1) The Truth of the Philippines Campaign, op. cit., p. 31. (2) Statement by Maj. Gen. Konuma, previously cited.

77 The force organized by the 2d Armored Division to assume this mission was designated the Shigemi Detachment. The detachment was to come under direct Area Army command as soon as it reached the north bank of the Agno River. 2d Armored Division Operations Order No. A-88, 7 Jan 45. ATIS Bulletin No. 1814, 27 Feb 45, pp. 1–4.

78 Summary of Operations and Battle Lessons, Luzon Campaign, op. cit., p. 31.

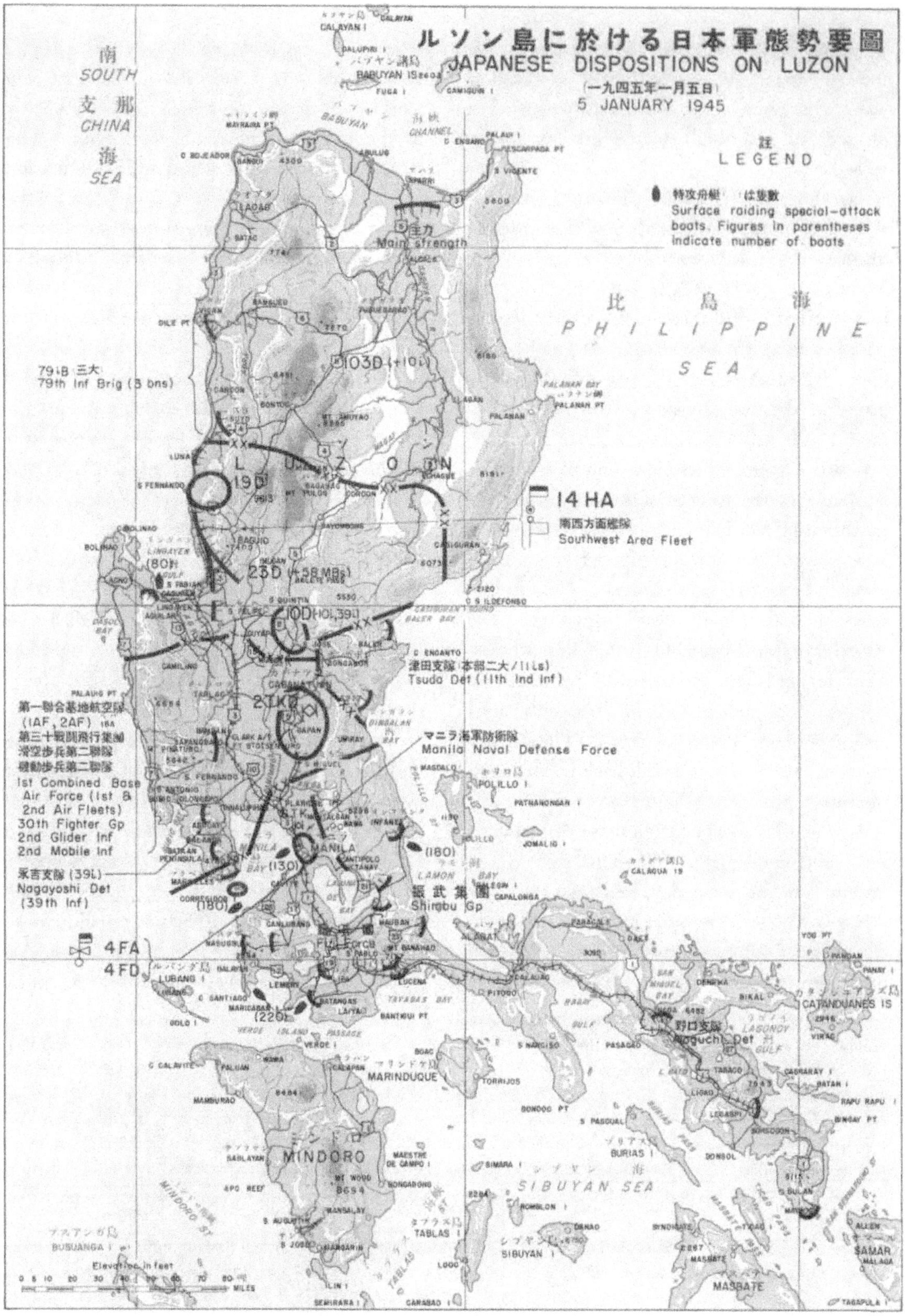

PLATE NO. 110

Japanese Dispositions on Luzon, 5 January 1945

while three other hits were claimed for suicide planes of the Fourth Air Army. Naval *Kamikaze* also struck at a follow-up group off Iba, with reported hits on five enemy ships.[79]

The heavy bombardment of the San Fernando sector on the 6th aroused great anxiety in General Yamashita's headquarters. An enemy landing in that sector would fall north of the main positions of the 58th Independent Mixed Brigade and create serious danger of a thrust toward Baguio over the Bauang-Naguilian Highway. The only means of meeting this potential threat was to change the planned employment of the 19th Division. The division was still in the San Fernando-Naguilian area due to postponement of its scheduled movement inland, and intense enemy air activity over the whole Lingayen area made it improbable that it could move in any case.[80] Consequently, on 7 January, the Area Army ordered the division to assume responsibility for the defense of the sector north of a line running through Santiago, Magungunay, Mt. Bilbil, and Asin. Its main strength was to be concentrated between Naguilian and Baguio.[81]

Retention of the 19th Division in the Naguilian-Baguio sector, however, weakened the planned dispositions for defending the vital south flank of the northern redoubt. To remedy this situation, General Yamashita decided to effect a gradual shift northward of part of the Shimbu Group forces from the area east of Manila. As the first step, the troops withdrawn from the Manila area were to move to the Cabanatuan sector, releasing the 2d Armored Division main body for operations to the west. Later, they were to transfer farther north and take up position behind the 10th Division to defend the passes leading into the northern redoubt. The 10th Division was also to broaden its deployment to take in the sector previously assigned to the 19th Division. On 8 January, the Area Army implemented its decision by ordering the following dispositions:[82]

1. The Shimbu Group commander will dispatch a force of approximately five battalions under the 105th Division commander to the Cabanatuan area. Maximum artillery and engineer units will be attached to this force. Elements of the force will relieve the 2d Armored Division of its mission of securing the Manila-San Miguel-Cabanatuan highway.

2. The main strength of the 2d Armored Division (less detached elements) will move to the Tarlac area and prepare to check an enemy advance toward the Clark Field sector.

3. The Shimbu Group will be responsible for the sector east of the line Calumpit-Pampanga River-Licab-Guimba-San Antonio, and the 2d Armored Division for the sector west of this line.

4. The 10th Division will deploy elements in the Triangle Hill[83] *and Tayug sectors to oppose an anticipated enemy advance from Lingayen toward San Jose.*

---

79 A total of 58 suicide aircraft and 17 escorting fighters took part in the strikes against enemy ships in Lingayen Gulf and off San Fernando and Iba during 6 January. (1) Philippine Area Naval Operations, Part II, op. cit., p. 103. (2) Philippine Air Operations Record, Phase Two, op. cit., Appended Chart 17. (American naval records show that one ship was sunk and 14 others damaged in these attacks. Seriously damaged ships included the battleship *New Mexico* and *California*.)

80 For a time after its debarkation at San Fernando on 27 December, the division was engaged in transporting its equipment and supplies from the beaches to an assembly point at Naguilian. It was then ordered to stand by to assist in unloading a fuel and ammunition convoy expected to reach San Fernando shortly. This convoy had not arrived by 6 January. The Truth of the Philippines Campaign, op. cit., p. 32.

81 Philippine Operations Record, Phase Three, op. cit. Vol. III, pp. 72–4.

82 Fourteenth Area Army Operations Order No. 368, 8 Jan 45. ADVATIS Bulletin No. 278, 15 Feb 45

83 The 19th Division had previously been scheduled to concentrate in the "Triangle Hill" sector. This sector, located south of San Leon, was so named by the Japanese because of three prominent hill features forming a triangle. Mt. Bangcay was at the southern corner of the triangle.

5. The Tsuda Detachment will assemble its main strength in the Bongabong area, leaving elements at coastal points.[84] *The roads leading from the coast to Bongabong will be destroyed.*

While the Area Army was making these last-minute changes in its dispositions, the pre-landing phase of enemy operations in Lingayen Gulf progressed swiftly. On 7 January the heavy naval bombardment units moved deeper into the gulf and began an intensive shelling of both coastal and inland targets in the Damortis and San Fabian sectors. Enemy carrier aircraft directly supporting the invasion force supplemented the naval gunfire by bombing attacks on 23d Division defense positions and ammunition dumps, while other air attack groups bombed and destroyed vital bridges in the Calumpit, Plaridel, and Manaoag areas. On 8 January the preparatory shelling continued, taking in both the San Fabian and Lingayen sectors.[85]

Despite renewed neutralization strikes by enemy task force aircraft, Fourth Air Army and First Combined Base Air Force planes continued their attacks on invasion shipping in Lingayen Gulf and off the west coast of Luzon on 7–8 January. Suicide planes inflicted additional damage,[86] but heavy losses sharply reduced the scale of these attacks. On 8 January the Second Air Fleet and First Combined Base Air Force headquarters were deactivated, and the First Air Fleet received orders the following day to retire to Formosa for reorganization. Fourth Air Army's decimated forces continued small-scale attack operations for a brief period, but rapidly mounting operational losses soon obliged them to retire to northern Luzon bases. Subsequent activity, for the most part, was limited to liaison and reconnaissance flights.[87]

The last valiant effort of the Japanese air forces in the Philippines had failed to repel General MacArthur's invasion of Luzon.

---

84 The Tsuda Detachment came under 10th Division command effective the same day as the Area Army order of 8 January. 10th Division Operations Order No. A-45, 8 Jan 45. I Corps ADVATIS Translation ICAET 0185, 25 Apr 45.

85 Sumary of Operations and Battle Lessons, Luzon Campaign, op. cit., pp. 31–2.

86 The number of sorties flown by army planes on 7–8 January is not known. Naval *Kamikaze* flew two missions with a combined total of 12 suicide planes and 11 escort fighters on 7 January, and only one sortie by a single suicide plane on the 8th. Nine *Kamikaze* and seven escorts made final sorties on the 9th. A total of 17 enemy ships were reported hit by Army and Navy suicide planes between 7–9 January. (1) Philippine Area Naval Operations, Part II, op. cit., pp. 103–4. (2) Philippine Air Operations Record, Phase Two, op. cit., Appended Chart 17. (According to American naval records, no ships were sunk by suicide planes during this period, but seven ships were damaged. One ship, HMAS *Australia,* was hit three times.)

87 Fourth Air Army headquarters transferred from Manila to Echague between 7 and 10 January preparatory to a general withdrawal of combat air units to northern Luzon bases for reorganization. These bases were found inadequate, however, and the headquarters withdrew to Formosa on 15 January under a plan to carry out the reorganization there. With the exception of one fighter company and reconnaissance elements which were to remain at northern Luzon bases, all combat air units on Luzon were to evacuate their remaining strength (mainly personnel) to Formosa. It was planned to complete the reorganization of three fighter regiments, one reconnaissance regiment, and one fighter-bomber regiment by 20 February, when these units would be redeployed to the Philippines if the situation permitted. (1) Philippine Air Operations Record, Phase Two, op. cit., pp. 122, 125–6, 130–3. (2) Statement by Col. Matsumae, previously cited.

CHAPTER XV

BATTLE ON LUZON

Launching of the American Invasion

At dawn on 9 January, three days after the appearance of the first Allied naval elements in Lingayen Gulf, outposts of the 23d Division observed two vast transport groups with a total of nearly 100 ships maneuvering into position off the southern and southeastern sectors of the gulf coast.[1] While naval units raked the shore with intense covering fire, landing craft loaded with assault troops began moving in toward the beaches. The American invasion of Luzon was under way.

By 1040 Tokyo time (0940 Philippine time) the assault waves of an estimated total enemy force of not less than one armored division and two infantry divisions were ashore in the San Fabian and Lingayen sectors.[2] Small outpost detachments of the 23d Division in these sectors made little or no attempt to defend the beaches against the overwhelming enemy assault, falling back to prepared inland positions after destroying key bridges near the landing points.

Opposition during the first day was consequently limited to harrassing fire thrown into the San Fabian beachhead first from mortars and then from artillery emplaced along the ridges back of the coastal plain.[3] In the Lingayen sector, where a major enemy landing had not been anticipated, the Japanese dispositions did not permit even artillery opposition.[4] (Plate No. 111)

The Area Army command estimated that the enemy would probably limit his initial operations to consolidation and expansion of the beachheads. However, the weakness of the defenses barring an advance into the central plain also aroused concern over the danger of a rapid enemy push inland either directly toward Manila or in the direction of San Jose.[5]

---

1 This chapter was originally prepared in Japanese by Maj. Hiroshi Komatsu, Imperial Japanese Army. For duty assignments of this officer, cf. n. 1, Chapter XIV. All source materials cited in this chapter are located in G-2 Historical Section Files, GHQ FEC.

2 *Senkun Tokuho Dai Yonjuyon-go; Ruson To Sakusen ni okeru Kyokun narabini Keika no Gaiyo* 戰訓特報第四十四號「ルソン」島作戰に於ける教訓竝に經過の概要 (Battle Lessons Special Report No. 44: Summary of Operations and Battle Lessons, Luzon Campaign) Imperial General Headquarters, Army Section, 29 Mar 45, pp. 34–5. (Japanese sources are either vague or in error with regard to the time of the enemy landings. Therefore, the time given here is based on Allied operational records.)

3 Statement by Lt. Col. Masaichi Takahashi, Staff Officer (Operations), 23d Division.

4 At the time of the Japanese invasion of the Philippines in December 1941, the beaches in the southern part of Lingayen Gulf around Dagupan and the town of Lingayen had been rejected as unsuitable for a landing because the marshy terrain of the coastal area would impede an advance inland. For the same reason it was thought unlikely that an American invasion force would execute major landings in this sector. (1) Cf. Chapter VI, Plate No. 17. (2) Unpublished memoirs of Lt. Gen. Akira Muto, Chief of Staff, Fourteenth Area Army: *Hitosen no Jisso* 比島戰の眞相 (The Truth of the Philippines Campaign) 15 Jun 47, p. 33.

5 Statement by Maj. Gen. Haruo Konuma, Deputy-Chief of Staff, Fourteenth Area Army.

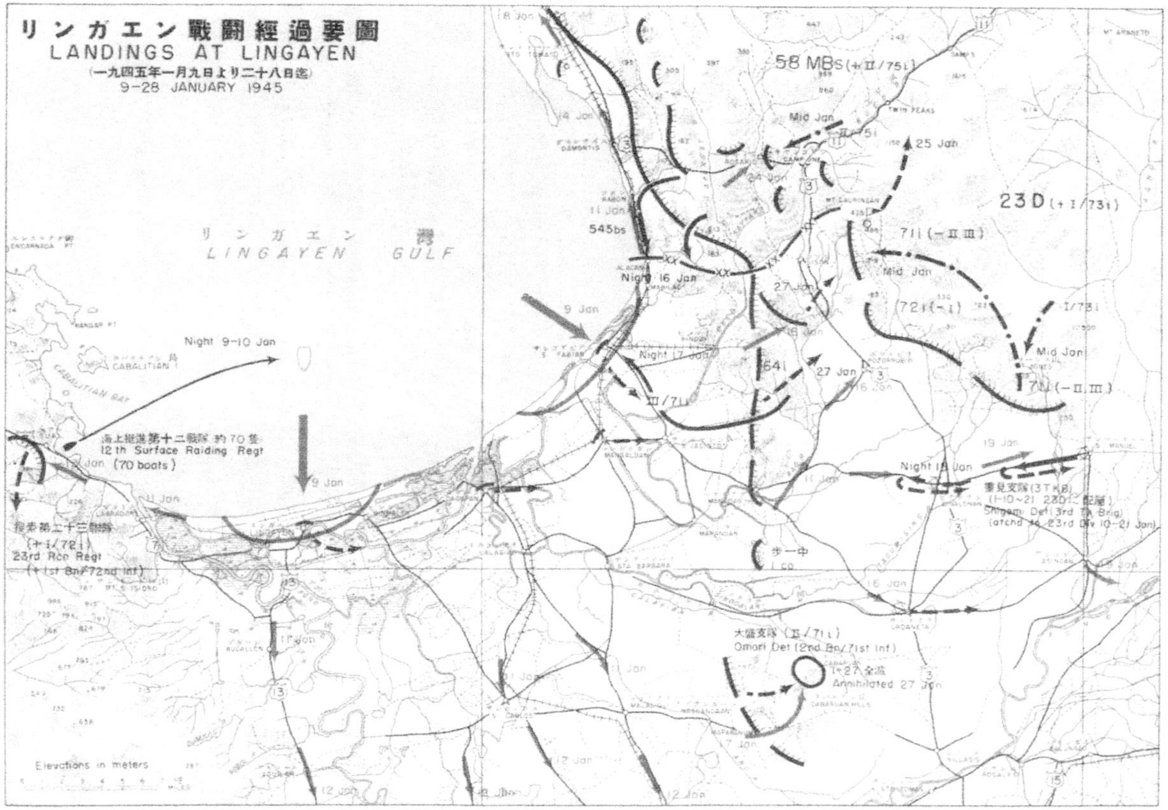

PLATE NO. 111

Landings at Lingayen, 9–28 January 1945

This danger was aggravated by the fact that the 2d Armored Division, due to the destruction of the highway bridge at Calumpit, could not without serious risk execute the Area Army order of 8 January directing the movement of its main strength from Cabanatuan to Tarlac.[6]

To cope with this situation, the Area Army on 9 January countermanded its previous order to the 2d Armored Division and instead instructed the division to move its main strength north to the line of the Agno River in the vicinity of Tayug.[7] General Yamashita's plan was now to employ the division in a counterattack from the north flank if the enemy overextended himself by too rapid an advance inland. If such an opportunity did not arise, the division's role was to become essentially defensive.

Back in Lingayen Gulf, the night of 9–10 January witnessed a determined but shortlived effort by the 12th Surface Raiding Regiment, based at Port Sual, to disrupt further enemy landing operations by an all-out suicide boat attack on ships standing off the beachhead areas. The full operational strength of the regiment, about 70 boats, sortied at midnight. Profiting from the element of surprise, the small explosive-laden craft stole in undetected to launch an attack of about two hours' duration. Although substantial damage was believed inflicted, nearly all the participating boats were lost in the attack, and the 12th Surface Raiding Regiment was rendered incapable of further operations.[8]

Despite this fleeting interference, the enemy continued to pour troops and supplies ashore. By 11 January the forces in the San Fabian and Lingayen beachheads had effected a juncture and were firmly in control of the coastal area extending from a point just south of Rabon to the mouth of the Agno River, west of Lingayen. At the same time, advance elements of the enemy left wing forces, supported by armor, were already starting to probe the 23d Division's first line defenses to the east of San Fabian.[9]

Developments elsewhere had meanwhile precipitated fresh debate in Area Army headquarters with regard to basic tactical policy. The 19–20 December plans called for a long-term delaying campaign in which the Area Army would avoid a large-scale counteroffensive except in the most favorable circumstances. However, on 9–10 January, just as the invasion was beginning, reports reached Baguio from various districts in northern Luzon to the effect that intensified guerrilla activities and enemy air attacks had virtually halted the collection and movement of essential food reserves into the key defense positions in the Baguio-

6 Cf. Chapter XIV, pp. 432–3. Most of the division main strength was to move via Calumpit because the highway network leading directly west from Cabanatuan to Tarlac included bridges too weak to support the heavier elements. An alternative route skirting around to the north via Talavera, Guimba and Gerona was ruled out because its use might expose the division to a flank attack in the event of a rapid enemy drive inland.

7 Following receipt of this order, the 2d Armored Division ordered the 6th Tank Regiment, which was still in the Manila area, to move north and rejoin the division. For a brief period the regiment had been attached to the Manila Defense Force but was returned to division control about 6 January.

8 On the basis of shore observation reports, 20 to 30 enemy ships were believed sunk in this attack. Summary of Operations and Battle Lessons, Luzon Campaign, op. cit., p. 35. (Actual results according to American naval records were two ships sunk and eight damaged.)

9 Despite the Area Army order of 5 January (Cf. Chapter XIV, p. 430.) directing the 23d Division to make these first line defenses its main line of resistance, the division by 11 January had augmented the forces in the forward line by only one additional infantry company and half a battery of artillery. The 64th Infantry, reinf., held the segment of the line running from southwest of Sison to Manaoag, while the 2d Battalion, reinf., 71st Infantry, designated the Omori Detachment, held the south flank anchor positions in the Cabaruan Hills.

Mankayan-Bambang triangle.

This obviously presented a serious threat to the Area Army plan of protracted resistance in the mountains of northern Luzon. So critical was the view taken by a majority of the staff that, on 10 January, recommendations were drawn up for submission to the Area Army Commander and Chief of Staff, urging that the policy of long-term delay be scrapped in favor of launching an early counteroffensive with maximum forces. These recommendations were placed before General Yamashita and Lt. Gen. Muto on 11 January.

After urgent consultations with Lt. Gen. Muto and Maj. Gen. Haruo Konuma, Deputy-Chief of Staff, the Area Army Commander firmly ruled against this proposal. His decision rested on two major grounds. First, the enemy's cautious development of the beachhead before starting any advance inland did not appear likely to create a situation in the near future, in which a large-scale counteroffensive could be mounted with reasonable chances of success. Second, although a shortage of food supplies would admittedly limit the Area Army's ability to wage a long-term defensive campaign, such a campaign still appeared to be the only feasible method of achieving the over-all strategic objective of delay.[10]

In line with General Yamashita's decision, the plan to move the main strength of the 2d Armored Division to the Tayug sector for a possible counterattack against the flank of an enemy thrust inland was now definitely abandoned. The Area Army, by a new order issued on the 11th, directed the division to assemble its main strength as speedily as possible at Lupao, about eight miles northwest of San Jose.[11] At the same time, the Shimbu Group commander was ordered to hasten the movement of the 105th Division main strength from the area east of Manila to Cabanatuan, the first stage of its projected transfer to the northern area.[12]

While firmly against risking a major counteroffensive, General Yamashita nevertheless felt that limited offensive action against the beachhead would not only give his forces the advantage of the initiative but provide the most effective means of delaying the enemy's advance inland.[13] This plan was studied by the Area Army staff on 11–12 January, and on the 13th Lt. Gen. Muto drove to the 23d Division command post east of Sison with an order directing the division to execute a strong raiding attack against the San Fabian-Alacan sector of the beachhead during the night of 16–17 January.[14] The primary objective was to effect maximum destruction of enemy weapons, supplies and key base installations.

On 14 January Lt. Gen. Fukutaro Nishiyama, 23d Division commander, issued an implementing order laying down the task organization and plan of attack. The 58th Independent Mixed Brigade and the 71st and 72d Infantry Regiments were each directed to organize a small, streamlined "suicide" raiding unit of hand-picked troops heavily armed with

10 The Fourteenth Area Army staff deliberations of 10–11 January regarding tactical policy are assumed on the basis of statement by Maj. Gen. Konuma, previously cited.

11 By the time this order was issued, the division headquarters had moved only as far as Talavera, a short distance north of Cabanatuan, while the advance elements were just north of San Jose.

12 *Shobu Sakumei Ko Dai Sambyakuhachijuyon-go* 尚武作命甲第三八四號 (Fourteenth Area Army Operations Order No. A-384) 11 Jan 45.

13 The Area Army estimated at this time that the enemy would start his advance inland momentarily and that it would take the form of a two-pronged drive, one prong heading south from Lingayen toward Manila, and the other southeast from San Fabian toward San Jose. (Statement by Maj. Gen. Konuma, previously cited.)

14 Ibid.

automatic weapons and reinforced by demolition squads. A fourth unit made up of a mobile infantry company and medium tank company was to be drawn from the Shigemi Detachment of the 2d Armored Division, now under 23d Division control.[15] These units were to penetrate the enemy beachhead perimeter simultaneously at different points between 0200 and 0400 on 17 January. After swiftly accomplishing their missions, they were to withdraw.[16]

From the first, however, the plan went awry. Serious enemy penetrations of the 23d Division first line positions on 15–16 January so hampered preparations for the attack that the 72d Infantry raiding unit was finally unable to participate. Another serious setback occurred during the night of 15–16 January, when the tank-infantry task unit of the Shigemi Detachment, moving up toward its assigned line of departure at Manaoag, encountered an enemy force just west of Binalonan and was compelled to withdraw eastward after a sharp engagement.

These developments left only two of the four raiding units called for in the attack plan capable of executing the operation. The attack was nevertheless launched at the scheduled time, with one raiding unit from the 58th Independent Mixed Brigade striking from the north toward Alacan, and the other from the 71st Infantry infiltrating toward San Fabian from the southeast. Both units penetrated almost to their objectives and succeeded in destroying some enemy weapons and equipment before heavy losses forced them to retire.[17] Due to its small scale, however, the operation had little effect toward impeding the enemy advance.

After this abortive effort, the 23d Division settled down to a stubborn defense of its prepared positions, limiting offensive action to small-scale night raids. The first-line defenses, however, had already been pierced at various points, and they now began to crumble rapidly under heavy and relentless ground, air and artillery assault. On 24 January, the Area Army finally recognized that a further defense of these positions was impossible and ordered the 23d Division to pull back the remaining strength of its forward elements to the secondary defense positions in the mountains southwest of Baguio.[18]

---

15 The Shigemi Detachment, ordered on 7 January to proceed to Urdaneta to bolster the 23d Division south flank, reached the vicinity of San Manuel by 10 January, when an Area Army order placed it under 23d Division control. On 12 January the 23d Division ordered the detachment to move up to Binalonan and Urdaneta, but this movement was delayed. Composition of the detachment was as follows:

Headquarters, 3d Tank Brigade
- 7th Tank Regt. (less 2d Tank Co.)
- 1st Bn. (less 1st Co. and one MG plat.), 2d Mobile Inf. Regt.
- 3d Bn. (less 8th Btry.), 2d Mobile Arty. Regt.
- 3d Co. (less one plat.), 2d Armored Division Engr. Unit
- 2d Co., 2d Armored Division Maint. Unit

16 23d Division Operations Order No. A-60, 1800 14 Jan 45. I Corps ADVATIS Translation ICAET 0027, 21 Jan 45.

17 Summary of Operations and Battle Lessons, Luzon Campaign, op. cit., pp. 53–5. (2) *Dai Nijusan Shidan (Asahi Heidan) Reika Butai Kodo Gaiyo* 第二十三師團（旭兵團）隷下部隊行動概要 [Summary of Operations, 23d Division (*Asahi* Force)] Home Depot Division, Mar 48, p. 26.

18 Fourteenth Area Army had already acted to strengthen these secondary defenses by drawing troops from the 19th Division to the north. On or about 11 January, the 2d Battalion, 75th Infantry, and the 1st Battalion, 73d Infantry, were ordered under 23d Division control for the purpose of reinforcing the defenses in the vicinity of Camp One. The 1st Battalion, 73d Infantry, was subsequently replaced by the 2d Battalion of the same regiment. (Statement by Maj. Gen. Konuma, previously cited.)

対戦車肉迫攻撃（リンガエン）　伊藤禎三　昭和二〇

Original Painting by Teizo Ito

PLATE NO. 112

Antitank Suicide Unit, Lingayen Gulf

The initial fighting had been costly. The 64th Infantry, when it began infiltrating to the rear on 27 January, was reduced to about one-third its original strength.[19] The Omori Detachment (2d Battalion, reinf., 71st Infantry) had been entirely wiped out in the defense of the Cabaruan Hill positions on the south flank. The 3d Battalion, 72d Infantry, had been reduced to less than half-strength by the losses incurred during the raid on San Fabian. The 23d Reconnaissance Regiment (reinforced by the 1st Battalion, 71st Infantry) had been driven into the Zambales Mountains behind Sual shortly after the enemy landing and was lost to the 23d Division for subsequent operations. The 58th Independent Mixed Brigade to the north had suffered less severely, but its forward positions in the southern part of the brigade sector had been breached, forcing a withdrawal to secondary positions north of the Damortis-Rosario road.

The close of January found the 23d Division and attached forces closely engaged all along the second line of defensive positions guarding the southwestern entrance to Baguio, heart of the northern defense bastion.

Defense of San Jose

Developments had meanwhile borne out the Area Army's estimate of probable enemy strategy. The invasion forces had unleashed two powerful drives from the beachhead—one aimed south toward Clark Field and Manila, the other directed east and southeast toward the key routes of access to the northern bastion from the central Luzon plain. These routes were the Villa Verde Trail, winding northeast from San Nicolas, and Highway 5, the main artery running north from San Jose over Balete Pass.

Of the two enemy drives, the second was viewed by the Area Army command as the more immediate and dangerous threat. From the strategic standpoint, seizure of either of the key routes would place the enemy in position to launch an early thrust into the northern bastion from the south, seriously jeopardizing the entire Area Army plan for a protracted defense of the northern Luzon area.

San Jose, which seemed certain to be a major enemy objective, was vitally important for other reasons as well. First, a large volume of supplies transported north from Manila had piled up at the San Jose railhead, whence its further movement into the northern bastion was proceeding slowly owing to inadequate motor transport and enemy air interference. It was therefore essential to hold San Jose until these urgently needed supplies could be moved north.

Second, there was the danger that a swift enemy penetration to the San Jose area would block execution of the plan to strengthen the defenses of the northern sector by shifting the 105th Division main strength up from the Manila area. On 15 January, when strong enemy elements broke through the 23d Division forward line into the sector northeast of Manaoag, signalling the start of the eastward push, the 105th Division had not yet completed the first stage of its northward movement to Cabanatuan.

The enemy breakthrough on 15 January spurred the Area Army command into swift action. An order issued the same day directed the 2d Armored Division, which had completed its concentration at Lupao, to advance immediately to the Tayug sector. Upon arrival, the division was to assume command of 10th Division elements already disposed north of Tayug and in the Triangle Hill sector to

19 General Yamashita cited the entire 64th Infantry for its valiant defense of the first line positions. Summary of Operations, 23d Division (*Asahi* Force), op. cit., pp. 13–4.

the south,[20] and the armored division itself was instructed to take up positions astride the trail north of San Nicolas and on both sides of the Ambayabang River. The mission of the combined forces was "to divert and contain" an enemy force attacking eastward from the beachhead.[21]

By the same order, the Area Army directed the main body of the 10th Division, which had completed its assembly in the San Jose area only about a week earlier, to occupy positions guarding the immediate approaches to San Jose. The division was to hold off an enemy attack while the 105th Division carried out its displacement through San Jose and occupied new positions to the north.[22]

A separate Area Army order dispatched the same day to the Shimbu Group stipulated that the 105th Division, upon reaching Cabanatuan, would pass under direct Area Army command. The division was directed to move its troops on from Cabanatuan to the north as speedily as possible and occupy positions behind the 10th Division at Minuli and north of Carranglan.[23]

Meanwhile, the Shigemi Detachment was all that stood between the enemy and Tayug, through which the 2d Armored Division was to move to its newly-assigned positions north of San Nicolas. To hold up the enemy advance until this move was completed, the Area Army ordered the detachment, through the 23d Division, to advance its main strength from San Manuel to Binalonan and defend that town to the last man.[24] Already at Binalonan were the remnants of the small armored task unit which had been beaten back by the enemy on the night of 15–16 January as it was moving up toward Manaoag for the raiding attack on San Fabian.

Before the main strength of the Shigemi Detachment had time to advance from San Manuel, however, the task unit in Binalonan was driven out by enemy forces which occupied the town on 18 January. The next day the enemy drove on toward San Manuel under powerful air cover. The Shigemi Detachment tried to stem this drive by strong resistance east of Binalonan but was forced to fall back on San Manuel, where it dug in for a desperate suicide stand.[25] (Plate No. 113)

The Area Army had meanwhile learned of two facts which further upset its calculations. The 2d Armored Division, after a preparatory reconnaissance of the area north of San Nicolas, reported back to Area Army headquarters at Baguio that the Villa Verde Trail was impassable to vehicular traffic. Since the division, if it occupied the positions directed by the Area Army on 15 January, would have to depend exclusively on this trail as its supply line from the rear and eventual route of withdrawal into the northern bastion, a change in plan was obviously necessary.

---

20 The main strength of the 10th Reconnaissance Regiment, 10th Division, was disposed in the vicinity of San Nicolas. Patrols of the same regiment were posted in the Triangle Hill sector.

21 *Shobu Sakumei Ko Dai Yonhyakujusan-go* 尚武作命甲第四一三號 (Fourteenth Area Army Operations Order No. A-413) 15 Jan 45.

22 Ibid.

23 *Shobu Sakumei Ko Dai Yonhyakunijuyon-go* 尚武作命甲第四二四號 (Fourteenth Area Army Operations Order No. A-424) 15 Jan 45.

24 Statement by Maj. Gen. Konuma, previously cited.

25 (1) Summary of Operations and Battle Lessons, Luzon Campaign, op. cit., p. 65. (2) An order issued by the Shigemi Detachment commander at San Manuel on 20 January declared: "The detachment will defend its present position to the death. The enemy must be annihilated." I Corps ADVATIS Translation ICAET 0037, 27 Jan 45.

Even more serious was the belated discovery that the 10th Division, which the Area Army believed to be guarding the approaches to San Jose, had in fact withdrawn northward to take up positions near Minuli and north of Carranglan. This mix-up, the result of which was to leave San Jose virtually undefended, came about through the following sequence of events.

Under the Area Army plans of 19–20 December, the 10th Division had been assigned to defense of the San Jose-Umingan-Natividad line.[26] It was to fulfill this mission with approximately two-thirds of its normal strength, the 39th Infantry Regiment and supporting elements of division troops having been transferred to Manila Defense Force command for the reinforcement of Bataan. Immediately afterwards, however, the strength available to the division was further reduced by the Area Army's sudden decision to divert the 10th Infantry Regiment to northern Luzon for employment in defense of the Cagayan Valley. This left the division with but five infantry battalions, plus the bulk of division supporting troops, to organize defensive positions over a front of more than 22 miles.[27]

In addition to the excessive length of the projected line, reconnaissance of the terrain convinced the division staff that the open country and relatively low hill features present in the area would not permit the organization of sufficiently strong positions. These objections, however, had not yet been communicated to Area Army headquarters by the time the main elements of the division completed their assembly in the San Jose area just prior to the enemy landings on Lingayen Gulf.

At this juncture, Maj. Shigeharu Asaeda, operations officer of the Area Army staff, visited San Jose to inspect defensive preparations in the 10th Division sector. The division command promptly seized the opportunity to point out that the reduced troop strength at its disposal, coupled with the unfavorable nature of the terrain, made impossible a strong defensive organization of the San Jose-Umingan-Natividad line. As an alternative, the division proposed to organize its main positions farther back in the mountainous terrain close to the vital passes to the north.

Inability to establish communications contact with Area Army headquarters at Baguio prevented Maj. Asaeda from obtaining immediate approval of this proposal. However, he was convinced that it would encounter no objection from the Area Army command, and at the same time the need for action was so urgent that he advised the division commander to begin moving his main strength immediately northward to organize positions in the vicinity of Minuli and north of Carranglan. A division order on 10 January initiated this movement.[28]

On 15 January, when the Area Army ordered the 10th Division to secure the approaches to San Jose, it was still unaware of the actual situation. By the 17th, however, Maj. Asaeda had succeeded in communicating the gist of

26 Cf. Chapter XIV, p. 422.

27 Under the 19–20 December plans, the 10th Division was to be reinforced by the attachment of the Tsuda Detachment (11th Independent Infantry Regiment, reinf., 26th Division), but this unit was to continue its mission of securing the Baler and Dingalan Bay sectors against an enemy landing. The main strength of the detachment was relieved of this mission and ordered to Bongabong by Area Army order on 8 January, but the detachment did not come under 10th Division command until 10 January.

28 *Tetsu* (10th Division) Operations Order No. A-54, 10 Jan 45. I Corps ADVATIS Translation ICAET 0185, 25 Apr 45. This order also directed organization of an Advance Force, composed of one infantry company, two artillery batteries, one and a half antitank companies, and one machine-gun platoon. This force was to remain in positions just north of San Jose.

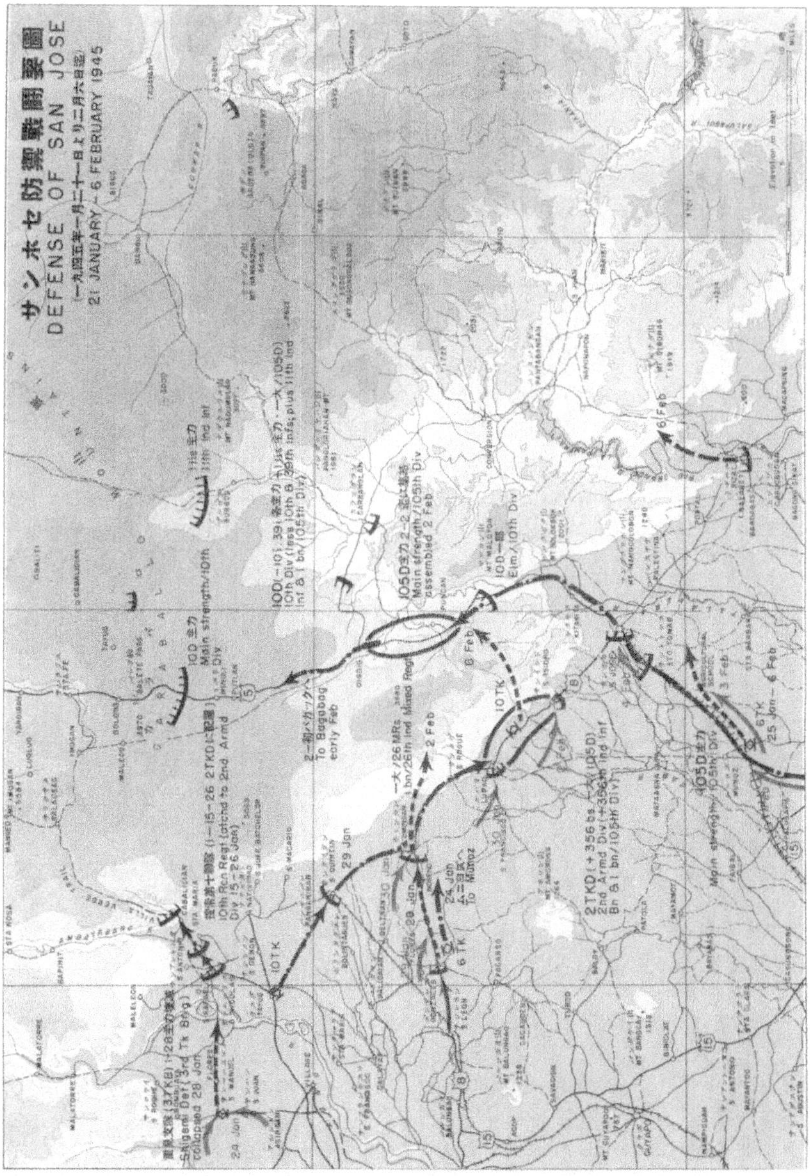

PLATE NO. 113
Defense of San Jose, 21 January—6 February 1945

his action to Area Army headquarters, and he was immediately ordered to return to Baguio by air to make a full report. Only upon his arrival from Bambang the following day did General Yamashita and his staff learn in detail of the new dispositions of the 10th Division.[29]

It was now doubly clear that the Area Army orders of 15 January must be drastically modified. On the one hand, the impassable condition of the Villa Verde Trail ruled out the previous plan to station the 2d Armored Division north of San Nicolas. On the other hand, the disclosure that San Jose had been left undefended except by a small residual force of the 10th Division made imperative new dispositions to secure that key point while the defense of the passes to the north was being strengthened.

General Yamashita decided that the only feasible course under these circumstances was to alter the deployment of the 2d Armored Division so as to throw it across the enemy line of advance toward San Jose for a series of frontal delaying actions. At the same time, the plan to move the entire main strength of the 105th Division north to defend the passes was modified to provide for the retention of two infantry battalions in the San Jose area to reinforce the 2d Armored Division and the residual elements of the 10th Division. On 20 January these decisions were implemented by an Area Army order, substantially as follows:[30]

1. The 2d Armored Division will deploy its main strength on the Tayug-Triangle Hill line and an element in the vicinity of Munoz. It will occupy positions in these sectors and endeavor to destroy the strength of enemy forces attacking from the Lingayen area across the central Luzon plain, especially in the direction of San Jose.

2. The main strength of the 10th Division will occupy positions in the vicinity of Minuli, north of Carranglan, and north of Gadeng, and will prepare both to defend these positions against an enemy attack from San Jose in the direction of Bambang and to counterattack toward San Jose. A strong element of the division will remain in positions just north of San Jose and check an enemy advance toward that point. A further element will be stationed in the Bambang area and secure that district. Division elements in the Tayug and Triangle Hill sectors will remain attached to the 2d Armored Division.

3. The main body of the 105th Division will advance north and assemble in the area south of Minuli as speedily as possible. One infantry battalion of the division will be placed under 2d Armored Division command at San Jose; and one infantry battalion will be transferred to 10th Division command to strengthen the advance positions just north of San Jose.

Steps to effect the new dispositions had barely started when enemy advance patrols on 21 January began probing the defenses set up by the Shigemi Detachment in and around San Manuel.[31] Despite relatively flat terrain and inadequate infantry strength, advance preparation of the ground had enabled the detachment to dig itself in firmly for a suicide defense of the town. Artillery positions were located on high ground immediately to the northwest, covering the road from Binalonan. The greater portion of the detachment's tanks (7th Tank Regiment less 2d Company) had been converted into stationary, fortified pill-boxes emplaced at key points around the perimeter of

29 Statements by Maj. Shigeharu Asaeda, Staff Officer (Operations), Fourteenth Area Army, and Maj. Gen. Konuma, previously cited.

30 *Shobu Sakumei Ko Dai Yonhyakusanjuyon-go* 尚武作命甲第四三四號 (Fourteenth Area Army Operations Order No. A-434) 20 Jan 45. (Plate No. 114)

31 Effective 21 January, the Shigemi Detachment reverted to 2d Armored Division command by order of Fourteenth Area Army. (Statement by Maj. Gen. Konuma, previously cited.)

the town.[32] Remaining tanks were held inside the town as mobile reserve together with the bulk of the infantry.[33]

The decisive phase of the battle for San Manuel began on 24 January when the enemy launched heavy and simultaneous attacks from both north and south. Although the defending forces inflicted heavy losses, the northern prong of the attack penetrated the detachment perimeter and gave the enemy a foothold inside the town. Fierce fighting continued for four more days, culminating in a final desperate counterattacks by the remnants of the Shigemi force on 27–28 January. The collapse of this attack ended all resistance. Only seven tanks and 400 troops of the detachment forces which had defended the San Manuel area succeeded in escaping to join the 10th Reconnaissance Regiment north of San Nicolas.[34]

Meanwhile, a new menace to San Jose had become apparent. On the left flank of the main enemy drive toward Manila, which had already enveloped Tarlac and Victoria, eight miles northeast of Tarlac, on 21 January, enemy tank-infantry patrols had begun probing northeastward in the direction of Guimba and Munoz. This appeared to foreshadow a move to outflank the 2d Armored Division's main positions on the Tayug-Triangle Hill line and strike at San Jose from the southwest. The armored division, in compliance with the Area Army order of 20 January, had already shifted the 6th Tank Regiment (less two companies), reinf. to the Munoz sector, but this force was all that stood in the way of the new enemy threat.[35]

The seriousness of the threat was only too clear. The 2d Armored Division main strength in the Tayug-Triangle Hill area was in danger of being cut off from its sole remaining escape route into the northern bastion via San Jose and Highway 5. Further, the 105th Division was just completing its movement through Cabanatuan on 25 January and faced the prospect of having its route northward barred by the enemy.

To meet the new situation, the Area Army on 26 January again ordered a change in the dispositions of the 2d Armored Division. Lt. Gen. Iwanaka was directed to pull back all remaining division strength from the Tayug-Umingan and Triangle Hill sectors with the exception of small outpost forces to be left at Gonzales and Umingan to delay an enemy advance from the northwest. The division was now to concentrate the bulk of its forces in a triangular-shaped area bounded by Lupao, Munoz and Rizal, with main positions on

---

32 Two factors were jointly responsible for the resort made by the 2d Armored Division to the practice of imbedding tanks in the ground as fixed defenses. The first was complete enemy control of the air, which made movement over the roads and highways next to impossible. The second was the fact that cross-country movement by armor was impossible in areas covered by rice-paddies. The employment of tanks as dug-in pill-boxes was considered highly effective on the basis of results achieved. (Statement by Lt. Col. Shigeo Kawai, Staff Officer (Operations), 2d Armored Division.)

33 Ibid.

34 (1) Ibid. (2) Radio, 10th Reconnaissance Regiment commander to Chief of Staff, 10th Division, 29 Jan 45. I Corps ADVATIS Translation ICAET 0184, 24 Apr 45.

35 The force dispatched to Munoz was commanded by the 6th Tank Regiment commander, Col. Kumpei Ida, and was designated the Ida Detachment. It began moving out of the Triangle Hill sector on 24 January and reached Munoz the following day. (1) Ida Detachment Operations Order No. 120, 24 Jan 45. I Corps ADVATIS Translation ICAET 0047, 3 Feb 45. (2) Ida Detachment Operations Order No. 122, 25 Jan 45. I Corps ADVATIS Translation ICAET 0049, 4 Feb 45.

Highway 8 between Lupao and San Jose and Highway 5 between Munoz and San Jose.[36]

These new dispositions had barely been effected when the enemy began attacking toward San Jose from both northwest and southwest. On 29 January enemy elements swept around the outpost force at Gonzales (Omuro Detachment) and cut its withdrawal route to Umingan, forcing the detachment to withdraw through the hills after destroying most of its tanks and all of its mechanized artillery.[37] The force at Umingan, one infantry battalion of the 26th Independent Mixed Regiment,[38] held out until 2 February, when it was finally compelled to retreat eastward.

Meanwhile, on the Munoz front southwest of San Jose, a battle even fiercer than that fought at San Manuel was in progress. The Ida Detachment had just occupied its positions in this sector on 25 January when the enemy began preliminary attacks to probe the detachment defenses.[39] Following these attacks, powerful enemy forces launched a full-scale assault on Munoz from the southwest on 1 February.

As at San Manuel, the Ida Detachment had entrenched itself in strong defensive positions with a large proportion of its tanks dug-in and reinforced to supplement the regular artillery. So firm was its resistance that the enemy assault made no progress on 1 and 2 February. On the 3rd, however, enemy forces by-passed Munoz on the southeast and cut in behind the Ida Detachment to sever Highway 5 in the vicinity of the General Luzon Agricultural School, between Munoz and Santo Tomas. The enemy then drove on up the highway toward San Jose. Antitank gun emplacements a short distance southwest of the town succeeded in stemming this advance, but enemy elements again swung around the Japanese positions to push into San Jose on 4 February. By the 6th, Highway 5 from the Agricultural School to San Jose was entirely in enemy hands.

The Ida Detachment cut off at Munoz had meanwhile been battling fiercely to hold its positions under renewed enemy attack. Col. Kumpei Ida, the detachment commander, had already been killed in action, and enemy artillery had poured into the Japanese positions as much as 15,000 rounds of fire in a single day.[40] By 6 February, however, the enemy had succeeded in gaining possession of only a portion of the town.

---

36 *Shobu Sakumei Ko Dai Yonhyakurokuju-go* 尚武作命甲第四六〇號 (Fourteenth Area Army Operations Order No. A-460) 26 Jan 45. (By the provisions of this order, the 10th Reconnaissance Regiment defending the entrance to the Villa Verde Trail north of San Nicolas reverted from 2d Armored Division to 10th Division control.)

37 Four tanks belonging to the Omuro Detachment succeeded in breaking through the enemy encirclement and escaped along the highway to Umingan. (Statement by Lt. Col. Kawai, previously cited.)

38 This unit was one of the five infantry battalions of the 105th Division being transferred to the north. Placed under 2d Armored Division command upon its arrival in the San Jose area, it was ordered to proceed immediately to Umingan, where it arrived on 29 January, just before the enemy attack.

39 Composition of the Ida Detachment was as follows:

Headquarters, 6th Tank Regt.
3d, 4th and 5th Cos., 6th Tank Regt.
8th Btry., 2d Mobile Arty. Regt.
356th Ind. Inf. Bn. (less 2 cos.)

(1) 2d Armored Division Operations Order No. A-88, 7 Jan 45, Annex 1. ADVATIS Bulletin No. 307, 20 Feb 45, pp. 3–4. (2) 6th Tank Regiment Order No. 135, 2 Feb 45. ATIS Bulletin No. 1971, 18 Apr 45, p. 5.

40 For its determined defense of Munoz, the Ida Detachment was subsequently cited by Fourteenth Area Army. (Statement by Lt. Col. Kawai, previously cited.)

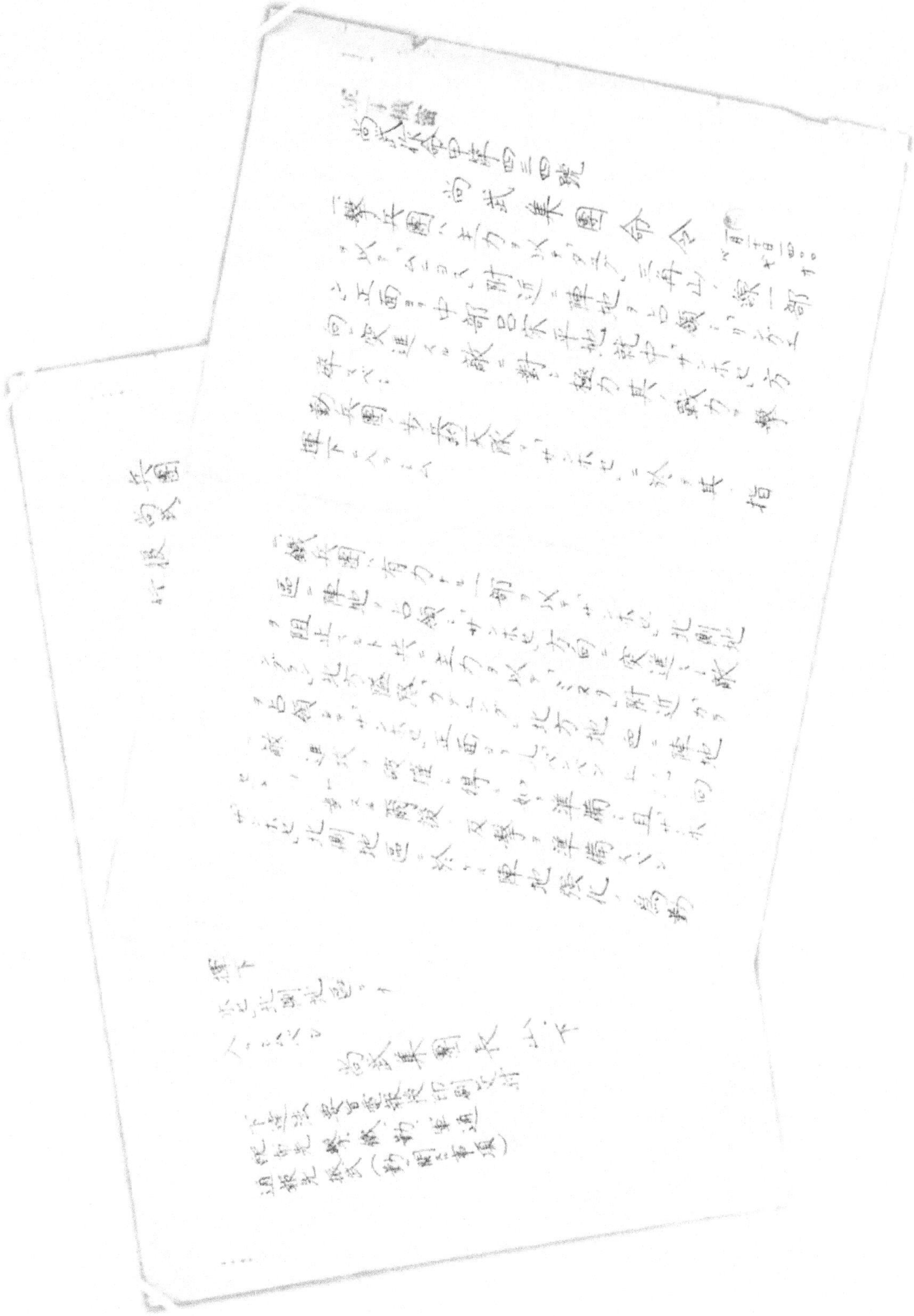

PLATE NO. 114
Facsimile of Fourteenth Area Army Operations Order No. A-434, 20 January 1945

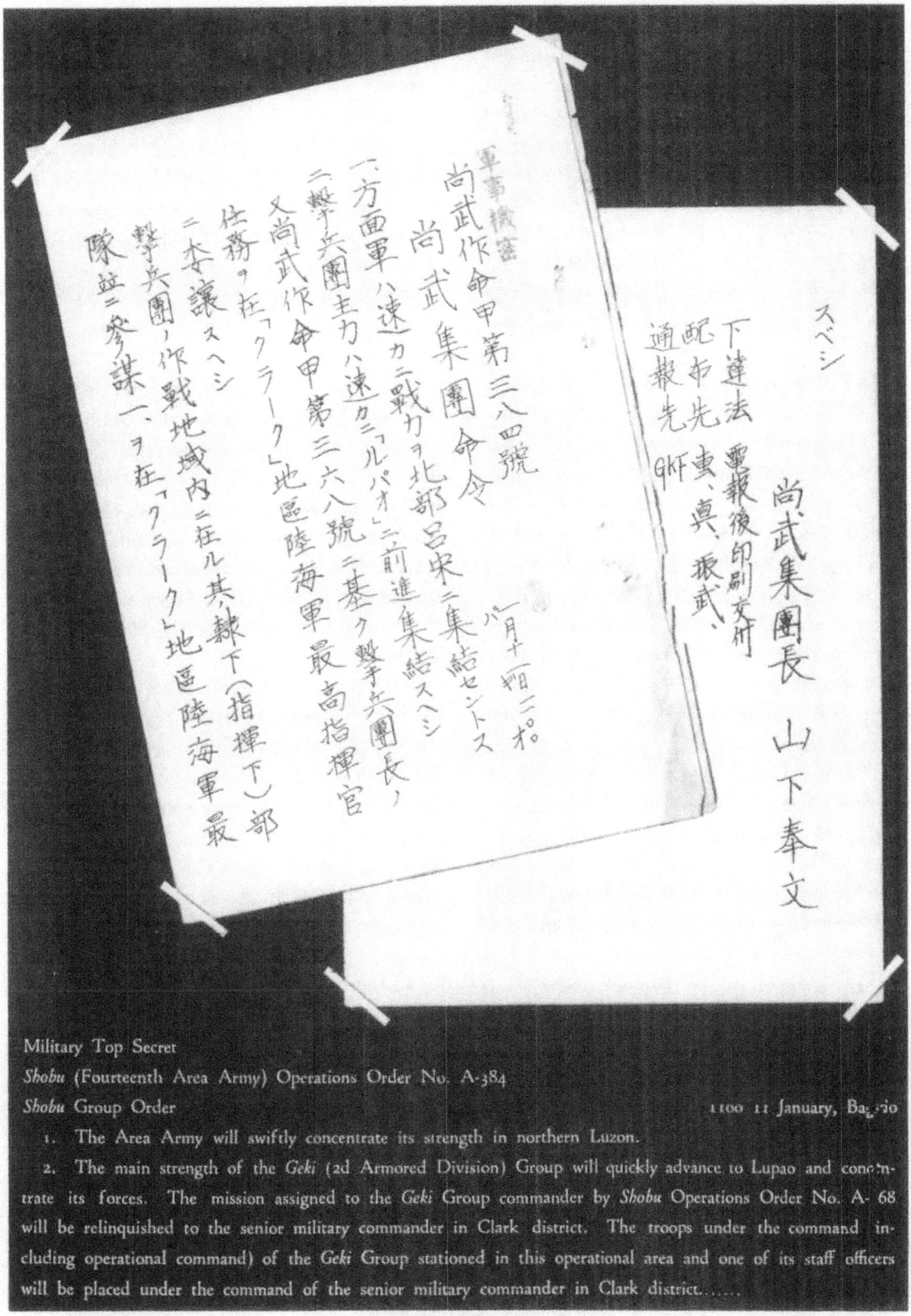

Military Top Secret
Shobu (Fourteenth Area Army) Operations Order No. A-384
Shobu Group Order　　　　1100 11 January, Baguio

1. The Area Army will swiftly concentrate its strength in northern Luzon.

2. The main strength of the *Geki* (2d Armored Division) Group will quickly advance to Lupao and concentrate its forces. The mission assigned to the *Geki* Group commander by *Shobu* Operations Order No. A-368 will be relinquished to the senior military commander in Clark district. The troops under the command (including operational command) of the *Geki* Group stationed in this operational area and one of its staff officers will be placed under the command of the senior military commander in Clark district......

PLATE NO. 115
Facsimile of Fourteenth Area Army Operations Order No. A-384, 11 January 1945

Under cover of the delaying actions fought by the 2d Armored Division, the main strength of the 105th Division had successfully slipped past San Jose prior to its capture by the enemy and reassembled along Highway 5 to the south of the main positions of the 10th Division.[41] On 4 February, therefore, the Area Army dispatched an order to the 2d Armored Division to break off action and withdraw through Balete Pass to Dupax for a period of rest and regrouping.[42]

This order actually did not reach 2d Armored Division headquarters until 6 February due to communications failure, but even at the time it was issued, the enemy seizure of San Jose had already sealed the escape route of the division forces still battling at Munoz and along Highway 8 in the Lupao-San Isidro area.[43] These forces now had but two alternatives; either to attempt to break through the enemy blocking their road of retreat, or to destroy their remaining tanks and mechanized equipment and withdraw cross-country into the northern hills.

The Ida Detachment chose to attempt a break-through. Covered by a diversionary attack, the remaining strength of the detachment pulled out of Munoz with what armor still was mobile at about midnight on the 6th and headed up Highway 5 toward San Jose. The attempt, however, was soon discovered. Between the Agricultural School and Santo Tomas, enemy artillery, antitank guns and automatic weapons raked the column with murderous fire which inflicted heavy casualties and destroyed virtually all tanks, trucks and mechanized artillery. Surviving personnel destroyed what vehicles remained and retreated off the highway to infiltrate around San Jose toward the advance positions of the 10th Division.

The 10th Tank Regiment and other 2d Armored Division elements defending Lupao and San Isidro, to the northwest of San Jose, decided against a similar break-through attempt. Destroying all tanks and mechanized equipment not already lost as a result of increasingly severe enemy attacks, the personnel of these units withdrew northeast from Highway 8 to make their way over the hills into the northern bastion.

By delaying the enemy's seizure of San Jose, the 2d Armored Division had successfully covered the transfer of the 105th Division to the north and the removal from San Jose of the bulk of supplies accumulated there. The cost, however, had been high. Of approximately 200 tanks which the 2d Armored Division possessed at the time of the Lingayen landings, 180 had been lost, together with

41 The 105th Division completed its reassembly in the area north of Puncan by about 2 February and immediately began preparing for a further move northward pursuant to an Area Army order issued on 29 January, which directed the division to proceed to the Bagabag area. The division now retained control of only three of its five infantry battalions, since one battalion had been released to the 2d Armored Division upon arrival in the San Jose area, and another had been transferred to the 10th Division to reinforce the latter's advance positions north of San Jose. Under the Area Army order of 29 January, one additional battalion was to be transferred to the 2d Armored Division, but the confused situation prevented the transfer from being carried out, and the battalion therefore remained with the 105th Division. To make up for the reduction of the division strength, the Area Army order of 29 January called for the transfer to 105th Division control of the 10th Infantry, which had arrived at Bagabag from Aparri on 25 January. (1) *Shobu Sakumei Ko Dai Yonhyakurokujuhachi-go* 尙武作命甲第四六八號 (Fourteenth Area Army Operations Order No. A-468) 29 Jan 45. (2) Statement by Lt. Col. Hideaki Kitagawa, Staff Officer (Operations), 105th Division.

42 Statement by Maj. Gen. Konuma, previously cited.

43 Enemy forces had begun attacking Lupao on 30 January and San Isidro on 2 February. (Statement by Lt. Col. Kawai, previously cited.)

corresponding proportions of the division's artillery and mechanized transport. About 2,000 division troops had lost their lives.[44] As an armored force, the division had been destroyed.

Clark Field and Bataan

Simultaneously with the enemy's crushing offensive along the northern rim of the central plain toward San Jose, his right wing forces to the west drove rapidly southward from the Lingayen beachhead toward Clark Field and Manila. Inadequate troop strength had forced General Yamashita to leave the area directly to the south of Lingayen Gulf virtually undefended. The enemy's advance consequently was impeded only by demolitions carried out by retreating Japanese patrols. By 21 January Tarlac had been overrun, and the enemy pushed on toward the outer defenses of the Clark Field sector in the vicinity of Bamban.

To deny the enemy the use of the vital airfields in this sector, Fourteenth Area Army had planned to organize the mountainous terrain immediately flanking Clark Field on the west and northwest as one of the three major defense strongpoints on Luzon. Right up until the Lingayen invasion, however, execution of this plan had been hampered by two factors. First, no unified command had been set up over the large number of Army and Navy Air force base personnel stationed in the area.[45] Second, Fourteenth Area Army had not been able to strengthen these forces with adequate numbers of regular ground combat troops.

Not until after General MacArthur's forces swarmed ashore in the Lingayen area was General Yamashita finally able to remedy the anomalous command situation. On 8 January, the day before the invasion, Maj. Gen. Rikichi Tsukada, commander of the 1st Airborne Raiding Group, had arrived at Clark Field from Japan with his headquarters. This at last made available a ranking general officer competent to direct ground operations.[46] On 11 January, General Yamashita placed Maj. Gen. Tsukada in command of all Army and Navy forces in west-central Luzon embracing not only the Clark Field area but almost all of Bataan Peninsula.[47] The mission assigned to these forces was :[48]

> *. . . .to check an anticipated penetration of the Clark Field sector, facilitate the operations of the air forces as far as possible, and as a last resort hinder enemy utilization of the airfields by operating from the strongpo nt west of Clark Field.*

In the Clark sector, Maj. Gen. Tsukada found himself in command of a heterogeneous collection of forces aggregating about 30,000 men. The 2d Mobile Infantry Regiment (less one battalion) of the 2d Armored Division and the 2d Glider Infantry Regiment were the only

---

44 Ibid.

45 A draft Army-Navy Local Agreement proposed by Fourteenth Area Army at the time of the formulation of the 19–20 December defense plans was never finalized. This draft stipulated that all Army and Navy airfield units in the Clark Field sector should be commanded by the ranking air commander present. Cf. Chapter XIV, p. 423.

46 Maj. Gen. Tsukada became the highest ranking commander in the Clark Field sector with the departure of the First Air Fleet Commander on 9 January.

47 Maj. Gen. Tsukada's forces were made responsible for operations west of the line Calumpit-Pampanga River-Licab-Guimba-San Antonio-Santo Tomas, including all of Bataan Peninsula except the southern tip around Mariveles. The 11 January order rescinded an earlier Area Army order of the 8th assigning this zone to the 2d Armored Division. Maj. Gen. Tsukada was to operate under direct Area Army command and not under Fourth Air Army. Fourteenth Area Army Operations Order No. A-384, 11 Jan 45, op. cit. (Plate No. 115)

48 Ibid.

regular infantry units. The remainder consisted mainly of Army and Navy airfield battalions, antiaircaft units, and miscellaneous base and service forces. Some of the base and service troops had been organized into provisional combat units.[49]

The fact that no Japanese forces of significant size stood in the way of an enemy drive south from Lingayen to the Clark Field area made it apparent that time was of the essence. Maj. Gen. Tsukada immediately ordered his troops to pour all their energies into the completion of defensive preparations. The formation of provisional combat units was hastened. Automatic weapons were dismounted from the numerous disabled aircraft littering the Clark Field strips and were emplaced in defensive positions to overcome the shortage of organic artillery. In the rugged mountain terrain to the west, work was accelerated on cave and trench positions.

On 17 January Maj. Gen. Tsukada ordered final dispositions to meet an enemy attack. All available Army combat units, regular and provisional, were tactically organized into four detachments with an aggregate strength of about 8,000 troops. All the naval forces, totalling about 15,000, remained integrated under the immediate command of Rear Adm. Ushie Sugimoto, 26th Air Flotilla commander. The remaining 7,000 Army service troops were placed under Maj. Gen. Tsukada's direct command. The composition of the four Army detachments was as follows:[50]

Takayama Detachment:
- *Hq, 2d Mobile Infantry Regt.*
  - *2d Bn, 2d Mobile Infantry Regt.*
  - *132d, 137th Airfield Bns.*
  - *25th Ind. Antitank Bn.*
  - *6th Btry., 2d Mobile Arty. Regt.*
  - *Misc. minor units*

Eguchi Detachment:
- *Hq, 10th Air Sector Unit*
  - *31st, 99th, 150th, 151st, 152d Airfield Bns.*
  - *Shibasaki Prov. Infantry Bn.*
  - *84th Field Antiaircraft Arty Bn.*
  - *Misc. minor units*

Takaya Detachment:
- *Hq, 2d Glider Infantry Regt.*
  - *2d Glider Infantry Regt.*

Yanagimoto Detachment:
- *Hq, 3d Bn., 2d Mobile Infantry Regt.*
  - *3d Bn., 2d Mobile Infantry Regt.*
  - *8th Ind. Tank Co.*

Maj. Gen. Tsukada assigned all but one of

---

49 Order of battle of Army and Navy forces in the Clark Field area as of 17 January (battalion level and above) was as follows:

Army Forces:
- Hq., 1st Airborne Raiding Gp.
- 2d Glider Inf. Regt.
- 2d Mobile Inf. Regt. (less 1st Bn.), reinf.
- 25th Ind. Antitank Bn.
- 84th Field Antiaircraft Arty Bn.
- Hq. 10th Air Sector Unit.,
- 31st, 99th, 132d, 137th Airfield Bns.
- 150th, 151st, 152d Airfield Bns.
- 22d Air Signal Regt.
- Shibasaki Prov. Inf. Bn.

Navy Forces:
- Hq., 26th Air Flotilla
- 141st, 153d, 201st, 221st Air Gps. (mainly ground personnel)
- 341st, 761st, 763d, 1021st Air Gps. (mainly ground personnel)
- 37th Naval Garrison Unit
- Elm., Northern Philippine Airfield Unit.

(1) 1st Airborne Raiding Group Operations Order No. A-24, 17 Jan 45. XIV Corps ADVATIS Translation XIVCAET 0038, 2 Feb 45. (2) *Hito Sakusen Kiroku Dai Sanki Dai Sankan Furoku Dai Ni Kuraku Chiku Kembu Shudan no Sakusen* 比島作戰記錄第三期第三卷附錄第二クラーク地區建武集團の作戰 (Philippine Operations Record, Phase Three, Vol. III Supplement 2: *Kembu* Group Operations in the Clark Sector) 1st Demobilization Bureau, Nov 46, pp. 14–18. (3) *Hito Homen Kaigun Sakusen Sono Shi* 比島方面海軍作戰其の四 (Philippine Area Naval Operations, Part IV) 2d Demobilization Bureau, Dec 48, pp. 46–7.

50 Outline of Troop Dispositions. XIV Corps ADVATIS Translation XIVCAET 0038, Separate Sheet, 2 Feb 45.

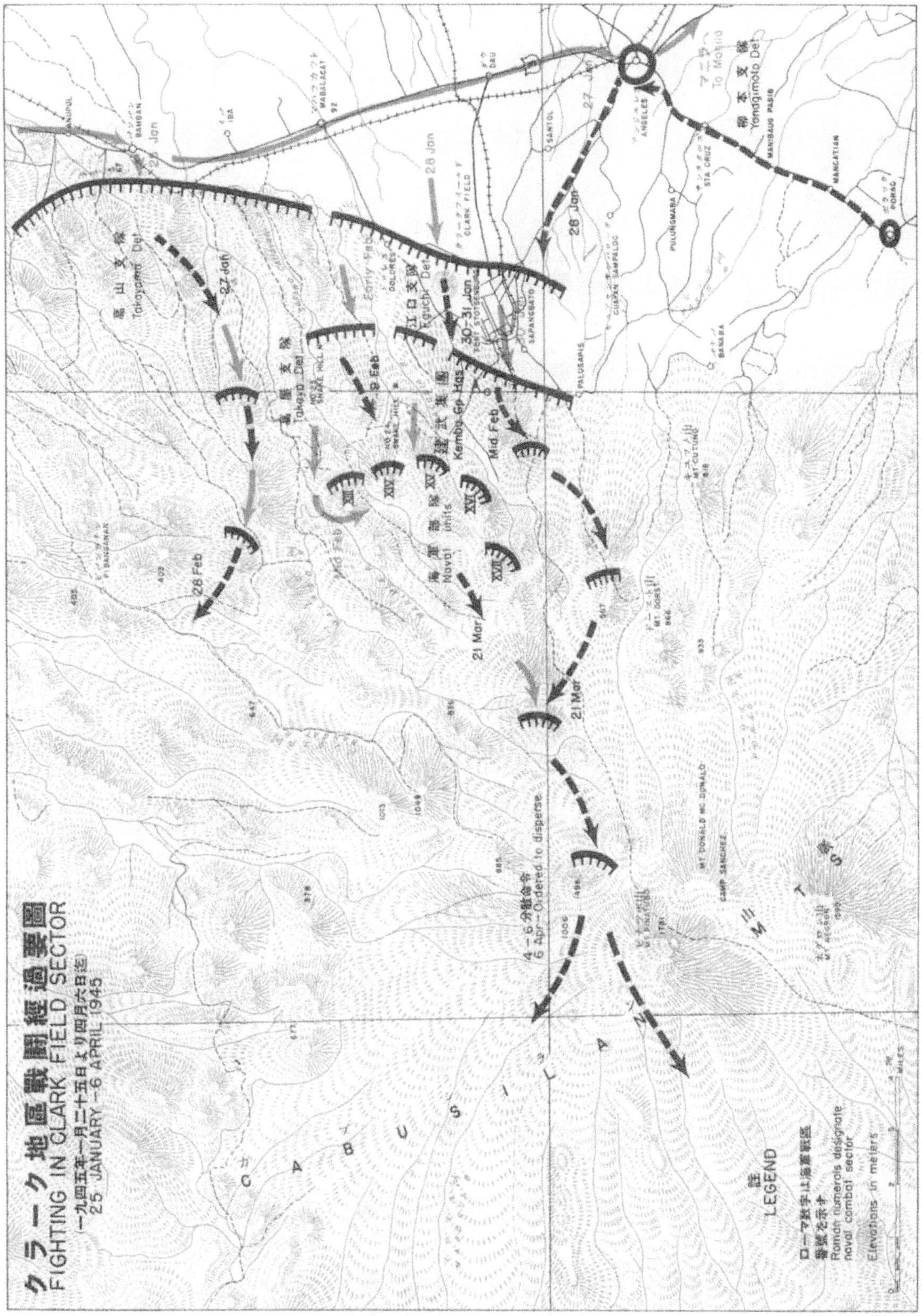

PLATE NO. 116
Fighting in Clark Field Sector, 25 January—6 April 1945

the Army detachments to the defense of his first and second-line positions. These positions faced east toward Highway 3, with the northern anchor in the hills west of Bamban and the south flank anchored in the vicinity of Sapangbato and Fort Stotsenburg. (Plate No. 116) The Takayama Detachment was deployed in the northern sector with its main strength in the forward hill positions overlooking Bamban. To the south, elements of the Eguchi Detachment occupied a forward line running through Dolores and Clark Field to a point southeast of Sapangbato, while the main strength of the detachment held strong mountain positions north and west of Fort Stotsenburg. The Takaya Detachment was deployed in the center of the second-line defenses. The Yanagimoto Detachment, not assigned to fixed positions, held its mobile units at Angeles and Porac in readiness to move against an enemy paratroop landing on the south flank of the Clark Field defenses.[51]

Rear Adm. Sugimoto's naval forces were assigned to the defense of positions in the rear of the two forward lines held by the Takayama, Eguchi and Takaya Detachments. Organized into five combat sectors, these forces dug themselves in on commanding heights to the northwest of Fort Stotsenburg.[52] While last-minute battle preparations were in progress, Fourteenth Area Army issued an order designating the forces under Maj. Gen. Tsukada's command as the Kembu Group.[53]

The enemy assault on the Clark Field sector defenses was not long in developing. On 22 January, the day after the fall of Tarlac, outpost patrols of the Takayama Detachment engaged in demolition work along the highway from Tarlac to Bamban were attacked by enemy spearhead elements and forced to withdraw southward. On the 23d the enemy invested Bamban itself and began probing the main defenses of the Takayama Detachment to the west of Highway 3.

The first major enemy attack on these defenses was launched on 25 January. The Takayama Detachment fought stubbornly in defense of each hill mass, but after two days of severe combat the detachment was forced to pull back toward its second-line positions.[54]

Other enemy forces had meanwhile driven on southward along Highway 3 to seize Mabalacat and Angeles. As this advance developed, the Yanagimoto Detachment had pulled back its elements from Porac and was preparing to engage the enemy when it received orders from Maj. Gen. Tsukada to evacuate Angeles and move into the main positions in the Clark Field-Fort Stotsenburg sector. The detachment therefore withdrew without making any serious defense of Angeles, which the enemy occupied on 27 January.

The Yanagimoto Detachment had barely moved back into the main positions when the enemy, wheeling west from Highway 3, launched a full-scale attack toward Clark Field and Fort Stotsenburg on 28 January. All available artillery firepower was thrown into an effort to stem this attack. By nightfall, however, the enemy had pressed up close to the first-line defenses of the Eguchi Detachment and was already in possession of part of Clark Field.

Elements of the Eguchi Detachment carried out a strong raiding attack on the enemy right flank during the early morning hours of 29 January but were obliged to withdraw before

---

51 1st Airborne Raiding Group Operations Order No. A-24, 17 Jan 45. XIV ADVATIS Translation XIVCAET 0038, 2 Feb 45.

52 Philippine Area Naval Operations, Part IV, op. cit., p. 49.

53 Statement by Col. Yasuji Okada, Chief of Staff, 1st Airborne Raiding Group.

54 Philippine Operations Record, Phase Three, op. cit. Vol. III Suppl. 2, Attached Map.

daybreak. The enemy then resumed his assault following a heavy artillery and mortar preparation. Although the tanks of the Yanagimoto Detachment were thrown into a determined counterattack in the late afternoon, this failed to regain any ground, and by the close of the day, a serious breach had been driven through the Japanese first-line positions.[55]

The enemy was now closing in rapidly on Fort Stotsenburg. Maj. Gen. Tsukada recognized that it would be futile to attempt to restore the shattered forward line and ordered the Eguchi Detachment to pull back to its main positions west and north of the fort. Kembu Group headquarters itself transferred to the rear during the night of 29–30 January. The surviving strength of the front-line forces withdrew to the main positions during the 30th and 31st, fighting stubborn rear-guard actions as they went.

Quick to exploit their advantage, the enemy forces overran Sapangbato and most of Fort Stotsenburg on 30 January and immediately pressed on to attack the Eguchi Detachment's mountain strongpoints to the west. Some of these were lost in the early days of February after heavy fighting. The detachment then pulled back to new positions farther to the rear.

In the Snake Hill North sector, the Takaya Detachment had meanwhile held firm against another prong of the enemy attack from the direction of Mabalacat. Early in February, however, pressure in this sector sharply increased, and on the 9th the detachment was ordered to move back into the rear positions occupied by Rear Adm. Sugimoto's naval troops.[56] This permitted the enemy to drive a wedge between the Takayama Detachment on the north flank and the rest of Maj. Gen. Tsukada's forces. After mid-February disintegration became rapid, and the Kembu forces were no longer able to do more than harass the enemy by small-scale raids.[57]

A startling new development had meanwhile taken place in another part of the Kembu operational zone. Just as the fighting in the Clark Field sector reached its climax on 29 January, General MacArthur landed a fresh invasion force near San Antonio, on the southwest coast of Zambales Province. The Nagayoshi Detachment,[58] responsible for the defense of Bataan Peninsula, had no forces in the landing area, and the enemy put his troops ashore without opposition. The invading force immediately began advancing along the highway leading southeast to Olongapo, its objective apparently being to cut across the base of Bataan Peninsula and join in a concerted assault on Manila.

---

55 Statement by Col. Okada, previously cited.

56 Philippine Operations Record, Phase Three, op. cit. Vol. III Suppl. 2, pp. 20–1.

57 During late February and March, the Japanese forces continued to execute a series of delaying actions which culminated in a withdrawal to the eastern base of Mt. Pinatubo. Here, on 6 April, the Kembu Group commander ordered his remaining troops to disperse and operate as guerrillas. Ibid., pp. 30–2.

58 Order of battle of the Nagayoshi Detachment as of the date of the enemy landing near San Antonio was as follows:

Hq., 39th Inf. Regt.
- 39th Inf. (less 1st Bn.)
- Elm., 10th Engr. Regt.
- Two prov. inf. cos.
- Misc. air-ground personnel

Main strength of the detachment (Hq., 3d Bn., 39th Infantry; one prov. inf. co.; and engr. troops) was disposed on northern Bataan, to the east of Olongapo. The 2d Battalion and one provisional infantry company were in the area east of Bagac, on the west coast. (1) Bataan Detachment Disposition Sketch, 17 Dec 44. XI Corps ADVATIS Translations XICAET 0032, 16 Feb 45. (2) Statement by Col. Sanenobu Nagayoshi, 39th Infantry Regiment commander.

At Olongapo, elements of the Nagayoshi Detachment fought a brief but brisk delaying action as the enemy moved in on 30 January. They then withdrew eastward to the detachment's main positions, which were organized in depth guarding the narrow mountain defile on the highway to Dinalupihan. The enemy advance was temporarily checked when it reached these defenses. Only after a sustained assault by forces estimated to total at least one division, supported by heavy firebomb attacks from the air, was the detachment forced to yield its forward positions.

The enemy now resorted to an encircling maneuver. Hostile elements swung around to the north of the Nagayoshi Detachment's main positions and attacked from the rear on 7 February. Under the combined weight of the frontal assault and the thrust from the rear, the detachment forces were driven from the last of their positions by 14 February. Col. Nagayoshi ordered the surviving troops to infiltrate southward to the eastern side of Mt. Natib.[59]

Battle Dispositions in the Shimbu Sector

As General MacArthur's divisions swept down the central Luzon plain, the forces of the Shimbu Group accelerated preparations for the defense of Manila and the key mountain positions to the east. In the latter part of January, Lt. Gen. Shizuo Yokoyama, Shimbu Group commander, and his staff estimated that the attack on Manila would develop about the middle of February, possibly involving a new enemy landing in Batangas Province to bring the Philippine capital under simultaneous assault from north and south.[60]

To meet the anticipated threat to Manila, Fourteenth Area Army, in close coordination with Southwest Area Fleet, had effected a sweeping reorganization and redeployment of the Army and Navy ground forces in the Manila area during January. By 31 January, all these forces, regardless of composition or service of origin, had been placed under the unified command of Lt. Gen. Yokoyama.[61]

In accordance with the Area Army's basic Luzon defense plans of 19–20 December, Lt. Gen. Yokoyama had withdrawn the bulk of the Army troops stationed in Manila and disposed them in the key mountain positions to the east. The Manila Naval Defense Force, in a coordinated move, had also transferred about 3,000 personnel and considerable quan-

59 After taking Dinalupihan, the enemy sent elements down the east coast of Bataan to Balanga. From Balanga a force advanced inland along the road crossing the peninsula to Bagac and attacked the Nagayoshi Detachment elements occupying positions to the east of Bagac. These elements withdrew northward to Mt. Natib, where they joined the remnants of the detachment main strength retreating from the north. (Statement by Col. Nagayoshi, previously cited.)

60 Statement by Col. Shujiro Kobayashi, Staff Officer (Operations), Shimbu Group. (This officer, previously senior operations staff officer in Fourteenth Area Army headquarters, was temporarily assigned to duty on the staff of Shimbu Group headquarters when it was organized late in December.)

61 Action to place the naval ground forces in Manila under Shimbu Group control for purposes of land combat was taken as early as 5 January. This was impelled by sudden concern that the objective of the enemy's Lingayen invasion convoy, then passing through Mindoro Strait, might be a point on southwestern Luzon close to Manila, and also by the fact that Southwest Area Fleet desired to transfer its headquarters immediately to Baguio to join the Area Army. A Southwest Area Fleet order was issued on 5 January to the Manila Naval Defense Force, specifying that it would henceforth act on orders from Lt. Gen. Yokoyama in all matters relating to ground operations. Upon the departure the same day of Southwest Area Fleet headquarters, a group of five staff officers, headed by Rear Adm. Kyugoro Shimamoto, Vice-Chief of Staff, remained behind in Manila to supervise operations of the Manila Naval Defense Force and to serve as liaison between the Force and Shimbu Group. (Statements by Col. Kobayashi, previously cited, Rear Adm. Kaoru Arima, Chief of Staff, Southwest Area Fleet, Comdr. Shigeichi Yamamoto, Staff Officer (Operations), Southwest Area Fleet and Lt. Comdr. Koichi Kayashima, Staff Officer (Operations), 31st Special Base Force.

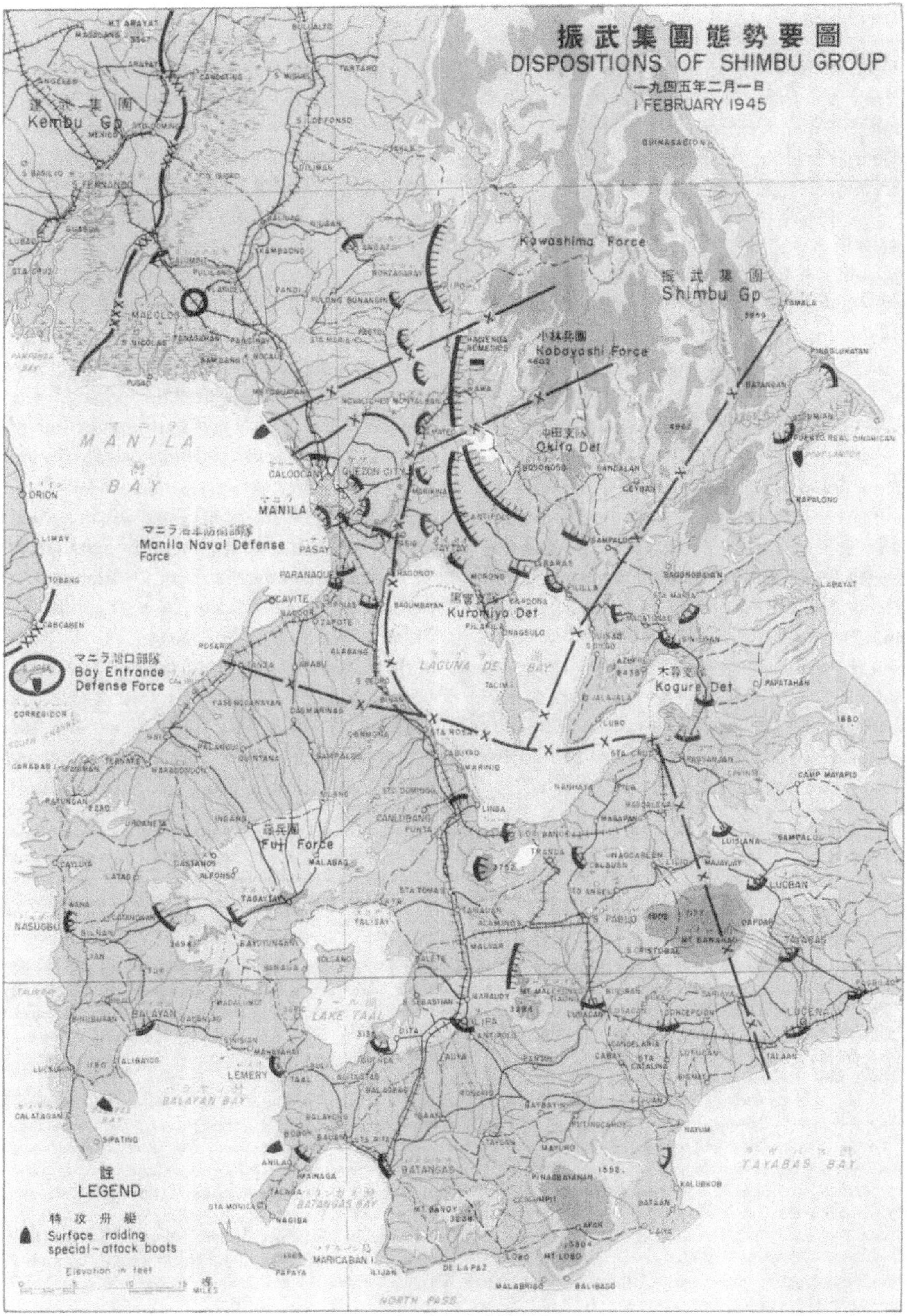

PLATE NO. 117
Dispositions of Shimbu Group, 1 February 1945

tities of food, ammunition and supplies to the Wawa-Bosoboso sector, east of the capital, but the major portion of its strength, numbering about 13,700, remained deployed in and around the city.

As finally organized to meet the enemy assault, the Shimbu Group zone was divided into two main operational areas. The Manila Naval Defense Force under Rear Adm. Sanji Iwabuchi, 31st Special Base Force commander, was made responsible for the defense of the Manila Bay islands and a mainland area bounded roughly by a line Caloocan-Novaliches-San Mateo-Pasig-Hagonoy-Binan-Tanza. Rear Adm. Iwabuchi's command included the entire naval garrison of Manila, the bay islands, and adjacent areas, as well as all remaining Army forces within the assigned zone.[62] The Army forces deployed in and around Manila numbered about 4,000. All Shimbu Group forces outside this zone, with a total strength of about 54,000, were under the direct tactical command of Lt. Gen. Yokoyama. The bulk of these latter forces were disposed in the vital Ipo-Wawa-Antipolo area east of Manila.

By the end of January, Shimbu Group headquarters had almost finished organizing the heterogeneous forces east of Manila into an effective combat task organization. (Plate No. 117) On the north of the line, in the vicinity of Ipo, was the Kawashima Force with an infantry component of four battalions, supported by the main strength of the 8th Field Artillery Regiment and reinforced by a considerable number of air force ground personnel. To the south, guarding the Wawa-Montalban area, was the main strength of the former Manila Defense Force, which had been moved out of the Philippine capital and renamed the Kobayashi Force after its commander, Maj. Gen. Takashi Kobayashi. This force was organized around a nucleus of four infantry battalions.[63]

The departure of the 105th Division for the nothern redoubt had opened a large gap in the Shimbu Group positions between the left flank of the Kobayashi Force and the north shore of Laguna de Bay. Although Lt. Gen. Yokoyama planned to fill in this weak point with the Noguchi Detachment,[64] the end of January found Maj. Gen. Noguchi's troops still engaged

62 The remaining Army troops in the zone assigned to the Manila Naval Defense Force had been placed under Rear Adm. Iwabuchi's command by a Shimbu Group order issued about 20 January. These troops comprised three provisional infantry battalions and some shipping troops. The provisional infantry battalion stationed on Corregidor had already been placed under Rear Adm. Iwabuchi's operational control late in December. (1) Statement by Lt. Comdr. Kayashima, previously cited. (2) Manila Naval Defense Force Operations Order No. 17, 21 Jan 45. XIV Corps ADVATIS Translation XIVCAET 0051, 9 Feb 45.

63 During late December and January, the Manila Naval Defense Force was strengthened by five provisional infantry battalions organized among the replacements, small combat elements, and other "loose ends" in the Manila area. Of nine such battalions organized, four were sent to the Kobayashi Force, one to the Kawashima Force, and four remained behind to be integrated into the Manila Naval Defense Force. The Kobayashi Force retained command over the 3d Surface Raiding Force located along the shores of Manila Bay northwest of the city. (Statement by Col. Kobayashi, previously cited.)

64 On or about 9 January, Lt. Gen. Yokoyama ordered the Noguchi Detachment to move from the Bicol Peninsula to the Manila area. By the end of January, the detachment headquarters, the 182d and the 185th Independent Infantry Battalions were en route, while the 1st Battalion, 26th Independent Mixed Regiment, scattered over northwestern Samar and the southern tip of Bicol, had not yet left for Manila. The 2d Battalion, 9th Infantry, newly attached to the Noguchi Detachment, was still on Samar. Maj. Gen. Noguchi planned to leave behind the 35th Air Sector Unit Hq., the 147th Airfield Battalion, and the 5th Surface Raiding Base Battalion to cooperate with the 35th Naval Garrison Unit in securing the Bicol Peninsula. (1) Statement by Maj. Gen. Susumu Noguchi, commander, 81st Infantry Brigade. (2) *Dokuritsu Konsei Dai Nijuroku Rentai Kodo Gaiyo* 獨立混成第二十六聯隊行動概要) (Summary of Operations, 26th Independent Mixed Regiment) Home Depot Division, Sep 47. (3) 81st Infantry Brigade (Noguchi Detachment) Operations Order No. A-11, 13 Jan 45. Sixth U. S. Army ADVATIS Translation 6AAET 0447, 11 Apr 45.

in the long march from the Bicol Peninsula. As a temporary measure the Okita Detachment, a five-battalion composite force organized around the 186th Independent Infantry Battalion, and the Kuromiya Detachment, a three-battalion force with the 181st Independent Infantry Battalion as its nucleus, were stationed in the Bosoboso-Antipolo area.[65]

While Lt. Gen. Yokoyama completed Shimbu Group preparations in the area east of Manila, Rear Adm. Iwabuchi, Manila Naval Defense Force commander, regrouped all the forces comprising the garrison of greater Manila.[66] By 1 February, the new organization for combat had been completed.

Three main operational sectors were estab-

---

65 Peripheral commands under the direct control of Shimbu Group included the Fuji Force in Batangas Province and the Kogure Detachment in the Lamon Bay area. The order of battle of the Shimbu Group (less the Manila Naval Defense Force) on 1 February was as follows:

- Hq., 8th Div. (Shimbu Group)
  - Kawashima Force
    - Hq. 82d Inf. Brig.,
    - 31st Inf. (less 3d Bn.)
    - 358th Ind. Inf. Bn.
    - Uno Prov. Inf. Bn.
    - 8th Fld. Arty. Regt. (less 1st and 4th Bns.)
    - 9th Air Intell. Unit
    - 12th Air Sig. Regt.
    - 18th Surface Raiding Base Bn.
    - Misc. Army Air-Ground Units
  - Kobayashi Force
    - Hq., Manila Defense Force
    - 4th, 7th, 8th, and 10th Prov. Inf. Bns.
    - 4th and 5th Med. Mortar Bns.
    - 3d Rocket Gun Bn.
    - Hq., 11th Air Sector Unit
    - Hq., 3d Surface Raiding Base Force
      - 11th, 17th, and 18th Surface Raiding Regts.
      - 11th, 17th, and 20th Surface Raiding Base Bns.
  - Okita Detachment
    - 186th Ind. Inf. Bn.
    - 7th and 8th Surface Raiding Base Bns.
    - Suzuki Prov. Bn.
    - 1st Composite Prov. Inf. Bn.
  - Kuromiya Detachment
    - 181st Ind. Inf. Bn.
    - Iwashita Prov. Inf. Bn.
    - Kuwazawa Prov. Inf. Bn.
  - Kogure Detachment
    - Hq., 1st Surface Raiding Base Force
      - 7th, 9th and 10th Surface Raiding Regts.
      - 9th and 10th Surface Raiding Base Bns.
  - Fuji Force
    - 17th Inf. (less 3d Bn.)
    - 3d Bn., 31st Inf.
    - 4th Bn., 8th Fld. Arty. Regt.
    - Hq., 2d Surface Raiding Base Force
      - 6th, 13th, 14th, 15th and 16th Surface Raiding Regts.
      - 6th, 13th, 14th, 15th, 16th, and 19th Surface Raiding Base Bns.
  - 8th Engr. Regt.
  - 8th Transport Regt.
  - 8th Rcn. Regt.
  - 22d Heavy Arty. Regt.
  - 20th Ind. Heavy Arty. Bn.
  - 26th Ind. Mixed Regt. (less 1st and 3d Bns.)
  - 3d Bn., 17th Inf.

(1) *Hito Sakusen Kiroku Dai Sanki Dai Sankan Furoku Dai Ichi Manira Toho Shimbu Shudan no Sakusen* 比島作戦記録第三期第三巻附録第一 マニラ東方振武集團の作戦 (Philippine Operations Record, Phase Three, Vol. III Supplement 1: Shimbu Group Operations East of Manila) 1st Demobilization Bureau, Oct 46, pp. 61–2. (2) Manila Defense Force Operations Order No. A-25, 10 Jan 45, and Kobayashi Force Operations Order No. A-45, 2 Feb 45. XIV Corps ADVATIS Translation XIVCAET 0060, 13 Feb 45, and XIVCAET 0075, 25 Feb 45. (3) Disposition Chart of Kuromiya Detachment, 31 Jan 45. XI Corps ADVATIS Translation XICAET 0049, 22 Mar 45. (4) Fuji Force Operations Order No. A-1, 1 Jan 45. XIV Corps ADVATIS Translation XIVCAET 0098, 20 Mar 45. (5) Statements by Maj. Gen. Osamu Kawashima, commander, 82d Infantry Brigade; Col. Kazuo Okita, commander, 186th Independent Infantry Battalion; Lt. Col. Nobutaka Kogure, commander, 1st Surface Raiding Base Force; and Lt. Col. Norio Tsutsumi, commander, 2d Surface Raiding Base Force.

66 This reorganization of forces in Manila was begun on 21 January, the same day that Rear Adm. Iwabuchi assumed operational control of Army forces in the city. On 3 February, Shimbu Group further strengthened Adm. Iwabuchi's hand by giving him absolute command of all Army forces in the Manila Naval Defense Force zone. (1) Manila Naval Defense Force Operations Order No. 17. XIV Corps ADVATIS Translation XIVCAET 0051, 9 Feb 45. (2) Shimbu Group Order, 3 Feb 45. Sixth U.S. Army ADVATIS Translation 6AAET 0376, 5 Mar 45.

lished for the defense of Manila. The Central Force, under Rear Adm. Iwabuchi's direct command, included all of metropolitan Manila south of the Pasig River (excluding Intramuros) and extending inland as far as Guadalupe. (Plate No. 118) In this sector were stationed the Headquarters and Headquarters Battalion, Manila Naval Defense Force, and three naval battalions. The Northern Force, under Col. Katsuzo Noguchi, was responsible for Intramuros on the south bank and for all of the city north of the Pasig. At Col. Noguchi's disposal were two provisional infantry battalions, a naval battalion, and a miscellaneous conglomeration of Army shipping units. The Southern Force, under Capt. (Navy) Takesue Furuse, included the Nichols Field and Fort McKinley sectors and all of the Hagonoy Isthmus. Capt. Furuse's command comprised one provisional infantry battalion and two naval battalions.[67]

The mission which the forces defending Manila were to carry out had meanwhile undergone a fundamental change from that laid down in the Area Army's basic plans of 19–20 December which had envisaged no decisive and prolonged defense of the capital itself by major forces. Preliminary talks to implement these plans took place late in December between representative of the Area Army and Rear Adm. Iwabuchi's command.[68] At this time the general policy laid down in the basic plans remained unchanged.

On 6 January, the day following the assignment of the Manila Naval Defense Force to the command of Lt. Gen. Yokoyama, Shimbu Group learned for the first time that, instead of only 4,000 naval troops in Manila, there were actually about 16,700. Pending further

67 The complete order of battle of the Manila Naval Defense Force on 1 February was as follows:

- Hq., 31st Special Base Force
  - Central Force
    - Hq. Bn., Manila Nav. Def. Force
    - 1st, 2d, and 5th Nav. Bns.
  - Northern Force
    - 2d and 3d Prov. Inf. Bns.
    - 1st Ind. Nav. Bn.
    - 3d Shipping Depot
    - Elms., 3d Shipping Trans. Hq.
  - Southern Force
    - 3d and 4th Nav. Bns.
    - Abe Prov. Inf. Bn.
  - Bay Entrance Defense Force
    - Elms., 31st Special Base Force
    - Ichinozawa Prov. Inf. Bn.

(1) Manila Naval Defense Force Operation Order No. 17. XIV Corps ADVATIS Translation XIVCAET 0051, 9 Feb 45. (2) Organization Chart of the Bay Entrance Defense Force. XI Corps ADVATIS Translation XICAET 0038, 26 Feb 45. (3) *Manira Boei Butai Sento Jokyo* マニラ防衛部隊戰鬪狀況 (Manila Naval Defense Force Operations) 2d Demobilization Bureau, May 47, p. 21.

68 The principal participants, Col. Shujiro Kobayashi, who at that time was still senior operations staff officer of the Area Army headquarters but who was already slated to be assigned temporarily to Shimbu Group headquarters, and Lt. Comdr. Koichi Kayashima, operations staff officer, 31st Special Base Force, disagree regarding some elements of the discussions. Col. Kobayashi states that it was agreed upon that the forces in Manila would be expected to stem an enemy assault on the capital only for a brief period of days, after which they would withdraw to the main Shimbu positions to the east. Lt. Comdr. Kayashima remembers no specific reference to the withdrawal of the naval forces after a brief defense. He did question Col. Kobayashi regarding the location where the Shimbu Group proposed to employ the Manila naval forces if they were subsequently evacuated from Manila. Col. Kobayashi replied that the area around Antipolo would be selected. The Manila Naval Defense Force considered the Antipolo area unsatisfactory. There being a possibility that the Manila forces might be withdrawn to the east, the naval forces, on their own initiative, began to transfer surplus rations from Manila to the Wawa-Bosoboso sector early in January. Shimbu Group later accepted this arbitrary decision of the naval forces. (Statements by Col. Kobayashi and Lt. Comdr. Kayashima, both previously cited.)

PLATE NO. 118
Dispositions in Manila, 1 February 1945

detailed study by the Shimbu Group staff and an exchange of views with regard to the employment of this unexpectedly large number of naval forces, Lt. Gen. Yokoyama on or about 8 January ordered the Manila Naval Defense Force " to continue its present duty, generally with existing dispositions ".[69]

Discussions began immediately thereafter between staff officers of the Shimbu Group and Southwest Area Fleet. These revealed that Southwest Area Fleet desired a firmer defense of Manila than was permitted in the Area Army's basic plans.[70] Although the Southwest Area Fleet staff officers had no authority to exercise control of the naval forces with respect to ground operations, the strong views they expressed influenced Lt. Gen. Yokoyama to modify his operational plan so as to provide for a firmer defense in Manila. An order issued to all subordinate commands within the Shimbu Group on 27 January included the following points.[71]

1. General Operational plan :

The Shimbu Group will concentrate its main strength in the key positions to the east of Manila and assemble maximum supplies in preparation for prolonged and self-sufficient operations. It will exploit the advantages of terrain and strongly fortified positions to crush enemy attacks on these positions and will seize tactical opportunities for carrying out strong raiding counterattacks. It will firmly defend Manila and Ft. McKinley and check their use by the enemy, at the same time destroying the enemy's fighting strength and preparing to counterattack the enemy rear from the main positions when a favorable situation arises.

2. Mission of Manila Naval Defense Force :

The Naval Defense Force will defend its already established positions and crush the enemy's fighting strength.

Defense of Manila

No sooner had this order been issued to the Manila Naval Defense Force than the Shimbu Group found itself subjected to the anticipated two-pronged attack. The speed with which the American forces, particularly the group approaching from the north, closed in upon Manila, however, caught Lt. Gen. Yokoyama's forces completely by surprise.[72]

The first knowledge in Shimbu headquar-

69 When the Manila Naval Defense Force was orgnaized on 20 December, it numbered only 4,000 troops because Southwest Area Fleet then had under study a plan to transfer the bulk of naval forces located in Manila to the Clark Field area. The following day or so, the Area Fleet notified the Area Army that only 4,000 naval troops would remain in Manila. . However, within a few days the Area Fleet decided against this plan in favor of incorporating the bulk of the naval forces then in Manila, into the Manila Naval Defense Force, raising the strength to 16,700. The Area Army, however, was not informed of this until about 6 January. The mission assigned on 20 December to the Manila Naval Defense Force was to defend the Nichols Field and Cavite sectors and those buildings and installations occupied by the naval forces in Manila. (1) Philippine Area Naval Operations, Part IV, op. cit., p. 11. (2) Statements by Col. Kobayashi, Rear Adm. Arima, and Comdr. Yamamoto, all previously cited. (3) Manila Naval Defense Force Operations Order No. 1, 22 Dec 44. XIV Corps ADVATIS Translation XIVCAET 0061, 14 Feb 45.

70 In the series of conferences between 8–13 January, opinion was divided among those staff officers of the Southwest Area Fleet remaining in Manila with regard to the degree of resistance in the city. Those who advocated a firm defense, however, argued their views more strongly than those who were willing to accept a brief defense. (Statements by Col. Kobayashi, Comdr. Yamamoto, and Lt. Comdr. Kayashima, all previously cited ; and Comdr. Takara Urabe, Staff Officer (Operations), Southwest Area Fleet, and Maj. Yorio Ishikawa, Staff Officer (Operations), Shimbu Group.

71 This order was communicated to staff representatives of all subordinate Shimbu Group commands at a conference at Lt. Gen. Yokoyama's headquarters. Lt. Comdr. Koichi Kayashima represented Rear Adm. Iwabuchi's command at the conference. Since leakage of the intention to withdraw after a brief defense might seriously impair the morale of the troops in Manila, this part of Lt. Gen. Yokoyama's plan was never included in formal orders issued by Shimbu commander. (Statements by Col. Kobayashi and Lt. Comdr. Kayashima, both previouslv cited.)

72 Although Shimbu Group was constantly in communication with Fourteenth Area Army headquarters at this time, very little information was obtained as to the progress of the operations in the north. Even less knowledge was available regarding the situation in the central Luzon plain and Clark Field sectors. (Statement by Col. Kobayashi, previously cited.)

ters of the imminence of the enemy assault was received 1 February. On that date a message from the Uno (Prov.) Infantry Battalion, Kawashima Force, stationed in the Plaridel-Calumpit area, reported that an enemy force, believed to be the advance guard of an estimated one division moving south from the direction of Clark Field, had crossed the Pampanga River near Calumpit.[73]

While the Uno Battalion, aided by the destruction of virtually the entire network of bridges northwest of Manila, delayed this advance, still another enemy thrust developed from the north. A task force, later identified as the 1st Cavalry Division, had crossed an undestroyed bridge at Cabanatuan, on 1 February. Speeding southward against little opposition, it reached the Angat River the following day.[74] After fording the river near Baliuag, this unit split into two columns one of which drove southeast towards Santa Maria while the other followed the Angat River to the east.

The Japanese outpost in Angat fought a delaying action against the latter force during the night of 2–3 February before withdrawing. The second enemy column, however, sped through the hills toward Novaliches and into Manila on the evening of 3 February.[75] (Plate No 119) Although the Kawashima Force immediately executed small-scale raids against the eastern flank of this column, it was unable to block the route Additional enemy troops and supplies continued to stream into Manila.

The southern prong of the enemy pincer had meanwhile been launched, 31 January, on Batangas Peninsula near Nasugbu. The Japanese detachment near the beach resisted the enemy landing force only briefly before retiring into the hills. During the initial phases of this invasion, however, special attack craft, operating from the base at Balayan, reportedly sank eight enemy vessels.[76]

Shortly after the American force moved out from the beachhead en route to Manila, it ran into the main strength of the 3d Battalion, reinf., 31st Infantry, Fuji Force, occupying positions in the narrow mountain defile about fifteen miles inland. This battalion delayed the advance until 3 February, when the enemy broke through the last barrier.

Simultaneously, the enemy launched an airborne invasion of the Batangas Peninsula. A force estimated at approximately one reinforced regiment dropped along Tagaytay Ridge and quickly overcame the scattered Japanese forces in that area. Following a juncture of the two

---

73 Ibid.

74 Although the Shimbu Group was charged with responsibility for the sector east of the line Calumpit-Pampanga River-Licab-Guimba-San Antonio, this was interpreted to mean protecting the Manila-Cabanatuan highway from the guerrillas. No elements of the group were posted north of the Angat River inasmuch as the movement of personnel from Manila and the 105th Division along the road was considered adequate. Moreover, members of the Ganap, pro-Japanese political organizations, patrolled the road between Baliuag and Cabanatuan. Ibid.

75 A few ground units had destroyed the bridges and parts of the roads leading along the shore northwest into the city. The enemy was forced to come into the city along the road from Novaliches. Some units had been stationed on the outskirts of the city to send up flares if the enemy was seen approaching. Whether the men were asleep or were killed in a surprise attack is not known, but no signals were ever seen. Santo Tomas University, which had been used as an internment camp, was one of the first installations overrun by this task force in Manila. Following negotiations with an American officer, Lt. Col. Hayashi, the Japanese commander, obtained permission to lead his unit out with what arms they could carry in exchange for the release of Allied internees held as hostages. They were taken by the American unit to the outskirts of Manila early 5 February and released. Lt. Col. Hayashi was subsequently killed in action. (Interrogation of Col. Kobayashi, previously cited.)

76 Summary of Operations and Battle Lessons, Luzon Campaign, op. cit., p. 88.

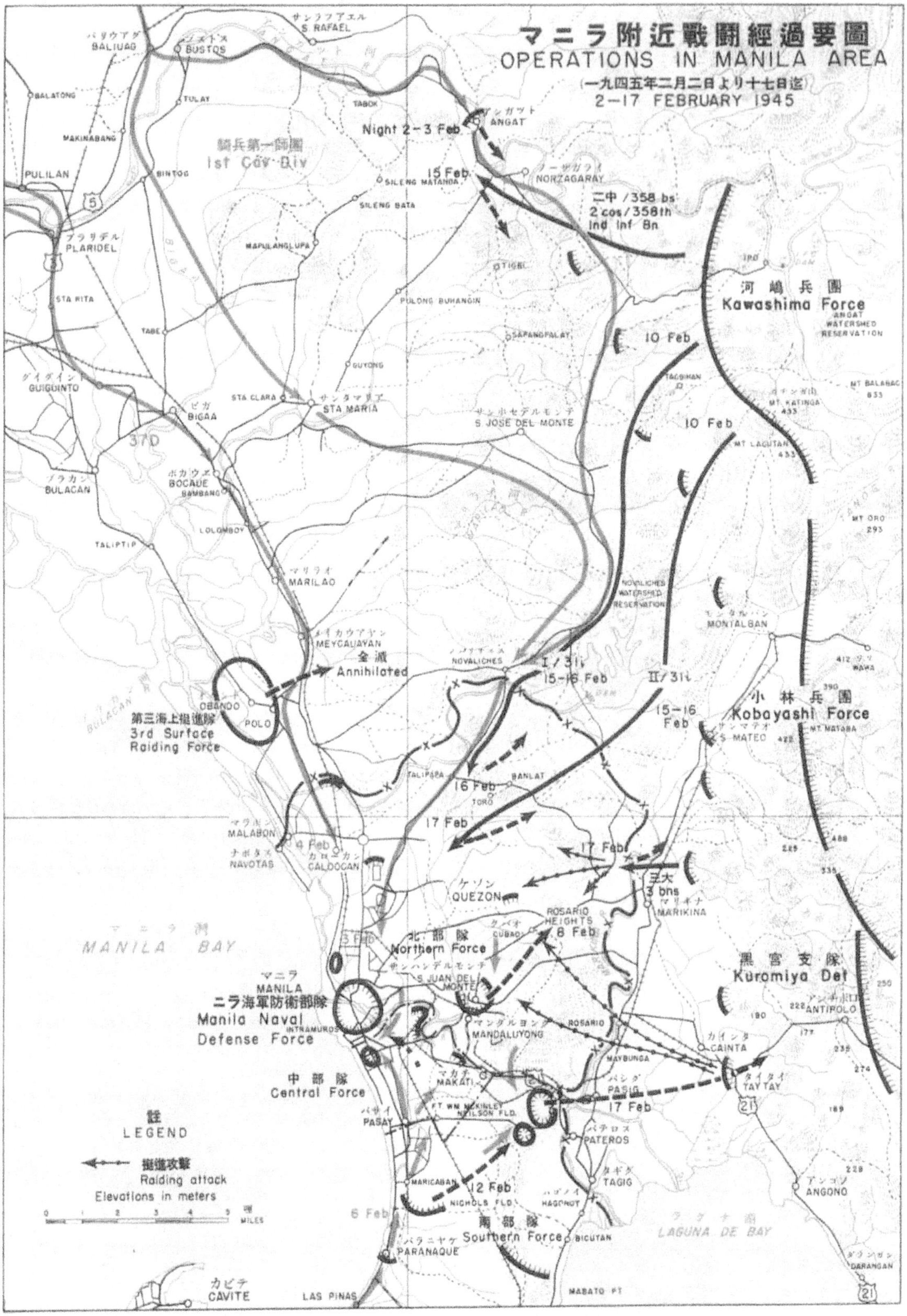

PLATE NO. 119

Operations in Manila Area, 2–17 February 1945

forces. the enemy pushed north rapidly until encountering outposts of the Southern Force, Manila Naval Defense Force, defending river crossings at Las Pinas and Paranaque where the bridges had been destroyed. As the Japanese withdrew towards Laguna de Bay the enemy resumed his advance and began attacking the western flank of the positions south of Nichols Field on 6 February.[77]

While the Southern Force parried this assault, the Northern Force completed withdrawing the major portion of its strength to the south bank of the Pasig River, leaving in north Manila small rearguard elements in addition to the 3d Surface Raiding Force along the Manila Bay sector[78] and the 1st Independent Naval Battalion in the vicinity of San Juan Del Monte. This latter force, under attack from the north since 6 February and threatened from the southwest by another enemy drive along the north bank of the Pasig, was compelled to retreat, 8 February, to Montalban.[79]

Simultaneously, the Shimbu Group began to formulate plans for a large-scale, coordinated raid against limited objectives occupied by the enemy. It was estimated that only about one American brigade was in northern Manila. Moreover, the successful delaying action being staged at this time by the Nagayoshi Detachment on Bataan and the erroneous belief that the enemy force which had landed at Nasugbu was still being delayed in Batangas by the Fuji Force emboldened the Shimbu staff into taking the offensive.[80]

Lt. Gen. Yokoyama hoped that such an operation, extending from Ipo Dam south to Antipolo, would not only break the enemy's offensive power by striking the vulnerable east flank before he had time to consolidate his positions but would weld the heterogeneous collection of Japanese personnel into a more effective, confident fighting force for subsequent operations.

In Manila, however, no such optimism prevailed. The Manila Naval Defense Force found itself under heavy attack from both north and south. The situation was rapidly reaching a point where a withdrawal had to be initiated if it was to be even partially successful. Rear Adm. Iwabuchi, with a staff party, therefore withdrew, 9 February, to Ft. McKinley. On that same day, he decided to dispatch one of the staff officers, Lt. Comdr. Koichi Kayashima, to Shimbu Group headquarters to report on the unfavorable situation existing in Manila and to submit a recommendation for withdrawing his forces to the east.

Upon arrival at Shimbu Group headquarters, 10 February, Lt. Comdr. Kayashima found the staff engrossed in the study of the counterattack plans. After he had received Lt. Comdr. Kayashima's report on the Manila situation,

---

77 The remnants of the Abe Provisional Infantry Battalion later withdrew south along Laguna de Bay and continued around the east shoreline to join the main body of the Shimbu Group in April. (Statement by Col. Hiroshi Hashimoto, Chief of Staff, Kobayashi Force.)

78 The 3d Surface Raiding Force, which had remained under command of Maj. Gen. Kobayashi and which had been by-passed by the enemy in his initial entry into Manila, now found itself under attack from the rear. The unit sought unsuccessfully to break through the enemy lines to join the main strength of the Shimbu Group in the hills east of Manila. During the fighting which followed, the entire unit was annihilated. Ibid.

79 About 880 troops completed this withdrawal. (Statement by Lt. Comdr. Kayashima, previously cited.)

80 By this date, 8 February, no reports had yet been received in Shimbu Group headquarters concerning the American attacks which had commenced against the Nichols Field positions on 6 February. (Statement by Col. Kobayashi, previously cited.)

which, although considerably worse than was known in Shimbu Group headquarters, did not yet appear critical to Lt. Gen. Yokoyama, the Shimbu Group commander directed that the attack plans be enlarged to include provision for the naval forces in Manila to carry out raiding attacks on a large scale in conjunction with the raids from the main positions.[81] The general outline of this plan was transmitted, 11 February, by radio to Ft. McKinley and Manila.[82]

The same day, however, Rear Adm. Iwabuchi had returned to Manila before the message was received from Shimbu Group.[83] His party was one of the last to pass from Ft. McKinley to Manila, the enemy effecting a juncture later the same day, 11 February, of the forces attacking from the south and the north in the vicinity of Nielson Field. The 3d Naval Battalion, still defending the Nichols Field area, was now threatened with complete encirclement and therefore withdrew during the night of 12–13 February to the northeast where it joined the naval forces defending Ft. McKinley.

With the enemy vigorously pressing his ground attacks, closely supported by armor and massed artillery fire, it was now apparent in Shimbu Group headquarters that the situation in Manila had reached a critical stage more rapidly than anticipated. Lt. Gen. Yokoyama therefore issued about 13 February an order directing the Manila Naval Defense headquarters to withdraw to Ft. McKinley as the initial step in evacuating the city.

By 15 February, however, when this order was received in Manila,[84] strong enemy forces were firmly in control of a three mile wide corrider between Ft. McKinley and the main strength of the naval forces in the city. Rear Adm. Iwabuchi replied on 16 February with the following message:[85]

Holding the strong point in the city is considered to be of great importance from the viewpoint of the general situation. Withdrawal of the headquarters from Manila would render difficult the execution of operations in Manila. Moreover, we did not succeed in reestablishing overland contact with Ft. McKinley. We are therefore unable to withdraw from Manila.

Lt. Gen. Yokoyama promptly ordered Rear Adm. Iwabuchi to withdraw from Manila on the night of 17–18 February in conjunction with the raiding attacks from the main positions.[86]

The Shimbu Group had meanwhile completed preparations for the extensive raids. The main attack was to be carried out by two

81 After learning of this decision of the Shimbu Group commander, Lt. Comdr. Kayashima made no formal recommendation to the Shimbu headquarters regarding the withdrawal of the Japanese forces from Manila. Lt. Gen. Yokoyama, intending to order the withdrawal of the Manila Naval Defense Force depending upon the outcome of the planned counterattack, still made no reference to the evacuation of Manila. (Statements by Col. Kobayashi and Lt. Comdr. Kayashima, both previously cited.)

82 Lt. Comdr. Kayashima departed on the night of 10–11 February to return to Ft. McKinley. Lt. Comdr. Kayashima's truck was destroyed, however, during an enemy air attack, and he was compelled to return to Shimbu Group headquarters on the 11th. The plan was immediately transmitted by radio to Ft. McKinley and Manila. (Statement by Lt. Comdr. Kayashima, previously cited.)

83 (1) Several naval officers in Manila and Col. Katsuzo Noguchi, who had assumed command of all Navy and Army troops in Manila when Rear Adm. Iwabuchi left for Ft. McKinley, urgently requested a naval staff officer to return and control the naval forces in Manila. Ibid. (2) General Yamashita, having previously learned that Rear Adm. Iwabuchi had transferred from Manila to Ft. McKinley, was very disturbed when, on 13 February he received the report that the naval commander had returned to Manila. He immediately directed Lt. Gen. Yokoyama to order Rear Adm. Iwabuchi to withdraw his forces from the city. The Truth of the Philippines Campaign, op. cit., p. 35.

84 Statement by Lt. Comdr. Kayashima, previously cited.

85 Philippine Area Naval Operations, Part IV, op. cit., p. 29.

86 Statements by Col. Kobayashi and Lt. Comdr. Kayashima, both previously cited.

battalions of the Kawashima Force attacking Caloocan Airfield and three battalions of the Kobayashi Force attacking Quezon and Banlat Airfield and the vicinity of Rosario. Smaller elements were to cover the extreme north and south flanks of this main attack.[87]

The 31st Infantry (less 3d Battalion), Kawashima Force, moved out of its assembly area in the vicinity of Novaliches on the night of 15–16 February.[88] The 1st Battalion moved along the highway to a point about three miles south of Novaliches before being halted by an American force. The 31st Infantry (less 1st and 3d Battalions), advancing along the east side of the Novaliches road during hours of darkness, was able to approach within about two miles of Caloocan Airfield before encountering enemy positions and retreating.[89]

In the center, the three provisional battalions of the Kobayashi Force had even less success. This composite force, inadequately prepared for such an attack, launched its assault against the west bank of the Marikina River on 16 February at a point about one mile above Marikina. The Americans had strengthened this line more quickly than had been anticipated, however, and the Japanese were unable to break through the enemy defenses. On the following day, when the attack was resumed, a small bridgehead was established on the west bank of the river. Casualties from enemy artillery proved to be so heavy, however, that the force began to withdraw on 18 February after having previously dispatched a raiding unit of several squads to infiltrate to the objectives.[90]

The covering forces on the flanks of the main attack registered only small, local gains before withdrawing. On the north, two companies of the 358th Independent Infantry Battalion succeeded in advancing as far as the south side of Angat before being checked by the enemy and forced to withdraw.

During these diversionary raids, only about 1900 troops, representing most of the surviving strength of the 3d and 4th Naval Battalions, withdrew from the Ft. McKinley area on the night of 17 February and entered the main positions of the Shimbu Group near Antipolo.[91] About 12,000 men remained in Manila with Rear Adm. Iwabuchi.[92] Although repeatedly ordered to evacuate, Rear Adm. Iwabuchi replied that it was now impossible to do so.[93]

Radio contact with the remnants in Manila

87 Statement by Col. Kobayashi, previously cited.

88 As a preliminary move in the attack, Lt. Gen. Yokoyama had ordered, 9 February, the Kawashima Force to move up the main force of the 31st Infantry to the vicinity of Novaliches. The regiment (less 3d Battalion) left its positions in the vicinity of Ipo on 10 February. (Statement by Col. Shimaji Kobayashi, commander, 31st Infantry Regiment.)

89 The 31st Infantry (less 3d Battalion) was preparing to renew the attack but, on 18 February, was ordered to withdraw to its main positions, completing the move on 21 February. Ibid.

90 Statement by Col. Hashimoto, previously cited.

91 Statement by Lt. Comdr. Kayashima, previously cited.

92 During the period 3–17 February, principally in the early phases of the Manila fighting, about 3500 additional miscellaneous naval personnel withdrew to Wawa. Ibid.

93 On 18 February, there having been no indication of an attempted withdrawal the preceding night, Lt. Gen. Yokoyama radioed the following to the Manila Naval Defense Force headquarters: "The counterattacks of the Kobayashi and Kawashima Forces will terminate on the night of 18 February. This will afford you a final opportunity to withdraw from Manila. Report direction and time of withdrawal." The reply of the same day stated, "With the intention of withdrawing from Manila we drafted a plan but later found that it was impossible to execute the withdrawal. If we attempt to withdraw we will be annihilated. On the other hand, if we remain in the city we will be able to hold the position for one more week. The most important problem is to hold your main positions as long as possible and to inflict maximum casualties on the enemy. It is the best way to defend to the last." Another message from Lt. Gen. Yokoyama on the same day stated as follows: "From experience gained during the Lingayen operations it is not difficult to penetrate the enemy's front lines during the night and retreat through his lines. Under unfavorable conditions the Manila Naval Defense Force headquarters, with some forces, should consider the possibillity of withdrawing by way of Hagonoy Isthmus or by boat across Manila Bay." Philippine Area Naval Operations, Part IV, op. cit.,pp. 30–2.

ceased on 24 February. Shortly thereafter Japanese resistance within the city was completely overcome.

Concurrently with the later phases of the fighting in Manila, the entrance to Manila Bay was brought under attack by the enemy. Following a prolonged sea and air bombardment, the Americans first sent troops ashore on 15 February near Mariveles, on the tip of Bataan Peninsula. Although the Japanese attacked this invasion fleet with about 50 special attack boats, reportedly damaging one cruiser, one destroyer, and two transports, the landing operations continued almost without interruption.[94] The small Army detachment defending that area from positions in the mountains northeast of Mariveles was soon forced to withdraw deeper into the mountains.

The garrison on Corregidor, numbering about 4,700, of whom about 800 were Army personnel,[95] was meanwhile being subjected to a particularly severe bombardment on the same day. The next day enemy paratroops descended on the western plateau simultaneously with an amphibious landing on the south beach. The enemy was further reinforced by another paratroop unit which dropped later the same day.

During the bitter fighting which followed, the Japanese utilized to a maximum the intricate system of cave and tunnel defenses traversing the island.[96] The terrific enemy superiority in tanks and flame throwers, however, finally succeeded in overcoming the last organized resistance on 27 February.

Two days later the American fleet had clear sailing into Manila Bay,[97] which by this time was littered with sunken vessels.

Fighting East of Manila, Phase I

Even before the intense fighting in Manila had begun to subside, the main strength of the Shimbu Group was subjected to increasing enemy pressure all along the line. By late February, most of the forward outposts between Ipo Dam and Antipolo had been driven back upon the main defensive positions.[98] (Plate No. 120)

Shimbu Group headquarters estimated that this preliminary probing signalled an imminent full-scale attack on the main positions. The northern end of the line, anchored in strong mountain positions near Ipo Dam, and the center, near Wawa and Montalban, were both considered to be strong enough to withstand heavy, prolonged attacks. These positions, furthermore, were well secured on the extreme north flank and in the rear by the trackless Sierra Madre Mountain Range.

The vulnerable southern flank, however, caused Lt. Gen. Yokoyama grave concern. Not only were the natural defenses weak, permitting attack from any of three directions, but the combat effectiveness of the provisional units manning that particular area was considerably below average.

Estimating that the main enemy assault

94 Ibid., p. 39.

95 Organization Chart of the Bay Entrance Defense Force. XI Corps ADVATIS Translation XICAET 0038, 26 Feb 45.

96 Following its capture in 1942, Corregidor had been left in ruins until the end of September 1944 at which time the Japanese began to reinforce the garrison and rebuild the fortifications. Philippine Area Naval Operations, Part IV, op. cit., pp. 36–7.

97 An American convoy of two cruisers, one destroyer, and 17 transports entered Manila Bay on 1 March. *Shukan Joho Dai Hyakujusan-go* 週間情報第一一三號 (Weekly Intelligence Reports No. 113) Imperial General Headquarters, Army Section, 10 Mar 45.

98 Philippine Operations Record, Phase Three, op. cit. Vol III Suppl. 1, p. 23.

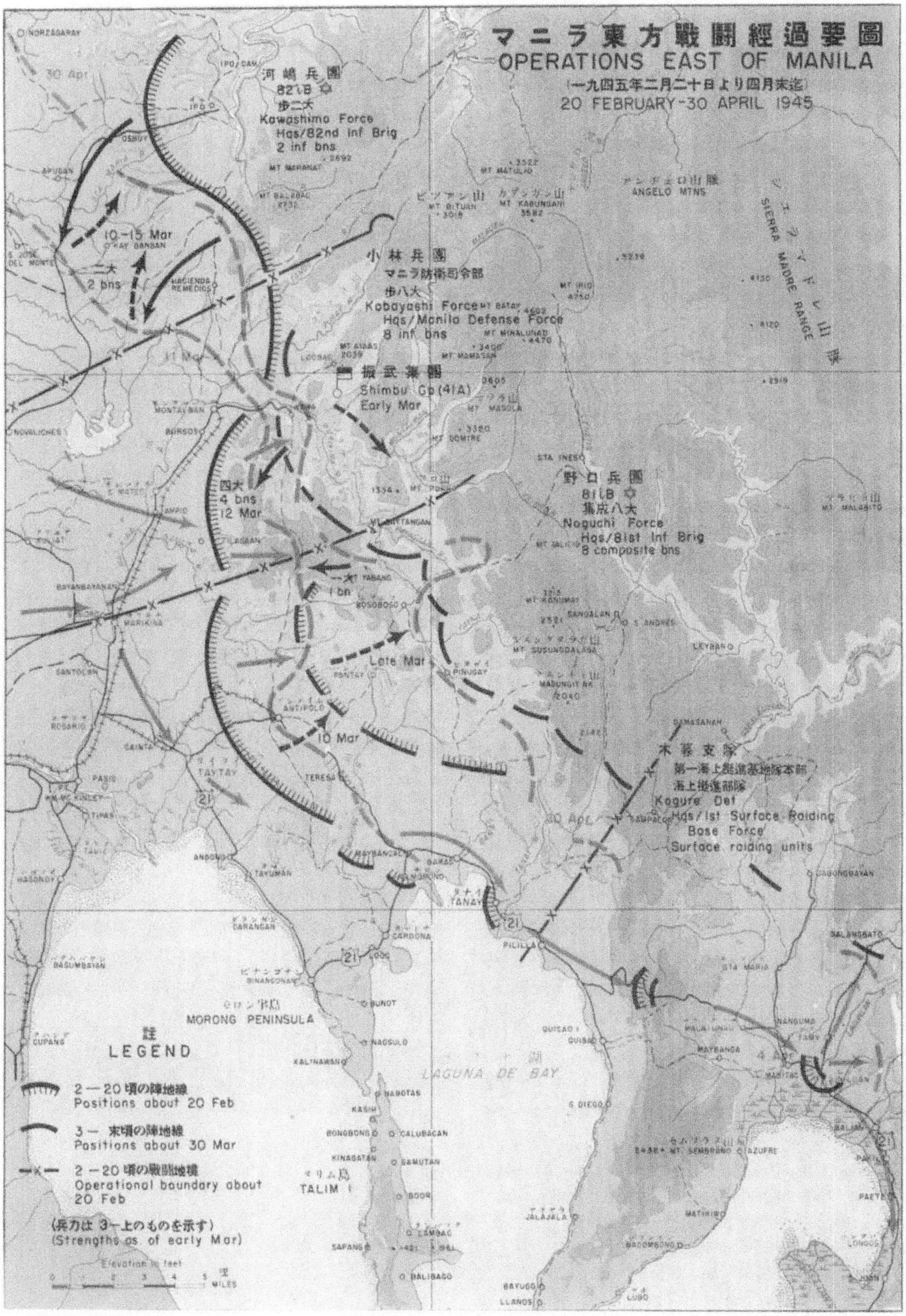

PLATE NO. 120
Operations East of Manila, 20 February—30 April 1945

would be directed at Antipolo and the hills immediately to the north, the Shimbu Group command believed that this attack would be accompanied by a powerful secondary attack. This might be directed at the north shore of Laguna de Bay after an amphibious move across the lake, or it might be an attack from the rear by way of Siniloan after launching new invasions near Infanta or on the Batangas Peninsula near Batangas or Lucena.[99]

To bolster this southern anchor Maj. Gen. Susumu Noguchi had been assigned on 12 February, the day after his arrival, to command the eight battalions between the Kobayashi and Kogure forces. The units which were under his direct command on Bicol Peninsula had begun closing in to the assembly area near Bosoboso about 18 February to further strengthen the main positions of the Shimbu Group.

To strengthen the rear, the naval forces which had escaped from Manila were ordered in late February to secure the Infanta Peninsula. On 27 February, Southwest Area Fleet appointed Capt. Takesue Furuse, commander Northern Philippine Airfield Unit, to command all naval forces in the hills east of Manila. Two days later the Eastern Naval Unit, comprising about 3,000 naval combat troops near Antipolo, and the Western Naval Unit, made up of about 6,000 naval service troops and civilians attached to the Navy in the Wawa-Bosoboso area, were organized. The former immediately began to transfer to Infanta under the personal command of Capt. Furuse.[100]

Shortly thereafter, on 6 March, the enemy commenced a terrific two-day artillery and air bombardment of the southern portion of the front lines. The ensuing ground attack first broke through south of Antipolo. Within a few days the first line had been penetrated at several points between Antipolo and Mt. Baytangan, compelling Maj. Gen. Noguchi to withdraw remnants of the first line units to the second line of defenses about 10 March.

The deepest and most dangerous enemy penetration, however, threatened to sever the main line in the center near Mt. Yabang. Lt. Gen. Yokoyama, therefore ordered the Japanese forces on about 10 March to counterattack the enemy salient, the main effort to be launched against the north shoulder of the enemy positions.[101]

Four battalions from the Shimbu reserve,[102] attached to the Kobayashi Force for this operation, launched the counterattack on 12 March from the area south of Wawa. However, they were soon stopped far short of their objective, Marikina. This assault was accompanied by a two-pronged attack south and southwest from the vicinity of Ipo by two battalions of the Kawashima Force and an attack west from the north side of Mt. Yabang by the 182d Independent Infantry Battalion of the Noguchi Force.

Hardly had this unsuccessful operation been terminated when the enemy launched another powerful attack, 17 March, aimed at the positions held by the Kobayashi Force west of Mt. Baytangan. The continued enemy pressure and the failure of the earlier counterattack made it necessary on about 20 March to order a withdrawal of the left flank of the Kobayashi

99 Ibid., p. 2.

100 Statement by Lt. Comdr. Kayashima, previously cited.

101 Philippine Operations Record, Phase Three, op. cit. Vol. III Suppl. 1, p. 25 and Attached Map No. 3.

102 These four battalions included the 31st Infantry (less 3d Battalion), which had been transferred from the Kawashima Force late in February and placed under direct command of the Shimbu Group, the 185th Independent Infantry Battalion, and one provisional battalion from a railway unit, both of which had just arrived from Bicol. (1) Kobayashi Force Operations Order No. A-124, 8 Mar 45. XI Corps ADVATIS Translation XICAET 0049, 22 Mar 45. (2) Statement by Col. Kobayashi, previously cited.

Force and the Noguchi Force to new positions east of the Bosoboso River.[103] This retirement, completed by the end of March, was followed shortly by an enemy drive southeast along the Pililla-Siniloan highway.

Concurrently with this withdrawal, the Western Naval Unit, was ordered to begin moving from the Wawa-Bosoboso area to the Infanta Peninsula.[104]

In the meantime, a change had been effected in the command status of the Shimbu Group. An Imperial General Headquarters order of 19 March redesignated the group as the Forty-first Army, giving Lt. Gen. Yokoyama for the first time complete command over all subordinate units.[105]

While the attacks north of Laguna de Bay eased off as the enemy forces regrouped for another attack, powerful enemy drives were launched against the widely scattered forces on the Batangas Peninsula. Here, the Fuji Force, commanded by Col. Masatoshi Fujishige, occupied key positions on the road net and commanding terrain south of Laguna de Bay. The first sustained attack against these positions was launched north and south around the east shore of Lake Taal in mid-March. The Japanese forces in this sector were soon compelled to withdraw east and take up new positions on the north slope of Mt. Malepunyo, five miles northeast of Lipa.[106]

As the enemy attacks turned eastward towards Lamon Bay and continued into April, Col. Fujishige decided to withdraw still further east and concentrate near the north base of Mt. Banahao, 17 miles northeast of Mt. Malepunyo. Moreover, by this time an enemy column moving along the south shore Laguna de Bay had already reached Pagsanjan, ten miles south of Siniloan completely severing all land communication with the main positions of the Forty-first Army. Following the assembly near Mt. Banahao during late April, the Fuji Force was incapable of conducting more than small raids.[107]

In the meantime, the remaining forces on the Bicol Peninsula were also being driven into the hills. An enemy force which landed near Legaspi, 1 April, soon overcame the resistance of the 35th Naval Garrison Unit,[108] forcing it about 7 April to withdraw to previously prepared positions about five miles north of Legaspi. Three weeks later the unit was again forced to retreat to the north. Just before reaching Naga, however, the remnants encountered an enemy force moving south and were dispersed in small groups into the nearby mountains. This was the final phase of organ-

---

103 Philippine Operations Record, Phase Three, op. cit. Vol. III Suppl. 1, pp. 25–6.

104 Philippine Area Naval Operations, Part IV, op. cit., pp. 62–3.

105 Shimbu Personnel Senior No. 50, 29 Mar 45. ADVATIS Bulletin No. 698, 5 Jan 45. This change had first been considered by Imperial General Headquarters about mid-December, but no decision was reached at that time. In January Fourteenth Area Army recommended to Imperial General Headquarters that the Shimbu Group be organized as an Army Headquarters. As a group commander appointed by General Yamashita, Lt. Gen. Yokoyama exercised only tactical control over troops assigned to the group. To have complete control, it was first necessary for the group to be recognized as an Army by Imperial General Headquarters, after which Lt. Gen. Yokoyama had to be appointed by the Emperor as commander of the Army.

106 Radio communication between Fuji Force and Forty-first Army ceased in late April. Philippine Operations Record, Phase Three, op. cit. Vol. III Suppl. 1, p. 32.

107 Ibid., p. 47.

108 This naval unit, composed of about 1,800 troops, was originally under command of the 31st Special Base Force in Manila. At the end of February, however, Southwest Area Fleet placed the 35th Naval Garrison Unit under command of the 33d Special Base Force on Cebu. (1) *Senkyuhyakuyonjuyon oyobi Yonjugo Nen Zai Hito Nihon Gun Heiryoku* 一九四四及四五年在比島日本軍兵力 (Strength of Japanese Forces in the Philippines, 1944–45) 1st and 2d Demobilization Bureaus, Nov-Dec 49. (2) Statement by Comdr. Kusumi, previously cited.

ized resistance on the Bicol Peninsula.[109]

Defense of Baguio

In northern Luzon, meanwhile, Fourteenth Area Army had estimated early in February that a major enemy drive on Baguio was imminent. By this date, moreover, guerrilla bands, operating along the Bagabag-Bontoc-Mankayan road, had already completely ruptured this one remaining vehicular route into the northern redoubt.[110] (Plate No. 121)

Although this threat on the north of the Baguio-Mankayan-Bambang triangle required prompt positive action by Fourteenth Area Army, no adequate reserves were immediately available. Only headquarters guards and the 16th Reconnaissance Regiment, 16th Division, consitituted the Area Army reserve.

Reserves were likewise unavailable from other sectors of northern Luzon. Although the enemy had not yet launched an invasion of the Aparri coast, it was estimated that such a move might occur at any time. It therefore appeared inadvisable to weaken this strategically vital sector, in which the 103d Division with three of its independent infantry battalions was deployed.[111]

Further south, the Cagayan valley defenses were already dangerously undermanned to meet a possible enemy airborne strike. Only one regular infantry battalion (177th Independent Infantry Battalion, 103d Division) was available for the defense of the naval air base at Tuguegarao. Elements of the Fourth Air Army had just assembled at Echague.[112] On 6 February, Area Army ordered the 4th Air Division to organize eight infantry companies

---

109 During the latter part of March and April, the main strength of the Bicol Detachment, which had been organized by Shimbu Group on 20 March 1945 and which included the Headquarters, 35th Air Sector Unit, the 147th Airfield Battalion and the 5th Surface Raiding Base Battalion, withdrew from the vicinity of Naga to previously prepared positions on Mt. Isarog, 11 miles to the east where the detachment remained until the end of the war. (1) 147th Airfield Battalion Operations Order No. 67, 22 Mar 45. XIV Corps ADVATIS Translation 165 LTD 0127, 12 May 45. (2) Philippine Operations Record, Phase Three, op. cit. Vol. III Suppl. 1, p. 48.

110 The only all weather route leading into Baguio, Highway 3 through Binalonan and Sison to the junction with Highway 11 at Camp One, had been severed during the first week of the ground fighting. Fourteenth Area Army had begun work early in January on a circuitous route winding about 60 miles through Trinidad to Inticak thence over a mountain trail to Kayapa where a carabao trail continued through Pingkian to Aritao. This particular trail was selected because it was believed to require the least effort to improve. Actually, however, the construction proved to be extremely difficult, particularly near the rice paddies and lowlands around Inticak and the steep grades southeast of Kayapa. No more than about one ton of supplies could be brought into Baguio each day at the beginning of February. Reserve supplies of food were already reaching a dangerously low level. It was mid-April before construction had progressed sufficiently to permit motor vehicles to travel the entire length. (1) Statement by Maj. Gen. Konuma, previously cited. (2) The Truth of the Philippines Campaign, op. cit., p. 35.

111 Another consideration for retaining a strong garrison at Aparri was the belief in Area Army that Japanese personnel might yet be transported to the Homeland aboard submarines, supplementing the air lift already operating at night from Tuguegarao and Echague. Imperial General Headquarters had dispatched a message to Fourteenth Area Army late in January ordering the evacuation to Japan of the maximum number of key personnel, such as pilots, engineers, and technicians, by every available means. On or about 10 February, three Japanese submarines left Takao, Formosa, for Batulinao, 30 miles east of Aparri. Two were sunk en route. The third returned to Formosa safely. This was the last submarine evacuation from the Philippines. (1) Statement by Col. Ichiji Sugita, Staff Officer (Operations), Imperial General Headquarters, Army Section. (2) *Hito Homen ni okeru Sensuikan Sakusen.* 比島方面に於ける潜水艦作戰 (Submarine Operations in the Philippine Area) 2d Demobilization Bureau, Dec 47, p. 36.

112 The Navy and Army were operating airlifts from Tuguegarao and Echague, respectively.

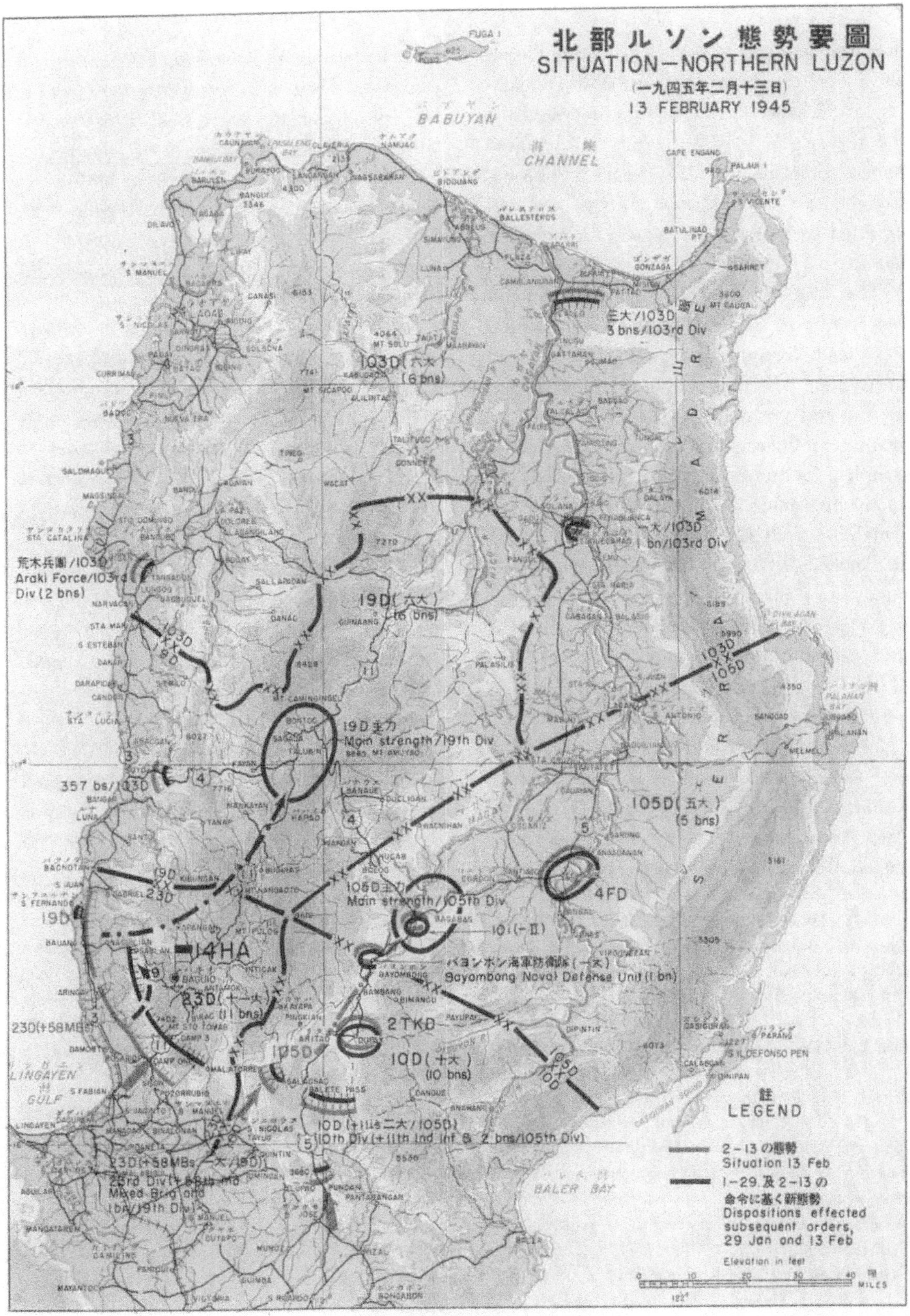

PLATE NO. 121

Situation—Northern Luzon, 13 February 1945

from the remnants of the 2d Parachute Group and miscellaneous flying personnel still remaining at Echague.[113] This unit, redesignated the *Takachiho* (高千穂) Unit[114] and placed under command of Col. Kenji Tokunaga, Commander, 2d Parachute Group, was not expected to be ready for combat for at least a month.

The Bagabag-Salacsac-Balete Pass sector was even more vulnerable. The 10th Division was already committed to the initial defense of the passes. The 105th Division had just begun to displace northward to Bagabag from the vicinity of Puncan and was not expected to complete its movement until early in March. In the meantime, the 10th Infantry (less 2d Battalion), 10th Division, which had arrived at Bagabag on 25 January and been placed under the control of the 105th Division four days later, was busy keeping the guerrillas under control in that vicinity. The survivors of the badly mauled 2d Armored Division were just beginning to assemble near Dupax.

These three divisions had been placed under the tactical control of Maj. Gen. Haruo Konuma, Deputy-Chief of Staff, Fourteenth Area Army, on 23 January, when he had been named chief of the newly organized Bambang Branch, Fourteenth Area Army headquarters.[115] Maj. Gen. Konuma, immediately set about the task of organizing provisional infantry battalions from the numerous miscellaneous air force ground personnel and service units. In view of the virtual isolation of the Baguio front, however, these provisional units were to be employed in the 10th Division sector and the Cagayan Valley.

Of almost 9,000 naval personnel in Bayombong, the majority were civilian employees of the Navy. One provisional battalion, with a strength of about 500, constituted the only regular combat force early in February.[116] Moreover, Southwest Area Fleet headquarters still exercised complete command over the entire group.

On the northwestern coast near Vigan, where the main strength of the Araki Force (79th Infantry Brigade, 103d Division) occupied positions, well organized and active guerrilla operations were believed to presage a secondary enemy landing on the nearby beaches. It was therefore decided not to transfer these Japanese forces to the Bontoc area to join the 357th Independent Infantry Battalion which was heavily engaged with the guerrillas between Suyo and Bontoc.

---

113 Following the departure of the Fourth Air Army headquarters for Formosa, 15 January, Fourteenth Area Army dealt directly with 4th Air Division. On 17 February, Imperial General Headquarters ordered the deactivation of the Fourth Air Army headquarters effective 28 February, at which time the 4th Air Division was incorporated into the Fourteenth Area Army order of battle. (1) *Dairikumei Dai Sennihyakugojuhachi-go.* 大陸命第千二百五十八號 (Imperial General Headquarters Army Order No. 1258) 17 Feb 45. (2) *Hito Koku Sakusen Kiroku Dai Niki.* 比島航空作戰記錄第二期 (Philippine Air Operations Record, Phase Two) 1st Demobilization Bureau, Oct 46, pp. 133–4.

114 *Takachiho* is the name of a mountain in Kyushu on top of which, according to a Japanese legend, Prince Ninigi, ancestor of Emperor Hirohito, was told to descend from heaven.

115 The Bambang Branch, Fourteenth Area Army headquarters, had been established 18 January, under the command of Lt. Col. Makoto Nakahara. The scope of the duties assigned to the branch at that time, however, was limited to establishing the Bambang area on a self-sufficient basis insofar as supplies were concerned. By the Area Army order of 23 January, the mission was broadened to include tactical control of the 2d Armored, 10th and 105th Divisions. Area Army, however, still reserved the prerogative of making the major decisions. (1) *Shobu Sakumei Ko Dai Yonhyakuyonjunana-go.* 尚武作命甲第四四七號 (Fourteenth Area Army Operations Order No. A-447) 23 Jan 45. (2) Statement by Maj. Gen. Konuma, previously cited.

116 (1) Philippine Area Naval Operations, Part IV, op. cit., p. 66. (2) Statement by Maj. Gen Konuma, previously cited.

After careful consideration, it was decided to contract the defense lines guarding Baguio and to transfer the 19th Division to the Bontoc area. The implementing order, substance of which was as follows, was issued on 13 February:[117]

1. The main strength of the 23d Division will establish strong positions along the perimeter Sablan-Mt. Apni-Mt. Lusen-Camp Three-Malatorre and will carry out fierce raiding tactics to inflict heavy losses on the enemy. Simultaneously preparations will be made to execute a counterattack at any time on the central Luzon plain.

2. The 19th Division, upon being replaced by the 23d Division, will shift to Bontoc and, while occupying as wide an area as possible, prepare to carry out a counterattack against the enemy in the direction of the Cagayan River. The division will also take immediate steps to establish peace and order and secure the main roads within the newly assigned area.

3. The boundary line, which will be the responsibility of the 23d Division, will extend from Mt. Pulog through Mt. Nangaoto and Kibungan to Bacnotan. The time for the change in responsibility will be published in a separate order.[118]

Another Area Army order of the same day defined the areas of responsibility of the 103d, 105th and 19th Divisions, and placed the 357th Independent Infantry Battalion of the Araki Force under the command of the 19th Division. The following boundaries, to become effective 2400 19 February, were established by this order:[119]

103d and 19th Divisions:

The line running through Santa Cruz—eastern boundary of Mountain Province—Mt. Camungingel—southern boundary of Abra Province—Hill 305 (four miles east of Santa Maria)—Santa Maria. The boundary will be the responsibility of the 19th Division.

105th and 19th Divisions:

The line running through Mt. Pulog—Hucab—Wacnihan—Santa Cruz. The boundary will be the responsibility of the 19th Division.

As the movement to new positions got under way the following day, the enemy resumed his offensive. (Plate No. 122) Two secondary drives began pressing in the direction of Pugo and along the coastal road toward Aringay, but the major effort was still in the vicinity of Camp One.

Although the last element withdrew in good order from Camp One, 18 February, the 58th Independent Mixed Brigade had borne the brunt of the heavy fighting in the Rosario-Camp One area since late January and the losses, consequently, were exceedingly high. A partial reorganization of the brigade was therefore necessary.[120]

117 This Area Army order provided for the transfer to the 23d Division of the Hayashi Detachment, which had been organized by the 19th Division about 7 January from the 544th Independent Infantry Battalion and miscellaneous shipping units stationed in San Fernando, and the 1st Battalion, 75th Infantry, 19th Division. Other provisions of the order stipulated that the 2d Battalion, 73d Infantry, and 2d Battalion, 75th Infantry, were to be returned to the 19th Division after the completion of the withdrawal. *Shobu Sakumei Ko Dai Gohyakujuroku-go.* 尚武作命甲第五一六號 (Fourteenth Area Army Operations Order No. A-516) 13 Feb 45.

118 The redeployment had progressed sufficiently by 28 February for the Area Army on that date to fix the time for the assumption of command in the newly assigned areas as 1200, 5 March. *Shobu Sakumei Ko Dai Gohyakugojuichi-go.* 尚武作命甲第五五一號 (Fourteenth Area Army Operations Order No. A-551) 28 Feb 45.

119 *Shobu Sakumei Ko Dai Gohyakujushichi-go.* 尚武作命甲第五一七號 (Fourteenth Area Army Operations Order No. A-517) 13 Feb 45.

120 The 546th Independent Infantry Battalion was depleted to such an extent that the remnants were absorbed into the 379th Independent Infantry Battalion. A provisional infantry battalion was integrated into the 545th Independent Infantry Battalion to bring that unit back up to approximately full strength. Summary of 58th Independent Mixed Brigade Organization, 20 Feb 45. I Corps ADVATIS Translation ICAET 0177, 21 Apr 45.

Early in March a new enemy drive began to materialize against the weak defenses of the Ambayabang River valley.[121] Although this attack, threatening the rear of the Baguio front and the uncompleted Baguio-Aritao road, fell within the boundary of the 2d Armored Division in Salacsac Pass, General Yamashita decided to shift the responsibility to the 23d Division in view of the desperate situation facing the Japanese forces in the Salacsac and Balete Pass area.[122] Accordingly, Area Army, on 10 March, shifted the 23d Division left boundary east to include the Ambayabang River, simultaneously assigning to the 23d Division the 16th Reconnaissance Regiment main strength to secure the upper reaches of the river.[123]

Shortly thereafter, about mid-March, Fourteenth Area Army began to recognize that the fall of Baguio might not be long delayed. Although stubborn defensive fighting forced the enemy to measure his gains in yards, the rate of attrition within the 23d Division was beginning to accelerate rapidly.[124] In view of the continuing deterioration, General Yamashita directed the Area Army staff to prepare a plan for prolonging the delaying action deeper in Mountain Province and the Bayombon plain.[125]

Meanwhile, the Japanese had been trying unsuccessfully since early March to clear the San Fernando area of guerrilla forces pouring in from the north. Concurrently, Maj. Gen. Bunzo Sato, Commander, 58th Independent Mixed Brigade, had been following with considerable concern the progress of the American force advancing north from Aringay. This enemy drive threatened to isolate the Japanese in the San Fernando area. On 10 March, therefore, Maj. Gen. Sato ordered the 1st Battalion, 75th Infantry, to withdraw and take up new positions in the vicinity of Naguilian. Simultaneously, the Hayashi Detachment was ordered to remain in the San Fernando area.[126]

As the withdrawal began, the Hayashi Detachment pulled in its outpost stationed in the town. The guerrillas quickly followed up this withdrawal to move into San Fernando on 14 March. Six days later, after the American forces had entered Bauang from the south on 18 March, the Hayashi Detachment was ordered to withdraw over secondary roads to Naguilian, thence to take up new positions further to the rear at Sablan.[127] The American col-

121 Only two guard companies under the command of the Inspectorate of the Line of Communications stood between the enemy force and the new road. Both of these companies were stationed in the vicinity of Santa Rosa. (Statement by Maj. Gen. Konuma, previously cited.)

122 For further details regarding the operations in Salacsac Pass at this time see page 478.

123 *Dokuritsu Konsei Dai Gojuhachi Ryodan (Mei) Butai Kodo Gaiyo* 獨立混成第五十八旅團(盟)部隊行動概要 [Summary of Operations, 58th Independent Mixed Brigade (Mei Force)] Home Depot Division, undated. Record of Operations and Movements of the 16th Reconnaissance Regiment. (pages unnumbered).

124 In addition to the mounting battle losses, for which the only replacements were untrained and inadequately armed service troops, the physical stamina of all troops was deteriorating sharply due to the progressive reduction in daily ration brought about by the shortage of food supplies. Moreover, the strain of continuous fighting against an enemy who had complete control of the air, unlimited supplies of artillery ammunition, and sufficient reserve units to permit rotation of the front line troops, was beginning to make itself felt among the Japanese troops. (Statements by Maj. Gen. Konuma, previously cited and Maj. Mitsusuke Tanaka, Staff Officer (Operations), Fourteenth Area Army.)

125 Philippine Operations Record, Phase Three, op. cit., Vol. III, Operations on Luzon, p. 164.

126 58th Independent Mixed Brigade Operations Order No. A-89, 10 Mar 45. I Corps-ADVATIS Translation ICAET 0177, 21 Apr 45.

127 Following the completion of this withdrawal, the Hayashi Detachment was dissolved. The 544th Independent Infantry Battalion thereupon took up positions in the main defense line at Sablan, replacing the 379th Independent Infantry Battalion which now concentrated on the north flank of the defense line. (Statement by Maj. Gen. Bunzo Sato, commander, 58th Independent Mixed Brigade.)

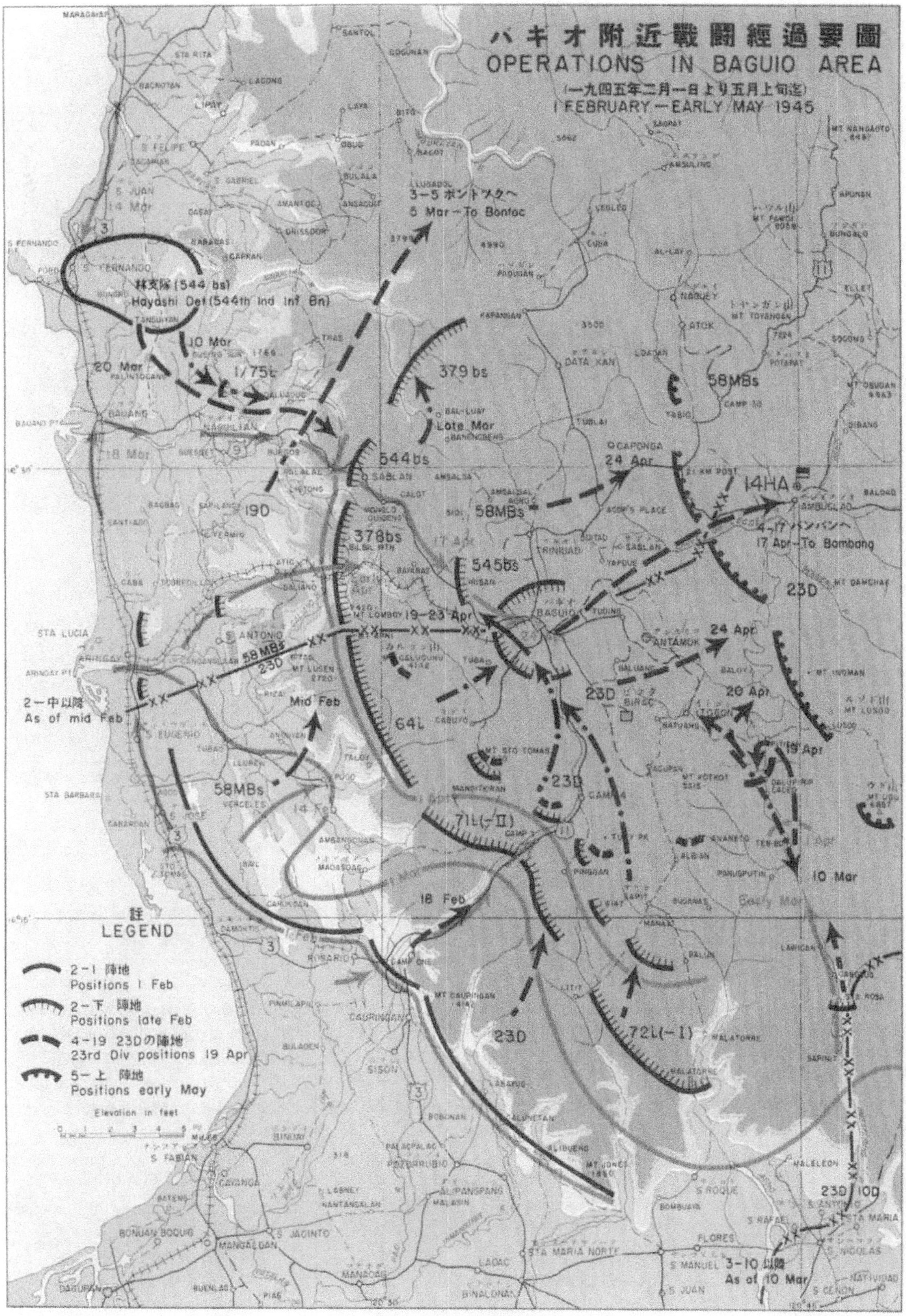

PLATE NO. 122

Operations in Baguio Area, 1 February—Early May 1945

umn, turning east at Bauang, overcame the Japanese defenses at Naguilian after three days of heavy fighting and pushed on late in March to Sablan where the Japanese lines finally held. In the meantime, another enemy assault had penetrated the outer defenses near Galiano, south of Highway 9.

Area Army now feared that the enemy's main effort against Baguio had been shifted from the southwest to the northwest. Long range artillery dropping on Baguio late in March from the northwest supported this contention.[128]

Preliminary steps to abandon the city were therefore initiated.[129] On 10 April, the Inspectorate of Line of Communications was ordered to establish service organizations and disperse supply dumps in the Cervantes-Mt. Pulog-Bayombon-Lubuagan area (22 miles northeast of Bontoc).[130]

General Yamashita decided, soon thereafter, to transfer his headquarters temporarily to Bambang, pending completion of the command post at Kiangan. Accordingly, the Area Army established on 13 April, the Baguio Branch, Fourteenth Area Army headquarters, to control tactical operations on the Baguio front following the departure of General Yamashita. Maj. Gen. Naokata Utsunomiya, Deputy-Chief of Staff, was simultaneously assigned chief of the branch.[131]

The defense of Baguio was now entering its final phase. An enemy tank-infantry task force, which had been halted late in March at Sablan, finally broke through the lines, 15 April, after first neutralizing the Japanese artillery emplaced near Monglo. Several tanks swept on through Irisan by the morning of the 17th. Maj. Gen. Sato, rallying every available unit including miscellaneous service elements from Baguio, immediately launched a determined counterattack. As the small enemy force withdrew, the brigade re-established the Japanese line through the northwestern section of Irisan.[132]

Word now arrived in Baguio that the road to Aritao was completed. General Yamashita, accompanied by a group of staff officers, thereupon departed the evening of the 17th for Bambang. Prior to leaving, the Area Army commander imparted last minute instructions to Maj. Gen. Utsunomiya relative to future steps to be taken by the forces on the Baguio front. The substance of these orders was as follows:[133]

---

128 Japanese civilians residing in Baguio were ordered to evacuate to Bayombong, 30 March. Fourteenth Area Army Line of Communication Staff Transportation Special Order No. 7, 30 Mar 45. ADVATIS Bulletin No. 702, 7 Jun 45, p. 1.

129 Maj. Gen. Haruo Konuma, who was conducting the defense of Balete and Salacsac Passes at this time, visited Fourteenth Area Army headquarters in Baguio on 4–5 April. Upon seeing how discouraged the Japanese were in this sector he proposed sending a provisional infantry battalion from the Cagayan Valley to aid in the defense of Baguio. He immediately dispatched a message to his command post in Bambang, ordering one provisional battalion to depart for Baguio. Because of the poor condition of the Baguio-Aritao trail, however, only the headquarters and one company completed the entire march, arriving at Baguio 16 April. Following the withdrawal of the Japanese from Baguio, this battalion was ordered by Area Army to come under the command of the 2d Armored Division. (Statement by Maj. Gen. Konuma, previously cited.)

130 Fourteenth Area Army Operations Order No. C–406, 10 Apr 45. in I Corps ADVATIS Translation 163d LDT 0205, 5 May 45.

131 *Shobu Sakumei Ko Dai Roppyakuyonjushichi-go* 尚武作命甲第六四七號 (Fourteenth Area Army Operations Order No. A–647) 13 Apr 45.

132 Statement by Maj. Gen. Sato, previously cited.

133 The Truth of the Philippines Campaign, op. cit., p. 46.

1. Prevent the enemy from entering Baguio as long as possible. When the situation makes it inevitable, withdraw to the line extending from the heights north of Point 21 Km to the heights southwest of Pingkian so as to maintain contact with the Balete front.

2. The main strength of the 23d Division will occupy new positions extending from Point 21 Km to the heights southwest of Pingkian. Care must be exercised to avoid enemy encirclement.

3. The 58th Independent Mixed Brigade will occupy new positions in the sector northwest of Point 21 Km.

Two days later, with the fierce battle still raging near Irisan, the Japanese defenses in that sector began to crumble. The 23d Division commander promptly issued an order calling for the transfer of elements from each of the three infantry regiments south and southwest of Baguio to reinforce the Irisan sector. The remainder of the 23d Division was to occupy the present positions along the line Mt. Calugong-Mt. Sto Tomas-Camp 4-Tuey Peak-Mt. Ananeto-Dalupirip.[134]

Although the reinforcements began arriving on the Highway 9 front that same day, the weight of the enemy assault forced the Japanese to retreat slowly towards Baguio.[135] By 24 April the enemy advance threatened to imperil the planned withdrawal deeper into the northern redoubt. Maj. Gen. Utsunomiya therefore directed the division to abandon the Baguio area and to take up new positions further north along a line running generally from Atok to Mt. Inoman.[136]

The same day, as the advance elements of the American forces entered the western approaches to the city, the 23d Division issued the implementing order and initiated the retreat. By 29 April most of the troops had successfully gained the north side of Baguio.

While the enemy paused in Baguio to reorganize, the remnants of the Japanese 23d Division and the 58th Independent Mixed Brigade took up positions along the new line in preparation for the final defense of the inner mountain stronghold.[137]

Battle for the Northern Passes

In the meantime, while the 23d Division had been delaying the American thrust into Baguio, the Japanese forces in the sector guarding Salacsac and Balete Passes were engaged in even a more bitter and protracted struggle to secure the southern flank of the northern redoubt.

The rugged landscape and dense forests of the Caraballo Mountains dominating this sector offset to a great extent the enemy's advantage of complete and unchallenged control of the air and his material superiority. Highway 5 was the single route over which vehicular traffic could move northward into Cagayan Valley. (Plate No. 123) The Villa Verde Trail, running from San Nicolas through steep mountains to join Highway 5 at Santa Fe, had been im-

134 23d Division Operations Order Nos. A-253 and A-254, 19 Apr 45. I Corps ADVATIS Translation 163d LDT 0206, 4 May 45.

135 One battalion of the 23d Division was thrown into the line about 19 April, followed two days later by another battalion of the 23d Division. Finally, about 22 April, the 16th Reconnaissance Regiment main strength, which on 20 April was ordered to withdraw from the vicinity of Tebbo, was committed with no more success. (Statement by Maj. Gen. Sato, previously cited.)

136 Statement by Maj. Gen. Naokata Utsunomiya, Deputy-Chief of Staff, Fourteenth Area Army.

137 During the evacuation of Baguio, the 378th Independent Infantry Battalion was almost annihilated. All of the artillery pieces of the 23d Division and the 58th Independent Mixed Brigade were lost during the withdrawal. (1) Statement by Maj. Gen. Sato, previously cited. (2) Philippine Operations Record, Phase Three, op. cit. Vol. III, p. 186.

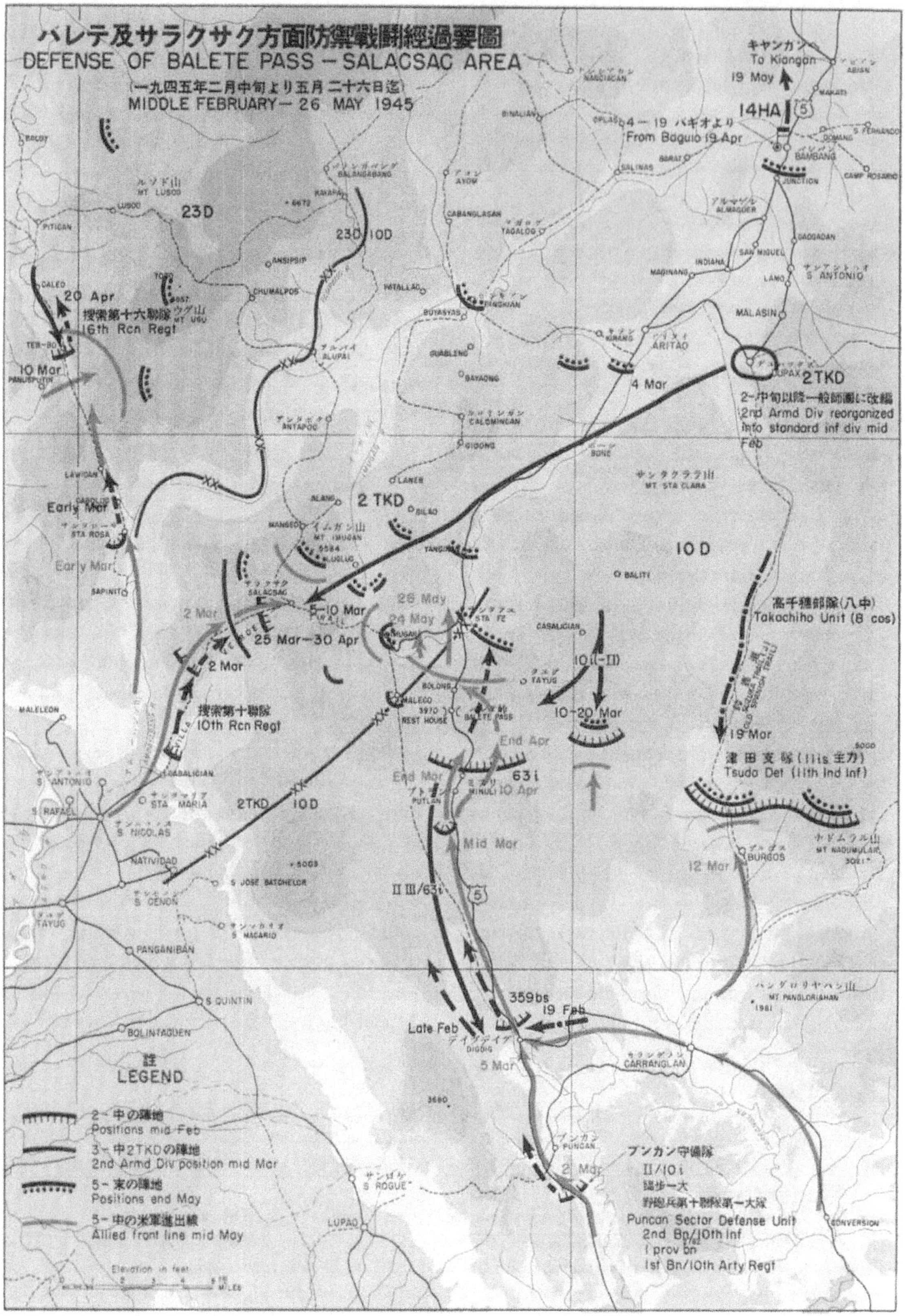

PLATE NO. 123

Defense of Balete Pass—Salacsac Area, Mid-February—26 May 1945

proved to some extent, but much of it could not be traversed by vehicles. East of Highway 5 several smaller trails passed north through the mountains. The most important, the Suzuka or Carranglan Trail,[138] led north from Carranglan through mountains considered less difficult to negotiate than those over which the Villa Verde Trail passed.

In mid-January, when the 10th Division first took up positions guarding these routes, it was estimated that the enemy main attack would be directed along Highway 5, with a secondary attack along the Suzuka Trail slightly more probable than one along the Villa Verde Trail.[139] Accordingly, the Tsuda Detachment with its two remaining regular infantry battalions was deployed to secure the area east of Highway 5, while the 10th Reconnaissance Regiment, with a total strength of less than a battalion of infantry, was ordered to continue its mission of guarding the entrance to the Villa Verde Trail. The main strength of the division, the 63d Infantry, was ordered to occupy and defend positions securing Balete Pass, through which Highway 5 led.[140]

During the following month, as the forces along Highway 5 and on the right flank were gradually reinforced from the reserves,[141] the division worked with feverish haste to improve the natural defenses guarding each of the terrain corridors. The surrounding area was literally honeycombed with caves, tunnels, gun positions and pillboxes. Stocks of food and munitions recently evacuated from San Jose were stored near the positions in preparation for a long and stubborn battle.

Although by mid-February the enemy had not yet launched a full-scale assault against either the Advance Force or the 10th Reconnaissance Regiment the tempo of his attacks nevertheless was increasing daily. At this juncture the Area Army, upon the recommendation of Maj. Gen. Konuma, ordered the 10th Division to defend its positions to the last man. Substance of this order, which had been previously imparted to the division by Maj. Gen. Konuma on 1 February and again on 11 February, was as follows:[142]

138 This same trail was referred to by the American forces as the Old Spanish Trail.

139 This estimate of the secondary attack was based on the premise that the enemy would attack up the more gradual slope north of Carranglan. Because of this more gradual slope the Japanese also considered the Suzuka Trail more difficult to defend. (Statement by Maj. Gen. Konuma, previously cited.)

140 Tetsu (10th Division) Operations Order No. A-67, 15 Jan 45. I Corps ADVATIS Translation ICAET 0185, 25 Apr 45.

141 The outposts in front of the 63d Infantry were first strengthened following a 23 January order by which the Inoue (Provisional) Battalion was added to the Advance Force immediately after being transferred from the 105th Division. Shortly thereafter, the 2d Battalion, 10th Infantry, was removed from the division reserve and also assigned to the Advance Force. Early in February, the 359th Independent Infantry Battalion, 105th Division, which had just been attached to the 10th Division from the 2d Armored Division, was ordered to take up positions mid-way between Digdig and Carranglan to block a possible enemy attack on Highway 5 from Carranglan. On 7 February, because of persistent though relatively small scale enemy attacks north of San Nicolas and Santa Maria, the division issued an order which included the following provisions: a. 1st Battalion, 39th Infantry, will take up positions in Salacsac Pass, behind the 10th Reconnaissance Regiment. b. The division artillery unit will place about one battalion in the vicinity of Imugan and have it prepared to cooperate in the Salacsac Pass fighting. c. Main strength of the Fujikuro Force (provisional battalion of the Tsuda Detachment) will proceed to the Ambayabang River valley about five miles north of Salacsac and pacify the guerrillas in that area. (1) Tetsu (10th Division) Operations Order No. A-76, 23 Jan 45. I Corps ADVATIS Translation ICAET 0185, 25 Apr 45. (2) Statement by Maj. Gen. Sakae Tsuchiya, Chief of Staff, 10th Division. (3) Tetsu (10th Division) Operations Order No. A-95, 7 Feb 45. I Corps ADVATIS Translation ICAET 0185, 25 Apr 45. (4) Cf. n. 41.

142 *Shobu Sakumei Ko Dai Gohyakunijuroku-go* 尚武作命甲第五二六號 (Fourteenth Area Army Operations Order No. A-526) 16 Feb 45.

1. Enemy strength must be reduced by carrying out vigorous raiding attacks deep within the enemy areas in the vicinity of San Nicolas, San Quintin, San Jose, Rizal, and to the south.

2. The entire area from the line running east to west immediately north of Puncan to the line running east to west through Balete Pass must be strongly defended to the last as the main position to crush the enemy. To accomplish this, counterattacks must be made and the enemy crushed on the spot with local forces or with powerful elements of the division main strength if the enemy attempts to infiltrate or advance into the rear through gaps.

3. The same tactics will be employed if the enemy attempts to infiltrate through the Ambayabang River or Mt. Nanaabung (13 miles northeast of Carranglan) flanks.

4. Depending on the situation, the main strength must be prepared to counterattack an enemy assault which may come from the San Nicolas, San Jose, or the Rizal sectors.

5. Withdrawal from the main positions will be only upon Area Army order.

The stage was now set for more than three months of the most violent and fanatical defensive fighting on all of Luzon. Near Puncan enemy forces were already moving through the hills on each side of Highway 5 to outflank the Japanese positions. On 19 February, Lt. Gen. Okamoto ordered the 359th Independent Infantry Battalion, to pull back its main strength closer to Digdig from where it could check an enemy advance from the south or east. Simultaneously the Advance Force was redesig nated the Puncan Sector Defense Unit.[143]

With operations progressing favorably at Puncan, Maj. Gen. Konuma, on 25 February, directed the 10th Division to counterattack the American forces in this area with a powerful element. Having no division reserve, Lt. Gen. Okamoto decided to employ the main strength of the 63d Infantry and immediately dispatched an implementing order to this effect.[144] Shortly after the designated units reached the hills northwest of Digdig, however, the enemy column overcame the Puncan Sector Defense Unit, 2 March, forcing the remnants to withdraw northwest into the hills. Elements of the 63d Infantry thereupon began moving back to the main positions.

In the meantime, a more dangerous enemy drive began gaining momentum on the right flank, where an estimated one division was forcing the 10th Reconnaissance Regiment back along the Villa Verde Trail. On 2 March, the Japanese were reported to be holding out with great difficulty in positions near Salacsac 2d Pass.[145] As a stop-gap measure, pending additional reinforcements from the Area Army, the 10th Division on 3 March placed the 1st Battalion, 39th Infantry, and the Fujikuro Force, both of which were in Salacsac 1st Pass, under the control of the 10th Reconnaissance Regiment.[146]

Maj. Gen. Konuma promptly cancelled his directive of 25 February to the 10th Division and issued, 3 March, in the name of the Area Army Commander, an order which included the following provisions:[147]

---

143 Tetsu (10th Division) Operations Order No. A-111, 19 Feb 45. I Corps ADVATIS Translation ICAET 0185, 25 Apr 45.

144 (1) Statement by Maj. Gen. Konuma, previously cited. (2) 63d Infantry Operations Order No. 83, 25 Feb 45. I Corps ADVATIS Translation ICAET 0185, 25 Apr 45.

145 Salacsac 2d Pass was the term applied by the Japanese to a saddle on the Villa Verde Trail between two prominent hills about two miles west of the village of Salacsac. The summit nearest Salacsac where the trail broke out into the open in the direction of Sante Fe was called Salacsac 1st Pass.

146 Tetsu (10th Division) Operations Order No. A-132, 3 Mar 45. I Corps ADVATIS Translation ICAET 0185, 25 Apr 45.

147 *Shobu Ha Sakumei Ko Dai Nijuichi-go* 尙武派作命甲第二一號 (Fourteenth Area Army, Bambang Branch, Operations Order No. A-21) 3 Mar 45.

1. The 2d Armored Division will depart from its present location on the evening of 4 March and advance to the vicinity of Salacsac Pass via the road through Aritao, Santa Fe, and Salacsac Pass. The division will occupy that area and crush the enemy on the west side of the pass by a systematic order of counterattacks. Local positions must be defended at all costs (withdrawals are prohibited) and the enemy must be destroyed before he reaches the rear perimeter (near Salacsac Pass) of the positions. Raiding assaults must be carried out not only on the front of local positions, but deep into the rear of enemy territory. Counterattacks with a large force will be carried out systematically and expediently at both flanks of the enemy salient.

2. The commander of the 10th Division will continue his present duties and transfer subordinate units (less units in the vicinity of Santa Fe) in the new operational area of the 2d Armored Division to the command of the 2d Armored Division commander when he arrives in the vicinity of Salacsac Pass. From the forces which were expected to counterattack the enemy in the vicinity of Puncan about three companies will immediately assemble in the vicinity of Imugan and come under the command of the 2d Armored Division.

3. The boundary, which will be the responsibility of the 2d Armored Division, will be the line running through Santa Fe—Hill 1210 (about five miles southwest of Santa Fe)—Natividad.

4. The 105th Division commander will dispatch the main strength (about one battalion commanded by the regimental commander) of the 10th Infantry and the Fourteenth Area Army Training Unit towards Aritao with objective of going into action near Salacsac Pass.

5. The 4th Air Division commander will dispatch the Takachiho Unit towards Aritao with the objective of going into action near Salacsac Pass.

Implementing orders by each of the divisions followed shortly thereafter. Advance elements of the 2d Armored Division began moving from Dupax on the evening of the 4th, and by 10 March the major portion of the division was deployed in positions securing the Salacsac 2d Pass.[148] Lt. Gen. Iwanaka immediately initiated a series of sharp counterattacks in an effort to regain lost positions in that area. Although these attacks succeeded in temporarily slowing the enemy drive, the 2d Armored Division was nevertheless finally forced to withdraw on 25 March to positions further to the rear near Salacsac 1st Pass, where the division began a stubborn defense.[149]

In the meantime, another threat had taken shape on the left flank of the 10th Division sector. On 12 March, it was reported that a strong American force, estimated to be another division, had concentrated along the southern approaches to the Suzuka Trail.[150] Upon receiving a further report two days later that this enemy division had begun to attack the Tsuda Detachment, Maj. Gen. Konuma immediately reinforced the 10th Division with one provisional infantry battalion and the *Takachiho* Unit,[151] which was expected to arrive at Dupax the following day.[152]

148 The 2d Armored Division had reorganized the 6th and 10th Tank Regiments and the division antitank unit into three infantry battalions. These were supported by the 2d Mobile Artillery Regiment, which by this time was reduced to a strength of approximately one artillery battalion. Total strength of the division numbered about 4,300. (Statement by Lt. Col. Kawai, previously cited.)

149 An attempt by the 2d Armored Division on 31 March to recapture the Salacsac 2d Pass failed to dislodge forward elements of the enemy force. Ibid.

150 *Shobu Ba Den Dai Sanbyakujushichi-go* 尚武バ電第三一七號 (Fourteenth Area Army, Bambang Branch, Telegram No. 317) 12 Mar 45.

151 With the assignment of the *Takachiho* Unit to the Suzuka Trail sector, Col. Kenji Tokunaga, who was senior to Col. Den Tsuda, took over command of all Japanese forces in this area. (Statement by Maj. Gen. Konuma, previously cited.)

152 When, on 18 March, it was learned that the original reports had overestimated the size of the enemy force attacking in the Suzuka Trail area, the newly added reinforcements, last elements of which arrived on 20 March, were nevertheless left in that area. Ibid.

Along Highway 5, meanwhile, the enemy continued to forge ahead. By mid-March he was attacking the forward positions of the 63d Infantry.[153] Although by the end of March the American forces had broken through the first line positions in one sector along the highway, they did not exploit the penetration. Instead, construction was begun on a new road east of, and parallel to, Highway 5.

This project was promptly recognized as foreshadowing a renewed enemy attack on the left flank of the 63d Infantry. The Japanese nevertheless were unable to adopt adequate measures to deal effectively with this new threat. Infiltration parties were sent out nightly to destroy construction equipment, but the Japanese could not break through the heavy infantry guard posted around the vehicles.

By this time, small groups of guerrillas were beginning to infiltrate through the front lines to occupy commanding terrain immediately in rear of the Japanese. Each guerrilla force subsequently was enlarged into a cancerous enemy island which attacked the Japanese from the rear in coordination with frontal assaults by the American forces.[154]

On 10 April, the enemy, supported by a terrific artillery bombardment and heavy air strikes, launched the anticipated attack against the left flank of the 63d Infantry.[155] In a desperate attempt to stem this drive with a decisive blow before it got well under way, Lt. Gen. Okamoto, on 15 April, ordered the 10th Infantry (less 2d and 3d Battalions) and an element of the Tokunaga Detachment to launch a coordinated counterattack against the enemy right flank. This attack failed to materialize, however, because the Americans struck the Japanese forces en route to the attack positions. By the end of April, despite the infusion of newly formed provisional battalions,[156] the enemy had secured positions from which he was able to control the heights east of Balete Pass.

The 2d Armored Division was also falling back slowly before the relentless enemy pressure in the Salacsac area. Despite repeated counterattacks and nightly infiltration raids, the Japanese were unable to check the American drive.[157] By 30 April the defenses in Salacsac 1st Pass had been penetrated at many points, leaving the 2d Armored Division in control only of the heights on either side of the summit.[158]

The enemy was now dangerously close to breaking through either the 10th Division or the 2d Armored Division. General Yamashita therefore issued on 5 May from his temporary command post at Bambang an " Outline of Future Operational Policy for Northern Luzon ",

153 In the meantime, the Area Army, after temporarily weakening the 10th Division by detaching three companies of the 63d Infantry to the 2d Armored Division, had augmented the strength, 8 March, by ordering the 10th Infantry (less 2d Battalion) to rejoin the 10th Division. Immediately upon arriving in the division area, 10–20 March, the 10th Infantry (less 2d and 3d Battalions) was assigned to the area between Highway 5 and the Suzuka Trail. The 3d Battalion was simultaneously attached to the 63d Infantry to further strengthen the defenses in Balete Pass. (1) *Dai Ju Shidan (Tetsu Heidan) Shijitsu Shiryo* 第十師團(鐵兵團)史實史料 [Historical Data, 10th Division (Tetsu Force)] Home Depot Division, 15 Feb 47. (2) Statement by Maj. Gen. Konuma, previously cited.

154 The Truth of the Philippines Campaign, op. cit., p. 50.

155 Statement by Maj. Gen. Tsuchiya, previously cited.

156 During the period from February to April the Bambang Branch, Area Army, had organized 24 provisional infantry battalions seven of which had been assigned to the 10th Division. (Statement by Maj. Gen. Konuma, previously cited.)

157 In the three month period ending 30 April the 2d Armored Division had been reinforced by the addition of six provisional infantry battalions. Ibid.

158 Statement by Lt. Col. Kawai, previously cited.

皇軍輜重兵隊の活躍　硲　伊之助

Original Painting by Inosuke Hazama

PLATE NO. 124
Japanese Ammunition Train in Action

summary of which follows:[159]

1. The Fourteenth Area Army will crush the enemy along a strategic line connecting the sector west of 21 Km. Point, the vicinity of Baloy, and Mt. Ugu, and also in the vicinity of Salacsac and Balete, the latter two areas being the present front line. For this purpose, Area Army reserve strength will be poured continuously into the 2d Armored Division and the 10th Division in order to maintain the battle strength of these divisions. At the same time, strength from the Carranglan front will be diverted to the Salacsac or Balete Pass fronts when necessary.

2. The Area Army will simultaneously hold a wide area, including the sector in the vicinity of Bontoc and the Cagayan Valley, making every effort to inflict heavy casualties on the enemy and maintain a source of supplies necessary for the execution of subsequent operations. For this purpose, the main strength of the 19th, 103d, and 105th Divisions and Araki Force will concentrate in the vicinity of Bontoc, Cauayan, Bagabag and Banaue, respectively.[160]

3. If the situation becomes unfavorable in the Baguio area, the Area Army, while covering the right flank of the Salacsac line, will crush the enemy by using its main force along the hills on the left bank of the Agno River and by using a part of its troops in the vicinity of Toyangan (15 miles northeast of Baguio).

4. Considering the possibility of an enemy penetration of the Salacsac-Balete line, preparations must be made south of Bambang to check the enemy, particularly enemy tanks. Every attempt must also be made to hold the area north of Bambang as long as possible. Furthermore, necessary preparations will be made rapidly in view of the possible loss of the Bagabag and the Cauayan plains. In the event the enemy penetrates the present Balete-Salacsac line, the 2d Armored Division will take up positions in depth in the direction Pingkian and Salinas and the 10th Division in the direction of Bagabag.

5. At all costs, the Area Army must hold the strategic area including the mountainous provinces in the vicinity of Mankayan, Bontoc, Bagabag, and Ambuclao (11 miles northeast of Baguio) and the sector at the head of the Agno River and attempt to hold out indefinitely.

Although the Japanese were still holding on doggedly to their positions in the passes, the struggle was rapidly reaching a climax. The 10th Division, in spite of reinforcing the 63d Infantry with every available unit, had been unable to retake any important positions or check the enemy's grinding offensive. Attacks on each side of Highway 5 continued to press through Balete Pass. The 2d Armored Division, by this time, was also unequal to the task.

With it now appearing just a matter of days until the American forces would break through the last defenses, General Yamashita decided, on 13 May, to transfer his command post to Kiangan. Maj. Gen. Naokata Utsunomiya,

159 Fourteenth Area Army Operations Orders, op. cit: Outline of Future Operational Policy for Northern Luzon, 5 May 45.

160 The following preliminary steps had already been taken by the Area Army to concentrate its scattered forces closer to the site of the final stand: a. The Araki Force, which had been placed directly under the Area Army on 16 March, was now assembled in the Cervantes area whence it had been ordered about 19 April; b The 103d Division had been ordered by Area Army, 27 April, to transfer its main strength to the Cauayan--Cabagan area, leaving in Aparri only the Yuguchi Detachment with a nuclear infantry strength of two battalions; c. The Bayombong Naval Defense Force had been placed under command of the 105th Division on 1 April for land operations. (1) *Shobu Sakumei Ko Dai Gohyakushichijushichi-go* 尙武作命甲第五七七號 (Fourteenth Area Army Operations Order No. A-577) 16 Mar 45. (2) *Dai Hyakusan Shidan (Shun) Butai Kodo Gaiyo* 第百三師團(駿)部隊行動概要 [Summary of Operations, 103d Division (Shun Force)] Home Depot Division, undated. (3) *Shobu Sakumei Ko Dai Roppyakukyuju-go* 尙武作命甲第六九〇號 (Fourteenth Area Army Operations Order No. A-690) 27 Apr 45. (4) 103d Division Operations Order No. A-283, 1 May 45. I Corps ADVATIS Translation 163 LDT 0274, 22 Jun 45. (5) *Shobu Sakumei Ko Dai Roppyakunijuichi-go* 尙武作命甲第六二一號 (Fourteenth Area Army Operations Order No. A-621) 31 Mar 45.

who had not yet arrived at Bambang from Baguio, was appointed Chief of the Bambang Branch the same day,[161] replacing Maj. Gen. Konuma who had been ordered by Imperial General Headquarters to return to the Homeland.[162]

Even before General Yamashita departed on 20 May, the enemy had broken through the 10th Division to seize Balete Pass on 13 May.[163] Other enemy forces had also pushed through the west side of the highway to capture positions northeast of Mt. Maleco and finally to sever the Imugan-Santa Fe road on 24 May.[164]

The 2d Armored Division, still defending the Villa Verde Trail from positions west of Imugan, was now compelled to shift its line of communications to the Imugan-Pingkian-Aritao road. Shortly thereafter, on about 26 May, the division withdrew to the north of Imugan. The pressure on this front thereupon eased off as the enemy continued along the Villa Verde Trail to Highway 5.

Having been driven from the passes, Fourteenth Area Army believed an opportunity might arise in the near future to attack the flanks of the enemy column pushing north along Highway 5 toward Bagabag. The Area Army therefore began to plan what might be the last coordinated attack on the American forces in northern Luzon.

Fighting East of Manila, Phase II

While the Japanese on the Balete-Salacsac front were still battling furiously to prevent a complete enemy breakthrough, the situation in the Forty-first Army area east of Manila continued to grow steadily worse. (Plate No. 125)

By early May Army headquarters had not yet been able to determine accurately the reason for the comparative lull which had prevailed north of Laguna de Bay for almost a month. It was believed that the Americans were either regrouping prior to resuming the offensive or were planning to contain the Forty-first Army with a small force while transferring the major strength to another sector. Regardless of the true situation, it was considered that a counterattack would achieve the maximum strategic delay envisaged in the basic Luzon defense plans.

Lt. Gen. Yokoyama issued orders accordingly about 3 May. Harassing raids were to be initiated from the north flank on the night of 10 May. Forty-eight hours later a large scale infiltration counterattack was to be launched by the 31st Infantry (less 3d Battalion but reinforced by the 3d Battalion, 17th Infantry) from the west side of Mt. Purro.[165] A provisional battalion of the Kobayashi Force was to attack simultaneously toward Montalban.[166] Further south, a maximum number of raiding units was to be dispatched in the direction of Antipolo by the Noguchi Force.

Shortly thereafter a renewal of the enemy offensive began to upset the plan. Fierce attacks against the center of the Forty-first Army line. 4 May, threatened to break through the Kobayashi Force near Wawa. Lt. Gen. Yokoyama promptly ordered the Kawashima Force

161 Fourteenth Area Army Operations Order No. A-750, 13 May 45. XIV Corps ADVATIS Translation 165 LDT 0143, 8 Jul 45.

162 Statement by Maj. Gen. Konuma, previously cited.

163 Historical Data, 10th Division (Tetsu Force), op. cit.

164 Statement by Lt. Col. Kawai, previously cited.

165 (1) Kawashima Force Operations Order No. A-146 (Shimbu Group Operations Order No. A-228) 4 May 45. XI Corps ADVATIS Translation 164 LDT 0080, 24 May 45. (2) Kobayashi Force Operations Order No. A-244, 4 May 45. XI Corps ADVATIS Translation 164 LDT 0082, 27 May 45.

166 Ibid., (Kobayashi Force Operations Order No. A-244.)

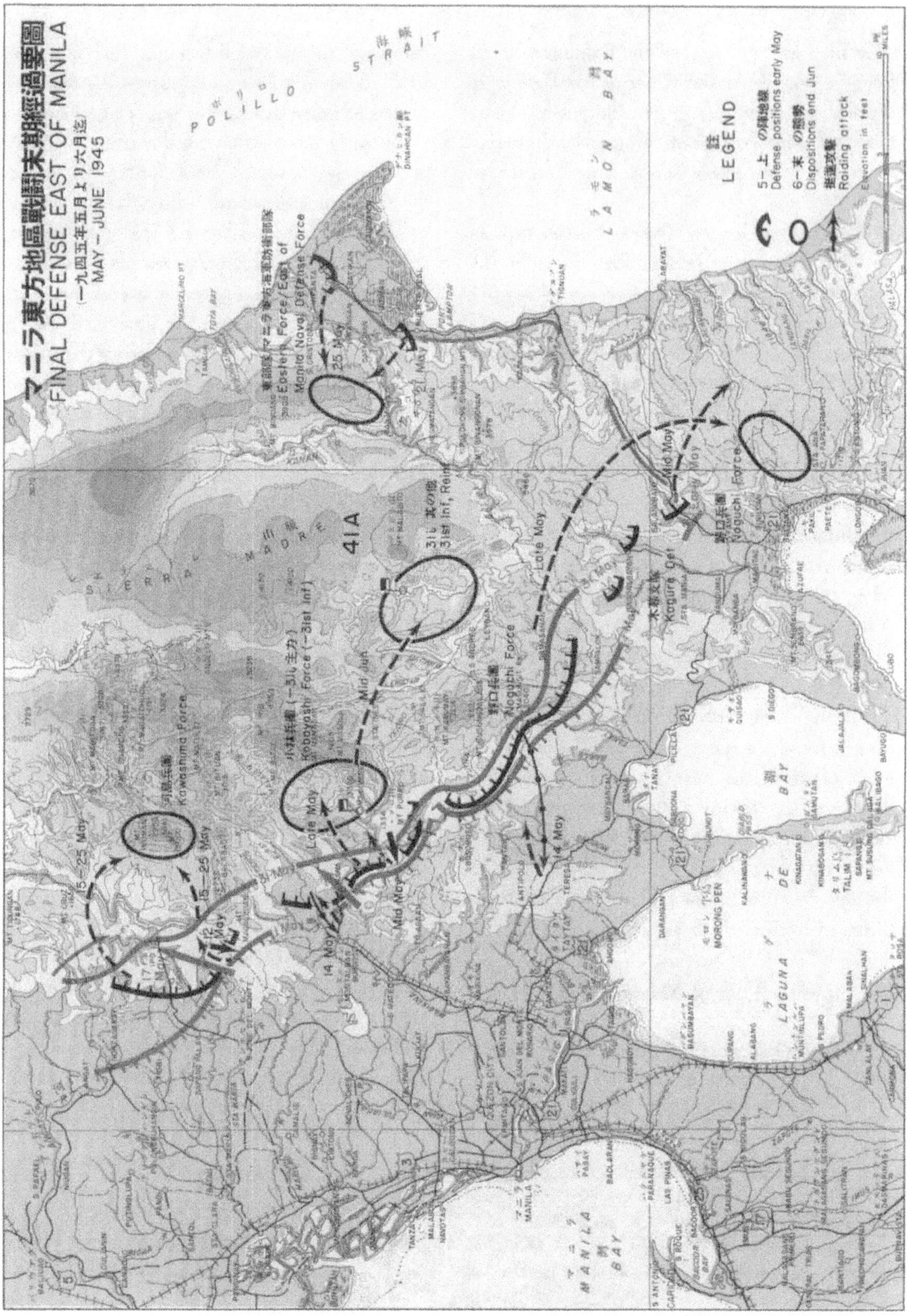

PLATE NO. 125
Final Defense East of Manila, May—June 1945

to initiate the infiltration raids immediately instead of on 10 May,[167] but before the raiding units jumped off the enemy offensive had spread to the north flank. The terrific artillery preparation laid on the Japanese positions guarding Ipo Dam on 6 May prevented the infiltration parties from moving forward, and the Kawashima Force was forced to suspend the planned operations.

The Army commander, although now fully cognizant of the weight of the enemy attacks in both sectors, still believed nevertheless that the major effort was concentrated against Wawa. He therefore ordered the Kawashima Force to launch a strong attack in the direction of Montalban to ease the pressure at the center of the line.[168]

The situation on the northern flank had now worsened to such an extent that Maj. Gen. Kawashima pressed the Army for cancellation of this latest attack plan. After this request was refused on the grounds that the situation on the Wawa front was still the most critical, preparations for the counterattack were immediately pushed by the Kawashima Force. The 358th Independent Infantry Battalion, which formed the nucleus of the Kawashima Force, was pulled out of the left center of the line on the evening of the 12th and moved southeast behind the left flank to a position from which the attack could be launched on the following morning.

Just at this critical moment an enemy attack was launched at the weakly held positions vacated the preceding night by the 358th Independent Infantry Battalion. Maj. Gen. Kawashima, having no reserves capable of stopping this assault, immediately ordered the battalion to reoccupy its original positions. By the time the battalion returned, however, the enemy had penetrated the line and, despite repeated counterattacks during the following two days, the positions could not be recaptured. The enemy forged ahead to capture Ipo Dam on 17 May as the Kawashima Force retreated deeper into the mountains to the rear.[169]

In the meantime, the other Japanese forces designated to attack in the original 4 May plan jumped off according to schedule. The 31st Infantry, reinf., proved unable to penetrate the enemy lines, however, and the main attack ended in failure. Small raiding units operated for several days behind the enemy lines near Antipolo and Montalban, but their effect was negligible.

The enemy now concentrated the full force of his overwhelming airpower in support of the ground attacks in the Wawa area. The combined weight of this bombardment and intense artillery fire took a heavy toll of the Kobayashi Force. By late May the defenses along the base of Mt. Purro crumbled, and the Japanese withdrew further east.[170]

South and to the rear of the main positions of the Forty-first Army, the Kogure Detachment had meanwhile been delaying an enemy column moving toward Lamon Bay from Siniloan. The detachment main strength, defending a narrow mountain defile near Salangbato, checked this drive from late April until mid-May when the American forces finally penetrated the last positions.[171]

This enemy column then sped on until it ran up against the positions of the Eastern

---

167 Statement by Maj. Noriaki Akutsu, Staff Officer (Intelligence), 105th Division, attached to Kawashima Force.

168 Ibid.

169 Ibid.

170 Statement by Col. Hashimoto, previously cited.

171 Statement by Lt. Col. Kogure, previously cited.

Naval Unit outside of Infanta.[172] The naval troops were able to stem the advance until 25 May, when Rear Adm. Furuse ordered the garrison to withdraw west,[173] leaving Infanta unguarded. Shortly thereafter, these Japanese troops broke up into small groups, the larger portion retreating deep into the Sierra Madre Mountains along the upper reaches of the Kanan River.[174]

The Forty-first Army was now crumbling rapidly. Heavy fighting on reduced rations had debilitated the survivors to the point where further organized resistance was impossible. To retain a semblance of organization, however, Army headquarters drafted a plan, 27 May, by which the remnants were to break up into small groups and organize defensive sectors deeper in the mountains, living off the land and fighting sporadically as guerrillas.[175] Following the issuance of implementing orders late in May, the Army's combat effectiveness was negligible.

End of the Luzon Campaign

The decimated remnants of the Fourteenth Area Army in the north were meanwhile being isolated by an unexpectedly swift enemy drive along Highway 5 which caught successive Japanese defenses completely by surprise. (Plate No. 126)

The Area Army had completed about 31 May the plans for a series of attacks against the flanks of the enemy column. The operation was scheduled to proceed substantially as follows:[176]

I General Plan

The Area Army's intentions include the following: The 23d Division will continue to secure the key line which runs from 21 Kilometer Point through the Baloy area to Mt. Ugu; the 10th Division and the 2d Armored Division will endeavor to delay the enemy forces advancing through Salacsac and Balete; the main strength of the 103d and 105th Divisions will concentrate rapidly in the Bayombong and Bagabag plain area and will destroy the enemy force advancing into that area; other units will adhere to the "Outline of Future Operational Policy for Northern Luzon," issued 5 May.

II Outline of Operations

1. The main strength of the 2d Armored Division will gradually withdraw from the present line and, after first assembling in Pingkian, will occupy a line running east and west through the Bato bridge by the early part of June. The division will attack the flank and rear of the column and destroy it.

2. A portion of the 10th Division will hold the present position east of the main road and will destroy the enemy by establishing points of resistance in depth. The main strength will simultaneously concentrate

172 In the meantime, the major strength of the Western Naval Unit had arrived at Infanta early in May. Food supplies were insufficient, however, to feed both naval forces. On 8 May, therefore, the Western Naval Force was ordered to disband and melt into the hills. Some of these service elements began drifting north along the coast while others moved south toward Laguna de Bay.

173 Rear Adm. Furuse's promotion from captain was effective on 1 May.

174 Philippine Area Naval Operations, Part IV, op. cit., p. 61.

175 The Noguchi Force was ordered to transfer its main strength to the mountainous region along the east shore of Laguna de Bay where some of its elements were already located. The Kobayashi Force was ordered to establish a defensive sector in the area north of Mt. Purro and east of Wawa. The 31st Infantry (less 3d Battalion) and miscellaneous troops of the 8th Division were ordered to take up positions north of the Lenatin River. The remnants of the Kawashima Force, already withdrawing from the Ipo Dam area, were ordered to move to the mountains east of San Miguel (22 miles northwest of Ipo) where food was more plentiful. This transfer later proved impossible and the main part of the Kawashima Force remained in the area centering about Mt. Inuman. Philippine Operations Record, Phase Three, op. cit. Vol. III Suppl. 1, pp. 36-8 and attached Map No. 4.

176 Philippine Operations Record, Phase Three, op. cit. Vol. III, pp. 228–30.

east of the Bato bridge on the left flank of the 2d Armored Division and will destroy the approaching enemy by attacking the flank and the rear.

3. The 105th Division will strengthen the defenses along the line extending from the vicinity of Mt. Azuma through Mt. Omori and the Farm School to Mt. Binhagan. If the enemy advances to this sector, the division will counterattack and destroy him.

4. The main strength of the 103d Division will occupy positions east of Bagabag and will counterattack and destroy the enemy if he presses forward to that area.

5. Other units will, in general, continue their present duties.

Almost from the start, however, the Area Army plan proved unworkable. Following receipt of the outline at the Bambang Branch, staff officers with implementing orders were dispatched on 3 June to rendezvous at Aritao with representatives of the 10th Division and the 2d Armored Division.[177] On the following day, the 2d Armored Division, which had not been heavily engaged since about 27 May, initiated a withdrawal to Pingkian pursuant to the new order. The enemy did not press the division, and the movement was therefore conducted in an orderly manner.

In the 10th Division sector along Highway 5, however, the situation was becoming desperate. Instead of relaxing the pressure following the capture of Balete Pass, the enemy had pushed ahead vigorously to capture Santa Fe on 26 May, meanwhile forcing the remnants of the 63d Infantry to fall back from the highway to the hills on the east. Moreover, the division had been compelled to commit the three provisional battalions which arrived by mid-May, leaving no elements to prepare defenses in depth along the highway.

The defenses along this artery were further weakened when the division proved unable to acquire satisfactory control over the two most recently assigned battalions, the stronger of which was the Tank Annihilating Unit.[178] Although both of these provisional units had gone into position north of Santa Fe shortly after arrival and were supported by a battalion of 10th Division artillery and 12 tanks of the 2d Armored Division,[179] they were not prepared to meet the full force of the American assault which bore down on them, 27 May. The Japanese, nevertheless, stemmed the advance until 2 June when the enemy broke through the last defenses. Two days later the enemy captured Aritao. As the American drive carried farther north, contact between the Area Army and the 10th Division was completely severed on 5 June.[180]

The only Japanese force now between the enemy and Bagabag was a battalion from the line of communication guard units in position south of Bambang. The enemy began assaulting these defenses on 5 June and on the following day broke through to continue north and enter Bagabag on 8 June.

177 Statement by Maj. Gen. Utsunomiya, previously cited.

178 The Tank Annihilating Unit, with a strength of about 400, was organized from the Fourteenth Area Army Training Unit. The mission of this unit was to make close-in attacks against enemy tanks with antitank mines. (Statement by Maj. Gen. Konuma, previously cited.)

179 Of the 20 tanks remaining in the 2d Armored Division after the fighting around San Jose these 12 were the only operational tanks by June. The guns had been removed from the other eight tanks and emplaced in the defenses in Salacsac Pass. (Statement by Lt. Col. Kawai, previously cited.)

180 Lt. Gen. Okamoto, whose staff officer did not return from the Aritao rendezvous, began moving his division, 10 June, on his own initiative toward Kasibu where he planned to establish a base from which the division could raid the enemy supply routes. After reaching Kasibu, however, the 10th Division was attacked by an enemy force advancing from the direction of Bambang late in July. The division again withdrew, this time to the upper valley of the Cagayan River. (1) Historical Data, 10th Division (Tetsu Force), op. cit. (2) Statement by Maj. Gen. Tsuchiya, previously cited.

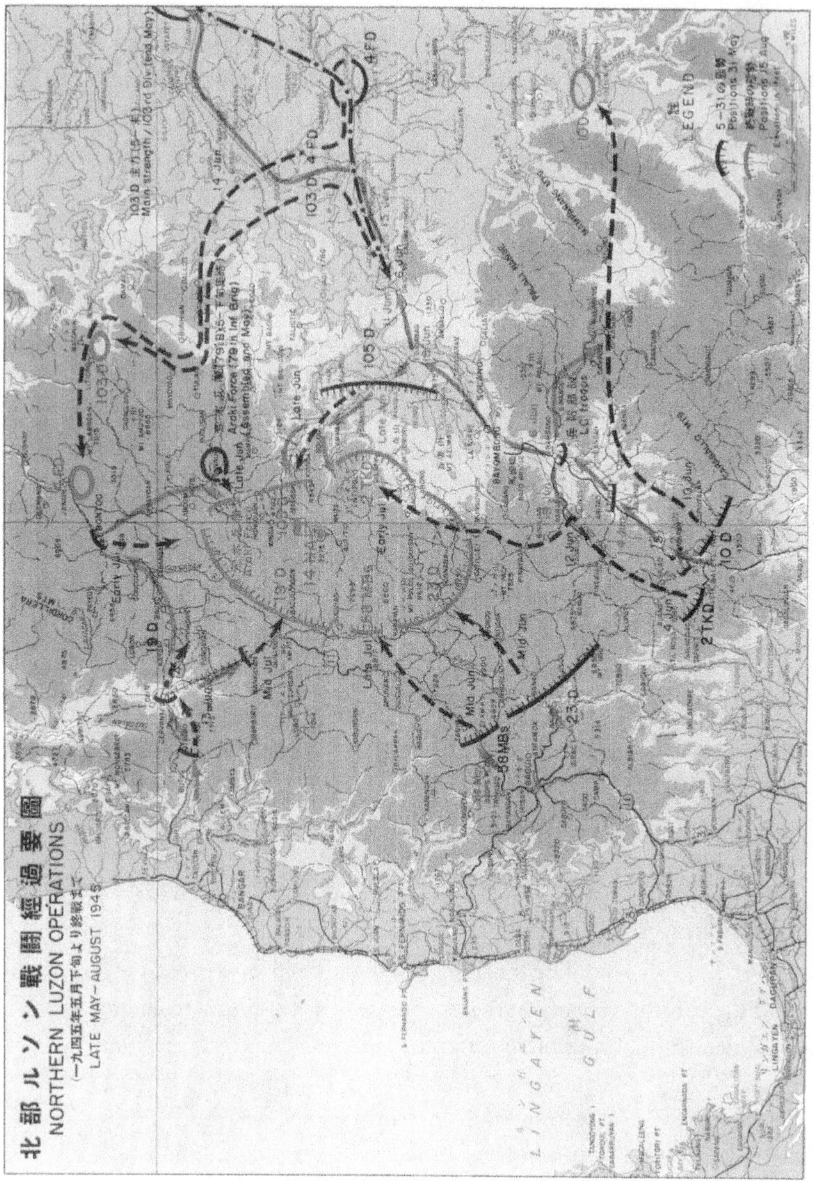

PLATE NO. 126
Northern Luzon Operations, Late May—15 August 1945

山岳地帯の退却　栗原　信

Original Painting by Shin Kurihara

PLATE NO. 127
Retreat in the Mountains

The rapidity of the enemy advance had caught the Japanese unprepared all along Highway 5. Those Area Army forces which were not already concentrated within the area selected for the final stand now began to disintegrate rapidly as the enemy exploited his advantage. The advance guard of the 103d Division, the 179th Independent Infantry Battalion, which had arrived at Orioung Pass on 6 June, was attacked on the 11th while the remainder of the division was strung out all the way back to Tuguegarao.[181] The division commander immediately rushed the nearest available units to reinforce the defenses in the pass, but the enemy penetrated the lines on 13 June. The division commander thereupon fled north from Santiago with part of the division headquarters. Efforts to re-establish contact between the division and Area Army were fruitless.[182]

At this juncture, General Yamashita issued a new operational plan, 15 June, substance of which was as follows:[183]

The Area Army, while occupying the upper reaches of the Asin River, will assemble its units as rapidly as possible and secure key points in Mt. Azuma, Kiangan, Banaue, Mankayan, the area west of Kabayan and Bokod sectors with the intention of sustaining resistance as long as possible. Efforts will be made to re-establish contact with the troops in the strongpoint east of Manila and in the Aparri area. Measures to reduce the enemy strength and to promote the self-sufficiency of all units will be strengthened.

An order conforming to this plan had already been issued to the 4th Air Division, 13 June, directing the remnants of that division to withdraw from Echague and join the main strength of the Area Army. If this proved to be impossible, the division was to engage in guerrilla operations in the area east of Echague. The division moved to Oscariz on the following day, but, after crossing the Magat River, it sought unsuccessfully to rendezvous with the Area Army.[184] By this date all Japanese forces remaining in the Cagayan Valley were isolated from the Kiangan stronghold.[185]

On 19 June, the 2d Armored Division, which had arrived in the Salinas area on 12 June preparatory to launching an attack against the flank of the enemy in Bambang but which had been attacked by the enemy that same day, was ordered to withdraw to the vicinity of Antipolo.[186]

In the meantime, the American forces had

181 The main strength of the 103d Division was disposed approximately as follows at this time:

Division Hq.—Santiago
179th Ind. Inf. Bn.—Orioung Pass
175th Ind. Inf. Bn.—San Luis (en route)
Arty. unit—Tuguegarao (en route)
Engr. unit—Oscariz (en route)

Summary of Operations, 103d Division (Shun Force), op. cit.

182 The division commander also lost contact with the major portion of his subordinate commands at this time. Ibid.

183 Philippine Operations Record, Phase Three, op. cit. Vol. III, pp. 240–3.

184 Statement by Col. Atsutaka Saruwatari, Chief of Staff, 4th Air Division.

185 The Yuguchi Detachment was ordered on 13 June to abandon Aparri and move to Tuguegarao and secure that area. Before the detachment could reach that vicinity, however, an enemy column moving north on Highway 5 attacked the 177th Independent Infantry Battalion, reinf., which was garrisoning that area, and drove it northeast into the hills. Here the remnants of the Tuguegarao garrison were joined by the Yuguchi Detachment. Summary of Operations, 103d Division (Shun Force), op. cit.

186 The 2d Armored Division subsequently arrived in the area south of Antipolo during the early part of July. (Statement by Lt. Col. Kawai, previously cited.)

begun closing in on the other flanks of the final redoubt. Since late May, the 23d Division had been under heavy attack, particularly in the sector assigned to the 58th Independent Mixed Brigade near 21 Kilometer Point. The steady pounding had forced the Japanese in mid-June to fall back along the road leading to Mankayan.

On the northwest, where the 19th Division had been securing the Bontoc-Mankayan-Tacbo area since early in March,[187] the Japanese began to fall back from the Tacbo pass on 13 June.[188] The 19th Division, however, was still able to hold the Bontoc-Mankayan sector against repeated forays by guerrilla bands.

The enemy column which had captured Bagabag on 8 June, meanwhile, had begun to attack the 105th Division deployed along the line extending from the hills behind Solano north through the Farm School. It was at this latter point that the enemy directed his strongest assaults during late June, forcing the division left flank to withdraw to the west of Highway 4.[189]

The divisions still remaining under control of Fourteenth Area Army now began to be slowly compressed deeper within the mountain wilderness between Mt. Pulog on the south and the upper valley of the Asin River. Although the Japanese were no longer capable of mounting large-scale attacks, they nevertheless continued to resist, hoping thereby to divert a portion of the enemy strength.[190]

As of late June, eight months had elapsed since General MacArthur's forces landed on Leyte. Since January Fourteenth Area Army had fought desperately to delay the enemy's conquest of Luzon and to force him to employ a maximum troop strength as long as possible. This terrific struggle was now terminated with the Area Army powerless to intercede against subsequent Allied operations.

187 In early April, the 73d Infantry, 19th Division, reinforced by the 359th Independent Infantry Battalion, 103d Division, succeeded in driving the guerrillas back from the Tacbo area. During April, however, the guerrillas were reinforced to such an extent that the 19th Division found it necessary to strengthen still further the defenses in that vicinity. Accordingly, the 76th Infantry (less 2d and 3d Battalions) was dispatched to the pass early in May. As the guerrilla strength continued to increase, the difficulty of supplying the Japanese forces also increased until they were finally compelled to withdraw in June. (Statement by Lt. Gen. Yoshiharu Ozaki, commander, 19th Division.)

188 3d Battalion, 73d Infantry, Operations Order No. A-65, 12 Jun 45. XIV Corps ADVATIS Translation 165 LDT 0150, 19 Jul 45.

189 Philippine Operations Record, Phase Three, op. cit. Vol. III, p. 245.

190 Statement by Maj. Gen. Utsunomiya, previously cited.

CHAPTER XVI

THE CENTRAL AND SOUTHERN PHILIPPINES

Situation, 22–25 December

Concurrently with Fourteenth Area Army's protracted delaying action in the key mountain redoubts of Luzon, Thirty-fifth Army, fully aware that there was no hope of obtaining further reinforcement, sought as best it could to carry out a similar policy of prolonged resistance in the central and southern Philippines.[1]

On 20 December General Yamashita dispatched an order to Thirty-fifth Army directing abandonment of the decisive battle on Leyte.[2] The substance of this order, delayed two days in transmission, was as follows:[3]

1. The Thirty-fifth Army will hereafter continue unbroken resistance in its assigned operational area and will support future counterattacks by Japanese forces.

2. The Army, in particular, will endeavor to secure the air bases at Bacolod, Cagayan and Davao and prevent their seizure and use by the enemy.

Shortly thereafter, Lt. Gen. Suzuki, Thirty-fifth Army Commander, ordered the 30th Division commander, who was still awaiting transportation to Leyte,[4] to remain on Mindanao and assume responsibility for defending that portion of the island east of Lanao Province.[5] Subsequently, Army headquarters transmitted to all subordinate units, other than those on Leyte, the same order which had been received from Fourteenth Area Army on 22 December.[6] (Plate No. 128). The interpretation and execution of this order was left entirely to the discretion of each individual commander in view of the belief prevailing in Army headquarters that it was unnecessary, at this time, to prepare detailed plans covering subsequent operations.[7]

In the first place, the Army Commander estimated that the enemy, rather than undertake the task of mopping up the Japanese forces in the remainder of the central and southern Philippines, would first employ all his forces in a massive assault against Luzon, and then continue his advance northward against the

1 This chapter was originally prepared in Japanese by Maj. Hiroshi Komatsu, Imperial Japanese Army. For duty assignments of this officer, cf. n. 1, Chapter XIV. All source materials cited in this chapter are located in G-2 Historical Section Files, GHQ FEC.

2 Statement by Maj. Shigeharu Asaeda, Staff Officer (Operations), Fourteenth Area Army.

3 (1) *Hito Sakusen Kiroku Dai Sanki Dai Nikan Furoku Reite Sakusen Kiroku* 比島作戰記錄第三期第二卷レイテ作戰記錄 (Philippines Operations Record, Phase Three, Vol. II, Supplement: Leyte Operations) 1st Demobilization Bureau, Oct 46, p. 190. (2) Statement by Maj. Gen. Yoshiharu Tomochika, Deputy-Chief of Staff, Thirty-fifth Army.

4 After the American landing near Albuera, Leyte, 8 December, Lt. Gen. Gyosaku Morozumi, 30th Division commander, transferred the division command post to Impalutao from Agusan because he estimated that the enemy might subsequently land at Macajalar Bay in conjuction with an airborne drop near Del Monte airfield. (Statement by Col. Hyobunosuke Yamazu, Chief of Staff, 30th Division.

5 30th Division Operations Order No. A-79 (Thirty-fifth Army Operations Order No. A-150), 25 Dec 44. Eighth Army ADVATIS Translation 8AAAE-T 75, 12 Apr 45.

6 Statement by Maj. Gen. Tomochika, previously cited.

7 Ibid.

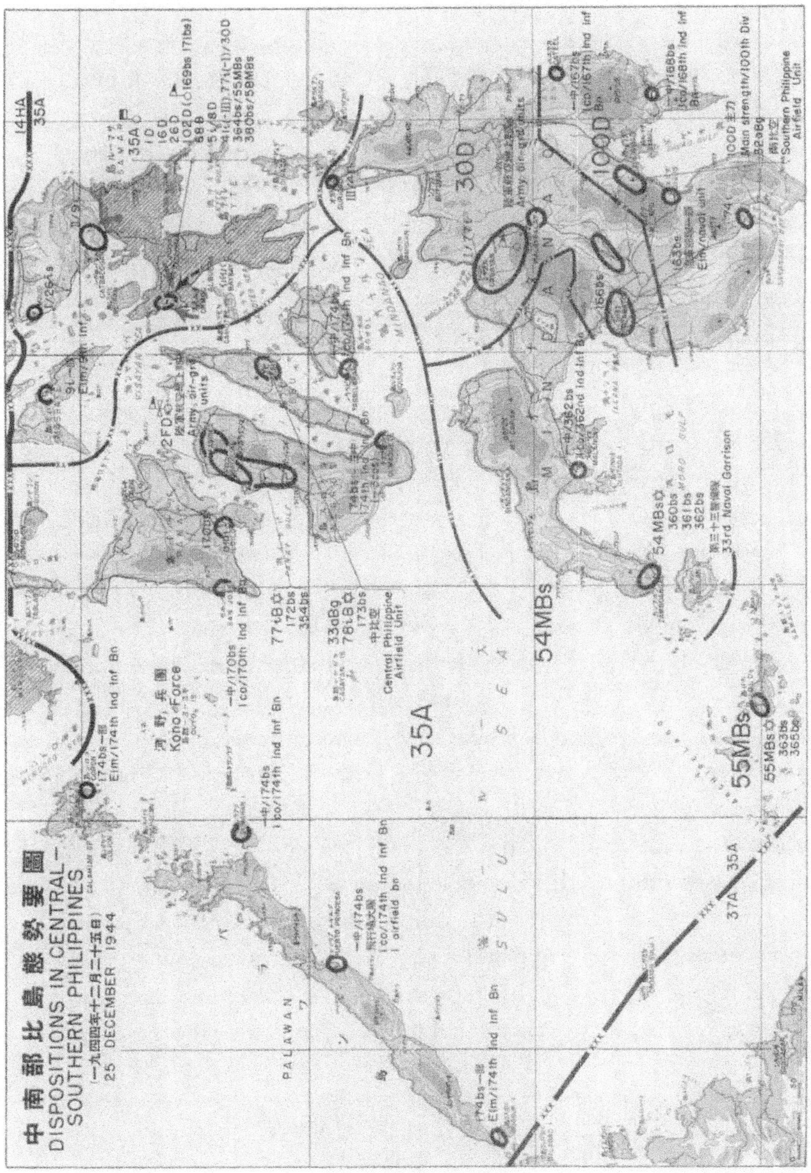

PLATE NO. 128
Dispositions in Central-Southern Philippines, 25 December 1944

Homeland,[8] leaving Thirty-fifth Army forces to wither on the vine like the by-passed Japanese troops in the southeast area.

Moreover, the enemy's progressively tighter sea blockade and air superiority throughout the central and southern Philippines clearly ruled out any long distance transfer of Thirty-fifth Army troops, even presuming that adequate surface transportation could be mustered.

In addition, the Army staff had been deeply engrossed in the conduct of the decisive battle on Leyte and consequently had given little consideration to the defense of other Army sectors. Even now that this battle was to be abandoned, the staff found it necessary to devote its full energies to regaining control over the badly disintegrated remnants of the Leyte forces.

The defensive potential of the island garrisons had meanwhile been seriously reduced during the Army's all-out efforts to win the Leyte contest. Many rear-area combat units had suspended battle preparations to work side by side with service troops in support of the Japanese forces on Leyte. Other units had been relieved from operational preparations to augment security details in an effort to restrict guerrilla activities, which had accelerated rapidly with the steady deterioration on Leyte and the enemy landing on Mindoro.

Of more far-reaching consequences, however, was the steady drain of combat troops during the decisive battle. The equivalent of about one full division had been withdrawn from other island garrisons, principally the 30th and 102d Divisions, and poured into Leyte,[9] reducing by almost one-fourth the total Army combat strength originally available for the defence of the other islands.

The largest concentration of forces was deployed along the south flank of the Thirty-fifth Army zone, on Mindanao.[10] The main strength of the unweakened 100th Division was still located in the Davao area. The 30th Division, its nuclear infantry strength reduced by almost one-half, was scattered from Surigao to Sarangani Bay, where the 74th Infantry was stationed. The western portion of Mindanao was defended by the 54th Independent Mixed Brigade, the main strength of which secured the Zamboanga area. Farther to the south, the 55th Independent Mixed Brigade (less one battalion on Leyte) occupied Jolo.

Dispersed throughout the central and western sectors of the Army area were the remaining

8 Ibid.

9 The larger units which had been transferred into Leyte from other sectors of the Army area included the following: (Cf. Chapter XIII.)

Hq., 102d Division	Elms., Division Engineer Unit
169th Ind. Inf. Bn.	41st Inf. (less 3d Bn.), 30th Division
171st Ind. Inf. Bn.	77th Inf. (less 1st Bn.), 30th Division
Division Artillery Unit	364th Ind. Inf. Bn., 55th Ind. Mixed Brig.

10 This concentration along the south flank of the Thirty-fifth Army area was the continuing result of the estimates made during August and September that the most probable target of the initial enemy invasion would be Mindanao. (Cf. Chapter XI, n. 116.) No redistribution of troops was made following the enemy landing on Leyte.

11 Until 14 November, the Japanese had considered this entire inland sector relatively secure behind the ring of defenses extending from Leyte through Surigao thence along the south coast of Mindanao to Zamboanga and down the Sulu Archipelago. On that date Lt. Gen. Suzuki, alarmed by the intensification of enemy surface activity in the Mindanao Sea, ordered the forces stationed on Cebu to accelerate defense preparations. It was not until the large American convoy sailed through the Sulu Sea on route to Mindoro, however, that the Japanese dream of "Safety Zone" was finally shattered. At that time the island garrisons began to construct defensive positions in earnest. (1) Statement by Lt. Col. Kotaro Katogawa, Staff Officer (Operations), Thirty-fifth Army. (2) Thirty-fifth Army Operations Order No. A-91, 14 Nov 44. *Dai Sanjugo Gun Hatchaku Bunsho Utsushi* 第三十五軍發着文書寫 (Document Files, Thirty-fifth Army Headquarters) Maj. Kohei Takahashi, Staff Officer (Operations), Thirty-fifth Army, Jan 46. This document was copied from original documents by Maj. Takahashi at the American Prisoner of War Camp on Leyte, January 1946.

elements of the 102d Division.[11] The strongest concentration was on Negros in the area around Bacolod. All Army ground forces on Palawan, Panay, Negros, and the small neighboring islands were under the command of Maj. Gen. Takeshi Kono, commander of the 77th Infantry Brigade, 102d Division.[12]

In addition to the Army ground forces, approximately 14,000 Navy ground combat troops and about 15,000 naval air and service personnel were stationed mainly at Cebu, Davao and Zamboanga.[13] Both the 33d Special Base Force at Cebu and the 32d Special Base Force at Davao and Zamboanga were under the command of Southwest Area Fleet. The commander of the 33d Special Base Force, Rear Adm. Kaku Harada, exercised operational control over Army ground units stationed in the vicinity of Cebu by virtue of an earlier agreement with Lt. Gen. Suzuki.[14]

The 2d Air Division, under command of the Fourth Air Army, continued to carry out intermittent special-attack sorties with the few planes still operable. Air ground units of this division were widely dispersed over Palawan, Panay, Negros, Cebu, Mindanao and Jolo.[15] The First Air Fleet had few planes in the area at this time, although base service elements of the Central Philippine Airfield Unit and of the Southern Philippine Airfield Unit were stationed at Cebu and on Mindanao, respectively.[16]

With the exception of the area near the city of Cebu, unified command over Army and Navy forces in each area would become operative only after an enemy landing, when the senior military commander would assume operational control over all forces present. Prior to such invasion, all forces merely co-operated in defensive preparations under the supervision of the senior commander.

Such was the over-all situation of the Japanese forces defending the central and southern Philippines on 25 December, when the enemy launched a new amphibious landing in the rear of the shattered Thirty-fifth Army remnants on northwestern Leyte. This assault, delivered in the Palompon area, virtually shattered all hope of prolonged resistance by Lt. Gen. Suzuki's forces.

Evacuation of Leyte

The disintegration of the Japanese remnants on Leyte now began to accelerate rapidly. Upon learning of the new enemy landing near Palompon, Lt. Gen. Suzuki quickly charged the defense of that area to nearby service troops, the only forces then available.

On the following day, 26 December,[17] Lt.

---

12 Maj. Gen. Kono had been placed in command of the Army ground forces on these islands in mid-November, when the 102d Division headquarters displaced from Bacolod to Leyte. (Statement by Maj. Gen. Tomochika, previously cited.)

13 (1) *Teikoku Kaigun Senji Hensei* 帝國海軍戰時編制 (Wartime Organization of the Imperial Navy) Navy General Staff, 15 Dec 44. (2) *Nairei Teiyo Bessatsu (Teiin Kankei)* 内令提要別冊(定員關係) (Naval Confidential Regulations Attached File: Naval Personnel), Secretariate, Navy Ministry, 1 Feb 44. (3) *Senkyuhyakuyonjuyon oyobi Yonjugo Nen Zai Hito Nihon Gun Heiryoku* 一九四四及四五年在比島日本軍兵力 (Strength of Japanese Forces in the Philippines, 1944–45) 1st and 2d Demobilization Bureaus, Nov-Dec 49.

14 Statement by Capt. Kenkichi Shigaki, Staff Officer (Operations), 33d Special Base Force.

15 Statement by Lt. Col. Shiro Nonogaki, Staff Officer (Operations), 2d Air Division.

16 Strength of Japanese Forces in the Philippines, 1944–45, op. cit.

17 On this date, Thirty-fifth Army received a further order from Fourteenth Area Army reiterating the provisions of the 20 December order. The order also directed Thirty-fifth Army to transfer the 2d Battalion, 9th Infantry (16th Division), from Catabalogan, Samar, to the northern tip of Samar, where it was to come under the command of the Noguchi Detachment on Bicol Peninsula, Luzon. Fourteenth Area Army Operations Order No. A-272, 25 Dec 44. Document Files, Thirty-fifth Army Headquarters, op. cit.

Gen. Suzuki revoked an earlier order for a general withdrawal into the Palompon area and drew up new plans to assemble the remnants of his forces farther north in the vicinity of Mt. Canguipot.[18] (Plate No. 129) Orders to this effect could be issued, however, only to units in the vicinity of Army headquarters. Other forces were to be notified upon arrival in the general area. On the same day the Army command post began displacing from Kompisao to the southern slope of Mt. Canguipot.[19]

No sooner had this change been effected than the enemy launched still another amphibious move, constricting still further the Leyte forces. This new landing was made on 27 December at San Isidro and Calubian, neutralizing the harbor facilities in that area. Only weak service elements and the rear echelon of the 68th Brigade defended this sector.

The Army had meanwhile redoubled its efforts to re-establish contact with the forces isolated northeast of Ormoc and with the 16th and 26th Division survivors in the hills along the Lubi trail. The remnants of the 1st and 102d Divisions, which had been isolated in the Mt. Catabaran area east of Highway 2, were now arriving in the Canguipot area after infiltrating across the enemy-controlled highway.[20]

While the general withdrawal continued, Thirty-fifth Army received from Fourteenth Area Army on 28 December another order reiterating the necessity of prolonging resistance in the Army area but also directing the gradual evacuation of Leyte. The substance of this order was as follows:[21]

1. Thirty-fifth Army will henceforth adhere to a policy of self-sufficiency and independent combat designed to prolong resistance.

2. Troops on Leyte will be evacuated gradually to other islands.

3. It is suggested that Thirty-fifth Army headquarters move to a point suitable for over-all command of the Army.

With confusion now increasing daily as the enemy pressed his attacks from the north, east and south, the Army staff began preparing plans to implement this new Area Army order. From the outset considerable doubt was entertained as to the probable success of such a mass evacuation. The enemy's tight sea blockade precluded daytime movement of ships in the waters off Leyte. Moreover, constant air interdiction foredoomed any large concentration of shipping.

Along the northeastern coast of Cebu and in the port of Cebu itself, however, there were many landing craft and other vessels which had been used to run the enemy blockade. It was estimated that about 40 landing barges plus numerous small boats and bancas, could be assembled for the evacuation.[22] Accordingly,

---

18 Statement by Maj. Gen. Tomochika, previously cited.

19 Ibid.

20 Both of these divisions commenced the withdrawal on the night of 21–22 December following the enemy's penetration of the 1st Division left flank. Although the Army order of 21 December had not yet reached either division, a withdrawal was permitted within the terms of the Army Commander's instructions of 19 December. (Cf. Chapter XIII, p. 403.) Philippine Operations Record, Phase Three, op. cit. Vol. II, Suppl., pp. 192–3.

21 Ibid., p. 198.

22 At this time there were about 30 landing barges, including those under repair, along the northeastern coast of Cebu and in the port of Cebu. In addition, ten large landing barges were expected to reach northern Cebu from Luzon early in January. (Statement by Maj. Manzo Yamazaki, Staff Officer (Operations), Thirty-fifth Army.)

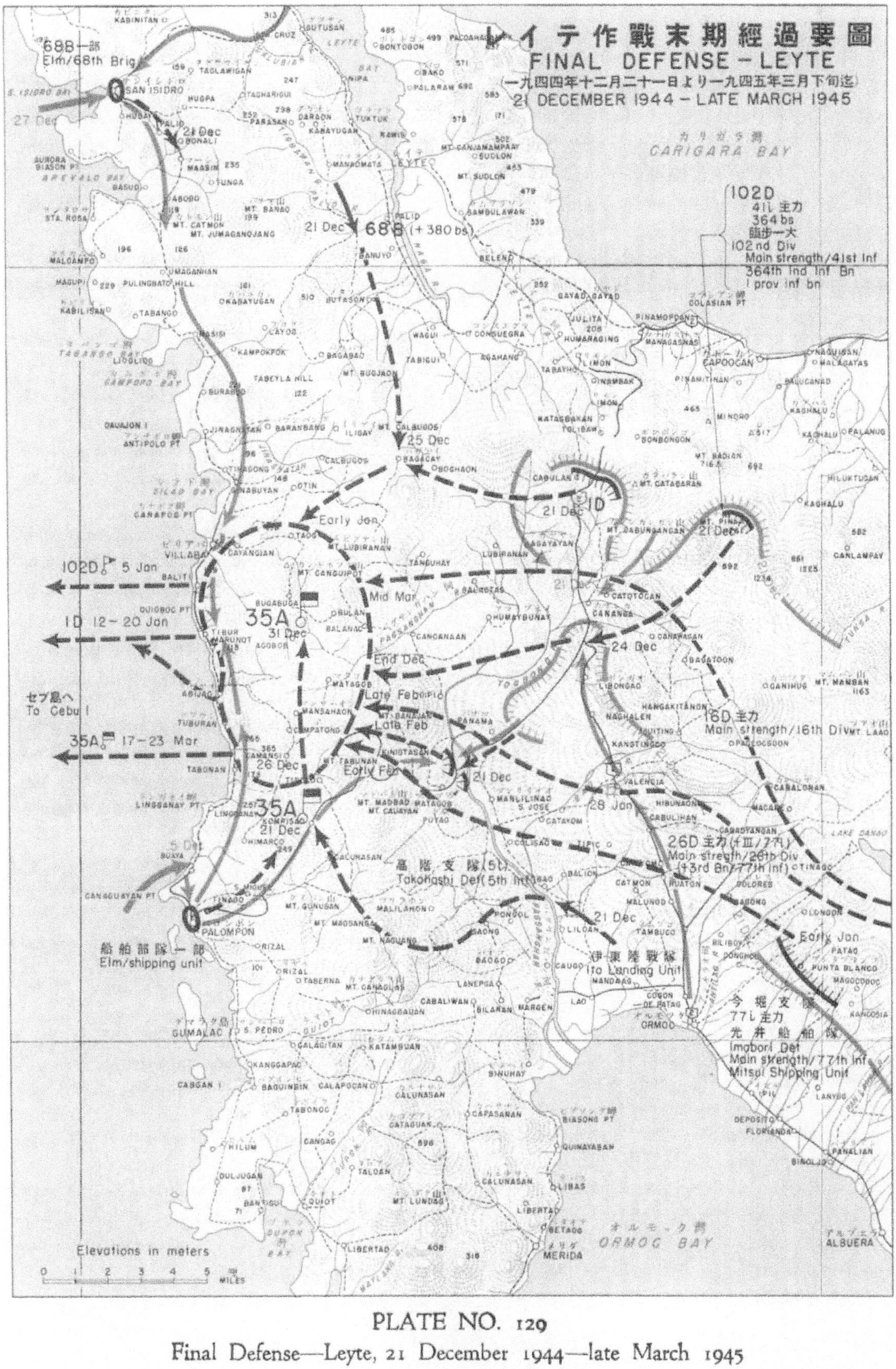

PLATE NO. 129

Final Defense—Leyte, 21 December 1944—late March 1945

the Army dispatched a staff group, designated the Rear Area Command Post, to Cebu on 4 January to procure and control all available shipping, as well as to arrange for the transport of rations and ammunition to Leyte on the eastbound trip.[23] The staff group was also charged with liaison between Army headquarters and Army troops stationed on other islands of the central and southern Philippines with regard to expediting operational preparations.[24]

The planning had progressed sufficiently by 5 January to initiate the evacuation on a small scale.[25] About 200 men of the 102d Division headquarters, including Lt. Gen. Shimpei Fukue, division commander, set out from Baliti in 25 bancas. Due to the unseaworthiness of some of the craft and the enemy's sea blockade, however, only 35 men, including the division commander, reached Tabogon, Cebu.[26]

On 8 January, Thirty-fifth Army completed an over-all plan covering the proposed evacuation.[27] The following provisions were included:[28]

1. The 1st Division, 5th Infantry, and 380th Independent Infantry Battalion will move to northern Cebu.

2. The 26th Division will transfer to the Bacolod sector of Negros.

3. The 41st and 77th Infantry Regiments will, if possible, return to the command of the 30th Division on Mindanao. Otherwise, these regiments will proceed either to Tagbilaran on Bohol, or to Dumaguete on Negros.

4. Elements of the 102d Division on Leyte will return to their original stations in the Visayan area.

5. The 16th Division, 68th Brigade and remnants of other units will remain on Leyte to cover the evacuation.

6. Army headquarters, with parachute regiments, will transfer to Cebu and subsequently to Mindanao, the withdrawal commencing after the majority of forces have withdrawn.

Implementing orders were to be issued according to the availability of transportation and

23 Lt. Gen. Takaji Wachi, Chief of Staff, Thirty-fifth Army, accompanied this staff group as far as Cebu on his way to Fourteenth Area Army headquarters in Baguio, where, late in January, he presented a report on the Leyte situation. On 20 Feburuary, Lt. Gen. Wachi was appointed Deputy-Chief of Staff, Southern Army. Simultaneously, Maj. Gen. Yoshiharu Tomochika was reappointed Chief of Staff, Thirty-fifth Army, the post he had held prior to the reorganization of Thirty-fifth Army headquarters on 14 November 1944. (Cf. Chapter XIII, n. 186.)

24 Statement by Lt. Col. Katogawa, previously cited.

25 The strength of the Japanese forces assembled in the Mt. Canguipot area as of 2 January 1945 totalled about 10,718, broken down as follows:

Hq., Thirty-fifth Army	300	380th Ind. Inf. Bn. / Camotes Detachment	1,200
1st Division	2,500		
102d Division	2,000	Takahashi Detachment	200
41st Infantry, 30th Div.	268	Ito Naval Landing Unit	250
68th Brigade	4,000		

In addition to these troops, there were an estimated 6,500 isolated northeast of Ormoc and along the Lubi trail, making a total strength of about 17,200. (1) Document Files, Thirty-fifth Army Headquarters, op. cit. (2) Statement by Maj. Yamazaki, previously cited.

26 *Dai Hyakuni Shidan Sakusen Keika Gaiyo* 第百二師團作戰經過概要 (Summary of Operations, 102d Division) Home Depot Division, Apr 46, p. 24.

27 In view of many indeterminate factors, the Army staff envisaged this plan primarily as a general guide. Moreover, the contents were not given wide distribution because of the harmful effect the disclosure might have on some of the Leyte forces. Those units which were most fit for subsequent operations were, in general, to have the highest priority in the evacuation. (Statement by Maj. Gen. Tomochika, previously cited.)

28 Philippine Operations Record, Phase Three, op. cit. Vol. II Suppl. pp. 209–11.

the progress of the assembly of units in the Canguipot area. The 1st Division was ordered on 8 January to board evacuation craft near Abijao, the first group embarking on the 12th. Enemy attacks in this area made it necessary for the next three echelons to load at Tibur on the 15th, 17th, and 20th.[29]

By this time the enemy had become aware of the attempted evacuation and subjected both embarkation and debarkation points to terrific air and surface attacks, destroying most of the Japanese craft. Concurrently, enemy ground forces launched an intensive clean-up campaign throughout northwestern Leyte to wipe out the remaining pockets of resistance. The evacuation therefore had to be temporarily suspended.[30]

Meanwhile, all Army units had finally received the order to withdraw into the mountainous area west of Highway 2. The remnants of the Imabori Detachment, Mitsui Shipping Unit, and 77th Infantry Regiment, numbering about 1,500, required almost a month to reach the Canguipot area, arriving early in February.[31] The 26th Division, after initiating the retreat from the Lubi trail area during mid-January, encountered the enemy at Valencia on 28 January. Following a brief skirmish, the Japanese withdrew and subsequently sifted through the enemy lines in small groups, approximately 800 survivors reaching the Canguipot area late in February.[32] The remnants of the 16th Division were still en route.

By late February, it had become apparent to Army headquarters that no further large-scale evacuations could be undertaken. Although the tattered forces on Leyte still controlled some of the commanding terrain along the shoreline and could, on occasion, infiltrate through enemy patrols to gain the coast, enemy air and torpedo boat attacks had all but wiped out Japanese surface transportation.

Nevertheless, in the middle of March, Lt. Gen. Suzuki decided to attempt the transfer of his headquarters in accordance with the plan of 8 January.[33] After dispatching an advance party on 17 March, Lt. Gen. Suzuki relinquished direct command over the Leyte remnants to Lt. Gen. Makino, commander of the 16th Division. The 300-odd survivors of this division were just arriving in the Canguipot area. The Army Commander, on 23 March, set out for Cebu by small boat.[34]

The weak and exhausted remnants still left on Leyte had long since ceased offensive operations. Lt. Gen. Makino, nevertheless, held his troops in the Canguipot area until late April when the shortage of food and relentless enemy clean-up campaigns forced them to retreat northeast to the vicinity of Bagacay. Shortly thereafter the order to disperse was issued, and centralized command ended.[35]

---

29 About 743 men were successfully evacuated from Leyte in the four echelons. No other 1st Division troops were evacuated. (Statement by Col. Junkichi Okabayashi, Chief of Staff, 1st Division.)

30 Philippine Operations Record, Phase Three, op. cit. Vol. II Suppl., pp. 203–6, 211–2.

31 (1) Ibid., p. 193. (2) Statement by Maj. Yamazaki, previously cited.

32 (1) Philippine Operations Record, Phase Three, op. cit. Vol. II Suppl., p. 194. (2) Statement by Maj. Yamazaki, previously cited.

33 One of the major factors influencing this decision was a request from the Rear Area Command Post early in March urging Lt. Gen. Suzuki to transfer his headquarters to a point which would permit him to resume control of the entire Army area. This request was prompted by reports of the enemy landing on Palawan on 28 February. (Statement by Maj. Gen. Tomochika, previously cited.)

34 Statement by Maj. Yamazaki, previously cited.

35 Following this dispersal, continuing enemy clean-up campaigns and the acute shortage of food resulted in further losses until, by the date of final surrender, only a few survivors were still alive.

Enemy Occupation of Palawan

Meanwhile, the enemy had already launched his campaign to clean out Japanese garrisons throughout the central and southern Philippines.[36] Palawan, the first to feel the impact of this mop-up,[37] had by late February lost much of its importance to the Japanese. In possession of the enemy, however, the island would provide an excellent base from which air domination of the South China Sea could be made more complete.

Extending almost 300 miles in a northeasterly direction from the northern tip of Borneo, Palawan had been used by the Japanese as a stepping-stone in sea and air traffic between the Philippines and Borneo.[38] Two airstrips, a seaplane base, and a good harbor at Puerto Princesa formed the hub of this important communications link. Moreover, the long coastline provided a sheltered passage and innumerable safe anchorages for the small supply and troop-carrying craft plying between Luzon and Borneo.

Enemy land-based aircraft operating from Leyte airfields had largely deprived Palawan of its value to Japanese communications. However, the heterogeneous collection of small Japanese units stationed in the area remained there since they could not be transferred elsewhere because of the lack of shipping.[39]

On 21 February, Maj. Gen. Takeshi Kono, 77th Infantry Brigade commander, dispatched a message to the Palawan garrison from his headquarters at Bacolod, ordering the various

---

36 The staff of the Rear Area Command Post had prepared, late in January, the following estimate of probable enemy intentions with regard to the Army area:

1. The enemy will probably initiate a clean-up campaign in the Visayan area about the time he gains control of Luzon.

2. At this time it cannot be determined whether or not the American forces will begin a Visayan campaign and a Mindanao offensive simultaneously. It is considered more probable, however, that operations will be undertaken against the former prior to launching an attack against Mindanao.

This estimate was promptly dispatched to all elements of the Army except those on Leyte. Army headquarters on Leyte had prepared in December an estimate not in agreement with this one, (Cf. p. 490) but it was not disseminated to subordinate commands because of the inadequate communication between Leyte and other islands at that time. (Statement by Lt. Col. Katogawa, previously cited.)

37 Earlier in February enemy forces cleared out the northern tip of Samar, which was within the Thirty-fifth Army area. The Japanese forces remaining on the island were not under the control of the Army, however. The 2d Battalion, 9th Infantry, was under command of the Noguchi Detachment on Bicol, and the 1st Battalion, 26th Independent Mixed Regiment, remained under command of its parent regiment, then en route to the Manila area. Both of these battalions had been ordered by mid-January to transfer to Luzon but were unable to make the move because of lack of transportation.

38 Statement by Maj. Gen. Toshio Nishimura, Deputy-Chief of Staff, Fourteenth Area Army.

39 Total strength of the Japanese forces in the Palawan area, 27 February, was about 1200. Order of battle was as follows:

Puerto Princesa sector:
131st Airfield Bn.
4th Co., 174th Ind. Inf. Bn.
1st Ind. Maint. Unit
Misc. Army air-ground personnel
Misc. Navy personnel

Pandanan Island (southern tip of Palawan):
One plat., 4th Co., 174th Ind. Inf. Bn.
En route to Puerto Princesa from Dumaran Island:
3d Co. (less two plats.), 174th Ind. Inf. Bn.
Coron:
One plat., 3d Co., 174th Ind. Inf. Bn.
Misc. Navy personnel

(1) Puerto Princesa Defense Operations Order No. 10, 22 Jan 45. Eighth Army ADVATIS Translation 8AAAE-T 36, 8 Mar 45. (2) Statements by 1st Lt. Toru Ogawa, commander, Supply Company, 131st Airfield Battalion, and Medical Probational Officer Toshimichi Urata, attached to 3d Co., 174th Independent Infantry Battalion.

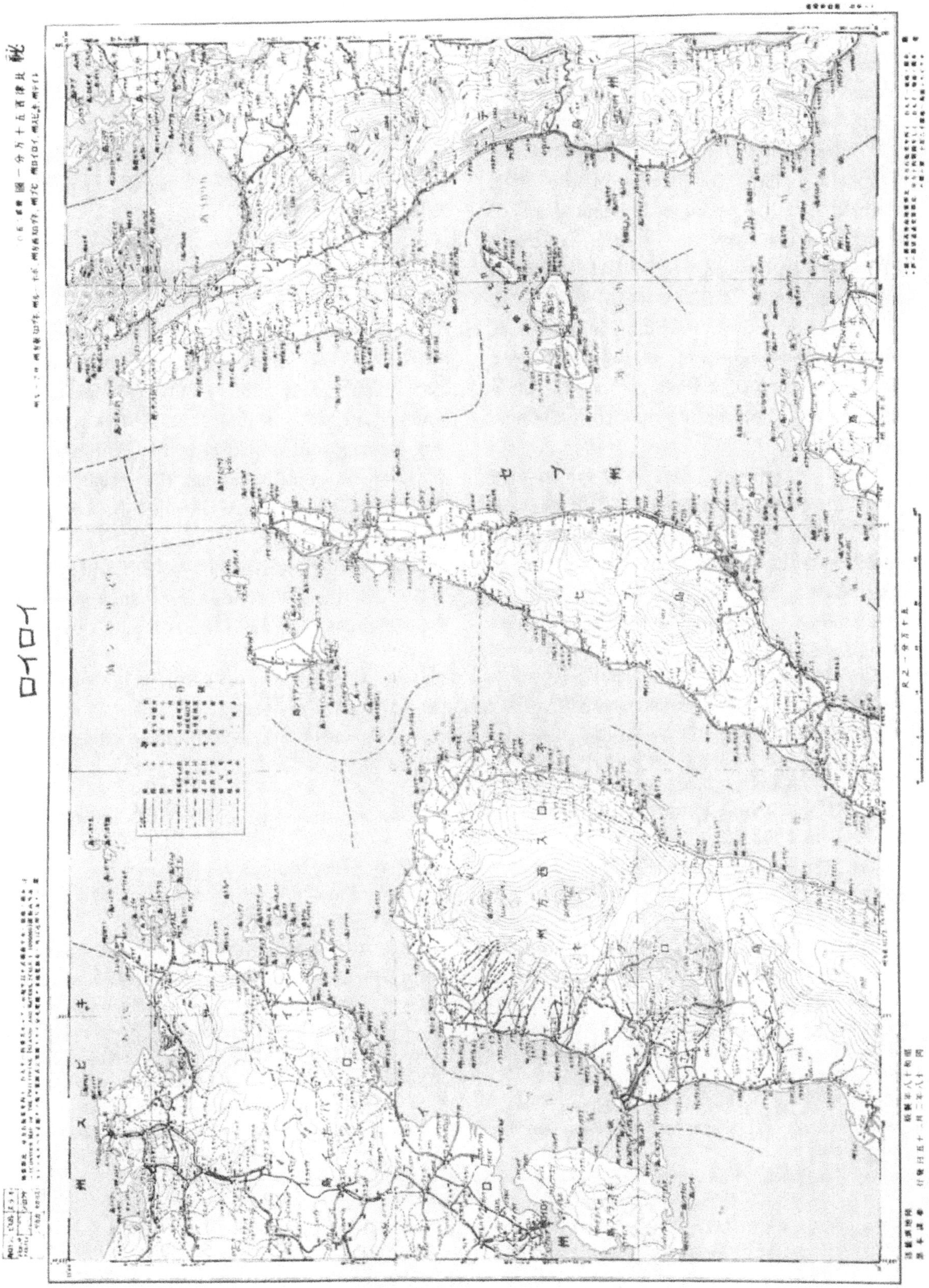

PLATE NO. 130
Map of Central Philippines Printed in 1943 by Land Survey Department, Imperial General Headquarters

Army and Navy units stationed there to be placed under unified command and to prepare for probable enemy invasion.[40]

Under the provisions of this order, Capt. Chokichi Kojima, commander of the 131st Airfield Battalion, assumed command of all Army and Navy forces at Puerto Princesa.[41] Defensive preparations, which had been started almost a month earlier,[42] were accelerated. The main defenses were being constructed in the mountains northwest of Irahuan, about six miles northwest of Puerto Princesa. Only small outposts were to be stationed in the port city.

These preparations were still in progress when the enemy commenced a terrific two-day air bombardment followed, on the morning of 28 February, by an intensive shelling of coastal installations. Later the same morning, American forces swarmed across the beaches unopposed,[43] the outposts having already withdrawn to the main positions. Shortly thereafter, on 2 March, the main defenses also began to crumble, and the surviving Japanese troops dispersed into the mountains to fight as guerrillas.[44]

Zamboanga—Sulu Archipelago

On the heels of the invasion of Palawan, other American forces swung south to invade Zamboanga and the Sulu Archipelago. Capture of the air and naval bases along this strategic link between the Philippines and Borneo would complete the isolation of the Philippines from the Japanese-held areas to the south.

Joint defense preparations of army and navy forces in the Zamboanga area were under the supervision of Lt. Gen. Tokichi Hojo,[45]

40 Radio Message Batsu 7 (77th Inf. Brig.), No. 104. Dispatched by Maj. Gen. Kono at Bacolod to Obayashi Unit, Palawan, 21 Feb 45. Eighth Army ADVATIS Translation 8AAAE-T 51, 21 Mar 45.

41 On 15 February, the Army guard units on Pandanan and Dumaran Islands were ordered to withdraw as soon as possible and assemble at Puerto Princesa. Palawan Sector Unit Order, 15 Feb 45. Eighth Army ADVATIS Translation 8AAAE-T 36, 8 Mar 45.

42 Puerto Princesa Defense Operations Order No. 10, 22 Jan 45. Eighth Army ADVATIS Translation 8AAAE-T 36, 8 Mar 45.

43 The American invasion force consisted of one escort carrier, two battleships, seven destroyers, three transports and over 40 LST's. The landing began at 1050 hours on 28 February. *Shukan Joho Dai Hyakujunigo* 週間情報第百十二號 (Weekly Intelligence Reports, No. 112) Imperial General Headquarters, Army Section, 3 Mar 45.

44 (1) The 3d Co. (less two plats.), 174th Ind. Inf. Bn., which had withdrawn from Dumaran Island shortly after receiving the order of 15 February, arrived at Fondeado Island, about 12 miles north of Puerto Princesa, on 28 February. After hearing the heavy bombardment then progressing in the Puerto Princesa area, the company landed on Palawan near Babuyan, four miles north of Fondeado Island. En route to Puerto Princesa, it fought off guerrillas and foraged for food until late June, when the company finally dispersed and melted into the mountains. (Statement by Medical Probational Officer, Urata, previously cited.) (2) The Japanese garrison on Pandanan Island could not comply with the order of 15 February to withdraw because of lack of transportation. (Statement by 1st Lt. Ogawa, previously cited.) (3) On 9 April an enemy task force invaded Coron and annihilated the small Japanese garrison on the island.

45 The official chains of command of the Japanese forces in the Zamboanga area were as follows:

54th Ind. Mixed Brig. —	Thirty-fifth Army (Leyte)
33d Naval Garrison Unit —	32d Special Base Force (Davao)
Navy air-ground personnel —	Southern Philippine Airfield Unit (Davao)
Army air ground personnel—	2d Air Division (Bacolod)

(Statement by Maj. Yasura Hanada, Staff Officer, 54th Independent Mixed Brigade.)

commander of the 54th Independent Mixed Brigade and senior officer in the area.[46] Main defenses were being prepared in depth in the hills between Recodo and a point north of Zamboanga.[47] (Plate No. 131)

Before these fortifications overlooking the beaches were completed, a large enemy task force appeared in Basilan Strait off Zamboanga. After heavily bombarding the coast, this force put assault troops ashore near San Mateo Point, about four miles west of Zamboanga, on the morning of 10 March. Small Japanese outposts stationed along that sector of the beach withdrew inland after a brief skirmish.[48]

The enemy quickly occupied Zamboanga airfield. On the following day, 11 March, attacks were initiated against the main defenses. The strongest thrust, beginning on 15 March, was aimed at the center of the line. By the 23d the Japanese were driven from these positions.

With communications now severed between the right and left flank positions, Lt. Gen. Hojo sought desperately to hold the positions on the east behind Zamboanga. During the following week, however, as the enemy intensified his attacks, the Japanese positions began to crumble rapidly. Finally, on 31 March, the brigade commander ordered a general withdrawal north along the peninsula.[49]

With the western tip of Mindanao in their possession, General MacArthur's forces now drove south into the Sulu Archipelago to complete the wedge between the Philippines and Borneo. Initial enemy landings were made on the virtually undefended islands of Sanga Sanga and Bongao, in the Tawitawi group, on 2 April.[50] A week later the enemy moved against Jolo.

The key Japanese position in the archipelago, Jolo, was garrisoned by the 55th Independent Mixed Brigade under Maj. Gen. Tetsuzo Suzuki.[51]

---

46 Order of battle of the forces in the Zamboanga area at this time was as follows:

54th Ind. Mixed Brig. (less elms.)	Misc. Navy air-ground personnel
33d Naval Garrison Unit	Misc. Army air-ground personnel

In addition, the 54th Independent Mixed Brigade posted the following detachments:

One co., 362d Ind. Inf. Bn. —	Malangas
Two cos., 361st Ind. Inf. Bn. —	Sindangan

Ibid.

47 Originally Lt. Gen. Hojo planned to defend positions on the beaches of Zamboanga. Upon receipt late in December of the Thirty-fifth Army order requiring sustained resistance, however, the positions were moved farther to the rear. Ibid.

48 Ibid.

49 The main strength of the brigade retreated via several routes through the mountains to Sibuco by mid-May. At this point the Japanese forces were again attacked by the enemy and once more forced to withdraw into the interior. When the enemy attacked the remnants on 26 May at Anungan, the brigade was finally forced to disperse. Ibid.

50 The Tawitawi group lay outside the Thirty-fifth Army zone of responsibility, coming under the Thirty-seventh Army on Borneo. Until early February the islands had been garrisoned by the main strength of the 25th Independent Mixed Regiment, which was then withdrawn to Borneo by order of Thirty-seventh Army. Sanga Sanga was entirely undefended, while Bongao was garrisoned by a small naval detachment of about 250 men of the 33d Naval Garrison Unit. On 27 April, survivors of the latter garrison escaped by boat to Sandakan. (Statements by Col. Shinshichi Iemura, commander, 25th Independent Mixed Regiment and Lt. Comdr. Yoshizumi Asano, Chief Paymaster, 33d Naval Garrison Unit.)

51 Total strength of the Jolo garrison at this time numbered about 3,000 Army personnel and 500 attached naval personnel. The order of battle was as follows:

55th Ind. Mixed Brig.	365th Ind. Inf. Bn. (less one co.)
363d Ind. Inf. Bn.	Misc. naval personnel

Dokuritsu Konsei Dai Gojugo Ryodan no Sento 獨立混成第五十五旅團の戰鬪 (Battle Report of 55th Independent Mixed Brigade) Nov 47. Unpublished memoirs of Maj. Tokichi Temmyo, commander, 365th Independent Infantry Battalion.

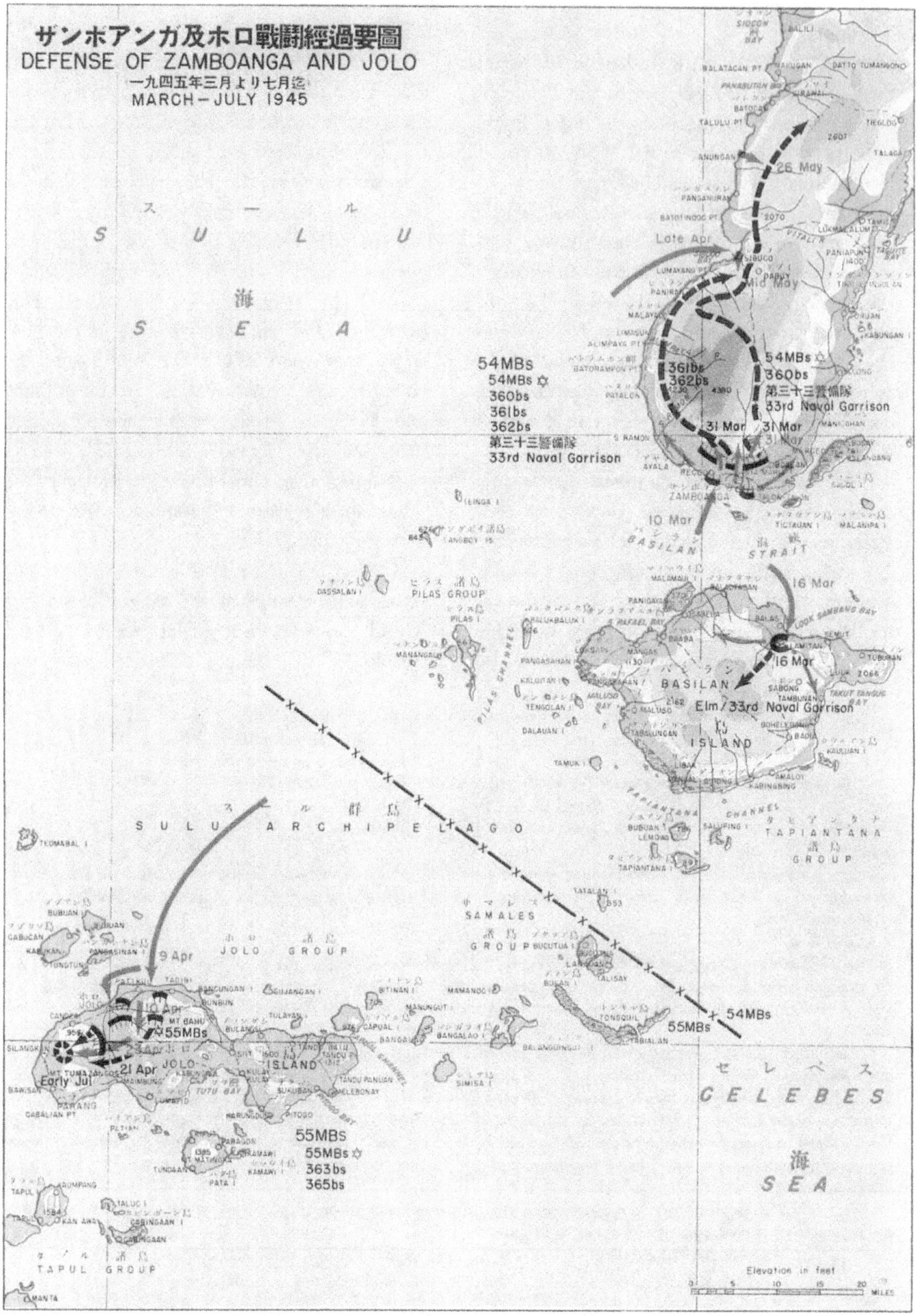

PLATE NO. 131
Defense of Zamboanga and Jolo, March—July 1945

Under the general policy of fighting delaying actions throughout the central and southern Philippines, Maj. Gen. Suzuki had abandoned earlier plans for defending the beaches and organized inland positions in depth back of the capital city of Jolo, on the north side of the island. Forward positions were prepared on scattered hills in the open terrain to the east of the city, covering the airfield, while the main defenses were farther to the rear in the more mountainous areas to the southeast and southwest.[52] The field fortifications were fairly well completed by early April, with sufficient rations stored near each position to last about one month.

During the night of 8–9 April, American naval craft stood into the coast, and on the following morning, after a heavy naval and air bombardment, an estimated 3,500 troops, supported by tanks, landed near Tagibi, eight miles east of Jolo. Before nightfall a smaller enemy detachment had effected a second landing nearer the city.

The Japanese forces were extremely hard pressed from the first because of the open terrain in which the forward positions were located. These were quickly overrun, and the enemy continued his frontal assaults in conjunction with guerrilla attacks from the rear. The day after the landing, the main enemy force began attacking the 365th Independent Infantry Battalion's main-line positions southeast of Jolo. Within ten days the battalion was driven from these positions, and on 21 April, Maj. Gen. Suzuki ordered it to withdraw to Mt. Tumazangos, about five miles southwest of Jolo.

Before the battalion could complete its movement, the enemy launched an attack against the Japanese positions in the Mt. Tumazangos sector. Enemy pressure eased after the early part of May, however, and the Japanese forces found time to reorganize their ranks.[53] Thereafter they were subjected only to sporadic guerrilla attacks.

Panay and Western Negros

Even before the conclusion of the enemy campaign to gain control of the strategic island links between the Philippines and Borneo, General MacArthur had launched another series of amphibious operations in the central Philippines. The objective of these operations was to wipe out remaining Japanese concentrations in the Visayan group of islands.

The largest concentration of army forces in the Visayan area was on northwestern Negros, formerly Fourth Air Army's most important base south of Luzon. Eight airstrips dotted the northwest part of the island, mostly centered around the town of Bacolod.[54] As of mid-March, Japanese strength in the area totalled almost 12,000, the major part of which was

52 Ibid.

53 Early in July, because of a shortage of rations and guerrilla activities in the area around Mt. Tumazangos, the Japanese forces moved to the west coast. One month later the remnants moved again, this time to Mt. Bahu, about seven miles southeast of Jolo. By the date of surrender, 16 September, only 87 Japanese still survived. Most of these were victims of various tropical diseases, and all were suffering from malnutrition. Ibid.

54 Of the eight airstrips on northern Negros, seven, located at Bacolod, Silay, Fabrica, Talisay, Tanza, Manapla and Salavia, were used by the 2d Air Division. The eighth, near Binalbagan, was used by the Navy.

made up of airfield troops and miscellaneous ground personnel of the 2d Air Division.[55] Regular ground combat troops were limited to two infantry Battalions of the 77th Infantry Brigade, 102d Division, although provisional combat units were rapidly being organized from the air ground and service units.[56] Lt. Gen. Takeshi Kono, brigade commander, directly controlled these troops in addition to exercising general command of Army ground forces on Panay and adjacent islands.[57]

Lt. Gen. Kono had recognized some time earlier that a prolonged defense of the airfields themselves was not feasible because of the flat terrain. To comply with the terms of the Thirty-fifth Army order requiring sustained resistance, he had decided to deploy the main strength of his troops in the nearby mountains, from which raiding squads could be dispatched.[58] (Plate No. 132) Small detachments were to be stationed near probable landing points, in towns, and near airfields[59] to delay the enemy's approach to the main defenses.

On the neighboring island of Panay, the Japanese garrison also prepared its main defenses in the mountains.[60] Simultaneously with an enemy landing, the main strength was to withdraw from Iloilo to the Bocari area, about 20 miles northwest of Iloilo.[61] Only small outposts were stationed along

55 Lt. Gen. Seiichi Terada, 2d Air Division commander, was the senior officer in the area at this time. He was expected, however, to transfer his headquarters at any time. The 2d Air Division had been placed under the command of the Third Air Army in Singapore about mid-February. Thereafter, the three-fold mission of the air division was to reconnoiter the enemy actions in the Leyte area, secure the airfields in the Visayan area, and transport air personnel to other areas. Permission for the transfer was not forthcoming until mid-March. Finally, on 27 March, Lt. Gen. Terada moved his headquarters to Mindanao, leaving all air ground units in the Negros area under the command of Lt. Gen. Kono. (Statement by Lt. Col. Nonogaki, previously cited.)

56 Order of battle of the Japanese forces on northern Negros at this time was as follows:

Hq., 77th Inf. Brig.	Hq., 6th Air Sector Unit
172d Ind. Inf. Bn. (less 2d Co.)	One AAA btry.
354th Ind. Inf. Bn.	Three airfield bns.
Elms. of 355th Ind. Inf. Bn.	Four airfield cos.
Transport Unit, 102d Division	Air supply and repair depots
Six cos., 102d Division Engineer Unit	Elms. of misc. air service units
Misc. service units	Elms. of naval ground units
Elms. of shipping units	

(1) Unit Disposition, undated. Eighth Army ADVATIS Translation 8AAAE-T 76, 13 Apr 45. (2) Yuge Force, Tokaido Sector Defense Plan, 10 Mar 45. Eighth Army ADVATIS Translation 8AAAE-T 74, 11 Apr 45. (3) Interrogation of Lt. Col. Shigekatsu Aritomi, Staff Officer (Operations), 102d Division. 10th Information and Historical Service, Headquarters Eighth United States Army.

57 Cf. p. 493. Kono was promoted to Lt. Gen. effective 1 March.

58 The outline of this defense plan had been completed during December, but because the guerrillas were in control of the mountain area selected, the Japanese were unable to proceed with the construction of fortifications. It was not until early in February, when the guerrillas were driven from the area, that the ground forces finally began preparation of defenses. (Statement by Maj. Kiyoshi Suzuki, Staff Officer (Supply), 2d Air Division.)

59 The Army airfields at Silay and Salavia were to be used as long as possible. 6th Air Sector Operations Order No. 290, 28 Mar 45. Eighth Army ADVATIS Translation, 8AAAE-T 72, 9 Apr 45.

60 After the American landing on Mindoro in mid-December, Maj. Gen. Kono's staff discussed the possibility of transferring to Negros the remainder of the Panay garrison which had been reduced by the dispatch of the 171st Independent Infantry Battalion to Leyte in October. Due to inadequate shipping facilities and enemy air interdiction, however, the project was dropped. (Statements by Lt. Col. Katogawa and Lt. Col. Aritomi, both previously cited.)

61 The Japanese forces on Panay at this time totalled about 2,500. The order of battle was as follows:

Iloilo sector:	San Jose sector:
170th Ind. Inf. Bn. (less 4th Co.)	4th Co., 170th Ind. Inf. Bn.
1st Co., 171st Ind. Inf. Bn.	Elms. of air ground units
Misc. units	

(Interrogation of Capt. Sadayoshi Ishikawa, attached to 170th Ind. Inf. Bn. 10th Information and Historical Service, Headquarters Eighth United States Army.)

the southern coast.

It was against Panay that the initial enemy assault was launched on 18 March. Japanese outposts west of Iloilo fell back as the enemy advanced along the coast to enter the capital city on the 20th. The small Japanese forces had already evacuated the city as planned and were withdrawing to the mountain redoubt, where they completed their assembly by 24 March.[62]

The enemy now turned to attack Negros. On 28 March, Lt. Gen. Kono's headquarters received radio reports from Japanese transport planes flying out of Bacolod that an enemy convoy had been sighted near Inampulugan Island in Guimaras Strait.[63] No further word of this enemy force was forthcoming until the following day, when it was reported that enemy tank-infantry teams were approaching Bacolod along the coastal highway from the south.[64]

Lt. Gen. Kono immediately issued orders for a series of counterattacks to be carried out on the night of the 30th against the enemy column entering Bacolod. The Japanese were unable to retard the advance, however, and the Bacolod detachment withdrew, followed on 2 April by the outposts in Talisay and Silay. Other small Japanese garrisons dispersed in the coastal area were quickly overcome and driven into the mountains. Lt. Gen. Kono had already ordered the Fabrica garrison on 1 April to displace south and join the main strength in the mountain redoubt east of Bacolod.[65]

While this concentration progressed, the enemy began attacking the prepared positions. After crushing the outpost line between Guimbalaon and Concepcion about 4 April, the enemy pressed on to assault the main defenses on 10 April. Bitter fighting continued in this area until late April, when the Japanese lines slowly crumbled under relentless attacks supported by overwhelming air and artillery bombardment.

During the following month, the Japanese were driven from successive defense lines until the enemy had overrun most of the prepared positions. By late May, food and supply shortages began to reduce sharply the combat efficiency of the Japanese forces. Organized resistance ended about 10 June, when the remnants began a general retirement deeper into the mountains of northern Negros to continue operating as guerrillas.[66]

The Panay garrison had meanwhile undergone, beginning about mid-April, intermittent enemy ground assaults supported by air and artillery bombardment. These attacks ceased altogether in late May. Thereafter the Japanese were harassed only by occasional guerrilla raids.

Southern Visayas

Concurrently with his drive into Panay and Western Negros, General MacArthur also moved clean-up forces against the southern

62 Shortly after this withdrawal was completed, the battalion commander ordered the forces near San Jose to assemble in the main positions. This detachment, after brief skirmishes with guerrilla forces, arrived in the Bocari area on 10 April. Ibid.

63 Statement by Lt. Col. Aritomi, previously cited.

64 Faulty communications between Lt. Gen. Kono's headquarters and the Japanese company stationed near Pulupandan, where the enemy landed, prevented transmission of reports of the enemy invasion. Ibid.

65 These forces completed the transfer by 5 April. (Interrogation of Lt. Col. Aritomi, previously cited.)

66 Ibid.

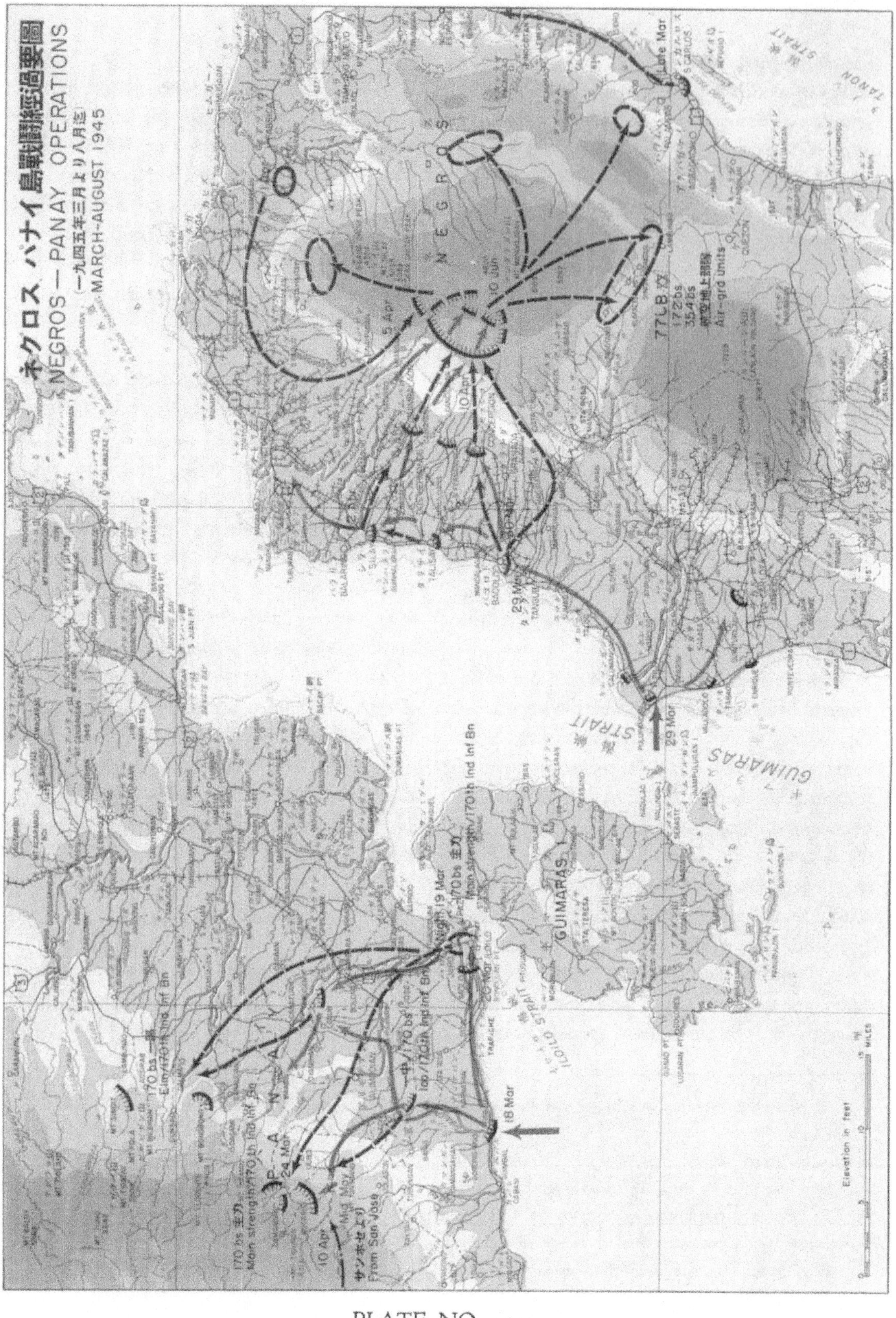

PLATE NO. 132
Negros—Panay Operations, March—August 1945

islands of the Visayan group. The most important and strongly defended of these was the island of Cebu.

Cebu had been the focal point of Thirty-fifth Army's rear support of the decisive battle on Leyte. Consequently, the Army forces stationed there were largely service elements, and little had been accomplished along the line of defensive preparations. The single combat unit of consequence, the 173d Independent Infantry Battalion, carried out guard duties during much of this period.

Following the termination of the Leyte battle, these troops assisted in the evacuation of the Leyte forces. When the evacuation was suspended late in January,[67] the Army forces accelerated the preparation of defenses near Cebu City and on the north end of the island, where about 1700 troops under command of the 1st Division were located.

The Navy's 33d Special Base Force had meanwhile been hard at work since mid-November completing operational preparations.[68] Rear Adm. Harada had selected the heights northwest of Cebu City to conduct the final stand, thereby hoping to deny the enemy early and unhampered use of the airfields and also to make unnecessary the early abandonment of the politically valuable capital city. (Plate No. 133)

The greatest effort was concentrated in constructing defenses in the Navy sector within the circular positions on high ground to the northwest of the airfields. Simultaneously, first-line positions were being prepared along the base of the heights.[69]

As the Army and Navy forces, aggregating about 14,500 on the entire island,[70] rushed to complete these defenses, the enemy, on 23 March, commenced a terrific and sustained aerial bombardment of the positions near Cebu City.[71] Three days later, following an intensive naval

---

67 The evacuation from Leyte of Lt. Gen. Fukue, 102d Division commander, created an unfavorable reaction in Army headquarters even though permission had been granted. The Army commander temporarily relieved Lt. Gen. Fukue of his command and placed him in restriction in the Cebu Army hospital. He was subsequently released on 24 March by the Army commander and restored to his command. (Statement by Maj. Gen. Tomochika, previously cited.)

68 Although Rear Adm. Harada had been informed by Lt. Gen. Suzuki on 1 November that he would command army units in the Cebu sector, he nevertheless was unable to relieve these units of their role in support of the decisive battle operations. Rear Adm. Harada therefore exercised very little actual control over the army units in the Cebu sector until February. (Statement by Capt. Shigaki, previously cited.)

69 First-line positions were originally prepared along the beach. In mid-February, however, it was decided to abandon these positions and construct a new first line farther inland.

70 (1) Sketch Map Showing Troop Dispositions on Cebu and Mactan Islands, owner and unit unknown. Eighth Army ADVATIS Translation 8AAAE-T 85, 29 Apr 45. (2) Statement by Col. Okabayashi, previously cited.

71 Order of battle of Japanese forces in the vicinity of Cebu city on 25 March was as follows:

Hq., 102d Division
- Hq., 78th Inf. Brig.
- 173d Ind. Inf. Bn.
- One composite co., 54th Ind. Mixed Brig.
- Elms. of miscellaneous service units

Hq., 33d Special Base Force
- 33d Special Base Force
- Misc. naval ground units

Central Philippine Airfield Unit

Shipping units
- Replacement Unit, 1st Shipping Engr. Unit
- 8th Surface Trans. Bn.
- Cebu Branch, Shipping Depot
- Elms. of miscellaneous shipping units

Elms. of Army air service units

(1) Summary of Operations, 102d Division, op. cit., pp. 28–9. (2) Sketch Map Showing Troop Dispositions on Cebu and Mactan Island, owner and unit unknown. Eighth Army ADVATIS Translation 8AAAE-T 85, 29 Apr 45.

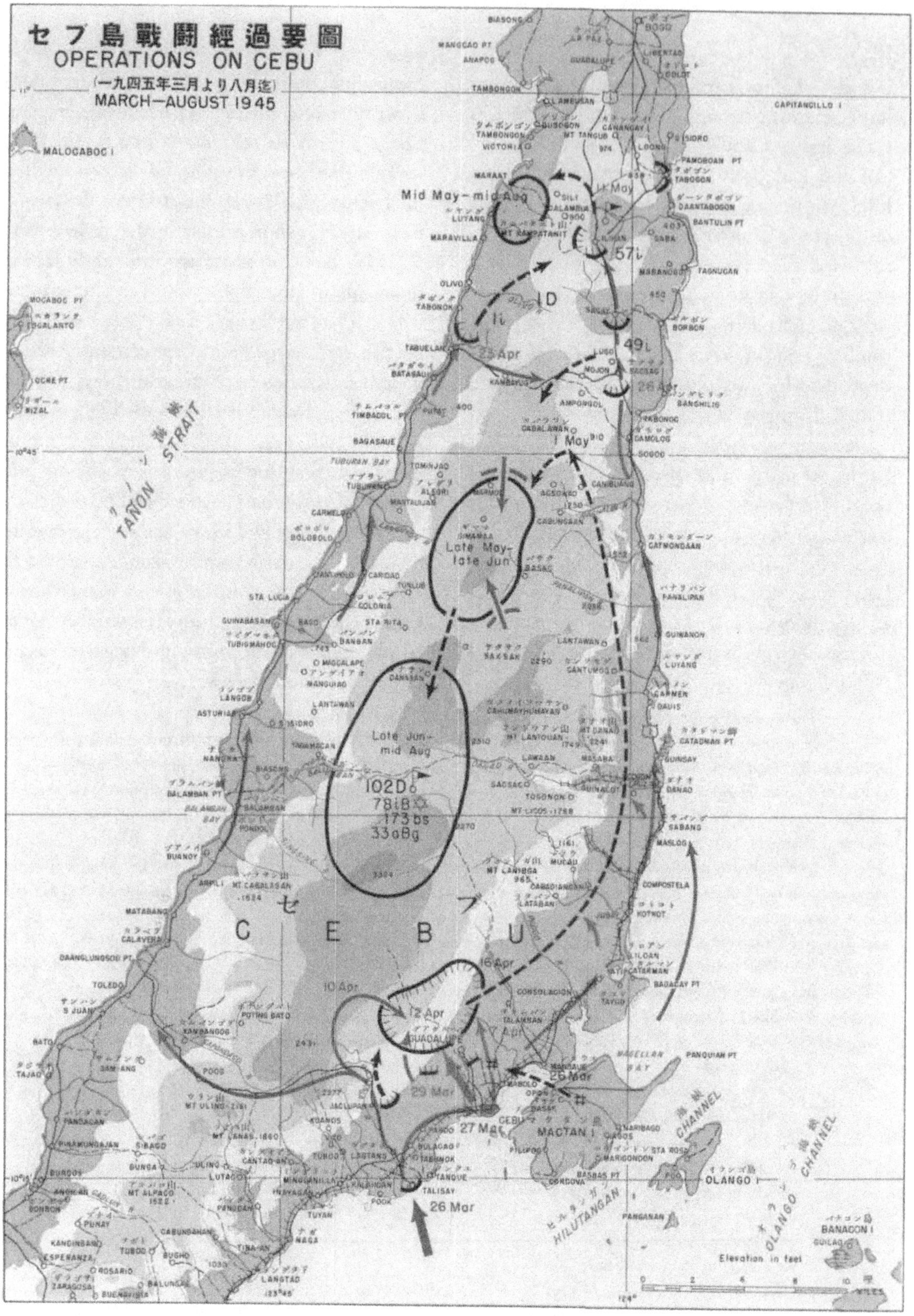

PLATE NO. 133
Operations on Cebu, March—August 1945

shelling, the enemy put ashore at Talisay an invasion force estimated at about one division.[72] The weak Japanese outposts were quickly overcome as the enemy moved out of the beachhead toward Cebu City.

On the same day that the landing took place, Lt. Gen. Suzuki, Thirty-fifth Army Commander, arrived in Cebu City en route from Leyte to Mindanao. Before departing for northern Cebu to continue his journey, Lt. Gen. Suzuki ordered Lt. Gen. Fukue, 102d Division commander, who had been placed in over-all command of the forces in the Cebu City area on 24 March,[73] and Rear Adm. Harada to defend the present positions in the heights sector as long as possible. Other provisions of this order included the following:[74]

1. When the continued defense of these positions becomes impossible, withdraw to northern Cebu and continue resistance.

2. After joining with the 1st Division, the headquarters of the 102d Division and its original subordinate units will move to Negros and command the units stationed on that island. Upon the departure of the 102d Division headquarters, the 1st Division commander will assume command of all units on Cebu.

Shortly after receiving this Army order, Lt. Gen. Fukue issued the following plan of operations for the defense of the positions northwest of Cebu City:[75]

1. The greatest effort will be made to secure the highest mountain position and the naval positions.

2. Withdrawal from these positions is tentatively scheduled for about 20 April. Rations for one month must be preserved for consumption following the withdrawal.

3. Movement from the present positions will be to the northeast via Liloan to northern Cebu.

4. Forces making the withdrawal will reassemble in the sector north of the Lugo-Tabuelan road.

The enemy had meanwhile entered Cebu City on 27 March. After overrunning the airfields, tank-infantry teams forged ahead to the newly-prepared first-line positions, which the Japanese defended briefly before retreating up the heights to the main positions.

Frontal attacks against these well-organized positions began on 29 March. Repeated assaults during the first week of April failed to score more than small local penetrations on the southeast flank defended by the naval forces. On the southwest flank, however, the Army positions were breached at several points. Unremitting air and naval bombardment supporting the enemy ground attacks also inflicted heavy damage on the Japanese positions and rear installations and seriously lowered the morale of the troops.[76]

Having registered only limited gains by his frontal attacks on the left flank, the enemy, on 7 April, began to encircle the naval positions on the northeast. This advance threatened to sever the route over which the Japanese planned to withdraw northward. The

72 The enemy landing on Cebu terminated the operation of Japanese midget submarines in the Mindanao Sea. Eight of these craft had been assigned to the 33d Special Base Force between October 1944 and January 1945. Using Dumaguete as an advance base, these midget submarines launched torpedo attacks against enemy convoys moving through the Mindanao Sea. Three were destroyed prior to the Cebu invasion, two were destroyed while attacking the Allied invasion convoy, and the remaining three were scuttled by the Japanese. (1) *Hito Homen ni okeru Sensuikan Sakusen* 比島方面に於ける潜水艦作戦 (Submarine Operations in the Philippine Area) 2d Demobilization Bureau, Dec 47, pp. 41–4. (2) Statement by Capt. Shigaki, previously cited.

73 Cf. n. 67.

74 Outline of Directives for Future Operations of the Thirty-fifth Army in the Visayan Sector, 23 Mar 45. Eighth Army ADVATIS Translation 166LD-T 87, 1 Jun 45.

75 Summary of Operations, 102d Division, op. cit., p. 35.

76 Statement by Capt. Shigaki, previously cited.

naval forces therefore launched a prompt counterattack, which temporarily halted the enemy drive.

An enemy force now moved around the southwest flank and gained the rear of the Japanese positions. Attacks were launched against the highest point of the defenses on 10 April. Although this particular sector had been recognized by the Japanese as the most vulnerable point because of unfavorable defensive terrain, shortage of combat troops had made it impossible to adequately defend the area. The enemy consequently broke through successive defense lines within a few days. By 12 April the defenses on the northwest flank of the circular position were almost completely overrun.

In view of the rapidly deteriorating situation, Lt. Gen. Fukue decided on 12 April that the withdrawal to northern Cebu should be initiated. Orders were issued the following day,[77] and the general retirement actually began on the night of the 16th. During the following two weeks, the Japanese remnants, harassed by enemy ground pursuit, air attack, and naval shelling, retreated to the north. By avoiding a pitched battle, the survivors reached the area south of the Sacsac-Tabuelan road by 1 May.[78]

Swift enemy columns had meanwhile raced ahead to occupy this road, preventing the Japanese retreating northward from effecting a juncture with the weak forces north of the road. The plan to transfer the 102d Division elements to Negros was therefore dropped.

When the enemy began attacking positions north of the road on 23 April, the 1st Division, which had been mainly preoccupied since its arrival from Leyte with fighting guerrillas, was ill prepared to meet the enemy drive. The defenses near the coast first gave way, followed two weeks later by the crumbling of the center of the line. By 11 May further defense of the area had become impossible. Lt. Gen. Tadasu Kataoka therefore ordered all units to disperse into the mountains west of Ilihan.[79]

Shortly thereafter the enemy extended his clean-up operations to the mountain area south of the road, where the remnants of the 102d Division had taken temporary refuge. By late June the survivors in this area were forced to disperse deeper into the mountains to the south.

Concurrently with the fighting on Cebu other enemy forces had launched two new invasions to complete the conquest of the neighboring island. The first landing was made on 11 April at Tagbilaran, Bohol, and was unopposed by the small Japanese garrison which had already withdrawn into the mountains.

The second landing was made shortly thereafter on 26 April near Dumaguete, eastern Negros, which had been an advance base for the midget submarines operating out of Cebu and a liaison point between the Visayas and Mindanao. This Japanese garrison was also too weak to permit more than a brief defense of the forward positions before withdrawing into the interior. In mid-June the survivors broke up into small groups to operate as guerrillas.[80] Organized resistance had now ceased throughout the Visayas.

---

77 Summary of Operations, 102d Division, op. cit., p. 38.

78 Ibid.

79 Statement by Col. Okabayashi, previously cited.

80 Interrogation of 1st Lt. Komei Fujitomi, 174th Independent Infantry Battalion. 10th Information and Historical Service, Headquarters Eighth United States Army.

Defense Preparations on Mindanao

With the enemy moves against Cebu and Bohol in late March and early April, the main portion of Mindanao to the east of the Zamboanga Peninsula remained the only important sector of the Philippines which had not yet felt the impact of General MacArthur's liberating forces. Here, Lt. Gen. Gyosaku Morozumi, in over-all command of army forces east of Lake Lanao, sped final preparations to meet invasion.

Army and navy forces deployed in this sector totaled about 58,000 far more than had been available for the defense of any other island. In addition to these regular forces about 17,000 Japanese civilians, of whom about 5,000 were employed as laborers, still resided in the large Japanese colony at Davao.[81]

In spite of this numerical strength, an inadequate road net dividing the area into two distinct and widely separated tactical sectors militated against a rapid transfer of troops to meet an enemy invasion.[82] In central Mindanao, extending from Cotabato, on the eastern shore of Moro Gulf, to Surigao on the north, Lt. Gen. Morozumi exercised direct command over the ground forces. The remnants of the Army air forces in this area were still under the command of Lt. Gen. Seiichi Terada, commander of the 2d Air Division, who had only recently transferred from Bacolod.[83] (Plate No. 134)

Defense of the remainder of the area was charged to Lt. Gen. Jiro Harada, commander of the 100th Division.[84] In addition to the

81 (1) Strength of Japanese Forces in the Philippines, 1944–45, op. cit. (2) Statements by Col. Yamazu and Lt. Col. Nonogaki, both previously cited, and Lt. Gen. Jiro Harada, commander, 100th Division.

82 That portion of Highway 3 between Kabacan and Kibawe was generally impassable to truck transportation even though Japanese engineers devoted considerable effort toward keeping it in repair. Less important roads were even more unfit for traffic, with frequent disruption caused by swollen streams or swampy areas. (Statement by Lt. Gen. Gyosaku Morozumi, commander, 30th Division)

83 Order of battle of the Japanese forces in central Mindanao in mid-April was as follows:

30th Division	19th Shipping Engr. Regt.
Hq., 30th Division	57th Fld. Rd. Const. Unit
3d Bn., 41st Infantry	22d Fld. Duty Unit
1st Bn., 77th Infantry	Misc. units
74th Infantry	2d Air Division
30th Fld. Arty. Regt. (less 3d Bn.)	Hq., 2d Air Division
30th Engr. Regt.	Hq., 31st Air Sector Unit
Division troops	32d, 33d, 37th, 102d, 103d, 133d Airfield. Bns.
166th Ind. Inf. Bn., 100th Div.	Misc. air units
7th and 9th Cos., 100th Div. Engr. Unit	

(1) 30th Division Operations Order No, A-80, 25 Dec 44. Eighth Army ADVATIS Translation. 8AAAE-T 75, 12 Apr 45. (2) Statement by Lt. Col. Nonogaki, previously cited.

84 Order of battle of the Japanese forces in the Davao and Digos areas in mid-April was as follows:

100th Division	
Hq., 100th Division	Division troops
75th Inf. Brig.	Misc. units.
163d, 164th, 165th Ind. Inf. Bns.	Hq., 13th Air Sector Unit
76th Inf. Brig.	8th, 12th, 14th, 26th Airfield. Bns.
167th, 168th, 352d, 353d Ind. Inf. Bns.	Misc. air units
100th Div. Arty. Unit	32d Special Base Force
3d Bn., 30th Fld. Arty. Regt., 30th Div.	1st, 2d, 3d, 4th (Prov.) Bns.
100th Div. Engr. Unit (less two cos.)	Southern Philippine Airfield Unit

(Statements by Lt. Gen. Harada and Lt. Col. Nonogaki, both previously cited and Comdr. Shinji Saji, Staff Officer (Operations), 32d Special Base Force.)

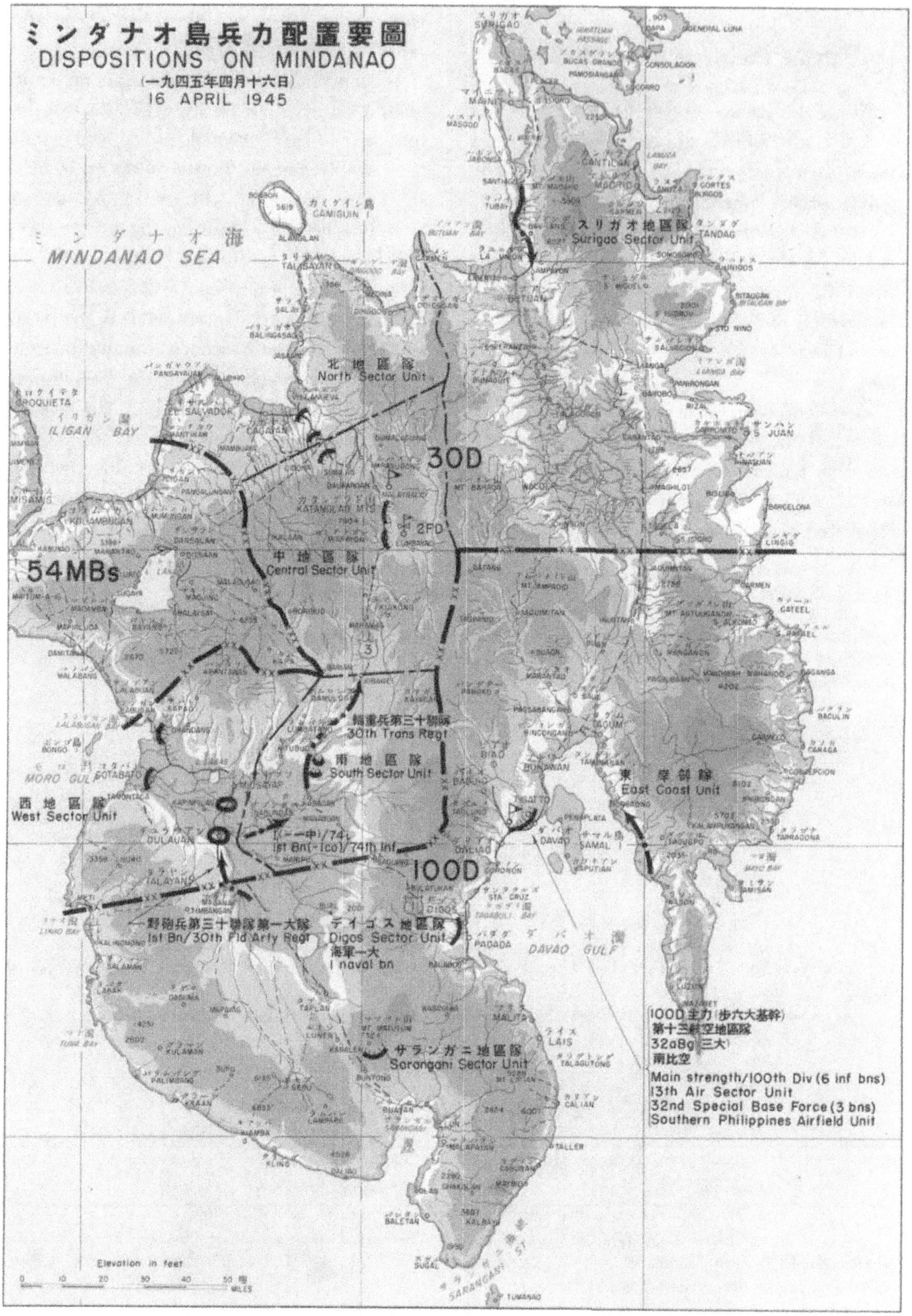

PLATE NO. 134
Dispositions on Mindanao, 16 April 1945

troops now under his command, about 7,700 Navy ground personnel were stationed in the Davao and Digos areas under Rear Adm. Naoji Doi, 32d Special Base Force commander, who had organized these troops into four provisional battalions. Also in this area were the 13th Air Sector Unit, under the 2d Air Division, and the remnants of the Southern Philippine Airfield Unit, still under Southwest Area Fleet control. All of these Navy and Air force units were to come under operational control of the senior Army commander in case of an enemy invasion.

The launching of the enemy clean-up campaign in the southern Visayas during late March and severe air raids on key points in Mindanao early the following month stimulated belief that an invasion of Mindanao was imminent. Lt. Gen. Morozumi estimated that the main enemy thrust would be delivered against either Cagayan or Davao, probably the former, with a possible secondary landing on a smaller scale in the Cotabato area.[85] No plans were entertained, however, for strengthening the Cagayan area at the expense of the defenses near Davao.

Operational preparations of the 100th Division, the main strength of which had been concentrated near Davao since September 1944, were well advanced by mid-April.[86] Lt. Gen. Jiro Harada still believed that the enemy would launch an amphibious invasion of the Davao area, the main assault falling just east of Daliao.[87] Positions were therefore constructed facing southeast on both sides of the Davao River. Disposition of the division at this time was as follows:[88]

Davao Area
- *Right Sector Unit (west of Davao River):*
  - *Hq., 76th Inf. Brig.*
  - *168th (less one co.), 352d, and 353d Ind. Inf. Bns.*
  - *Two engr. cos.*
- *Left Sector Unit (east of Davao River):*
  - *Hq., 75th Inf. Brig.*
  - *164th and 165th Ind. Inf. Bn.*
  - *One arty. btry.*
  - *One engr. co.*
- *Arty. Unit:*
  - *100th Division Arty. Unit (less one btry.)*
- *Reserve:*
  - *One co., 163d Ind. Inf. Bn.*
  - *167th Ind. Inf. Bn. (less two cos.)*

Digos Sector Unit: (89)
- *163d Ind. Inf. Bn. (less one co.)*
- *3d Bn. (less two Btrys.), 30th Fld. Arty. Regt., 30th Div.*

Sarangani Sector Unit: (90)

85 This represented a slight modification of the estimate made by Lt. Gen. Morozumi in December 1944, at the time of the American landing on Mindoro. At that time he had considered it slightly more probable that the main enemy landing in an invasion of Mindanao would be made in the Davao area rather than on the north coast around Cagayan. A secondary landing at Cotabato or along Sarangani Bay was also anticipated in this estimate. (Statement by Col. Yamazu, previously cited.)

86 By the end of April all division units had completed strong field fortifications. (Statement by Col. Muneichi Hattori, Chief of Staff, 100th Division.)

87 Statement by Lt. Gen. Harada, previously cited.

88 Ibid.

89 The Digos Sector Unit was disposed to counter an enemy amphibious invasion of the beaches in that area. In the event of such a landing, the Japanese were to prevent the enemy's unrestricted use of Padada airfield southwest of Digos. Ibid.

90 In the period following receipt of the Army order late in December, the Japanese garrison in the Sarangani Bay area was progressively reduced in view of the revised estimate that the enemy would not now launch a major assault in that area. The 74th Infantry, 30th Division, was pulled out during late December and January, the 100th Division replacing that regiment with the 167th Independent Infantry Battalion (less one co.) in January. About 20 March this battalion (less two cos.) was ordered back to the Davao area. (Statements by Lt. Gen. Harada and Col. Yamazu, both previously cited.)

One co., 167th Ind. Inf. Bn.
East Coast Unit (en route to Davao):[91]
One co., 167th Ind. Inf. Bn.
One co., 168th Ind. Inf. Bn.

Plans had been prepared to incorporate the naval forces into the existing tactical grouping of the 100th Division. The 1st and 2d Naval Battalions were to be attached to the Left Sector Unit, the 3d Battalion to the Right Sector Unit, and the 4th Battalion to the Digos Sector Unit.[92]

In central Mindanao, where there were fewer troops to cover a far greater area, operational preparations of the 30th Division had been virtually stalemated because of other urgent requirements. An acute shortage of foodstuffs prevailing in the upper highlands along Highway 3 since August 1944 made it necessary to employ troop units to gather provisions, particularly rice in Cotabato Province, and to transport them over the inadequate road net to the Malaybalay sector.

With the realization early in April that an enemy attack was imminent and that the danger to Cagayan was increasing, Lt. Gen. Morozumi called a halt to these foraging activities and speeded up the division regrouping which had been started more than three months previously. By mid-April, three widely scattered elements were moving toward the northern part of central Mindanao.

The 3d Battalion, 41st Infantry, had recently arrived in the Ampayon area en route to Balingasag from Surigao.[93] The main strength of the 1st Battalion, 74th Infantry, had just been ordered to suspend the collection of supplies in the Dulawan area and was near Kabacan en route north to join the main strength of the regiment. Farther south, the 1st Battalion, 30th Field Artillery Regiment, was in Dulawan on its way north from Sarangani Bay.[94] The tactical grouping of the 30th Division was now as follows:

Surigao Sector Unit:
3d Bn., 41st Inf.
North Sector Unit:
30th Rcn. Regt.
1st Bn., 77th Inf.
Central Sector Unit:
74th Infantry Regt.
Under Division Control:
30th Fld. Arty. Regt. (less 3d Bn.)
30th Engr. Regt. (less 2d and 3d Cos.)
30th Trans. Regt., reinf.
Misc. division troops
South Sector Unit:[95]
One co., 1st Bn., 74th Inf.
Two field duty cos. (temporary)
Misc. units
West Sector Unit:
166th Ind. Inf. Bn.

Fighting in Central Mindanao

The northward shift of 30th Division troop strength to meet an expected invasion of the Cagayan area was in full swing when the enemy

91 Both companies in the East Coast Unit began withdrawing to Davao late in February. Lack of transportation and guerrilla interference prevented their arrival before the surrender. (Statement by Lt. Gen. Harada, previously cited.)

92 One other provisional battalion, mainly comprised of the Southern Philippine Airfield Unit, was to remain under the direct command of Rear Adm. Doi. (Statement by Comdr. Saji, previously cited.)

93 *Dai Sanju Shidan Hyo Butai Kodo Gaiyo* 第三十師團(豹)部隊行動概要 (Summary of Operations, 30th Division (Hyo) Force) Home Depot Division, undated.

94 This battalion had been ordered, on or about 20 March, to move to Malaybalay where it was to revert to regimental control. Ibid. Cf. n. 90.

95 Early in April the 30th Engineer Regiment commander, who was then in command of the South Sector Unit, was ordered to proceed north and supervise the construction of defenses in the Dalirig area. Simultaneously, four engineer companies were attached to the 30th Transport Regiment which was then engaged in transporting supplies from Kabacan to the Malaybalay area. (Statement by Col. Yamazu, previously cited.)

struck suddenly at the division back door with an amphibious landing near Parang on 17 April. (Plate No. 135) Outposts of the 166th Independent Infantry Battalion had already withdrawn from Parang to positions farther inland.

Before reports of this latest American move reached 30th Division headquarters, a firm beachhead had been established in the Cotabato area. As the enemy forces advanced east along the highway and up the Mindanao River, Japanese detachments destroyed the highway bridges and withdrew. The main strength of the 166th Independent Infantry Battalion also withdrew from the Midsayap area to prepared positions on the north flank of the enemy column.[96]

Only small miscellaneous Japanese forces now stood between the enemy and Kabacan, the vital road net junction from which Highway 1 continued southeast to the Davao area and Highway 3 ran north through central Mindanao. These elements engaged the enemy south of Kabacan but were quickly overcome and driven north of the Pulangi River on 23 April.[97]

By this time 30th Division headquarters had swung into action in an attempt to stem the enemy advance before it moved north into central Mindanao. Col. Koretake Ouchi, commander of the 30th Engineer Regiment, was ordered about 22 April to take command of operations along the southern approaches and to annihilate the enemy south of the east-west line running through Omonay. For this mission he was given command of the South Sector Unit and the equivalent of about a reinforced infantry battalion.[98]

On 23 April, Maj. Gen. Tomochika, Chief of Staff of Thirty-fifth Army, arrived at the 30th Division command post at Impalutao from Agusan, where he had arrived on 21 April from Cebu. Maj. Gen. Tomochika immediately conferred with Lt. Gen. Morozumi regarding implementation of the Army policy of protracted resistance. As a result of this discussion, Lt. Gen. Morozumi concluded that it was necessary to strengthen the Malaybalay area. He therefore ordered the transfer of the 1st Battalion, 77th Infantry, from the North Sector Unit to the Central Sector Unit and the movement of this battalion to new positions near Malaybalay. About the same date Lt. Gen. Morozumi ordered the Surigao Sector Unit to move from Ampayon to the Waloe area, where it was to assist in carrying out the division's self-sufficiency program.

Although the 30th Division commander thus adopted measures to strengthen the area of final resistance, he still considered the landing near Parang a secondary invasion and estimated that the main enemy landing would soon be launched in the Macajalar Bay area.[99] The preponderance of division strength was therefore retained in the north.

With regard to the protracted defense of Mindanao, the following plan, based on the policy under study by the Army Commander

---

96 *Dai Hyaku Shidan (Kyo) Butai Kodo Gaiyo* 第百師團(據)部隊行動概要 (Snmmary of Operations, 100th Division (Kyo) Force) Home Depot Division, undated.

97 The 1st Battalion, 30th Field Artillery Regiment en route to Malaybalay was cut off by the rapid enemy advance and, on the night of 18 April, began withdrawing on its own initiative to the Sarangani Bay sector. In early May, this battalion was again placed under the 100th Division and merged into the Sarangani Sector Unit. (Statement by Maj. Makoto Takatsuka, commander, 1st Battalion, 30th Field Artillery Regiment.)

98 The forces, in addition to the South Sector Unit, included the main strength of the 1st Battalion, 74th Infantry, the 3d Company, 30th Engineer Regiment, and two companies, 100th Division Engineer Unit. 30th Division Operations Order No. A-106. X Corps ADVATIS Translation XCAET 0058, 29 Apr 45.

99 Statement by Col. Yamazu, previously cited.

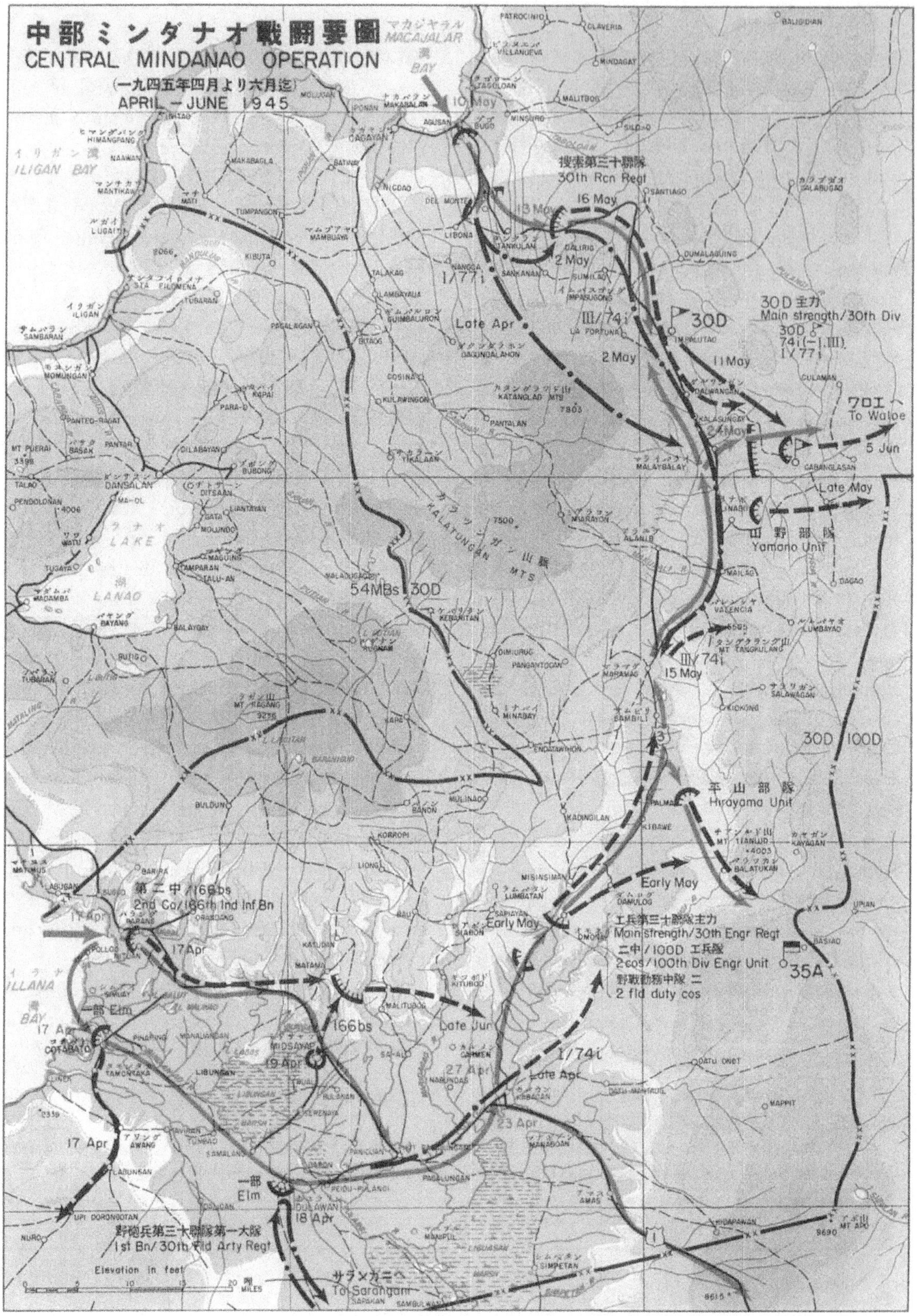

PLATE NO. 135
Central Mindanao Operation, April—June 1945

since the Leyte campaign, was evolved on 28 April:[100]

I. General Policy

The Army will repel the invading enemy. When this is no longer possible, the forces will remain in the southern Philippines, strive to reduce the enemy's combat strength and prepare to support future Japanese counterattacks, thus contributing to the over-all operations.

II. Outline of Operations

1. A portion of the 30th Division will cooperate with the 100th Division in operating against the enemy advancing from Cotabato Province to the Davao area. The main body of the 30th Division will endeavor to reduce the combat strength of the enemy penetrating into central Mindanao and will establish a primary self-supporting region in the area east of Malaybalay. Preparations will be made later to establish a second self-supporting area near Waloe. The forces preparing defenses in the north near Del Monte will be reduced as much as possible.

2. The 100th Division, concurrently commanding the Army air ground and service troops and Naval forces, will attempt to crush the enemy invading the Davao plain from its present positions. It will establish strong self-supporting regions in the areas north of Davao, north and west of Tamugan, and on the east slope of Mt. Apo. The mouth of the Tagum River and its vicinity will be secured by portion of its strength.

3. The Yamano Unit (Army air ground service troops in central Mindanao) will establish a self-supporting region in the area east of Mailag.[101] *It will then expand the region to the south and will cooperate with operations of the 30th Division.*

4. The Hirayama Unit (57th Field Road Construction Unit and attached miscellaneous elements) will defend the rear of the 100th Division and the western front of the Army self-supporting region against American forces expected to penetrate into the Davao area along the Palma-Tamugan road.

After implementing orders were issued by Lt. Gen. Morozumi, Maj. Gen. Tomochika moved to Basiao to establish the Thirty-fifth Army Command Post and await the Army commander, who had not yet arrived from Cebu.[102]

Meanwhile the enemy paused only briefly at Kabacan before launching a drive southeast along Highway 1 toward Davao on 23 April. Four days later another column resumed the advance north along Highway 3 in what appeared to be an offensive in considerable strength. Lt. Gen. Morozumi promptly decided upon a further shifting of troops in the north. He ordered the 3d Battalion, 74th Infantry, to withdraw from Dalirig to Malaybalay and the 30th Reconnaissance Regiment to pull back from Del Monte airfield to occupy the positions near Dalirig vacated by the 3d Battalion.

The forces under Col. Ouchi were by this time proving inadequate to cope with the enemy column advancing along the highway. Successive road blocks and delaying detachments were overrun as the enemy pushed rapidly north. To reinforce this sector with additional infantry strength, the division commander, on 2 May, ordered the 3d Battalion, 74th Infantry, which had just arrived at Impalutao, to continue marching south to Maramag and block the enemy advance. Shortly after arriving, the battalion, on 15 May, launched a counterattack against the head of the enemy column but the Japanese were quickly pushed aside. The road

100 Philippine Operations Record, Phase Three, op. cit. Vol. II, Suppl., pp. 220–1.

101 Shortly thereafter, on 3 May, the 2d Air Division headquarters transferred to Singapore in compliance with a Third Air Army order. On 25 May the headquarters was deactivated. (Statement by Lt. Col. Kazuyoshi Obata, Staff Officer (Operations), 2d Air Division.)

102 Lt. Gen. Suzuki, Thirty-fifth Army Commander, left Cebu Island on 10 April with Maj. Gen. Tomochika and other members of Army headquarters in five bancas. The bancas became separated while en route to Mindanao. After the surrender it was learned that the Army Commander had been killed during a strafing attack. (Statement by Maj. Gen. Tomochika, previously cited.)

to Malaybalay from the south now lay open.

The 30th Division had, in the meantime, come under attack from the north. An enemy force, estimated at about one regiment, landed at Agusan on 10 May and pushed rapidly inland, opposed only by small outposts. The following day, Lt. Gen. Morozumi ordered the division headquarters, which had been preparing to move, to transfer from Impalutao to the strongpoint east of Malaybalay.

The main defenses of the 30th Reconnaissance Regiment near Dalirig were brought under attack on 13 May. Three days later, after the enemy had overrun the last of these positions, Lt. Gen. Morozumi ordered the remnants to withdraw to the east of Malaybalay.[103] Shortly thereafter, on 23 May, the two enemy columns from the north and south effected a juncture at Impalutao.

On the following day the first-line defenses east of Malaybalay were subjected to a terrific artillery bombardment. The enemy followed this with a strong ground assault which the Japanese were able to check only temporarily. By 5 June, the division was driven back to positions near the Pulangi River. On that date Lt. Gen. Morozumi ordered a withdrawal to the Waloe area.[104] The 30th Division was thereafter unable to mount more than small-scale raiding forays.[105]

Defense of the Davao Area

Meanwhile the enemy column driving southeast from Kabacan had pushed swiftly along Highway 1 before the Japanese could organize effective delaying action and entered Digos on 27 April. Outposts of the Digos Sector Unit delayed the enemy force briefly before retreating to the main positions in the mountains to the northwest.[106] (Plate No. 136) A smaller force, dispatched from the Right Sector Unit in Davao to positions near Astorga, was overcome following a short engagement. The enemy column moved on into Davao on 3 May.

The enemy had meanwhile begun attacking the first-line positions of the Right Sector Unit on both banks of the Talomo River on 1 May.[107] As more enemy elements closed into the Davao area, these frontal attacks spread to the east to include the Left Sector Unit on 3 May.

After moving division headquarters from Mintal to Lapuy, Lt. Gen. Harada took steps on

103 On 1 June, a report was received by 30th Division that the regiment had arrived in the vicinity of Dalwangan. No further word, however, was forthcoming from this unit. (Statement by Lt. Gen. Morozumi, previously cited.)

104 By 5 June the 30th Division had only sufficient rations to last about two weeks. Nevertheless, it was believed that the trek over the mountains could be made in about two weeks. Actually, the survivors who finally reached the Waloe area marched almost two months through the trackless mountain wilderness. When they finally arrived, they found the 3d Battalion, 41st Infantry, under heavy attack by an enemy force which had moved south. Shortly thereafter, the remnants of the Japanese force began to move to the east coast but were able to march only eight miles southeast before the surrender. Ibid.

105 The Yamano Unit had meanwhile withdrawn to the eastern bank of the Pulangi River following a brief engagement near Linabo late in May. These air ground service troops remained in the mountains until the surrender. Philippine Operations Record, Phase Three, op. cit. Vol. II, Suppl., pp. 227-8.

106 In accordance with a 100th Division order, the main strength of the Digos Sector Unit had displaced from Padada to the vicinity northwest of Digos between 23 and 25 April. On 1 May these newly organized defenses were attacked, and four days later the Digos Sector Unit began to withdraw northward toward the Japanese main positions. (Statement by Col. Rinzo Shizuru, commander, 163d Independent Infantry Battalion.)

107 The 32d Special Base Force and the Army air service troops in Davao had meanwhile been placed under the operational control of the 100th Division. (Statement by Lt. Gen. Harada, previously cited.)

3 May to reinforce the defenses in the vicinity of the Talomo River, where the major enemy effort appeared to be aimed. The 167th Independent Infantry Battalion (less two companies) was removed from division reserve and attached to the Right Sector Unit. Simultaneously, the 168th Independent Infantry Battalion was ordered to pull back from its forward positions near Bayabas to a line slightly north of Mintal.[108]

During the following week, the enemy attacks grew steadily heavier, requiring additional measures to bolster the right flank. On 11 May, Lt. Gen. Harada ordered Maj. Gen. Muraji Kawazoe, commander of the 75th Infantry Brigade, to relinquish command of the Left Sector Unit to Rear Adm. Doi and redeploy a maximum number of his troops to new positions in the vicinity of Ula. Accordingly, the 164th and 165th Independent Infantry Battalions, each leaving approximately one company in the original positions, were pulled out of the line and moved across the Davao River to the newly-assigned area on 14–15 May, followed on the 16th by Maj. Gen. Kawazoe's headquarters. Two days later the forces on the right bank of the Davao River were reorganized into the Right and Left Front Units.[109]

In the meantime, the fighting near Mintal had continued with unabated fury. The Japanese were driven from the forward positions on 11 May. On the following day, however, the 353d Independent Infantry Battalion launched a determined counterattack and recaptured the defenses. The battle continued until 24 May when the American forces again reoccupied the positions. This time they could not be driven off.[110] On the night of the 25th the Left Front Unit withdrew to new positions between the Right Front Unit and the Left Sector Unit.

Four days later Lt. Gen. Harada, foreseeing eventual independent action by the Left Sector Unit because of the terrain, issued the following order to cover subsequent operations of the Japanese forces on the left bank of the Davao River:[111]

> *Rear Adm. Doi will remain in command of all Army and Navy forces on the left bank of the Davao River and will destroy the enemy in that area. Should the situation require a withdrawal, the forces remaining on the left side of the river will move to the upper*

108 Ibid.

109 Following this transfer the tactical grouping of the 100th Division was as follows:

Left Sector Unit:
- 32d Special Base Force
- Elms., 75th Inf. Brig.
- 13th Air Sector Unit

Left Front Unit:
- Hq., 76th Inf. Brig.
- 167th Ind. Inf. Bn. (less two cos.)
- 352d, 353d Ind. Inf. Bns.
- 12th, 14th Airfield. Bns.
- One btry., 100th Div. Arty. Unit
- Two cos., 100th Div. Engr. Unit
- 3d Naval Bn.

Right Front Unit:
- Hq., 75th Inf. Brig.
- 164th, 165th Ind. Inf. Bns. (less elms.)
- One co., 166th Ind. Inf. Bn.
- 168th Ind. Inf. Bn. (less one co.)
- One btry., 100th Div. Arty. Unit
- One co., 100th Div. Engr. Unit

Under division control:
- 100th Div. Engr. Unit (less elms.)
- Division troops

Ibid.

110 The Digos Sector Unit had meanwhile arrived in the Wangan area on 21 May. The naval forces which had been operating with this unit joined the naval units under Rear Adm. Doi. The Digos Sector Unit (less naval elements) was ordered to occupy positions southeast of Wangan and northwest of Ula and come under the command of the Right Front Unit. Ibid.

111 Ibid.

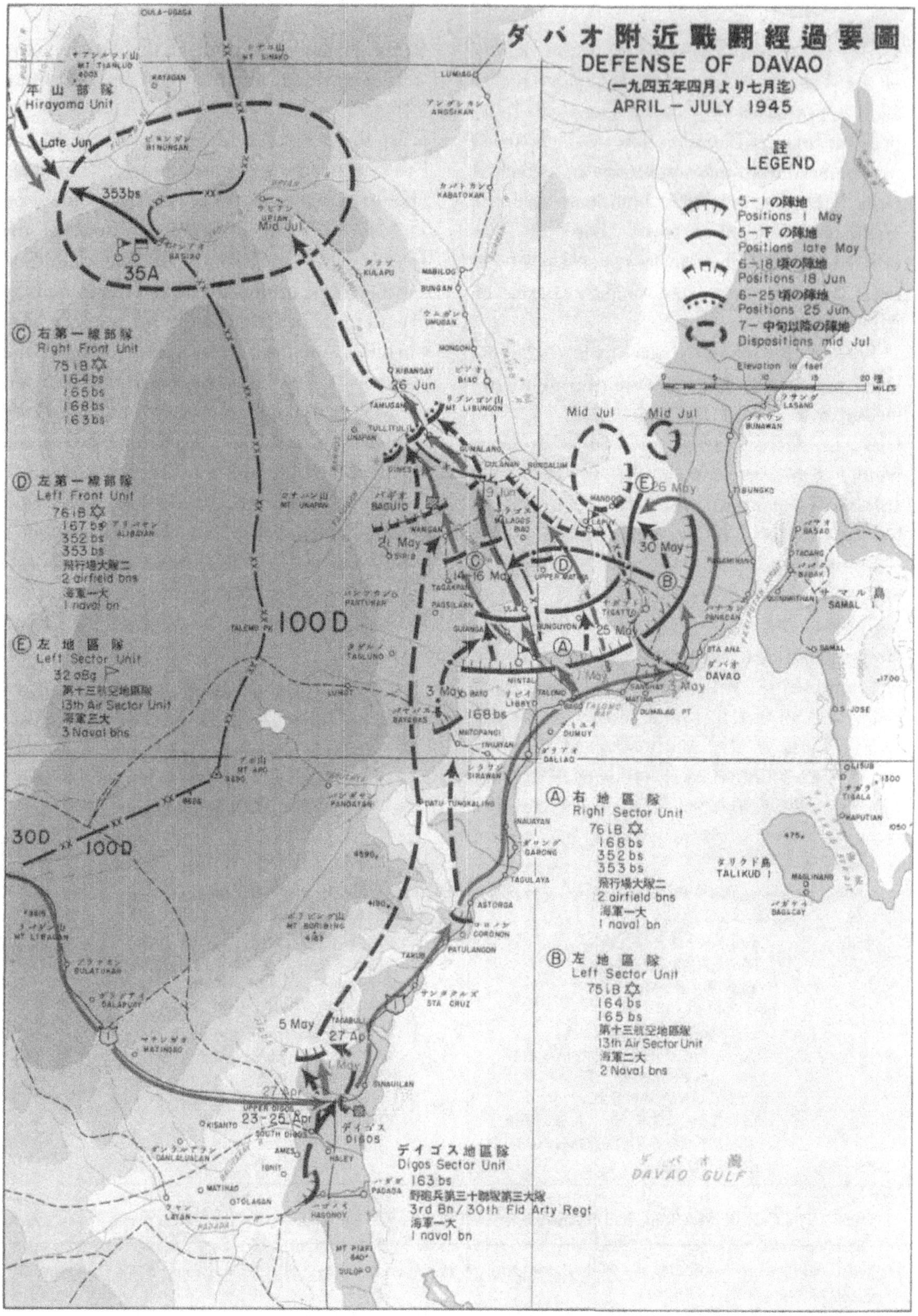

PLATE NO. 136
Defense of Davao, April—July 1945

reaches of the Davao River, where they will carry out self-sustaining operations.[112]

Meanwhile, the enemy had quickly followed up the withdrawal of the Left Front Unit to launch attacks against the Japanese positions near Ula on 27 May. By 1 June these positions were overrun. The enemy pressed forward to capture Wangan on the 16th and Culanan on the 19th. All of the prepared defenses were now in the hands of the American forces.

While the 100th Division remnants assembled along the line of the Tamogan River, Lt. Gen. Harada decided on 26 June to withdraw still farther in order to take advantage of more favorable terrain. Accordingly, he ordered the 76th Infantry Brigade to take up positions about four to six miles north of the Tamogan River and the 75th Infantry Brigade to occupy positions farther north near the Bannos River.[113]

As the withdrawal to the new positions got under way, a fresh enemy threat to the division rear from central Mindanao began to take shape. The Hirayama Unit, defending the area southeast of Palma,[114] was forced back by late June to the juncture of the Pulangi and Kulaman Rivers. Lt. Gen. Harada, pursuant to an Army order, promptly ordered the 353d Independent Infantry Battalion to move to the south bank of the Kulaman River and defend that area.[115]

About 11 July it was reported that the American forces had been replaced by large guerrilla forces. Although the enemy pressure eased thereafter, Lt. Gen. Harada issued orders to disperse and forage for food in the vicinity of Upian, Basiao and the southern bank of of the Kulaman River.[116] The remnants began to move to the designated areas on 15 July. Soon thereafter, enemy attacks ceased altogether.[117]

Recapitulation

By late June, scarcely eight months after the invasion of Leyte, the once powerful Fourteenth Area Army in the Philippines was completely broken. The remnants, consisting of small isolated groups of survivors holed up in mountain areas, were more concerned with foraging for food than with offensive operations.

The strategic consequences of the loss of the Philippines were disastrous in the extreme. The sea lanes between the industrial area in the Homeland and the resources of the southern regions, which had been slowly withering under the relentless attacks of American submarines and long range aircraft,[118] were now

112 On 9 June, the enemy launched an attack against Mandog which the Japanese evacuated two days later. After 14 June, however, the enemy appeared to withdraw from this sector and thereafter no large scale engagements were fought on the left bank of the Davao River. (Statement by Comdr. Saji, previously cited.)

113 Statement by Lt. Gen. Harada, previously cited.

114 The remnants of the 1st Battalion, 74th Infantry, and the South Sector Unit, 30th Division, which had been driven into the mountains along Highway 3 by the swift enemy advance in central Mindanao had meanwhile been attached to the Hirayama Unit early in June.

115 This battalion arrived in the newly assigned area early in July. (Statement by Lt. Gen. Harada, previously cited.)

116 Ibid.

117 The Sarangani Sector Unit, in the vicinity of Mt. Matutum, about 20 miles northwest of Sarangani Bay, was attacked in mid-July. On 1 August, this unit withdrew deeper into the mountains to the northeast. Summary of Operations, 30th Division (Hyo) Force, op. cit.

118 The American carrier task force which sortied through the South China Sea in January 1945 further disrupted the tenuous sea lines of communication. (Statement by Capt. Atsushi Oi, Staff Officer (Operations), General Escort Command.)

completely severed by air and surface patrols operating from Philippine bases. Only the rapidly dwindling stockpiles in Japan and the limited amount of raw materials which could be obtained from China and Manchuria could hereafter feed the flagging war potential.[119] Moreover, the Japanese armed forces helplessly isolated in the southern areas must henceforth depend entirely on the supplies already within those areas.

The Japanese High Command had thrown into the battle for the Philippines the maximum ground, sea and air strength which could be mustered in a desperate but futile attempt to halt the American advance grinding toward the heart of the Homeland. The decisive battle phase ended in failure when the enemy landed on Mindoro, 15 December 1944.

Thereafter the only choice left to these Japanese isolated in the Philippine Islands was to inflict the greatest possible losses and delay on the American forces in the hope of postponing the invasion of the Homeland itself.

The defeat had cost the Japanese military machine dearly. The Navy surface forces suffered such a crippling blow in the battle for Leyte that the Combined Fleet was thereafter able to conduct only small-scale forays.

Losses in ground forces had totalled more than 500,000.[120] Fourteen divisions, four independent mixed brigades, one regular brigade, about 50 provisional battalions organized from miscellaneous Army service units,[121] and numerous naval and air elements had been destroyed in the bitter ground fighting since the enemy landing on the Leyte beaches on 20 October 1944.

The tremendous losses in aircraft and trained flying personnel had influenced the adoption of special-attack, or *tokko*, methods by the Army and Navy air forces. This type of warfare had by now been accepted by the Japanese armed services as an essential feature of the plans for the defense of the Homeland.

---

119 Cf. Chapter XVIII.

120 Breakdown of the Japanese ground forces employed in the defense of the Philippine Islands, 1944–45, is as follows:

Army Ground Troops:
- Fourteenth Area Army Hq.
- Troops under direct Area Army command
- Area Army Line of Communication troops
- Thirty-fifth Army Hq.
- 3d Shipping Transport Hq.
- 1st Division
- 2d Armored Division
- 8th Division
- 10th Division
- 16th Division
- 19th Division
- 23d Division
- 26th Division
- 30th Division
- 100th Division
- 102d Division
- 103d Division
- 105th Division
- 1st Airborne Raiding Group
- 54th Independent Mixed Brigade
- 55th Independent Mixed Brigade
- 58th Independent Mixed Brigade
- 61st Independent Mixed Brigade
- 68th Brigade

Navy Ground Troops:
- Southwest Area Fleet Hq.
- 3d Southern Expeditionary Fleet Hq.
- 31st Special Base Force
- 32d Special Base Force
- 33d Special Base Force

Strength of Japanese Forces in the Philippines, 1944–45, op. cit.

121 Data regarding the number of provisional battalions organized by the Japanese in the Philippine Islands is included in the statement by Maj. Gen. Haruo Konuma, Deputy-Chief of Staff, Fourteenth Area Army. (Cf. Chapter XV, p. 49, 65 and 67.)

CHAPTER XVII

TOKKO "SPECIAL-ATTACK"

Evolution of *Tokko*

The resort to suicide weapons and tactics on an organized and ever-increasing scale in the battle for the Philippines was grim evidence of Japan's mounting desperation in the struggle to halt General MacArthur's penetration of the inner defenses of the Empire. Such methods reflected the strong determination of the Japanese High Command and fighting services to overcome, at any cost, the growing disparity in material fighting power.[1] This type of warfare *à outrancé*, which so vividly expressed the ardent patriotism of the Japanese soldier, was termed *tokko*, or "special-attack", a generic name covering all suicide operations on land, sea, and in the air.[2]

Although special-attack methods had been seriously discussed by Army and Navy field commanders as early as the fall of 1943, they were not employed to any significant extent until the American invasion of Leyte in October 1944. By the end of the Philippines campaign, however, suicide attack, particularly in the air, had become a major combat technique which later played an even more prominent role in the battle for Okinawa and in the defense plans for the Homeland.

The concept of suicide attack developed gradually under the impact of the Army's successive reverses in the Solomons and in New Guinea between 1942 and 1944 and the Navy's inability to halt the enemy's offensive across the Central Pacific. By the time of the invasion of Leyte, the war situation had become so critical that the Japanese High Command was finally forced to sanction and then actively to push the organization o *tokko* units as virtually the last effective means o. stemming the Allied tide.

"Special-attack" found partial expression during the Papuan Campaign in the winter of 1942–43, when raider squads were thrown into night assaults in a desperate attempt to stall the Allied drive on Buna.[3] The ground troops

1 This chapter was originally prepared in Japanese by Col. Takushiro Hattori, Imperial Japanese Army. Duty assignments of this officer were as follows: Chief, Operations Section, Imperial General Headquarters, Army Section, 1 Jul 41—14 Dec 42; Secretary to the War Minister, 14 Dec 42—20 Oct 43; Chief, Operations Section, Imperial General Headquarters, Army Section, 20 Oct 43—12 Feb 45; Commander, 65th Infantry Regiment, 12 Feb 45—15 Aug 45. All source materials cited in this chapter are located in G-2 Historical Section Files, GHQ FEC.

2 The term *tokko* is an abbreviation of *Tokubetsu Kogeki* (特別攻撃) or "special-attack." *Tokkotai* (特攻隊) or "special-attack unit" was the official designation given to any group organized to undertake suicide missions. The term *Kamikaze* was originally limited to naval air units using suicide attack methods but later came to be applied by the western world to special-attack air units of both services.

3 These squads were known as *Teishintai*—raiding units—and *Giyutai*—volunteer units. Initially they consisted of about 20 men and were assigned to such missions as destroying artillery, munition dumps, and other important installations behind enemy lines. The Eighteenth Army in May 1943 activated two provisional raiding companies which subsequently were employed in the Ramu Valley and in the Finisterre Mountain campaigns of September—December 1943. In the summer of 1944, the Inspectorate-General of Military Training compiled a "Raiding Attack Training Guide", based on experience up to that time, for distribution to the entire Army. (Statement by Lt. Col. Kengoro Tanaka, Staff Officer (Operations), Eighteenth Army.)

レイテ島に於ける薫空挺隊の奮戦　藤田嗣治　昭和二〇

Original Painting by Tsuguji Fujita

PLATE NO. 137
Airborne Raiding Unit on Leyte

resorted to this type of offense when weakness in numbers and weapons rendered conventional methods of attack ineffectual. These raider units, however, were not suicide forces in the strict sense; they were hit-and-run "commando" groups which were ordered out with the possibility and expectation that at least some would return. The compelling reason for their employment was nevertheless the same as that which later produced the *Kamikaze*—the necessity of achieving maximum results at a minimum cost in materiel and men.

By early 1943, the Japanese Air forces had lost their initial superiority. Slowly weakened by irreplaceable losses in skilled pilots, they found themselves increasingly incapable of effective opposition. The air war over New Guinea underlined the ever-widening difference in the strengths of the opposing forces, and emphasized the superiority of American aircraft and combat training. Before the end of 1943, the Air forces were definitely on the defensive,[4] and the Japanese found it impossible to resume the initiative at any point.

Under these conditions, front-line airmen who were under constant Allied assault began to feel that orthodox combat tactics were futile against the enemy's more powerfully built and more heavily armed aircraft, particularly the B-17 "flying fortress." Pilots arrived at the conclusion that, barring the unlikely development of a radically new scientific weapon, a superior fighter plane, or a novel defensive technique, the only solution lay in reliance on suicide tactics.[5]

Influenced by this widespread feeling of desperation at the front, the Army Section of Imperial General Headquarters, in May 1944, initiated informal discussions on the subject of organizing a special-attack corps within the Army Air force. At the same time, however, technical studies were begun with a view to devising improved combat techniques and ordnance, and every effort was made to perfect the use of existing weapons.

Following the disastrous Philippine Sea Battle of 19–20 June 1944, Navy Air force officers also began to study special-attack techniques more seriously. This stunning defeat not only cut heavily into the reservoir of skilled fliers which the Navy was trying to rebuild but also gave a fresh demonstration of the weakness of the Japanese air arm. An increasing number of pilots, and on one occasion a flag officer, urged the Navy High Command to consider the adoption of suicide-attack methods, but their recommendations were not acted upon at that time.[6]

---

4 An indication of this defensive attitude was the fact that the emphasis of Army aircraft production in the Homeland, by the autumn of 1943, had shifted from bombers to fighters. More and more fighters were being turned out in proportion to bombers to meet the huge demand by the Army Air forces for interceptors. (Statement by Lt. Col. Kotaro Katogawa, Military Affairs Bureau, War Ministry.)

5 Not only was the fire power of the Japanese fighter planes no match for enemy B-17's and B-24's, but insufficiently trained replacements, frequent breakdowns in aircraft supply, poor quality of aircraft and equipment, and high combat losses contributed to the feeling of helplessness and desperation among front line units. This state of progressive deterioration on New Guinea and in the Solomons led the Japanese to believe that the only way to stop the enemy in the air was for their aircraft to crash deliberately into the invading air formations. (Statement by Lt. Col. Koji Tanaka, Staff Officer (Air Operations), Imperial General Headquarters, Army Section.)

6 Immediately following the Battle of the Philippine Sea, Rear Adm. Obayashi, Commander of the 3d Carrier Division, volunteered to organize special attack units and requested Vice Adm. Ozawa, then Commander of the First Mobile Fleet, for authorization to do so. Ozawa unofficially informed Admiral Toyoda, Commander-in-Chief of the Combined Fleet, but no positive action was taken on the grounds that the formation of special-attack units had to remain on a volunteer basis and could not be ordered. United States Strategic Bombing Survey (Pacific), Naval Analysis Division, *Interrogations of Japanese Officials*. Vol. I, p. 152.

Experimentation with *Tokko* Tactics

With the formulation of the *Sho-Go* Operation plans in July 1944, the Army quietly moved ahead with blueprints for the creation of special-attack air units. By early August plans covering the organization, training, and tactics of such units were completed. Volunteers were being recruited from Army air schools to receive instructions at special training centers[7] and aircraft especially designed for use by *tokko* units were in various stages of planning and production.[8]

The Navy Air forces, meanwhile, continued to adhere to orthodox attack methods. Some Navy fighter pilots were being trained in skip-bombing technique, and the most competent pilots of regular bombing and torpedo attack squadrons were placed in the "T" Attack Force, a newly organized unit designed to carry out attacks on enemy ships at night or under adverse weather conditions.

Beginning in September 1944, the Army and Navy began a series of studies to determine the amount of air power and the type of tactics required to smash an Allied invasion fleet. The Army Air force shortly thereafter began actual tests in home waters. Staff studies were made simultaneously by Army and Navy commanders in the Philippines, where the next enemy blow was expected to fall.

The tests conducted by the Army showed conclusively that flying personnel were inadequately trained to make orthodox high-level bombing an effective means of attack. Actual combat experience had also demonstrated that although skip-bombing obtained far better results, it involved prohibitive losses in planes and pilots because the enemy's naval antiaircraft defenses were so highly perfected. Skip-bombing, moreover, required experienced air crews, of which the Air forces were painfully short.

These studies and experiments supported a growing conviction that suicide or crash attack was the best means of ensuring maximum results at minimum cost in planes and pilots. The effectiveness of this method had already been demonstrated on a number of isolated occasions prior to the summer of 1944.[9] Many high staff officers were convinced that it was

7 In late July the directors of air training schools and local Air force commanders were ordered by the Inspectorate-General of Army Aviation and the War Minister to submit lists of volunteers for special-attack training. Fifty flight personnel were recruited in the first group, their training to be completed in September. Abont sixty flight personnel (for four groups) were later recruited in August and September. Instructions were given to the recruits at Army air schools at Hamamatsu and Hokoda for heavy and light bomber pilots, and at Hitachi and Akeno for fighter pilots. (Statement by Col. Monjiro Akiyama, Chief, Training Section, Office of the Inspectorate-General of Army Aviation.)

8 The Army began conversion of a number of new Type-IV heavy bombers and Type-99 twin-engine light bombers into special-attack aircraft. The heavy bombers were armed with two 800-kilogram demolition bombs and a contact fuse built into the fuselage. The light bombers were similarly armed with one 800-kilogram bomb. Fighters and fighter-bombers were also being equipped with devices for carrying a greater load of bombs to permit instant conversion into special-attack planes. The Navy meanwhile speeded the secret development of a new type of special-attack ordnance, the *Oka* (dubbed "Baka bomb" by the Americans), a one-man rocket-powered glide bomb invented by a Navy ensign. It was put into production in September. The bomb was attached to the underfuselage of a mother plane and was released at a point most effective for a glide and dive on the target. This was usually 11 miles from the objective at an altitude of 3,500 meters or at a point not liable to interception by enemy carrier fighters. About 70 of these bombs were launched against Allied ships during the Okinawa campaign from March to June, 1945. (Statements by Lt. Col. Koji Tanaka, previously cited, and Rear Adm. Katsuhei Nakamura, Chief, General Affairs Bureau, Naval Aeronautical Department.)

9 In April 1943, a Sgt. Ono of the 6th Air Division, rammed his plane into a B-17 north of Madang, New Guinea, while on transport escort duty. In May 1944, four fighter planes led by Maj. Takada of the 5th Air Regiment, launched a crash attack against an Allied warship off southern Biak, New Guinea. During the Midway naval battle of 5 Jun 1942, a Lt. Tomonaga executed a suicide dive on an enemy ship. (Statements by Lt. Col. Koji Tanaka, previously cited, and Capt. Mitsuo Fuchida, Staff Officer (Operations), Combined Fleet.)

the only solution in view of the insufficient training which most pilots were then receiving. Crash tactics, moreover, required relatively little instruction.[10]

While these views generally prevailed in lower Army and Navy echelons,[11] the employment of suicide techniques on a large scale by the Air forces was not yet officially sanctioned by either the Army or Navy High Commands; as late as mid-October, Imperial General Headquarters had not yet issued any formal orders directing the use of *tokko* tactics.[12]

This was the situation when the incursion of an enemy carrier task force into the Ryukyus—Formosa area precipitated the critical air battle off Formosa. On 15 October, in the closing phase of this engagement, Rear Adm. Masabumi Arima, Commander of the 26th Air Flotilla in the Philippines, set an example of suicidal bravery by attempting to crash his torpedo bomber into an enemy carrier.[13] This act of self-sacrifice by a high flag officer spurred the flying units in forward combat areas and provided the spark which touched off the organized use of suicide attack in the battle for Leyte.

Tokko Air Units

The underlying compulsion which brought the first organized air special-attack units into being, however, came from the crucial state of unpreparedness in which the enemy invasion of Leyte caught the Air forces in the Philippines. Mass air strikes against the invasion forces had been envisaged under the *Sho-Go* Operation plans, but the First Air Fleet found itself with fewer than 50 operational aircraft, mostly fighters and a few bombers, as a result of enemy carrier attacks. The Fourth Air Army had less than 100 planes, and its strength was widely dispersed. The Second Air Fleet, which contained the bulk of the Navy's Air forces, had suffered heavy losses in the Formosa air battle and had not yet begun its redeployment to the Philippines.

On 17 October, as the enemy fleet began moving into Leyte Gulf, Vice Adm. Takijiro Onishi arrived in Manila to assume command of the First Air Fleet. On the following day, the *Sho* No. 1 Operation was activated. Its success hinged on a decisive employment of air power. Determined to secure maximum results

10 "To achieve *Kamikaze*, the ordinary technique of the pilot is sufficient, no special training methods are necessary. . . . But to pilots who have had short training and least flight experience we give the essence of the *Kamikaze* attack in the shortest period possible." (Statement by Capt. Rikihei Inoguchi, Staff Officer, First Air Fleet, USSBS *Interrogations of Japanese Officials*, op. cit. Vol. I, p 61.)

11 By mid-September, the Fourth Air Army in the Philippines began independently to form *tokko* units for special missions. On 21–22 September, three Type-I Army fighters, organized for such a mission, carried out an attack on an enemy carrier force. (Statement by Lt. Col. Katsuo Sato, Staff Officer (Operations), Fourth Air Army.)

12 While the Army and, to a lesser degree, the Navy had proceeded with the physical formation of *tokko* units and the production and equipment of special-attack planes, Imperial General Headquarters issued no general orders placing such units on a compulsory basis. The Army and Navy High Commands still considered special-attack a method of coping with local situations and not an over-all policy. Large-scale special-attacks in the Philippines Campaign were counted upon, but no orders to impress flying personnel into *tokko* units were issued because numerous volunteers were available. (Statement by Lt. Gen. Torashiro Kawabe, Deputy-Chief, Army Aeronautical Department.)

13 Although Rear Adm. Arima's attempted crash attack was reported successful by returning Japanese aircraft, U. S. Navy sources indicate that his plane crashed into the sea. No carrier or any other ship was hit by a suicide plane in the period 14–18 October. USSBS (Pacific), Military Analysis Division, *Japanese Air Power*, July 46, p 62.

with the handful of planes at his disposal, Admiral Onishi decided late in the night of 19 October, a few hours before formally assuming command, to create a suicide corps on a volunteer basis within the First Air Fleet.

This corps, christened the *Kamikaze* Special-Attack Corps, was originally made up of 24 volunteer fliers selected from the 201st (Fighter) Air Group, stationed in the Manila area.[14] Assembled at Clark Field, the fliers were organized into four units, each composed of six pilots and bearing a special name designation. The aircraft allotted for their use were Zero fighters stripped of radio equipment, armament, and all but essential flying instruments and fitted to carry a 250-kilogram demolition bomb underneath the fuselage. On attack missions, the suicide units were to be led to the target by guide planes and covered against enemy interception by fighter escort. Crash attacks were to be executed only when reasonably certain of hitting enemy carriers or major task force units.[15]

The initial sorties of the *Kamikaze* units on 21 October were largely unsuccessful. Three of the units took off from Mabalacat, near Clark Field, to attack enemy task forces reported operating to the east of Luzon, but were unable to locate the targets. The fourth, which had advanced to a base on Cebu the preceding day, was preparing to sortie against enemy naval forces in the vicinity of Leyte Gulf when a sudden strike by enemy planes on the morning of the 21st destroyed all six suicide aircraft on the ground. Three reserve fighters were undamaged, however, and it was decided to carry out the mission with these planes. Two of the fighters were unable to execute attacks because of unfavorable weather and were forced to return to their base, but the third was believed to have crash-dived into an enemy warship.[16]

The first real display of *Kamikaze* effectiveness was given on 25 October, the second day of the all-out air offensive launched in conjunction with the surface force attack toward Leyte Gulf. One unit flew from Davao to attack an enemy carrier group east of Surigao Strait; another shortly thereafter struck at the same carrier force which had just escaped destruction by Vice Adm. Kurita's warships in the battle off Samar. In the latter strike, four *Kamikaze* planes were reported to have executed successful crash attacks. On 26 October, an enemy carrier group east of Surigao Strait was the target of yet another suicide attack, in which two carriers were reported hit.[17]

The small *Kamikaze* corps organized by Vice Adm. Onishi was originally intended as a temporary expedient to meet an express emergency. The unfavorable trend of the Philip-

14 The commander of this corps was a regular Navy officer. The other 23 airmen were graduates of naval pre-flight training courses and had received some special training in skip-bombing. (Statement by Capt. Fuchida, previously cited.)

15 The technique of approaching the target was of great importance. This varied according to circumstances and the skill and preference of the pilot, but was generally calculated to achieve surprise and evade or confuse radar detection, visual spotting, and fire control. During the Leyte operations, *tokko* planes generally used a high or medium-altitude approach up to 6,000 meters, although on occasion a low-level approach using cloud cover was employed. Over the target each plane was on its own, one plane usually being assigned to one ship. Pilots were instructed to attack enemy carriers by diving at the flight deck at as steep an angle as possible, or if this was not feasible, to aim at the side of the vessel. (Statements by Lt. Col. Koji Tanaka and Capt. Fuchida, both previously cited.)

16 Since there is no record of any other suicide attack on ships in Leyte Gulf during 21 October, it appears probable that this aircraft was the one which, according to Allied naval records, hit and damaged the HMAS *Australia* on that date. This was the first successful suicide attack in the Leyte campaign.

17 *Hito Homen Kaigun Sakusen Sono Ni* 比島方面海軍作戦其の二 (Philippine Area Naval Operations, Part II) 2d Demobilization Bureau, Oct 47, p. 101

特攻隊(空)內地基地ヲ進發

Original Painting by Sentaro Iwata

Photograph by U. S. Army Signal Corps

PLATE NO. 138
Departure of Special-attack Unit from Homeland Base

pines Campaign, however, forced an increasing reliance upon suicide tactics. This tendency was spurred by heavy aircraft losses, an inadequate flow of air replacements, and a steady deterioration in the quality of pilots arriving at the front.

As the original *Kamikaze* corps was expended, new units were quickly organized in the First Air Fleet with volunteers drawn from regular tactical units and replacement groups.[18] Soon any flight of planes on a suicide mission was called a special-attack unit. Usually the planes went out in groups of two or three, accompanied by an equal number of fighter escorts and a single guide plane. They took off at dawn and at other hours when they were the least liable to interception.

By the end of October, special-attack units of the First Air Fleet had carried out twenty separate missions with from one to eight planes participating. Twenty-eight aircraft were credited with having crashed into their objectives, principally enemy naval vessels in the waters east of the Philippines.

The Second Air Fleet, which deployed to the Philippines on the eve of the big air offensive of 24–25 October, at first did not resort to the formation of *Kamikaze* units, but as combat losses pared down its strength and orthodox methods failed to achieve adequate results, it soon found itself obliged to follow the example of the First Air Fleet. By the end of the first week in November, planes of the Second Air Fleet's 221st Air Group were executing crash attacks on enemy ships in Leyte Gulf. Between 1–15 November, the First and Second Air Fleets—at this time operating together as the First Combined Base Air Force—carried out 21 suicide missions using a total of 71 *Kamikaze* planes. During this period, 19 aircraft were reported to have crashed successfully into their objectives.[19]

Although the lead had been taken by the Navy Air forces, the Army was not long in following suit. With the activation of the *Sho* No. 1 Operation, Imperial General Headquarters ordered to the Philippines the first of the specially trained and equipped Army *tokko* units which had been organized in the Homeland during the summer of 1944. The Banda and Fugaku units, each with twelve special-attack bombers, reached Philippine bases too late to participate in the 24–25 October air offensive and were held in reserve by Fourth Air Army for use only on the most vital missions.[20] They finally went into action for the first time on 12–13 November.

Eighteen additional special-attack units of the same type, with a total of 216 aircraft, were ordered to the Philippines during November. Meanwhile, under High Command authorization given early in November, Fourth Air Army organized provisional *tokko* units within its regular tactical forces already in the Philippines. The strength and composition of these units were determined by the regimental commanders, and they were to use ordinary fighter and fighter-bomber aircraft with bombs attached to the underfuselage.

Although the appearance of the *tokkotai* exerted no appreciable effect on the outcome of the battle for Leyte, the reason did not lie in a lack of effectiveness. Reported results for the Leyte campaign showed that 30 per cent of

18 Pilots, as well as aircraft of various types, were drawn from the 153d, 601st and 761st Air Groups. The initial corps used only bomb-equipped fighters. However, after the reorganizations of the *Kamikaze* corps, all types of aircraft were used. (Statement by Capt. Fuchida, previously cited.)

19 Philippine Area Naval Operations, Part II, op. cit, pp. 101–2,

20 The Banda unit used a twin-engine bomber manned by a crew of two and armed with a single 800-kilogram bomb. The Fugaku unit used a Type-IV heavy bomber, which carried two 800-kilogram bombs.

the special-attack planes which actually attempted crash attacks hit their objectives, a rate of accuracy far higher than that achieved by orthodox bombing. The ratio of aircraft losses to damage inflicted on enemy ships was also lighter. During the first three weeks, 30 Army *tokko* planes were lost in attacks which reportedly sank or damaged 23 enemy vessels, as compared with 24 aircraft lost in regular bombing attacks which were reported to have sunk or damaged only 6 Allied ships.[21]

But even suicide methods, however destructive, could not overcome the enemy's decisive superiority in numbers. His carrier task forces were still able to send preponderant air strength against Japanese bases throughout the Philippines. About one-third of the planes, sent as reinforcements from the Homeland, were destroyed on the ground even before they could go into action.[22] Moreover, the element of surprise which heightened the effectiveness of the *tokkotai* in the initial period of their activity was soon gone. The enemy took prompt steps to strengthen the air cover and antiaircraft defenses of his ships, so that more and more special-attack aircraft and escorting fighters had to be sent out on each mission in order to achieve worthwhile results.

Special-attack nevertheless remained the sole means whereby the Japanese Air forces could pit their inferior strength against the enemy with any degree of success. By 15 December, the day the enemy invaded Mindoro, more than half the air strength deployed in the Philippines was being committed to *tokko* combat. By the end of the same month, Army special-attack units were frequently flying to the target without fighter escort, and virtually anything that could leave the ground, including trainer aircraft, was being fitted for crash missions.

The *tokkotai* made their final mass effort of the Philippines Campaign in January 1945 against the enemy's Luzon invasion forces. On 5 January, sixty Army special-attack planes struck the huge invasion convoy as it moved west of Subic Bay and inflicted considerable damage. The effort reached maximum intensity on the following day, when both Army and Navy aircraft joined in day-long attacks on enemy ships moving into Lingayen Gulf. By 7 January, however, the Air forces had sustained such heavy losses that their offensive power was spent.

For a brief period thereafter, a small number of Army suicide planes continued to operate from hidden fields on northern Luzon, sortieing in groups of two or three planes for sporadic attacks on enemy re-supply shipping. By mid-January even this meager activity had ceased, and the Japanese air effort in the Philippines came to an end.

During the entire period of the Philippines Campaign, Fourth Air Army sent out about 400 special-attack planes on 61 missions, while *Kamikaze* of the Navy's First Combined Base Air Force flew 436 planes on 107 missions. The results officially reported for these attacks were 154 vessels of all types hit by Army planes and 105 by naval *Kamikaze*.[23] While the frequent difficulty of ascertaining exact results

21 Statement by Lt. Col. Koji Tanaka, previously cited.

22 Between 15 October 1944 and the end of January 1945, a total of 1,970 Army aircraft were lost in the Philippines theater. Of this total, about 700 planes were destroyed on the ground because of the inability of the air defense to protect Japanese bases from enemy air strikes. Attempts were made to reduce losses by dispersal to concealed airfields at various points on Luzon, but these efforts were largely ineffectual. Top priority was also given to moving air replacements from the Homeland. However, ferrying losses were extremely high due to poor quality of aircraft and inadequately trained pilots. (Statement by Lt. Col. Sato, previously cited.)

23 (1) *Hito Koku Sakusen Kiroku Dai Niki* 比島航空作戰記錄第二期 (Philippine Air Operations Record, Phase Two) 1st Demobilization Bureau, Nov 46, Attached Chart XVII. (2) Philippine Area Naval Operations, Part II, op. cit., pp. 101–2.

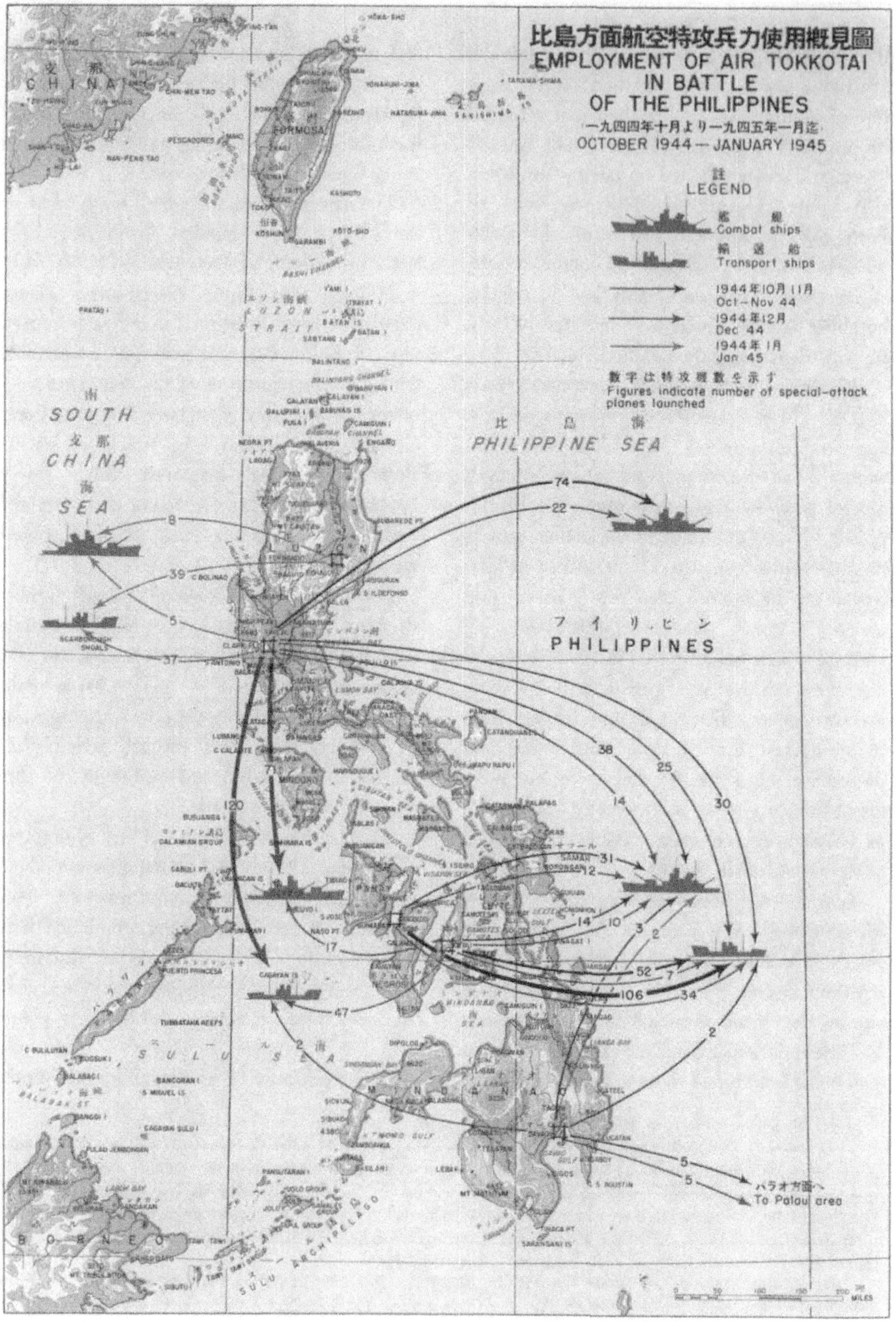

PLATE NO. 139

Employment of Air *Tokkotai* in Battle of Philippines, October 1944—January 1945

特攻舟艇敵艦に突入　福澤一郎　昭和二〇

Original Painting by Ichiro Fukuzawa

PLATE NO. 140
Surface Raiding Unit Dares Desperate Ramming

made these figures subject to unavoidable error, there was no question that the sharp sting of the special-attack forces had hurt the enemy.[24] (Plate No. 139)

Other Forms of Special-Attack

Air *tokko* was by far the most spectacular and extensive form of special-attack to emerge during the struggle for the Philippines, but it was not the only manifestation of the suicide technique whereby the Japanese sought desperately to change the tide of battle. Both the Army and Navy resorted to other forms of special-attack, all of which were designed to obtain maximum results with rapidly dwindling resources.

The use of small airborne raiding units in the abortive attempt of November 1944 to regain control of the Burauen airfields on eastern Leyte was the only example of the application of special-attack principles to airborne ground troops during the Philippines Campaign.[25] Air force weakness barred further operations of this kind in the battle for Luzon, but the Burauen airborne assault furnished the prototype for a similar suicide operation carried out in May 1945 during the final stage of the Okinawa Campaign.[26]

The Luzon battle witnessed the first use of another special-attack weapon designed to offset Japan's poverty in regular surface combat forces. This new weapon was the suicide crash boat, a small craft of light plywood construction manned by a single navigator and fitted with depth charges to be dropped directly alongside an enemy vessel or rammed into it without release from the attacking craft.[27] Such boats were specially designed for close-in attacks on enemy invasion craft anchored off landing beaches.

Surface raiding units equipped with crash boats were organized and trained by both the Army and Navy during the summer of 1944 in preparation for the defense of the Philippines. Plans formulated in August called for the construction of Navy surfac eraiding bases at

24 A check of the results officially claimed by the Japanese Air forces against the figures of actual losses and damage given in American naval records covering the Philippines Campaign shows the following: Only 19 Allied ships, including two escort carriers, were actually sunk by *Kamikaze* attack. However, 121 suicide planes hit their targets, inflicting varying degrees of damage, and 53 scored damaging near-misses. One third of the recorded hits were on carriers, battleships and cruisers. USSBS, *Japanese Air Power.*, op. cit., p. 63.

25 Cf. Chapter XIII, pp. 393–397.

26 On the night of 24 May 1945, the Giretsu (Army) Airborne Raiding Unit crash-landed on *Kita* (north) and *Naka* (central) airfields which had been occupied by the enemy, and succeeded in doing extensive damage to parked aircraft before the unit was wiped out. The Navy planned a similar raiding attack on American B-29 bases in the Marianas in July 1945, but the 25 long-range Navy bombers assigned to the mission were destroyed by enemy carrier raids on Misawa airfield, in northern Honshu, on 14–15 July, and the plan had to be abandoned. (Statements by Lt. Col. Koji Tanaka, previously cited, and Comdr. Yoshimori Terai, Staff Officer (Operations), Imperial General Headquarters, Navy Section.)

27 The Navy gave its crash boats the designation *Shinyo*, while those used by the Army's surface raiding regiments were known as *Renraku-tei* or "liaison boat." The boats were essentially of the same type, 18 feet in length and powered by automobile engines capable of driving them at a maximum speed of 26 knots. The *Shinyo* carried a single 250-kilogram explosive charge in the bow, while the *Renraku-tei* loaded two 120-kilogram depth charges, one on each side of the cockpit. *Shinyo* crews were instructed to crash headlong into the target vessel, while *Renraku-tei* navigators were originally trained to approach the side of an enemy ship abreast the engine compartments at high speed and drop their charges while executing a U-turn. The charges were fixed to detonate six seconds after release. This allowed virtually no leeway for crash boat to get away. Both the Army and Navy planned to concentrate as many boats as possible for a surprise attack, but actually no more than 70 boats were ever employed on a single raid. (1) *Shusenji ni okeru Nihon Kantei* 終戦時に於ける日本艦艇 (Japanese Naval Vessels at the End of the War) 2d Demobilization Bureau, 25 Apr 47. (2) Statement by Lt. Col. Tsugunori Kuriya, Staff Officer (Shipping), Fourteenth Area Army.

Davao and Sarangani Bay on Mindanao, at Tacloban on Leyte, and at Lamon Bay on Luzon.[28] The boat units and maintenance personnel began arriving on Luzon early in September, but the projected bases in the southern and central Philippines were not completed in time to permit deployment before the invasion of Leyte. On Luzon the surface raiding forces were concentrated at four main points of anticipated invasion—Lingayen Gulf, Manila Bay, Batangas, and Lamon Bay.[29]

On only two occasions during the defense of Luzon did these forces go into action on a large scale, and, in both instances, they proved ineffectual. About 70 Army surface raiding boats launched a mass attack against enemy invasion shipping in Lingayen Gulf on the night of 9–10 January 1945, but were unable to disrupt landing operations.[30] The Navy's *Shinyo* boats based in Manila Bay went into action for the first and only time on the night of 15 February against enemy warships anchored in Mariveles Bay, on Bataan Peninsula. All participating craft were expended in both these raids.

Although devised long before the general adoption of special-attack weapons and less suicidal in nature than either the *Kamikaze* plane or the crash boat, the Navy's midget submarines also fitted naturally into the pattern of *tokko* warfare.[31] Ten of these craft, operating from bases in the central and southern Philippines, harassed northbound enemy convoys in the Mindanao and Sulu Seas during December 1944 and January 1945 until they finally met their doom.

The Navy's *Kaiten*, a one-man piloted torpedo which was first put into production in September of 1944, was more clearly a full-fledged suicide weapon. These torpedoes were designed to be carried on the deck of a parent submarine, generally four to one parent craft, and released within range of the target. The operator, seated in the center of the torpedo, then steered his explosive missile directly to the objective and certain death. The *Kaiten* were first thrown into action on 20 November 1944, when 8 were employed in a surprise attack on enemy carriers and battleships at Ulithi Anchorage. In January and February 1945, 28 more were used in operations against enemy naval bases at a number of scattered points in the Central and south Pacific.[32]

Although about 450 *Kaiten* were produced,

28 Philippine Area Naval Operations, Part I, op. cit., pp. 40–1.

29 The Army surface raiding forces in the Philippines were organized into boat regiments, each regiment with an authorized strength of 100 boats and 104 navigating personnel, and supported by a base battalion of about 900 men. Boat regiments and base battalions stationed in the Manila Bay, Batangas and Lamon Bay areas were commanded by three Surface Raiding Force headquarters, one for each of these areas. One boat regiment and a base battalion, commanded by a local division commander, were stationed in Lingayen. Most of the boat regiments were below authorized strength because of the large proportion of craft lost in transit to the Philippines. At the time of the Allied invasion of Luzon, about 150 surface raiding boats were based at Sual, on Lingayen Gulf. (Statement by Lt. Col. Kuriya, previously cited.)

30 Cf. Chapter XV, p. 436.

31 Midget submarines were first used in the Pearl Harbor attack in December 1941, all five participating craft being lost in the operation. (Cf. Chapter V, p. 71.) During 1942 they were used in small-scale attacks on shipping inside the harbors of Sydney, Australia, and Diego Suarez, on Madagascar Island, and also operated in the Solomons area during the fight for Guadalcanal. Limited cruising range, low speed, difficulty in launching from the mother submarine, and inadequate power of the torpedoes carried so reduced the effectiveness of these craft that a new type of midget submarine, with a cruising range of 300–350 miles, was developed. Sixteen of this new type, designed to operate from shore bases, were built between 1942 and 1945. (Statement by Capt. Toshikazu Ohmae, Naval Affairs Bureau, Navy Ministry.)

32 Seven submarines, each carrying four *Kaiten*, were used in attacks against enemy ships in Ulithi Anchorage, Apra Harbor on Guam, Kossol Passage and Humboldt Bay between 20 November 1944 and 12 January 1945. Enemy vessels off Iwo Jima in the Bonins were also attacked by eight *Kaiten* in late February 1945, but the results are not known. (Navy Ministry's Report to Col. F. P. Munson, 22 November 1945.)

including some improved models, the Navy's plans for using them extensively never materialized. Between May 1945 and the end of the war, they had been employed on only eleven occasions, achieving a total of 18 reported hits on enemy vessels.[33]

The institution of special-attack weapons and methods in air and sea warfare was paralleled by no striking innovations in the field of regular ground combat. Nevertheless, as the Philippines Campaign progressed, there was a tendency toward an increasing reliance on infiltration and raiding tactics of a suicidal nature. During the defense of Luzon, Fourteenth Area Army also employed suicide antitank units which endeavored to destroy enemy armor by close-in attacks with grenades and lunge mines.

The *tokkotai* came into being during the final phase of the Pacific War because of a compelling necessity to devise some counterweight to the overwhelming enemy superiority in all aspects of military power. The resort to suicide methods on so large a scale, however, would never have been possible but for a single factor—the deeply inculcated spirit of patriotic self-sacrifice which characterized the Japanese soldier and sailor and which was reflected in the psychology of the nation as a whole.

It was this spirit, the product of centuries of religious and moral training, that brought forth in ample number volunteers for every special-attack unit or mission; that led Japanese troops, often starving and cut off from all hope of succor, to fight on rather than surrender; and that spurred the nation to still greater efforts despite successive defeats, mounting hardships and growing doubt of ultimate survival.

Although the *tokkotai* had not proved equal to the task of turning back the enemy in the Philippines, their heroic example bolstered the determination of the people at home and gave an impetus to the still wider adoption of special-attack methods. These suicide techniques were to play a large role in the subsequent defense of Okinawa, and under plans which the High Command began formulating early in 1945, were to constitute the backbone of the final battle to turn back the anticipated invasion of the Homeland.

---

33 Ibid.

CHAPTER XVIII

HOMELAND DEFENSE: BASIC PLANS AND PRELIMINARY OPERATIONS

Strategic Situation—January 1945

As the year 1945 began, the Japanese nation faced disaster.[1] The great decisive battle in the Philippines had been lost, and the enemy was moving on to invade Luzon, wresting from the Japanese their last operational base in the Philippines and extending the Allied air perimeter far out over Formosa, the China coast and southern Japan. Behind General MacArthur's forces hundreds of thousands of trained Army, Navy, and Air Force personnel were isolated, unable to contribute directly to the prosecution of the war. Ahead of the enemy, the road to the Japanese Homeland lay wide open.[2]

The war potential of the nation was at a record low. The backbone of both the fleet and the air forces had been broken in the Philippines, while the Army's ability to concentrate for a final effort was hampered by its commitments to many scattered and now isolated theaters. More serious than even the long chain of tactical defeats which had all but destroyed Japan's armed forces, however, was the rapid disintegration of the national war economy at its base.

Largely as a result of submarine interdiction of the South China Sea, the nation had already been deprived of essential strategic raw materials. Only in the early part of the war had Japanese shipbuilding facilities been able to keep up with sinkings by Allied submarines and planes. By January 1945, 69% of Japan's merchant fleet, including 59% of her vital oil tankers, had been sent to the bottom. Total gross cargo loadings in the quarter ending in December 1944 were a mere 33% of the wartime peak.[3]

As a result of the crisis in shipping, the flow of coal, iron ore, non-ferrous ores and concentrates, salt, and chemicals from the Asiatic continent had been materially reduced, while the supply of oil, rubber, nickel, chromium, and

1 This chapter was originally prepared in Japanese by Lt. Col. Iwaichi Fujiwara, Imperial Japanese Army. Duty assignments of this officer were as follows: Staff Officer (Intelligence), Imperial General Headquarters, Army Section, Aug 39—Dec 41; Staff Officer (Intelligence), Southern Army, Dec 41—Mar 43; Staff Officer (Operations and Intelligence), Fifteenth Army, Mar 43—Feb 45; Instructor (Military Tactics), Army War College, Feb 45—Mar 45; Staff Officer (Operations), Second General Army, Mar 45—Jun 45; Staff Officer (Operations), Fifty-seventh Army, Jun 45—Aug 45. All source materials cited in this chapter are located in G-2 Historical Section Files, GHQ FEC.

2 The term Homeland includes (from north to south) Karafuto (Sakhalin) and the Chishima (Kurile) Islands, Hokkaido, Honshu, Shikoku, Kyushu, the Ryukyu (Nansei) Islands, and the Ogasawara (Bonin) Islands.

3 *Tokei Chosa Hokoku* 統計調査報告 (Statistics and Analysis Reports), Japanese Research Group, G-2 Historical Section, GHQ FEC, 30 Nov 50, Report No. 1: Gross Tonnage of Japanese Merchant Shipping Losses through Allied Action, Dec 41–Aug 45.

aluminous concentrates from the southern area had slowed to a mere trickle.[4] General MacArthur's seizure of Luzon and operations of the American carrier task force in the South China Sea gave notice that even this slender lifeline was about to be decisively severed.[5]

The collapse of Japan's overseas supply of raw materials was immediately reflected in the production figures of her basic industries. By early 1945, the iron and steel industry was operating at only about 49% of its wartime peak production, the three most important nonferrous metals, copper, aluminum, and magnesium, were off 49% from their peak, and a similar situation obtained in the output of ferro-alloys. The chemical industry was producing at 62% of the first quarter of 1944, while the total production and import of crude and refined petroleum was down to a disastrous 26% in the quarter ending in December 1944. From the long-range strategic viewpoint, the situation had become hopeless.[6]

For their part, the Allies seemed to be in an extremely favorable position. In the European theater, the failure of the Germans' great Ardennes offensive presaged the defeat of Japan's distant ally, and it was a certainty that termination of hostilities in that area would be followed by the redeployment of overwhelming land, sea, and air power against Japan. Intelligence reports disclosed that the United States was rushing production of all types of amphibious

4 Imports of oil from the southern area in fiscal year 1944 were only 34% of the wartime peak, of bauxite 42%, and of rubber 55%. There were no imports at all of rice from the southern area after March 1944. In order to make up for the loss, larger quantities of substitute foods from the continent were imported. Due to the shipping shortage and the diversion of shipping to this food haulage program, imports of heavy coking coal from Manchuria, North China, and Karafuto fell to only 36% of the wartime peak and those of iron ore to 33%. (1) *Gomu Kogyo no Hatten* ゴム工業の發展 (Development of the Rubber Industry) Compiled by Mr. Katsumi Ikeo, Chemical Bureau, Munitions Ministry, 25 Nov 48, pp. 49, 70–2. (2) *Gikai Hokokusho* 議會報告書 (Reports to the Diet) Japanese Government Ministries ; Summary of Staple Food Supply and Demand Factors Accompanying the Prolongation of the War, Sep 45. (3) U. S. Strategic Bombing Survey (Pacific), Over-All Economic Effects Division, *The Effects of Strategic Bombing on Japans' War Economy*, Dec 46, pp. 108–9, 121, 135. (4) "The whole South Seas plan to make good Japan's deficiencies . . . . was based on the erroneous assumption that ship tonnage . . . . would remain level. Because of the tremendous loss in shipping . . . . the South Seas were important for only the first year. After that the economy of Japan was based on China, Manchuria, and Japan Proper." U. S. Strategic Bombing Survey (Pacific), Naval Analysis Division, *Interrogation of Japanese Officials*, 1945, Interrogation No. 155. (Interrogation of Mr. Aichi, Cabinet Planning Board, Greater East Asia Ministry.)

5 In January, the U. S. carrier task force sortied into the South China Sea and began a campaign of destruction against ports and shipping. On 12 January, a convoy of 9 fully loaded tankers was almost wiped out off Quinhon, French Indo-China. On the 15th, a southbound convoy off Hongkong suffered a similar fate. (Statement by Capt. (Navy) Atsushi Oi, Staff Officer (Operations), General Escort Command.

6 The hopeless situation in raw materials and basic industry was not yet fully reflected in the important fabricating industries. Aircraft were still being produced at 69% of wartime rate, shipping at 47%, and Army munitions at about 72%. (1) Statistics and Analysis Reports, op. cit., Report No. 2 : Output of Strategic Materials in the Japanese Homeland ; Report No. 3 : Japanese Army Ordnance Production, January 1945. (2) Reports to the Diet, op. cit : Summary of Production of Aircraft and Loss of Capacity through Air Raids, 1 Sep 45. (3) *Nihon Kaiun no Hembo* 日本海運の變貌 (Situation of the Japanese Sea Transportation), Mr. Genko Tsuboi, Sea Transport Board, Ministry of Transportation, 20 Aug 46, Attached Chart No. 11.

craft, heavy bombers, tanks, artillery, new rocket weapons, and fleet carriers.[7] In the Philippines and the Marianas the enemy had excellent staging bases from which to launch the next effort.

In this disheartening atmosphere, Imperial General Headquarters undertook to determine the probable course of future Allied action. In general, it seemed clear that the United States would desire to bring the war to an early conclusion, and that the ultimate goal of enemy strategic policy would be the annihilation of the Japanese Army in the Homeland and the occupation of Japan. To prepare for the ultimate invasion, the High Command felt that the enemy would take the following course of action.[8]

1. Complete the occupation of key areas on Luzon as soon as possible.

2. Take one of the following courses of action given in order of probability:

a. A two-pronged advance into the Homeland defense perimeter to secure advance bases and tighten the blockade of Japan Proper. The eastern advance would move from the Marianas into the Ogasawara (Bonin) Islands in February or March. On the west, a much heavier offensive would move from the Philippines to one or more key points bordering on the East China Sea including Formosa, the Nansei (Ryukyu) Islands, and the lower reaches of the Yangtze River, this operation to commence in March or April. The two prongs of the advance would converge in the final assault on the Homeland in the fall of 1945 at the earliest.

b. An advance from the central Pacific into the Ogasawara (Bonin) Islands and thence directly to the Homeland, (this operation to begin in February or March,) invasion of Japan Proper, in this case, to be in June or July.

3. In conjunction with operations on the southern approaches to the Homeland, the enemy would probably launch a secondary offensive from the Aleutians against the Chishima (Kurile) Islands in order to capture advance air bases.

A direct attack on Japan Proper without these preliminary steps was regarded as a remote possibility. Within the framework of the most probable Allied strategic plan, it was believed that Iwo Jima in the Bonin Islands would be the next invasion target, with Okinawa in the Ryukyus following soon thereafter.

An inventory of the forces available to meet the expected onslaught gave the High Command no cause for optimism. The basic composition and deployment of forces adopted for the *Sho-Go* operations still survived in principle, weakened and modified however by the heavy commitment to the *Sho* No. 1 front in the Philippines. Under Imperial General Headquarters there were two major commands responsible for the defense of the Homeland. The General Defense Command, under General Prince Naruhiko Higashikuni, was responsible for the air and ground defense of Honshu, Shikoku, Kyushu, and the Izu Islands, and for a portion of the long-range air operations against enemy invasion convoys and bases. The Combined Fleet, under Admi-

7 (1) Statement by Lt. Gen. Seizo Arisue, Chief, 2d Bureau (Intelligence), Imperial General Headquarters, Army Section. (2) *Daihonyei Rikugun Tosui Kiroku* 大本營陸軍統帥記錄 (Imperial General Headquarters Army High Command Record) 1st Demobilization Bureau, Nov 46, p. 294.

8 (1) *Sojo Sho Sambo Socho Gunreibu Socho* 奏上書參謀總長軍令部總長 (Report to the Throne by the Chiefs of Army and Navy General Staffs) 19 Jan 45. (2) *Saiko Senso Shido Kaigi Tsuzuri* 最高戰爭指導會議綴 (Proceedings of the Supreme War Direction Council) Report No. 10, 15 Feb 45. (3) Statement by Col. Takushiro Hattori, Chief, Operations Section, Imperial General Headquarters, Army Section.

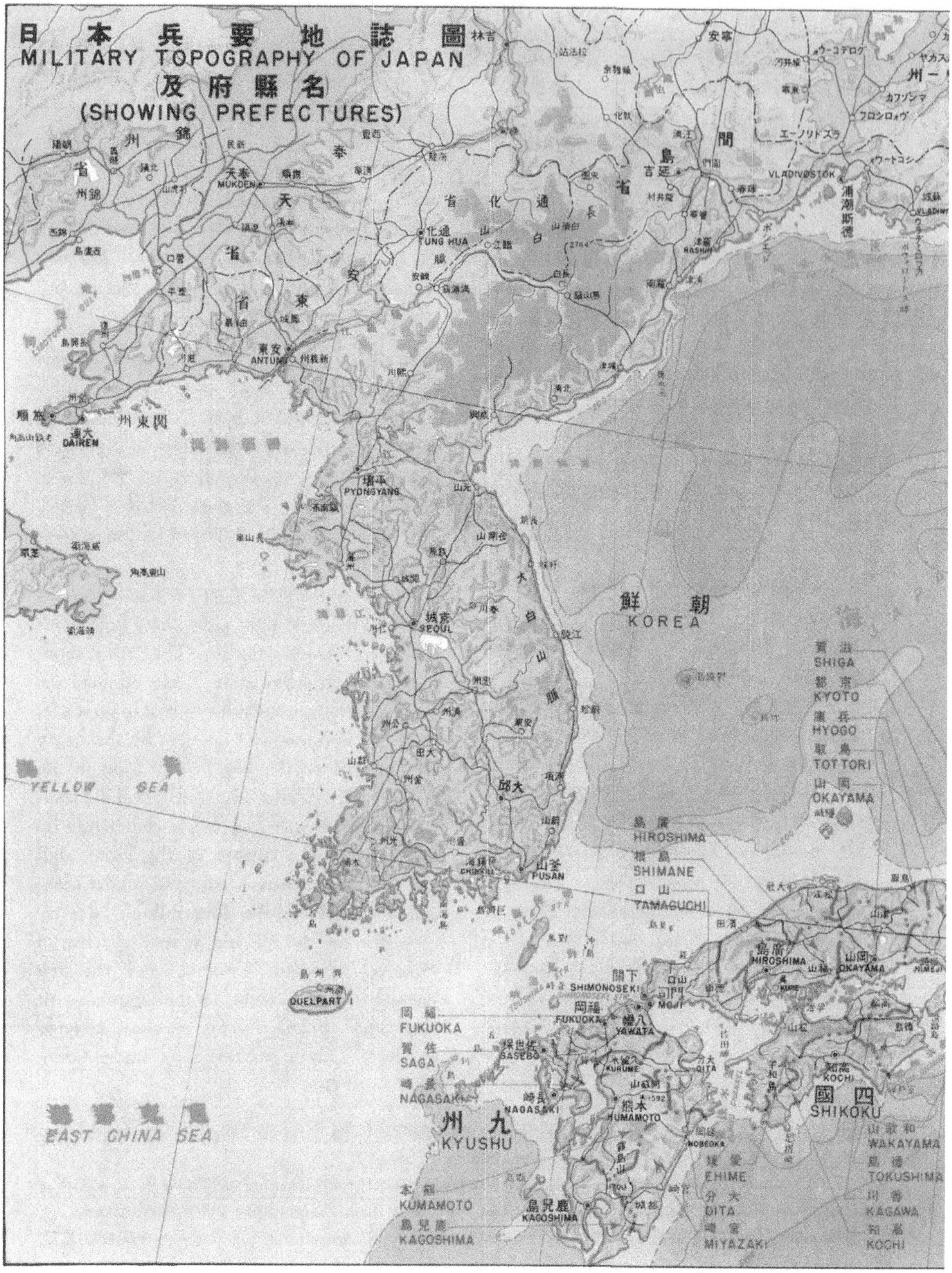
日本兵要地誌圖
MILITARY TOPOGRAPHY OF JAPAN
(及府縣名)
(SHOWING PREFECTURES)
MUKDEN
TUNG HUA
RASHIN
VLADIVOSTOK
ANTUNG
DAIREN
PYONGYANG
SEOUL
KOREA
YELLOW SEA
CHINKAI
PUSAN
QUELPART I
SHIMONOSEKI
MOJI
FUKUOKA
YAWATA
SASEBO
KURUME
OITA
NAGASAKI
KUMAMOTO
NOBEOKA
KAGOSHIMA
KYUSHU
EAST CHINA SEA
HIROSHIMA
OKAYAMA
KOCHI
SHIKOKU
SHIGA
KYOTO
HYOGO
TOTTORI
OKAYAMA
HIROSHIMA
SHIMANE
YAMAGUCHI
FUKUOKA
SAGA
NAGASAKI
KUMAMOTO
KAGOSHIMA
EHIME
OITA
MIYAZAKI
WAKAYAMA
TOKUSHIMA
KAGAWA
KOCHI

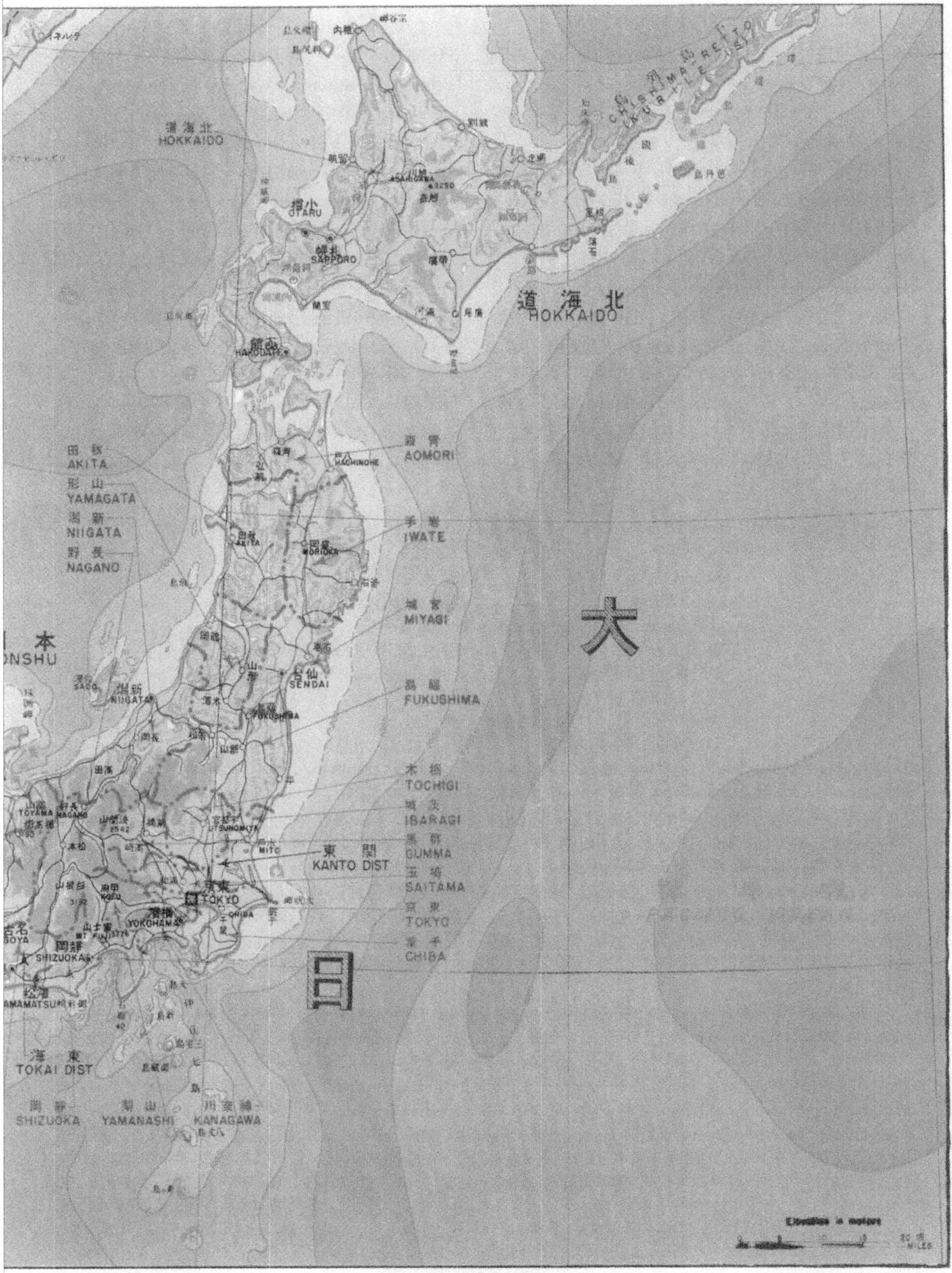

PLATE NO. 141
Military Topography of Japan

ral Soemu Toyoda was responsible for all surface operations and the bulk of the long-range air operations over the approaches to the Homeland. (Plate No. 142)

Peripheral commands under Imperial General Headquarters included the Fifth Area Army under Lt. Gen. Kiichiro Higuchi, responsible for Hokkaido, the Kurile Islands, and Karafuto;[9] Lt. Gen. Tadamichi Kuribayashi's Ogasawara Group in the Bonin Islands; and the Tenth Area Army under General Rikichi Ando, responsible for Formosa and the Ryukyu Islands.

For the land and air defense of Japan Proper (excluding Hokkaido and certain naval stations), the Commander-in-Chief, General Defense Command, had at his disposal the Western District Army (Lt. Gen. Isamu Yokoyama), responsible for Kyushu and southwestern Honshu, the Central District Army (Lt. Gen. Masakazu Kawabe) in central Honshu and Shikoku, the Eastern District Army (General Keisuke Fujie) in northeast Honshu and the Izu Islands, the Thirty-sixth Army (Lt. Gen. Toshimichi Uemura) as a mobile reserve in the Kanto-Shizuoka area, and the newly-activated Sixth Air Army.[10]

To execute the Combined Fleet mission, the Commander-in-Chief had available Vice Adm. Seiichi Ito's Second Fleet, now engaged in

---

9 Forces available to the Fifth Area Army for the defense of the Northeast Area were as follows:

Hokkaido	Kurile Islands
Ground	Twenty-seventh Army
77th Division	91st Division
7th Division	42d Division
Air	3d and 4th Amphibious Brigades
Twelfth Air Fleet	43d Independent Mixed Brigade
1st Air Division	69th Independent Mixed Brigade
Karafuto	Chishima 1st Brigade
Karafuto Mixed Brigade	Chishima Naval Base Force

Lt. Gen. Higuchi exercised a combined operational command including naval units. Since stress had been laid on the defenses of the Northeast Area ever since the loss of Attu in May 1943, a high state of preparedness had been reached. On 27 March 1944 the headquarters of both the Twenty-seventh Army and Fifth Area Army had been activated, the latter replacing the Northern Army which had previously absorbed the Northern District Army on 11 February 1943. There were now, however, only 36 operational aircraft in the 1st Air Division and Twelfth Air Fleet combined, mostly long-range reconnaissance planes. (1) *Hondo Sakusen Kiroku* 本土作戰記錄 (Homeland Operations Record), Vol. IV: Fifth Area Army, 1st Demobilization Bureau, Oct 46, pp. 4–6, 16–8, 24–6. (2) *Dairikumei Dai Senhachijuyongo* 大陸命第千八十四號 (Imperial General Headquarters Army Order No. 1084) 24 Jul 44. (3) *Hoppo Homen no Sakusen* 北方方面の作戰 (Northern Area Naval Operations) 2d Demobilization Bureau, Mar 49, p. 74. (4) Report to the Throne, op. cit.

10 Sixth Air Army under Lt. Gen. Michio Sugawara had the basic mission of long-range air attacks against enemy invasion bases and convoys. However, most of the aircraft were unsuitable for such operations and were detached from the Air Army for air defense. Of the approximately 100 long-range planes available, slightly more than half were assigned to the 7th and 98th Bomber Regiments, these units being attached to the 11th Air Flotilla in Kyushu for operations. This left the effective air strength of Sixth Air Army at less than 50 planes. In mid-January, Imperial General Headquarters issued a policy memorandum to Sixth Air Army indicating its desire that emphasis be placed on the speedy organization of *tokko* units to facilitate the execution of its basic mission. (1) *Dairikumei Dai Sennihyakujunigo* 大陸命第千二百十二號 (Imperial General Headquarters Army Order No. 1212) 26 Dec 44. (2) Report to the Throne, op. cit. (3) *Iwo Jima oyobi Nansei Shoto Homen Koku Sakusen Kiroku* 硫黄島及南西諸島方面航空作戰記錄 (Iwo Jima and Ryukyu Islands Air Operatisns Record) 1st Demobilization Bureau, Aug 46, p, 12. (4) Statements by Capt. Mitsuo Fuchida, Staff Officer (Operations), Combined Fleet, and Lt. Col. Takashi Kagoshima, Staff Officer (Air Operations), Imperial General Headquarters, Army Section.

repair and training activities in the Hiroshima-Kure area[11] and the Sixth Fleet under Vice Adm. Shigeyoshi Miwa, comprising Japan's 52 remaining submarines. For long-range offensive air operations Admiral Toyoda had at his disposal the Third Air Fleet under Vice Adm. Kimpei Teraoka in the home islands, the First Air Fleet under Vice Adm. Takijiro Onishi in Formosa,[12] and Rear Adm. Chikao Yamamoto's 11th Air Flotilla, an independent unit, on Kyushu. Of these, only the latter was at this time in a state of real combat readiness.

With the disintegration of the Japanese surface forces, the burden of intercepting invasion threats against the Homeland fell almost exclusively on the air establishment. The air situation, however, was extremely discouraging. Only about 550 aircraft were available for offensive purposes. Distribution of strength was as follows:[13]

Combined Fleet
- *First Air Fleet*
  - *Deployment: Formosa*
  - *Strength: 50 aircraft*
- *Third Air Fleet*
  - *Deployment: southwest Honshu, Shikoku, Kyushu and the Ryukyus*
  - *Strength: 200 aircraft*
- *11th Air Flotilla*
  - *Deployment: southern Kyushu*
  - *Strength: Navy—200 aircraft; attached Army—54 aircraft*

General Defense Command
- *Sixth Air Army*
  - *Deployment: central and southwest Honshu*
  - *Strength: about 50 aircraft*

The situation in regard to defensive air strength, vitally needed to guard the Homeland against large-scale enemy air raids from newly-won bases, was equally bad.[14] Available

11 On 1 January 1945, the organization of the Second Fleet was as follows:

1st Battleship Division	1st Carrier Division	2d Destroyer Squadron
Yamato	*Amagi*	*Yahagi* (CL)
Nagato	*Katsuragi*	Ten DD
Haruna	*Ryuho*	
	Junyo (XCV)	

Nihon Kaigun Hensei Suii oyobi Heiryoku Soshitsu Hyo 日本海軍編制推移及兵力喪失表 (Table Showing Organizational Changes and Losses of Japanese Naval Forces) 2d Demobitization Bureau, Oct 49, pp. D-3-5; J-44, 47.

12 Due to the situation in the Philippines, Combined Fleet decided that any further commitment of naval air forces to the Philippine area would be impracticable. First Air Fleet was accordingly transferred on 9 January to Formosa where it replaced the deactivated Second Air Fleet. (Cf. Chapter XI, p. 309 and Chapter XIV, p. 433) *Okinawa Homen no Kaigun Sakusen* 沖縄方面の海軍作戦 (Okinawa Area Naval Operations) 2d Demobilization Bureau, Oct 47, pp. 15–6.

13 (1) Ibid., pp. 12–3. (2) Report to the Throne, op. cit. (3) In addition to the units mentioned, the 8th Air Division operating off Formosa under Tenth Area Army command was available for limited attack operations over Homeland approaches. Although its 126 aircraft were primarily fighters organization of *tokko* units and recoupment of bomber strength was proceeding rapidly. (Statement by Lt. Col. Kanichi Ishikawa, Staff Officer (Operations), 8th Air Division.)

14 Although a few scattered raids had been conducted on northern Kyushu in the summer of 1944 by enemy B-29's operating from bases in China, the Superfortress campaign against Japan did not begin in earnest until November 1944. Operating from bases in the Marianas, the B-29's flew a total of 500 sorties against a variety of industrial targets in Tokyo and Nagoya during November and December, one raid involving about 100 enemy aircraft. The raids steadily mounted in frequency and intensity so that during January an additional about 600 sorties were flown, Kobe and Osaka being added to the target list. While these raids, using primarily high explosives, did little damage, they did serve to bring home to the people the desperate plight of the nation. (1) Imperial General Headquarters Army High Command Record, op. cit., p. 322. (2) *Jiji Nenkan* 時事年鑑 (Jiji Year Book) Jiji Press Co., Tokyo, 5 Jan 47, p. 151.

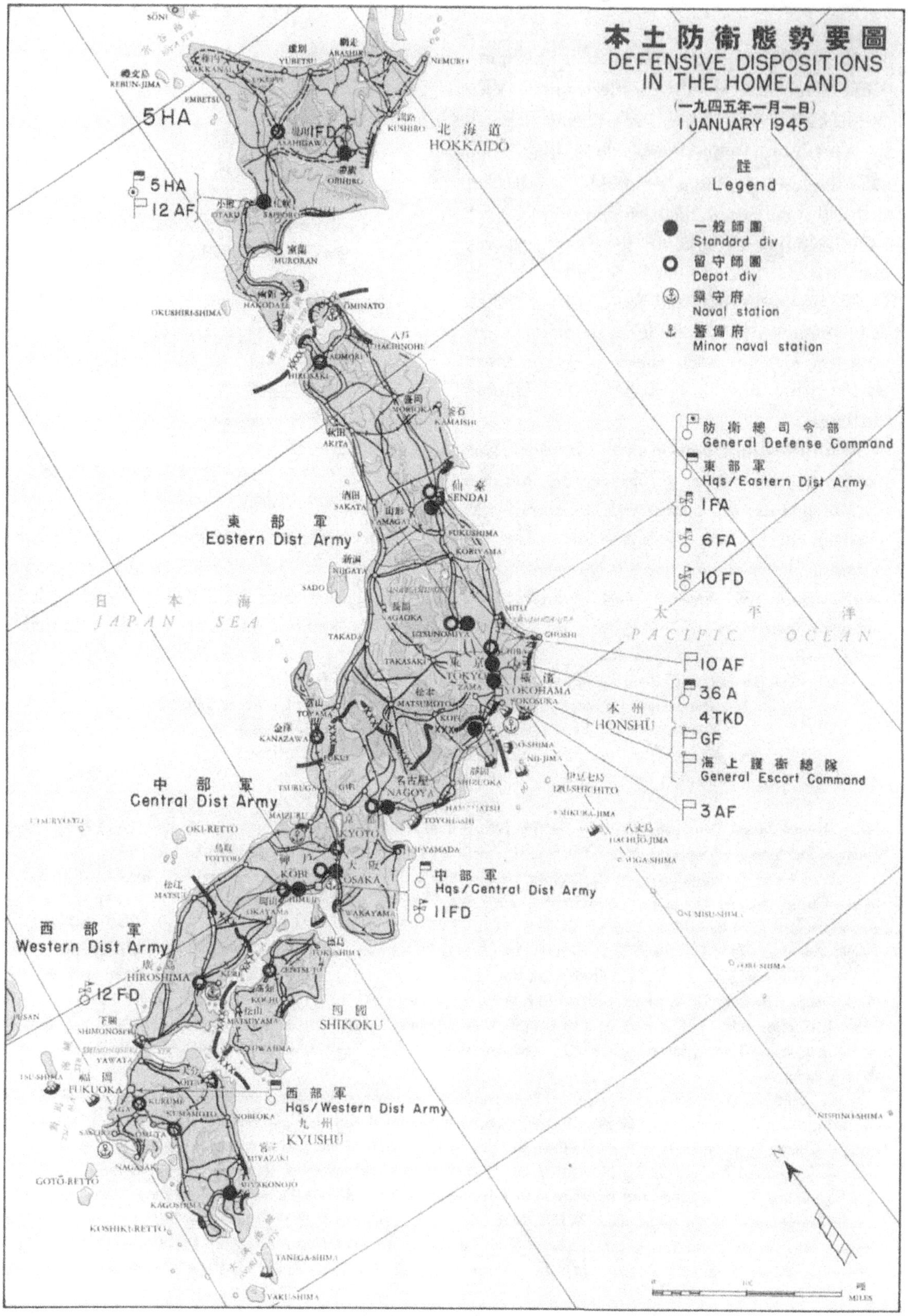

PLATE NO. 142
Defensive Dispositions in the Homeland, 1 January 1945

for this purpose was a total of about 770 aircraft and 1200 antiaircraft guns for all of Japan Proper (excluding Hokkaido). Air defense operations were decentralized, each District Army having under its command an air division for interceptor operations and a number of antiaircraft units. Disposition of the air strength attached from Sixth Air Army was as follows:[15]

Western District Army—12th Air Division (111 aircraft)
Central District Army—11th Air Division (267 aircraft)
Eastern District Army—10th Air Division (288 aircraft)

In addition, the 302d, 332d, and 352d Naval Air Groups were attached to the Eastern, Central, and Western District Armies respectively for interceptor operations. This added about 100 fighters to the air defense system. These Army and Navy forces were concentrated principally in the Tokyo-Yokohama, Osaka-Kobe, and Shimonoseki-Moji areas, and around vital naval installations such as Yokosuka, Kure, and Sasebo.

Primary emphasis in the *Sho* No. 3 plan, formulated in the summer of 1944, had been placed on sea and air operations.[16] As a result, the build-up of ground combat forces in Japan Proper during late 1944 had been on a very modest scale. A total of only eight divisions was available for operations in Japan Proper (excluding Hokkaido).[17] The command system and deployment were as follows:[18]

Western District Army
86th Division—Miyakonojo, Kyushu
Central District Army
44th Division—Osaka
73d Division—Nagoya
Eastern District Army
3d Imperial Guards Division—Tokyo
72d Division—Sendai
65th, 66th, and 67th Independent Mixed Brigades —Izu Islands
Thirty-sixth Army
81st Division—Utsunomiya

15 (1) Homeland Operations Record, op. cit. Vol. I: General Defense Command, pp. 32–33. (2) Report to the Throne, op. cit. (3) *Daikairei Dai Sanju-go* 大海令第三十號 (Imperial General Headquarters Navy Order No. 30) 19 Jul 44. (4) Table Showing Organizational Changes and Losses of Japanese Naval Forces, op. cit., pp. J-15, 19.

16 (1) *Dairikumei Dai Senhachijuichi-go* 大陸命第千八十一號 (Imperial General Headquarters Army Order No. 1081) 24 Jul 44. (2) *Dairikushi Dai Nisenhyakushichijuroku-go* 大陸指第二千百七十六號 (Imperial General Headquarters Army Directive No 2176) 22 Sep 44. (Cf. Chapter XI, pp. 295–301 and Chapter XII, pp. 328–30.)

17 All eight of these divisions had been activated during July 1944 as part of the preparation for *Sho* No. 3 operation. In addition to the indicated units, the 1st Imperial Guards Division, which had been formed on 14 May 1943 from the Imperial Guards Depot Division, was stationed in Tokyo under command of the Eastern District Army, but was considered as not being available in connection with the defense of the Homeland in view of its mission of securing the Imperial Palace, the role formerly executed by the Imperial Guards Depot Division. Moreover, the 84th Division, also activated in July 1944, was also in the Homeland waiting at Himeji for transportation to Okinawa. (Cf. n. 34) Also in the Homeland to aid the combat divisions in their preparations were the staff and overhead of 13 depot divisions located at Tokyo, Sendai, Utsunomiya, Kanazawa, Hirosaki, Nagoya, Osaka, Kyoto, Himeji, Zentsuji, Hiroshima, Kumamoto, and Kurume. (1) *Hengo Butai Gaiken Hyo* 編合部隊概見表 (Unit Organization Tables) Army General Staff. (2) Statment by Col. Hiroshi Fuwa, Staff Officer (Operations), First General Army. (3) Homeland Operations Record, op. cit. Vol. I., pp. 49–50.

18 (1) Unit Organization Tables, op. cit. (2) Homeland Operations Record, op. cit. Vol. I, pp. 49–50. (3) *Dairikumei Dai Senhachijuroku-go* 大陸命第千八十六號 (Imperial General Headquarters Army Order No. 1086) 26 Jul 44.

93d Division—Mt. Fuji area

4th Armored Division—Chiba and Narashino

Under the *Sho* No. 3 plan the mission of these units had been mainly the construction of coastal defense works in southern Kyushu, the Kanto plain, the Toyohashi-Hamamatsu area in central Honshu, and the Hachinohe district at the extreme northeastern tip of Honshu.[19] At the beginning of the year, construction of semi-permanent artillery positions was on schedule. Construction of infantry positions, on the other hand, was lagging seriously. Only in the Ariake Bay area of southern Kyushu were they more than 40% complete. In the Toyohashi and Hachinohe districts, completion had only reached about 10%, while in the Kanto area, construction had just been started.[20]

In contrast to the small beginnings that had been made on the defenses of Japan Proper, strong ground formations had been disposed on the Homeland defense perimeter. On Formosa and the Ryukyus was the Tenth Area Army with a powerful force of eight divisions, seven independent mixed brigades, and an air division.[21] The Ogasawara Islands were garrisoned by the Ogasawara Group, a heterogeneous combat formation built around the 109th Division.[22]

New Plans for Homeland Defense

By mid-January 1945, the progress of events clearly indicated the need for a new and far-

19 *Sho* No. 3 plan called for the employment of the following nuclear infantry and heavy artillery strength in these coastal defense positions:

	Infantry Battalions	Heavy Artillery Guns
Southern Kyushu	8½	20
Kanto Plain	11	70
Toyohashi-Hamamatsu	5	30
Hachinohe	3	0

Homeland Operations Record, op. cit. Vol. I, p. 53.

20 Because defense policy had not yet been definitely formulated by the High Command, most of these fortifications, following as they did the tactical idiosyncrasies of the local commanders, were later found to be inadequate and in some cases useless. (1) Ibid., pp. 51–4. (2) Homeland Operations Record, op. cit. Vol. III; Second General Army, pp. 6–7.

21 Tenth Area Army forces were disposed as follows:

Tenth Area Army
- 8th Air Division—Formosa
- 66th Division—Formosa
- 76th Independent Mixed Brigade—Formosa
- Fortieth Army—Formosa
  - 9th, 12th, 50th, and 71st Divisions
  - 75th Independent Mixed Brigade
- Thirty-second Army—Ryukyus
  - 24th, 28th, and 62d Divisions
  - 44th, 45th, 59th, 60th and 64th Independent Mixed Brigades

(1) *Dai Sanjuni Gun Shijitsu Shiryo* 第三十二軍史實資料 (Historical Data, Thirty-second Army) Home Depot Division, Mar 47, pp. 7–10. (2) *Taiwan Homen Sakusen Kiroku* 臺灣方面作戰記錄 (Formosa Area Operations Record) 1st Demobilization Bureau, Aug 46, pp. 105–12. (3) Unit Organization Tables, op. cit.

22 Attached to the 109th Division were the 145th Infantry Regiment, the 12th and 17th Independent Mixed Regiments, and the 26th Tank Regiment. (1) *Chubu Taiheiyo Homen Sakusen Kiroku* 中部太平洋方面作戰記錄 (Central Pacific Operations Record) 1st Demobilization Bureau, Nov 46, Vol. I, 98, 103–4. (2) *Dairikumei Dai Sensanju-hachi, Senyonjugo-go* 大陸命第千三十八、千四十五號 (Imperial General Headquarters Army Orders No. 1038 and 1045) 26 Jun and 1 Jul 45.

reaching strategic plan to replace the now-defunct *Sho-Go* plan. In the pessimistic and confused official atmosphere of this period, further complicated by inter-service differences, it was difficult to formulate even the most general of policy directives.[23] However, the need for action was imperative, and Imperial General Headquarters, on 19 January, submitted for Imperial sanction the draft of a general policy directive known as the "Outline of Army and Navy Operations." This directive, having been approved by the Emperor, was officially promulgated on 20 January and became the basis for all future Homeland defense planning. Its essence was as follows:[24]

1. General Policy

a. The final decisive battle of the war will be waged in Japan Proper.

b. The armed forces of the Empire will prepare for this battle by immediately establishing a strong strategic position in depth within the confines of a national defense sphere delineated by the Bonin Islands, Formosa, the coastal sector of east China, and southern Korea.

c. The United States will now be considered Japan's principal enemy. Operational planning of all headquarters will be directed toward interception and destruction of American forces, all other theaters and adversaries assuming secondary importance.

2. Preparation and Conduct of Operations

a. Resistance will continue in the Philippines so as to delay as long as possible the enemy's approach to the Homeland defense perimeter.

b. Key strongpoints to be developed within the perimeter defense zone include Iwo Jima, Formosa, Okinawa, the Shanghai district, and the south Korean coast. The main defensive effort will be made in the Ryukyus area. Preparations in the perimeter defense zone will be completed during February and March 1945.

c. When the enemy penetrates the defense zone, a campaign of attrition will be initiated to reduce his

---

23 The chief controversy centered around air operations over the Homeland defense perimeter. The Army insisted that a maximum effort be planned over the East China Sea area, even to the extent of stripping eastern Japan of its strength. The Navy, on the other hand, was extremely chary of undertaking any large scale air operation until May, wishing to avoid a premature, piecemeal commitment while rebuilding its shattered air forces. The Army point of view finally prevailed and became High Command policy. (Statements by Rear Adm. Sadatoshi Tomioka, Chief, First Bureau (Operations), Capt. Toshikazu Ohmae, Staff Officer (Operations), and Comdr. Yoshimori Terai, Staff Officer (Operations), all of Imperial General Headquarters, Navy Section; and Col. Ichiji Sugita, Staff Officer (Operations), and Lt. Col. Takashi Kagoshima, Staff Officer (Air Operations), both of Imperial General Headquarters, Army Section.)

24 (1) *Daikairei Dai Sanjushichi-go Bessatsu Teikoku Rikukaigun Sakusen Keikaku Taiko* 大海令第三十七號別冊帝國陸海軍作戰計画大綱 (Imperial General Headquarters Navy Order No. 37, Separate Table, Outline of Army and Navy Operations) 20 Jan 45. (Plate No. 143 is a facsimile reproduction of a portion of this document.) (2) Report to the Throne, op. cit. (3) Statements by Lt. Gen. Shuichi Miyazaki, Chief, First Bureau (Operations), Col. Hiromu Hosoda, Staff Officer (Operations), both of Imperial General Headquarters, Army Section, and Col. Takushiro Hattori, previously cited. (4) To support the operational plans the Supreme War Direction Council, on 25 January, laid down the basic national logistics plan. Top priority was given to aircraft, special-attack weapons, antiaircraft, petroleum products, food, and ships. Raw material and basic commodity goals for 1945 were set as follows: Carbonsteel–3,000,000 metric tons; coal–55,000,000 metric tons; liquid fuel (all types)–15,750,000 barrels; and shipping–2,040,000 metric tons. In view of the state of Japanese economy at the time, these goals were completely unrealistic. (Cf. pp. 533–4) Proceedings of the Supreme War Direction Council, op. cit., Report No. 9, 25 Jan 45. (5) On 20 January, Imperial General Headquarters planned a last emergency effort to transport vital strategic raw materials from the southern area. This was to be a blockade-running operation along the China coast, involving a strong naval and air escort. The plan, designated the *Nan-Go* (南號) Operation, called for the transport of 750,000 barrels of oil and quantities of rubber and bauxite to the Homeland during March. *Daikaishi Dai Gohyaku-go* 大海指第五百號 (Imperial General Headquarters Navy Directive No. 500) 20 Jan 45.

preponderance in ships, aircraft, and men, to obstruct the establishment and use of advance bases, to undermine enemy morale, and thereby to seriously delay the final assault on Japan. The air forces will make a maximum effort over the perimeter defense zone. Enemy troops that succeed in getting ashore at points on the Homeland defense perimeter will be dealt with by those ground forces on the spot without reinforcement from other theaters.

d. Emphasis in ground preparations will be laid on Kyushu and Kanto. Strong air defenses will be established along key lines of communication, such as the Shimonoseki and Korea Straits, and at important ports and communications centers such as Tokyo, Nagoya, Osaka-Kobe, and the Moji-Kokura-Yawata area.

e. During the delaying operations in the forward area, preparations for the decisive battle will be completed in Japan Proper by the early fall af 1945.

f. In general, Japanese air strength will be conserved until an enemy landing is actually underway on or within the defense sphere. The Allied invasion fleet will then be destroyed on the water, principally by sea and air special-attack units.

Pursuant to this basic policy directive, the Army Section of Imperial General Headquarters assigned missions to its major subordinate commands as follows:[25]

1. The General Defense Command will secure the Homeland by destroying any enemy invading force. Emphasis in operational preparations will be on the Kyushu, Kanto, and Tokai[26] districts. Efforts will be made to annihilate the Allied forces on the sea through the vigorous application of special-attack tactics.

2. The China Expeditionary Army will shift to a two-front campaign and concentrate forces in the east and south China coastal sectors. It will secure the continental key area by destroying any enemy invading force. Main defensive effort will be in the lower reaches of the Yangtze River.

3. Southern Army will secure key points in the southern area and check the advance of the enemy toward the Homeland defense perimeter, thus facilitating the over-all operations.

4. Tenth Area Army will secure Formosa and the Ryukyus by destroying any enemy invading force.

5. Seventeenth Area Army[27] will shift operational emphasis from northern to southern Korea and secure key sectors by destroying any enemy invading force.

On 30 January, the Navy Section of Imperial General Headquarters conducted a conference in Tokyo of the chiefs-of-staff of all fleets and naval districts. The "Outline of Army and Navy Operations" was presented to the conference and preliminary plans and general missions promulgated.

Immediately following the publication of the "Outline of Army and Navy Operation," joint conferences were held to iron out difficulties in the actual implementation of the plan. The first and most important problem was to formulate a sound and workable air policy. Efforts in that direction culminated on 6 February in the drafting of a Joint Army-Navy Air Agreement for the first half of 1945. This was to be subject to ratification by the two services

25 *Dairikumei Dai Sennihyakunijuhachi, Sennihyakusanjuroku, Sennihyakuyonjuni, Sennihyakuyonjugo-go* 大陸命第千二百二十八、千二百三十六、千二百四十二、千二百四十五號 (Imperial General Headquarters Army Orders No. 1228, 1236, 1242, and 1245) 22 Jan, 27 Jan, 3 Feb, and 6 Feb 45.

26 Although it was not mentioned in the "Outline of Army and Navy Operations," the Army Section felt that there was a chance the enemy might attempt a secondary landing in the Tokai district around Nagoya. Defense of this district was therefore added to the missions of the General Defense Command. Special concern was felt for the coastal areas on both sides of the entrance to Ise Bay.

27 The operational missions of the Korea Army were assumed by the Seventeenth Area Army which was activated on 6 February as part of the reorganization plan. (Cf. p. 546.) Since the transfer, 20 November 1944, of the 19th Division to the Fourteenth Area Army in the Philippines (Cf. Chapt. XIII, n. 159), the principal operational strength of the Korea Army consisted of three depot divisions.

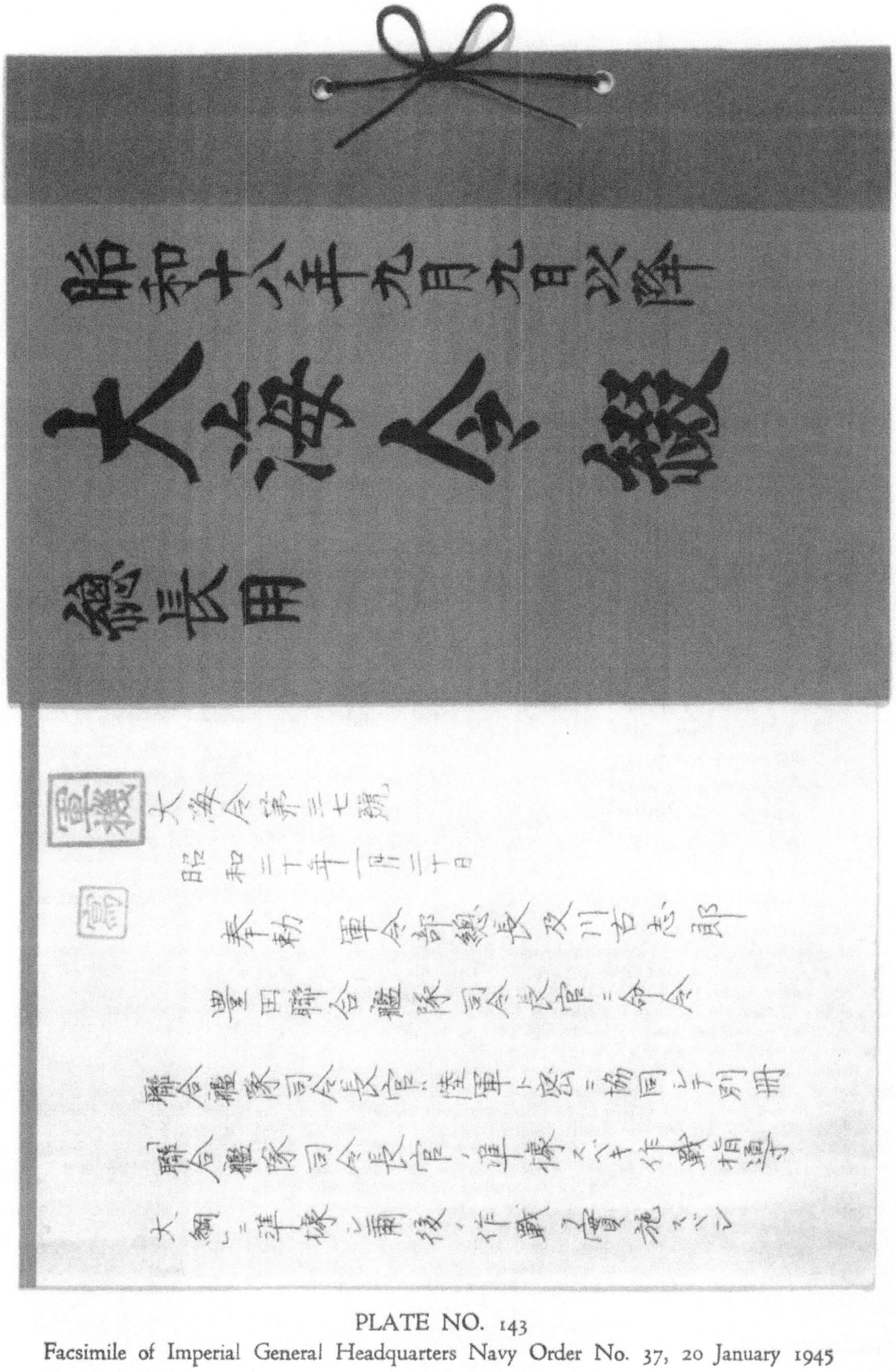

PLATE NO. 143

Facsimile of Imperial General Headquarters Navy Order No. 37, 20 January 1945

after study on the operational levels. In it essentials, the draft was as follows:[28]

1. All Army and Navy air forces in the Homeland (less air defense and training forces) will be concentrated in the East China Sea area (Formosa, the Ryukyus, east China, and Korea) during the months of February and March 1945. This concentrated air strength, together with air units already in Formosa (First Air Fleet and 8th Air Division), plus certain reinforcements from other theaters, enumerated below, will crush any enemy attempt to invade points within the aforementioned area.

2. Primary emphasis will be laid on the speedy activation, training, and mass employment of air special-attack units.

3. The main target of Army aircraft will be enemy transports and of Navy aircraft, carrier task forces.

4. Scheduled Strength (to be assembled by 1 April)

Army

Basic Force—1175 aircraft

Sixth Air Army—735

8th Air Division—440

Reinforcements—215 aircraft

from Fifth Air Army (China)—175

from Third Air Army (SE Asia)—40

Navy (Tentative)

Basic Forces—400–580 aircraft

Third Air Fleet and Eleventh Air Flotilla —350–480

First Air Fleet—50–100

Reinforcements—125–175 aircraft from China Area Fleet and Thirteenth Air Fleet (SE Asia)

5. Command System: The basic command relationship will be one of inter-service cooperation. Coordination will be effected through Combined Fleet for the Navy air forces and through the General Defense Command and Tenth Area Army for the Army air forces.

On the same day that the draft of this joint agreement was issued, the Army Section of Imperial General Headquarters published an implementing directive entitled "Outline of Air Operations in the East China Sea Area." In addition to the provisions of the joint agreement on which it was based, this plan contained detailed directions for the preservation and replenishment of existing units, strengthening of bases, and the redeployment necessary to bring the required air power into the theater by 1 April. This plan, designated the *Ten-Go* (天號) Operation Plan, became the basis for all future Army air operations over the East China Sea Area.[29]

To prepare for the forthcoming operations,

28 (1) Imperial General Headquarters Army High Command Record, op. cit., pp. 359–65. (2) Statement by Lt. Col. Katsuki Mizumachi, Staff Officer (Operations), Sixth Air Army. (3) Although the Army air forces at this time were weaker numerically than the Navy's, the Army High Command had a very optimistic outlook on the production and training situation. It was relatively easier for the Army to train pilots for attack on enemy transports since the Navy's mission of attacking carriers and other combat types required more skillful personnel. As far as the Navy was concerned, it was extremely difficult to foresee the actual availability of strength at the end of March. It therefore assumed a conservative attitude toward making its commitment under the agreement final. Although Army ratification was immediate, the Navy delayed final action until 1 March while studying the actual trends of production and training. (Statements by Capt. Ohmae and Col. Sugita, both previously cited). (4) The Naval High Command nevertheless recognized that positive measures should be adopted in an effort to delay the enemy's next operation as long as possible. Accordingly, plans were completed by mid-February for a force of Kyushu based long-range bombers to conduct a surprise raid against an enemy carrier group in Ulithi anchorage. This attack, designated the Second *Tan* Operation was executed on 11 March but was assessed as completely unsuccessful on the basis of subsequent photographic reconnaissance. Only 11 of the original 24 planes arrived over Ulithi, and these arrived after sunset, rendering accurate bombing impossible. Okinawa Area Naval Operations, op. cit., pp. 22–7. (American Editors Note: In early August 1944, The Combined Fleet planned to carry out an air attack on the Allied carrier groups anchored in the Majuro Atoll (Marshall Islands) in order to delay the next Allied invasion. This plan, designated the First *Tan* (丹) Operation, called for a long range, one way flight, by Navy land based medium bombers; however, the plan was never put into effect.)

29 (1) Iwo Jima and Ryuku Islands Air Operations Record, op. cit., p. 15. (2) Statement by Lt. Col. Mizumachi, previously cited.

the Navy, on 10 February reorganized its air units within Japan Proper. The Third Air Fleet, with headquarters in the Kanto district, was relieved of all further responsibility for the Kyushu-Ryukyus area and assigned exclusively to operations in central and eastern Honshu. The Fifth Air Fleet was activated on Kyushu, absorbing the 11th Air Flotilla and all former Third Air Fleet units in the area. It assumed responsibility for future naval air operations in the East China Sea area under the joint agreement.[30]

For further implementation of the "Outline of Army and Navy Operations," Imperial General Headquarters quickly realized that reorganization and redeployment of the ground establishment was also imperative. Of first priority was the reorganization of the major commands in Japan Proper. This was accomplished by a War Ministry order of 22 January and an Imperial General Headquarters order of 6 February, which transferred the operational missions of the old district armies to a number of new area Army headquarters.[31] New district Army commands were established in each area Army zone to assume responsibility for logistics and administrative matters.[32] The General Defense Command was now constituted as follows: (Plate No. 144)[33]

Northeast Honshu
Eleventh Area Army—Lt. Gen. Teiichi Yoshimoto
Northeastern District Army Command—Sendai
East-Central Honshu
Twelfth Area Army—Gen. Keisuke Fujie
Eastern District Army Command—Tokyo
West-Central Honshu
Thirteenth Area Army—Lt. Gen. Tasuku Okada
Tokai District Army Command—Nagoya
Western Honshu and Shikoku
Fifteenth Area Army—Lt. Gen. Masakazu Kawabe
Central District Army Command—Osaka
Kyushu
Sixteenth Area Army—Lt. Gen. Isamu Yokoyama
Western District Army Command—Fukuoka

---

30 (1) Okinawa Area Naval Operations, op. cit., pp. 13–4. (2) It was intended that, should a decisive air battle be fought over the East China Sea, the Third Air Fleet would constitute a reinforcement pool for Fifth Air Fleet. (Statement by Capt. Ohmae, previously cited).

31 During the early period of the war the principal functions of the district armies, centering around the activation, organization and training of newly formed units, had been carried out directly under the command of Imperial General Headquarters. The General Defense Command existed as a headquarters solely for the purpose of exercising operational control over the district armies in matters connected with internal security and air defenses. (Cf. Plate No. 10) In May 1944, however, the General Defense Command was invested with complete operational control over the Eastern, Central and Western District Armies, each of the latter continuing to deal directly with the War Ministry in all logistical matters. Shortly thereafter, pursuant to the provisions of *Sho* No. 3 plan, these district armies were assigned operational control over the newly organized divisions in their respective areas. (1) Homeland Operations Record, op. cit., Vol. I, pp. 1–13. (2) Imperial General Headquarters Army High Command Record, op. cit., pp. 242–3. (3) Unit Organization Tables, op. cit.

32 The commander of each Area Army was also the commander of the District Army Command and therefore held a dual responsibility, reporting to the Commander-in-Chief, General Defense Command, for operations and to the War Ministry for logistics and administration.

33 (1) Imperial General Headquarters Army High Command Record, op. cit., p. 325. (2) *Dairikumei Dai sennihyakuyonjuyon-go* 大陸命第千二百四十四號 (Imperial General Headquarters Army Order No. 1244) 6 Feb 45. (3) The "Outline of Army and Navy Operations" had in effect called for a shift in emphasis from the Northeast Area to central and western Japan. The requirements of the Fifth Area Army were accordingly scaled down in subsequent reorganizations and redeployments. This process began on 23 January with the transfer of Twenty-seventh Army headquarters to Sendai where it was subsequently reorganized as Headquarters, Eleventh Area Army. *Dairikumei Dai Senninyakunijuku-go* 大陸命第千二百二十九號 (Imperial General Headquarters Army Order No. 1229) 22 Jan 45.

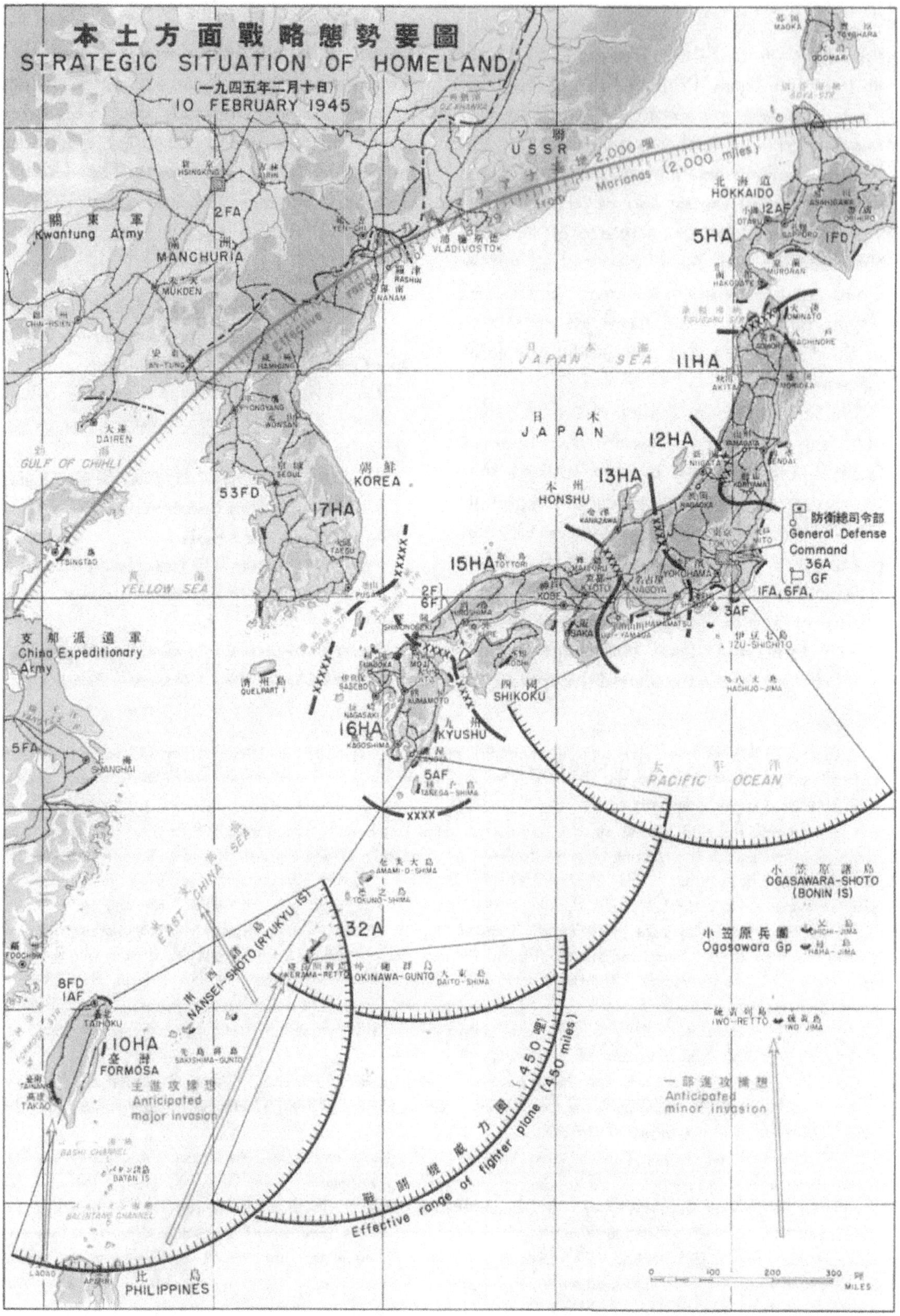

PLATE NO. 144
Strategic Situation of Homeland, 10 February 1945

Mobile Reserve—Kanto Plain

Thirty-sixth Army—Lt. Gen. Toshimichi Uemura

The reorganization of the high-level headquarters in Japan Proper was of course only a first step toward building up the ground combat forces. Although one division had been added to the Homeland forces on 22 January,[34] there were still too few troops to garrison even the most critical points. Concurrently with the activation of the Homeland area armies, therefore, the Army High Command ordered the immediate organization of four independent mixed brigades as an emergency mobilization to fill the most critical gaps as rapidly as possible. These were as follows:[35]

95th Independent Mixed Brigade—Eleventh Area Army

96th Independent Mixed Brigade—Twelfth Area Army

97th Independent Mixed Brigade—Thirteenth Area Army

98th Independent Mixed Brigade—Sixteenth Area Army

Following this emergency measure, the Army undertook to establish a firm troop basis for the future Homeland defense armies. Discussions between the War Ministry and the High Command culminated on 26 February in the adoption of a plan calling for the mobilization of 42 divisions, 18 independent mixed brigades,[36] and six tank brigades, the bulk of this force to be added to the one armored and eight line combat divisions and seven independent mixed brigades at that time active in the General Defense Command. These units, together with the required logistic and administrative support elements, would contain a total of about 1,500,000 men. Mobilization of this enormous force was to be accomplished in three stages, the first from late February to early April, the second during April, and the last by the end of September. To provide the necessary high level headquarters, it was also planned to activate nine Army headquarters for tactical field command and two general Army headquarters to exercise command at Army group level. The activation schedule for units in the Homeland (less Hokkaido) was as follows:[37]

---

34 The 84th Division was at Himeji staging for Okinawa. (Cf. n. 17, p. 541.) On 22 January, on the eve of its scheduled departure, Imperial General Headquarters directed that the division remain in the Homeland. Lt. Gen. Shuichi Miyazaki, Chief, First Bureau (Operations), Imperial General Headquarters, Army Section, felt that the forces on the defense perimeter were already sufficient without the addition of this division, and that the extremely weak state of the Homeland defense forces justified its diversion. On 6 February, it was assigned to Fifteenth Area Army, pending its final deployment. The withholding of this division from Okinawa was to have grave consequences. (Cf. n. 64) (1) Unit Organization Tables, op. cit. (2) Statement by Col. Sugita, previously cited. (3) Historical Data, Thirty-second Army, op. cit.

35 In addition, three infantry brigades, the 1st, 2d and 3d Tokyo Garrison Brigades, were organized and placed under the Eastern District Army Command with a specific mission of carrying out guard duty only. Of the mixed brigades, the 95th was stationed at Hachinohe, the 96th in Chiba Prefecture, the 97th in the Ise Bay area, and the 98th in the Ariake Bay area. In addition, two divisions were activated, the 79th in Korea and the 96th in Kyushu, for subsequent assignment to the Seventeenth Area Army on 28 February. (1) Unit Organization Tables, op. cit. (2) Imperial General Headquarters Army High Command Record, op. cit., p. 308. (3) *Dairikumei Dai Sennihyakugojuku-go* 大陸命第千二百五十九號 (Imperial General Headquarters Army Order No. 1259) 18 Feb 45.

36 Of these, two divisions and two independent mixed brigades were to be activated in Korea, one division and one independent mixed brigade on Hokkaido, and two divisions in the Kuriles and on Karafuto, the former from the 43d Independent Mixed Brigade and the 69th Independent Mixed Brigade, the latter from the Karafuto Mixed Brigade. (1) Statement by Col. Hosoda, previously cited. (2) Unit Organization Tables, op. cit. (3) Imperial General Headquarters Army High Command Record, op. cit., pp. 305–12.

37 In addition one Army headquarters was to be activated in Korea. (1) Imperial General Headquarters Army High Command Record, op. cit., pp. 308–10. (2) Homeland Operations Record, op. cit., Vol. II: First General Army, pp. 8–11, and Attached Maps 1–2; Vol. III: Second General Army, Attached Charts 1–3; Vol. IV, pp. 11–8, 28–38 and Attached Maps 1–2; Vol. V; Seventeenth Area Army, pp. 10–21. (3) Unit Organization Tables, op. cit. (4) *Dairikumei Dai Sennihyakukyujushichi, Sensambyakunijuhachi, Sensambyakugojuichi-go* 大陸命第千二百九十七、千三百二十八、千三百五十一號 (Imperial General Heaquarters Army Orders Nos. 1297, 1328 and 1351) 8 Apr, 10 May and 19 Jun 45. (5) Statement by Col. Hosada, previously cited.

1. *First Mobilization*
    - *13 coastal combat divisions*[38]
    - *1 independent mixed brigade*
2. *Second Mobilization*
    - *2 general Army headquarters*
    - *8 Army headquarters*
    - *8 line combat divisions*
    - *6 tank brigades*
3. *Third Mobilization*
    - *7 line combat divisions*
    - *9 coastal combat divisions*
    - *14 independent mixed brigades*

In addition, during the period of the second mobilization, three line combat infantry divisions and one armored division were to be moved from Manchuria to the Homeland. This redeployment, together with the planned mobilizations, brought the projected total of Homeland (less Hokkaido) defense ground forces to a basic combat strength of 26 line combat divisions, 22 coastal combat divisions, and 21 independent mixed brigades.[39] Armored composition was fixed at two armored divisions and six tank brigades.

The problem of supplying weapons and equipment to the new units was a serious one. However, when the mobilization plan was drafted, the total amount of weapons in the hands of Army ordnance supply, together with those to be produced during February and March plus weapons slated for transfer from the continent, was fortunately sufficient to permit a large enough initial issue to enable the units to begin their training.[40] Provided production could be held at present levels, it was felt that the ordnance industry could fill the remaining combat equipment requirements by September. On this basis, the War Ministry set the following ordnance production targets for the period April through September:[41]

Rifles	*523,200*
Light machine guns	*9,360*
Machine guns	*1,260*
Infantry cannon	*2,160*
Antiaircraft guns	*606*
Mortars	*3,300*
Self-propelled cannon	*168*
Light artillery	*178*
Heavy artillery	*50*

The production schedule for sea special-attack weapons was simultaneously set as follows:[42]

38 The coastal combat division was a type organization developed to engage in close-quarter fighting against enemy landings. It was characterized by lower troop strength. In infantry fire power it was superior to the regular line combat division, particularly in grenade dischargers and light and heavy machine guns. In artillery fire power it was markedly inferior. Due to its mission, the coastal combat division was practically immobile having only about one-fourth the horse and vehicular capacity of the average line combat division, this being used primarily for logistic support. Divisions bearing numbers in the 100 and 300 blocks were coastal combat divisions with a few exceptions, and all others were line combat divisions. *Dai Iichiji oyobi Dai Sanji Doin Shidan Shuyo Heiki Genkyo Hyo* 第一次及第三次動員師團主要兵器現況表 (Actual Condition of Ordnance Equipment for Divisions in the 1st and 3d Group Mobilizations) Ordnance Section, Fifteenth Area Army Headquarters, 20 Jun 45.

39 The 321st Division, scheduled for activation in the third mobilization, was to absorb the 65th Independent Mixed Brigade stationed on the Izu Islands. Unit Organization Tables, op. cit.

40 Typical of the partial issue of available weapons were rifles, 50 %, light machine guns, 23 %, infantry cannon, 28 %, and antitank guns, 74 %. Some categories (e. g. machine guns and light artillery) were already well over 75 % available. (Statements by Maj. Tosaku Hirano and Maj. Yoshimichi Furukawa, Staff Officers (Organization and Equipment), Imperial General Headquarters.)

41 Statistics and Analysis Reports, op. cit., Report No. 17: Production of Japanese Army Ordnance.

42 (1) *Showa Niju Nendo Zenki no Kokuki, Tokko Heiki no Seisan Keikaku to Seisan Jisseki* 昭和二十年度前期の航空機、特攻兵器の生產計畫と生產實績 (Planned and Actual Production of Aircraft and *Tokko* Weapons in the First Half on 1945) 2d Demobilization Bureau, 15 Jul 49. Attached Chart II. (2) Statements by Maj. Furukawa, and Maj. Hirano, both previously cited.

Army	
Renraku-tei (crash boats)	3,000
Navy	
Shinyo (crash boats)	3,840
Koryu and Kairyu (midget submarines)	1,440
Kaiten (human torpedo)	660

Aircraft production presented a special problem due to the concentrated attention being given the industry by enemy air units. Based on the assumption that production could be increased slightly over the average output in December and January,[43] the Supreme War Direction Council adopted a figure of 16,000 planes as the goal for the period April through September.[44] Production during the quarter ending in March was to be allocated exclusively to the *Ten-Go* Air Operation.

Attacks on the Homeland Defense Perimeter

Long-range preparations for the defense of the Homeland had barely gotten underway when enemy attacks on the defense perimeter began. On 19 February, a powerful U. S. amphibious task force launched an invasion of Iwo Jima, keystone of the Ogasawara Island sector of the perimeter.

At the same time, the enemy violently accelerated the aerial offensive against Japan Proper, striking heavily at industrial targets and airfields in Tokyo, Osaka-Kobe, and Nagoya. From a total of 598 sorties flown over Japan Proper in January, Allied air activity rocketed to 3,193 sorties in February.[45]

To slow up the mounting American air offensive against Japan, it was imperative that the enemy be denied forward bases on the Homeland defense perimeter. Iwo Jima consequently assumed an importance far out of proportion to its actual size and facilities, because from its airfields enemy fighter planes would be able to fly both escort and attack missions over almost all of Japan Proper south of Sendai. (Plate No. 146)

Despite the importance of Iwo Jima, the Japanese were unable at this time to make any large-scale air commitment to its defense. The Ogasawara Islands did not afford a good chain of mutually supporting bases, and they were too far from Japan Proper. In any case, the Japanese air establishment was still in too debilitated a state to seek a decisive battle in the area. The defense of Iwo Jima therefore fell exclusively to the ground forces. For over a month Lt. Gen. Tadamichi Kuribayashi's Ogasawara Group (109th Division) held out against immensely superior American forces, fighting a tenacious battle in the many excellent caves and field fortifications on the island. By 22 March, despite this heroic resistance, Iwo Jima fell completely into enemy hands.[46]

As the campaign on the Homeland defense perimeter began, the enemy unleashed a series

43 Production in December was 2,200 and in January 1,900, the loss being due to raw material shortages and air raids. In February, production declined still further to 1,260 mostly as a result of the program of dispersal of the industry carried out in anticipation of heavy enemy air raids. As soon as plants were well established in their new locations, the High Command expected the figure to go back up to the January level. Reports to the Diet, op. cit: Summary of Production of Aircraft, Loss of Capacity through Air Raids, 1 Sep 45.

44 Proceedings of the Supreme War Direction Council, op. cit., Report No. 9.

45 On 16 February, about 1000 U. S. carrier planes staged a raid on Tokyo. Attacks of the same scale came on the 17th and 25th. During the last two weeks in February B-29's struck five times at industrial targets in the Tokyo, Nagoya, and Osaka-Kobe districts. The 25 February raid on Tokyo consisted of at least 130 aircraft, the largest number of B-29's recorded up to that time. (1) Imperial General Headquarters Army High Command Record, op. cit., p. 322. (2) *Dai Niji Sekai Taisen Ryakureki Otsu* 第二次世界大戰略歷(乙) (Abridged Chronicle of World War II, B) 2d Demobilization Bureau, Mar 46. Part III, pp. 30–1. (3) Jiji Year Book (1947), op. cit., p. 151.

46 (1) Central Pacific Operations Record, op. cit., Vol. I, p. 121. (2) *Iwo Jima Sakusen* 硫黃島作戰 (Iwo Jima Operations) 2d Demobilization Bureau, Jan 48, pp. 4–5.

of air attacks on large urban areas in the Homeland which rocked the nation to its very foundations. Switching from high explosives to incendiaries, the B-29's began these operations on the night of 9-10 March with a heavy raid on Tokyo. The new tactics caught the Japanese completely off-guard, and the results were indescribably horrifying. Well over 250,000 houses were destroyed, rendering more than a million persons homeless, and 83,793 were burned to death.[47] Between 10 and 17 March, raids on the same pattern were flown against Osaka, Kobe, and Nagoya. B-29 raids from November through February had been little more than an annoyance to the Japanese, but with the beginning of the fire-bomb campaign, enemy strategic bombing loomed for the first time as a threat to the entire social and economic fabric of the nation.

With the invasion of Iwo Jima and the sudden acceleration of the aerial offensive against Japan, it was clear that a move into the East China Sea area could not be far distant. The Navy, which had originally entertained considerable reservations about an air operation in that area before May, suddenly took steps to meet the emergency.

First of all, the Army-Navy Air Agreement of 6 February, which until now had been considered a highly tentative document, was formally ratified by the Navy High Command on 1 March. At the same time, plans were made to strengthen naval air participation in the forthcoming campaign. The 11th, 12th, and 13th Combined Air Training Groups were converted into operational units. To take over the new groups, Tenth Air Fleet was activated on 1 March in the Kanto district, Vice Adm. Minoru Maeda assuming command. The Third Air Fleet, whose mission under the original plans had been to garrison eastern Honshu and provide a reinforcement pool for the East China Sea air battle, was now informed that, upon the activation of the operation, it would immediately displace to Kyushu in full strength. Both the Third and Tenth Air Fleets were ordered to begin intensive training in special-attack methods. This training was to be completed by the end of April, when Tenth Air Fleet was also to move to Kyushu bases. The planned strength of Navy air units in the East China Sea operation was now fixed as follows:[48]

Fifth Air Fleet	*520 aircraft*
Third Air Fleet	*510*
Tenth Air Fleet	*2000*
First Air Fleet	*85*

On 20 March, the Navy High Command issued an over-all policy directive which clearly set forth the basic concept of the Navy's participation in the Homeland defense campaign. This document, entitled the "Imperial Navy Outline Plan of Immediate Operations," contained the following general provisions:[49]

47 (1) *Kushu Saigai Jokyo Ichiran Hyo* 空襲災害狀況一覽表 (Chart of Damages by Air Raids) Metropolitan Police Board of Tokyo, 1949. (2) "The raid was in an area off the Sumida River criss-crossed by many rivers and canals .... This was all encircled with fire .... People were unable to escape. They were found later piled up on the bridges, roads, and in the canals, 80,000 dead and twice that number injured. We were instructed to report on actual conditions. Most of us were unable to do this because of horrifying conditions beyond imagination." U. S. Strategic Bombing Survey (Pacific), Morale Division, *The Effects of Strategic Bombing on Japanese Morale*, Jun 47, pp. 36–7. (Statement by an unidentified official of the Police Bureau, Ministry of Home Affairs.)

48 (1) Okinawa Area Naval Operations, op. cit., p. 11. (2) *Daikaishi Dai Gohyakuju-go* 大海指第五百十號 (Imperial General Headquarters Navy Directive No. 510) 1 Mar 45.

49 (1) *Daikaishi Dai Gohyakujusan-go* 大海指第五百十三號 (Imperial General Headquarters Navy Directive No. 513) 20 Mar 45. (2) The Navy was by this time convinced that the East China Sea air operation would be decisive as regards the future defense of the Homeland and planned a maximum effort. The Navy's concept differed from the Army's in that the Army regarded the operation as a perimeter defense or delaying action, rather than as decisive. *Bakugeki Chosa Dan Kaigun Bunkakai Dai Shichi-go ni taisuru Kaito* 爆擊調査團海軍分科會第七號ニ對スル回答 (Reply of the Navy Ministry to Questionnaire No. 7, U.S. Strategic Bombing Survey (Pacific), Naval Analysis Division) 18 Oct 45.

民防空陣の活躍　鈴木　誠　昭和二〇・[illegible]

Original Painting by Makoto Suzuki

PLATE NO. 145

Civilian Air Raid Defense Activity: Women Fire-Fighters

1. The Ryukyu Islands are designated as the focal point of the decisive battle for the defense of the Homeland.

2. Special emphasis will be placed on the destruction of enemy ships by the prompt and vigorous use of sea and air special-attack forces.

3. Hit-and-run raids will be conducted against forward American attack bases to delay the launching of an enemy invasion of the East China Sea area.

4. At the same time, the defenses of the Homeland will be strengthened with emphasis on Kyushu and the Kanto area. Defenses at important straits and bay entrances will be strengthened. Sea routes to the continent will be protected.

Meanwhile, the preliminaries of the East China Sea air battle were already underway, giving the Japanese no time to complete their long-range preparations. Early on the morning of 17 March, Imperial General Headquarters learned through intelligence channels that sizeable American fleet elements had quit Ulithi atoll in the western Carolines. It was assumed that the enemy task force was on its way to attack the Kyushu area. The attack was expected the following day.

Combined Fleet immediately alerted Fifth Air Fleet in Kyushu. The Air Fleet commander, Vice Adm. Matome Ugaki, was ordered to attack the enemy task force only if it contained invasion transports and to refrain from an engagement if it was composed exclusively of warships.[50] Vice Adm. Ugaki, however, feared that, if he did not attack, he would lose his entire command on the ground. He therefore forwarded a strong recommendation to Tokyo that counteraction be taken regardless of the composition of the task force. The High Command forthwith released the local commander from the earlier injunction and instructed him to use his own judgement.[51]

At 2300 on the 17th, a Fifth Air Fleet search mission detected by radar a large enemy task force barely 250 miles southeast of the southern tip of Kyushu.[52] Vice Adm. Ugaki immediately issued orders for a dawn attack in force. Even as this strike mission was taking off, enemy carrier planes were swarming in to hit air bases in southern Kyushu and Shikoku. Heavy damage was sustained.[53]

During the next four days, a violent air battle raged over the southern Homeland. Fifth Air Fleet threw into the battle a total of 193 aircraft, including 69 *tokko* planes. Losses were staggering, amounting to 161 planes, or 83 per cent of the total aircraft committed. These losses, together with the widespread havoc caused by enemy attacks on airfields, left the Air Fleet powerless to participate in further large-scale action for several weeks to come. On the other hand, the damage

50 This was in line with the policy contained in the Army-Navy Air Agreement of hoarding aircraft against the date of the actual invasion. (Cf. pp. 544–5) (1) *Dai Go Koku Kantai Sakusen Kiroku* 第五航空艦隊作戰記錄 (Fifth Air Fleet Operations) 2d Demobilization Bureau, Aug 46, p. 19. (2) Okinawa Area Naval Operations, op. cit., pp. 27–8.

51 (1) Okinawa Area Naval Operations, op. cit., pp. 28–9. (2) Fifth Air Fleet Operations, op. cit., pp. 19–20. (3) The Japanese were again faced with the dilemma so familiar in the Philippines. If the air forces were now committed against enemy carriers, it would be difficult, if not impossible, to make a maximum effort against the amphibious convoy when it later appeared. On the other hand, if the air forces were not committed against the carriers, the Japanese ran the risk of having them completely destroyed on the ground. (Statement by Rear Adm. Tomioka, previously cited.)

52 Subsequently three other separate enemy task groups were discovered in the vicinity. It was believed that none of these contained transports. Fifth Air Fleet Operations, op. cit., pp. 20–1.

53 (1) On 18 March, the enemy carriers flew off 1400 sorties mainly against airfields in central and southern Kyushu. Shikoku and the coast of Wakayama Prefecture were also hit. Ibid., pp. 21–4. (2) As this battle began, Sixth Air Army commenced its planned move to Kyushu, the headquarters reaching Fukuoka on 18 March. Operational units were still scattered throughout the Home Islands. Iwo Jima and Ryukyu Islands Air Operations Record, op. cit., pp. 26–7, 31–2.

inflicted on the enemy was believed so severe that any immediate invasion threat to the East China Sea area was considered drastically reduced.[54]

Preparations to meet eventual invasion were nevertheless carried forward vigorously. Every effort was made to assemble the planned strength of 4,500 aircraft in the battle theater as rapidly as possible. On 21 March, Sixth Air Army, which was still engaged in displacing its main strength to Kyushu, was placed under the command of Combined Fleet. The Commander-in-Chief, Combined Fleet, now exercised unified command over the bulk of all Army and Navy air units operating in the East China Sea area.[55]

While the Japanese strove to complete these preparations, the enemy continued his air and sea blockade of the home islands, hampering deployment for battle and constantly whittling down the overall war potential of the nation. Enemy B-29's continued their immensely destructive incendiary campaign, now concentrating on Nagoya, Japan's chief aircraft production center.[56]

Late in March, the Superforts began a surprising new series of attacks which wrought havoc in the Homeland rear area. On the 27th and the 31st, they carried out attacks on the Kyushu bases of the Fifth Air Fleet, closing down each field several days for repair.[57] On the 27th and 30th, they sowed thousands of aerial mines in Shimonoseki Strait and the western Inland Sea, closing that vital supply artery for an entire week.[58]

Submarines and aircraft also continued to take their murderous toll of Japanese shipping. By the end of March, 74% of the nation's total merchant tonnage had been sent to the bottom, including 75% of the tanker fleet.[59] As the

---

54 Fifth Air Fleet reported that five carriers, two battleships, three cruisers, and one unidentified vessel had been sunk in these attacks. Imperial General Headquarters was inclined to seriously doubt the accuracy of these figures, but the Combined Fleet felt that the enemy had been dealt a heavy blow even if the report were discounted by half. (1) Fifth Air Fleet Operations, op. cit., p. 26 (2) Statements by Rear Adm. Tomioka and Capt. Fuchida, both previously cited.

55 (1) An exception was the 8th Air Division on Formosa which remained under the command of Tenth Area Army. (2) *Dairikumei Dai Sennihyakushichijuhachi-go* 大陸命第千二百七十八號 (Imperial General Headquarters Army Order No. 1278) 19 Mar 45 and *Daikaishi Dai Gohyakujuni-go* 大海指第五百十二號 (Imperial General Headquarters Navy Directive No. 512) 19 Mar 45. (3) The Navy, on 20 March, officially adopted the Army code name *Ten-Go* to apply to all air operations over the East China Sea area. Imperial General Headquarters Navy Directive No. 513, op. cit.

56 Despite their destructiveness, these raids had no immediate effect on aircraft production. As a result of the completion of the dispersal program in the aircraft industry, the output of aircraft actually increased to 1900 in March from February's low of 1260. Summary of Production of Aircraft, Loss of Capacity through Air Raids, op. cit.

57 (1) *Senkyo Shubo* 戰況手簿 (Daily Record of the War Situation) Operations Section, Army General Staff. (2) In addition, the U.S. carrier task force joined in the attack on Kyushu air bases on 28 and 29 March. On 31 March carrier planes struck in great force at airfields on Tokunoshima in the Ryukyus, causing extensive damage. This forward air base, a little over 100 miles north of Okinawa was an installation vitally important to the successful conduct of the *Ten-Go* operation. Its interdiction was a severe blow. Iwo Jima and Ryukyu Islands Air Operations Record, op. cit., pp. 40, 42.

58 (1) Daily Record of the War Situation, op. cit. (2) Statement by Capt. Oi, previously cited.

59 Statistics and Analysis Reports, Report No. 1, op. cit. The *Nan-Go* Operation (Cf. n. 24) for the emergency transportation of material from the southern area was ended on 18 March, having fallen far short of expectations. The submarine threat plus the enemy air patrols being flown off Clark Field, Manila, and Puerto Princesa airdromes closed down the one-time southern lifeline of the Empire completely. There was no further trade after this date. (1) Statement by Capt. Oi, previously cited. (2) *Daikaishi Dai Gohyakujuichi-go* 大海指第五百十一號 (Imperial General Headquarters Navy Directive No. 511) 16 Mar 45.

shipping crisis deepened, transport of essential supplies to the southern Homeland battle theater became almost impossible.

On 23 March, a strong Allied carrier task force suddenly attacked the island of Okinawa in the Ryukyus. This attack immediately appeared to cast doubt on the Fifth Air Fleet's battle claims of 18–21 March. The High Command, however, took the sanguine view that this was merely a minor operation, undertaken by the enemy while en route back to Ulithi in retaliation for the losses he had suffered off the Kyushu coast.[60]

This serious misjudgement was quickly exposed on 25 March when the enemy began to put ashore a landing force on the Kerama Islands, a small group about 25 miles southwest of Okinawa. On the same day, Combined Fleet issued an alert for the *Ten-Go* air operation.[61]

The situation was now precarious. The commitment of the Fifth Air Fleet on 18 March, far from having delayed the enemy invasion, had actually resulted in a premature and largely ineffective expenditure of Japanese air strength in Kyushu. The Sixth Air Army and the Third Air Fleet had not yet completed their westward displacement, while the Tenth Air Fleet was just beginning its specialized training.[62] Air reaction to the landings in the Ryukyus was therefore negligible. On 1 April, when the enemy extended the amphibious offensive to Okinawa, the burden of defending that important island fell initially on the ground forces alone.[63]

The battle on Okinawa went badly from the very beginning. Handicapped by insufficient troop strength,[64] Lt. Gen. Mitsuru Ushijima, commander of the Thirty-second Army, had concentrated the bulk of his forces on the southern portion of the island, where the terrain was relatively favorable to a strong defense. This, however, left the central sector of the island, containing two valuable airfields, virtually unguarded. The enemy landed directly in this weakened sector and quickly mopped up the small security detachments stationed there. Lt. Gen. Ushijima's forces to the south dug themselves in on strong battle positions in the Naha-Shuri area. By the end of the second week of the campaign, enemy forward elements began to approach this line preparatory to a show-down battle.

Although continually harassed by enemy air raids on the Homeland, Combined Fleet finally succeeded during the first week in April in effecting a partial concentration of air forces in Kyushu. Execution of the *Ten-Go* operation then began in earnest.

The decisive air battle began on 6 April with

60 (1) Okinawa Area Naval Operations, op. cit., p. 34. (2) Fifth Air Fleet Operations, op. cit., p. 28.

61 On the next day, 26 March, Combined Fleet ordered all forces to execute the *Ten-Go* operation. Okinawa Area Naval Operations, op. cit., p. 36.

62 On 26 March, Third Air Fleet and Tenth Air Fleet were placed under the command of Fifth Air Fleet, thus unifying the operational command of all naval air forces in the Homeland. Ibid., p. 36.

63 Historical Data, Thirty-second Army, op. cit., p. 36.

64 Until November 1944, the Thirty-second Army on Okinawa had a combat strength of three divisions and one mixed brigade. On 13 November, however, Imperial General Headquarters ordered the transfer to Formosa of the 9th Division, the strongest of the three divisions on Okinawa, to replace the 10th Division, which was being shipped from Formosa to the Philippines. The 9th Division displaced from Okinawa to Formosa in mid-January 1945. Imperial General Headquarters planned to assign the 84th Division to Thirty-second Army to replace the 9th, but late in January this plan was abandoned in favor of retaining the 84th Division in the Homeland. (Cf. n. 34) The permanent loss of one division forced Thirty-second Army to weaken drastically its plans for the defense of Okinawa. (1) *Okinawa Sakusen Kiroku* 沖繩作戰記錄 (Okinawa Operations Record) 1st Demobilization Bureau, Aug 46, pp. 3–5. (2) Imperial General Headquarters Army High Command Record, op. cit., p. 357. (3) Historical Data, Thirty-second Army op. cit., pp. 8, 15–20.

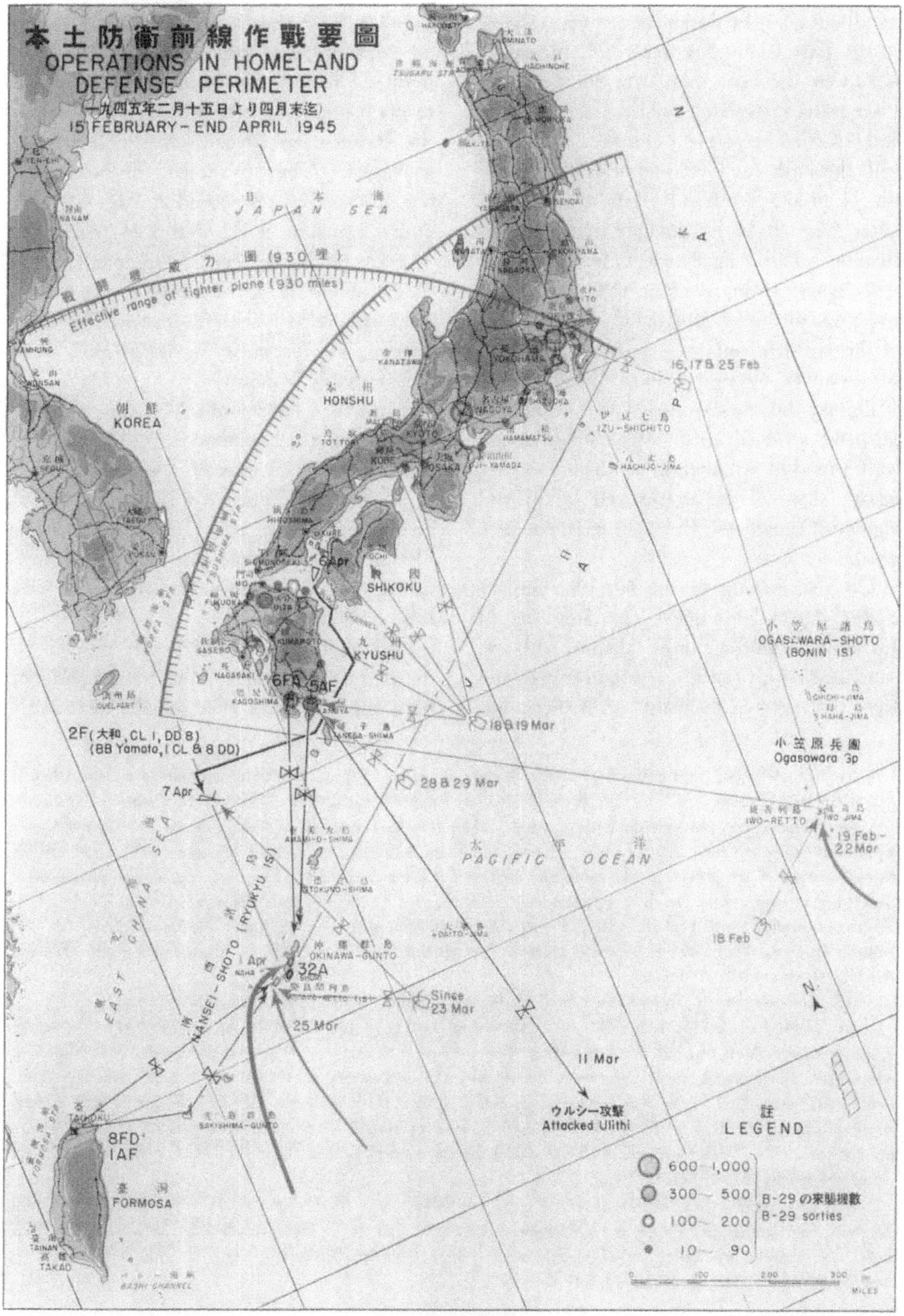

PLATE NO. 146
Operations in Homeland Defense Perimeter, 15 February—7 April 1945

an all-out effort by the air forces then available in the East China Sea sector. Participating units were the Sixth Air Army and Fifth Air Fleet in the Homeland, and the First Air Fleet and 8th Air Division on Formosa. Operating with the Fifth Air Fleet were attached units of the Third and Tenth Air Fleets, although the latter were in an exceedingly low state of training. Following the commitment of these units against enemy invasion vessels off Okinawa, it was planned to send out the last remnants of the Japanese surface fleet from the Inland Sea to attack survivors in the anchorage area.[65]

During the two-day period 6–7 April, the Japanese attacked enemy ships off Okinawa with a total of 699 aircraft, of which 355 were *tokko* planes. These attacks were boldly and vigorously carried out, and reported results were good.[66]

On the evening of the 6th, the Surface Special-Attack Force under Vice Adm. Seiichi Ito sortied from the Bungo Channel with one super-battleship (*Yamato*), one light cruiser, and eight destroyers in company. On the morning of 7 April, this force was spotted by enemy air reconnaissance while still over 400 miles north of Okinawa. A large part of the enemy's carrier plane strength, apparently undeterred by the Japanese air offensive, was unleashed against Vice Adm. Ito's force. In two strikes, one at 1240 and the second at 1345, the light cruiser and four of the destroyers were sunk. *Yamato*, hit by ten aerial torpedoes, went to the bottom at 1417, the second of Japan's two great 64,000-ton battleships to succumb to air attack. The four surviving destroyers returned independently to Japan.[67]

Between 12 April and 4 May, four additional large-scale air offensives were launched against the enemy fleet off Okinawa, the last of which was coordinated with the final all-out ground offensive of the Thirty-second Army During these operations, the Japanese air units flew a total of 1653 sorties, of which 761 were *tokko* missions.[68]

Although great successes were claimed by the participating units,[69] the attacks did not have any noticeable effect on the local battle

---

65 (1) Okinawa Area Naval Operations, op. cit., pp. 42–3. (2) U.S. Strategic Bombing Survey (Pacific), Naval Analysis Division, *The Campaigns of the Pacific War*, Appendix 103 (Action Report of 2d Destroyer Squadron).

66 In this attack, two battleships, three cruisers, eight destroyers, three minesweepers, twenty one transports, and 25 unidentified vessels were reported sunk. In addition, 61 other vessels were claimed damaged. This attack was the first of a series of ten general attacks conducted between 6 April and 22 June. Due to attrition, they progressively diminished in size, and the last three were relatively minor affairs. These general attacks were heavily interspersed with small-scale attacks, which kept the enemy almost continuously engaged. Statistics and Analysis Reports, op. cit., Report No. 4: Commitment and Losses of Japanese Aircraft and Damages Inflicted on Allied Forces in the Okinawa Air Operations, 25 Mar–22 Jun 45.

67 This was the last Japanese naval sortie of the war. Okinawa Area Naval Operations, op. cit., pp. 52–5.

68 General attack Nos. 2, 3, and 4 were executed on 12–13, 15–16, and 21–28 April respectively. A total of 1,272 sorties were flown, of which 565 were *tokko* missions. General attack No. 5, which was coordinated with the ground offensive of Thirty-second Army, was carried out on 4–5 May, with a total of 381 sorties, of which 196 were *tokko*. During all of this period, enemy B-29's continued to blast Kyushu airfields, seriously disrupting the mounting of general attacks in planned strength, interrupting the flow of reinforcements into the theater, and reducing supply and maintenance installations. The Japanese were never able to deploy the 4,500 aircraft called for in the *Ten-Go* plan. Statistics and Analysis Reports, op. cit., Report No. 4.

69 Two carriers, one (carrier or cruiser), six battleships, four (battleships or cruisers), three cruisers, one destroyer, one (cruiser or destroyer), six transports, two patrol craft and seventeen unidentified vessels were reported sunk. In addition, 45 vessels, including two carriers, were reported damaged. (1) Ibid. (2) Fifth Air Fleet Operations, op. cit., pp. 45, 49–50, 59–60.

situation. The Thirty-second Army ground offensive of 4–5 May made limited gains against exceedingly strong resistance, faltered, and fell back after a bloody two-day fight. Following this, Thirty-second Army was no longer capable of effective offensive action. Although the *Ten-Go* air operation was sustained for another six weeks, the enemy was securely in possession of Okinawa as a forward base after the failure of the 4 May offensive.[70] All of Japan south of Tokyo, as well as Korea and the lower Yangtze, were now brought within easy range of enemy land-based fighters.

General Plan of Operations

Just as the battle for control of the East China Sea area began, the Japanese High Command learned through intelligence channels that Soviet Russia had begun to redeploy troops from the European theater to the Far East. In Europe, the crossing of the Rhine by the Allied armies and the catastrophic defeats suffered by the Germans in the Saar and Ruhr valleys indicated that final collapse of Hitler's forces was imminent. Japan faced the dread of fighting the war alone.

Against the background of these alarming strategic developments, Homeland defense planning was hurriedly continued. On 20 March, Imperial General Headquarters Army Section transmitted to all major subordinate commands the preliminary draft of a voluminous operations plan to meet an invasion of the home islands. This document, based upon the general policy directive of 20 January, covered in considerable detail objectives, tactics and technique, troop movements, internal security, transportation, communications, and logistical administrative support. The draft was studied by subordinate headquarters, revised by the Army Section, and submitted to the Navy and to various governmental agencies for concurrence.[71] On 8 April it was formally disseminated to the field commands. Its essential points were as follows:[72]

1. The forthcoming decisive operation in the Homeland and adjacent areas will be referred to as the **Ketsu-Go** (決號) *Operation. Designations of the component operations will be as follows:*

Ketsu	*No. 1*	*Hokkaido, Karafuto, and the Kurile Islands*
	No. 2	*Northern Honshu*
	No. 3	*Kanto District*
	No. 4	*Nagoya-Shizuoka area*
	No. 5	*Western Honshu and Shikoku*
	No. 6	*Kyushu*
	No. 7	*Korea*

70 During the climax of the *Ten-Go* operation (general attacks Nos. 1–5), the Japanese lost 1,711 aircraft, seriously reducing the combat potential of the air forces. However, a total of 159 enemy ships were reported sunk (including those sunk in minor operations) and 141 damaged. It was not until after the war that access to Allied records revealed to the Japanese that only 17 vessels were actually sunk though a considerable number was damaged in varying degrees.

71 The Navy did not at this time issue a detailed operations plan for Homeland defense. Navy general policy had already been announced in the "Outline Plan of Immediate Operations" of 20 March. (Cf. pp. 551–2.) Intensely preoccupied with the *Ten-Go* operation, the Navy delayed detailed planning for *Ketsu-Go*. Little action was called for by the High Command in any case, since all *Ketsu-Go* planning had been delegated to Combined Fleet. (Statement by Capt. Ohmae, previously cited.)

72 (1) *Dairikumei Dai Sennihyakukyujukyu-go* 大陸命第千二百九十九號 (Imperial General Headquarters Army Order No. 1299) 8 Apr 45. (Plate No. 147 is a facsimile reproduction of a portion of this document.) (2) *Dairikushi Dai Nisenyonhyakusanjuhachi-go* 大陸指第二千四百三十八號 (Imperial General Headquarters Army Directive No. 2438) 8 Apr 45, Annex: Outline of Operational Preparations for Operational Preparations for Operations *Ketsu*: Logistics Plan. (3) Imperial General Headquarters Army High Command Record, op. cit., pp. 337–47.

The original *Ketsu-Go* plan was designed only for the initial direct invasion. No second battle was envisaged or considered in drafting these plans, though, of course, it was recognized that a defense of Kanto would be executed if Kyushu were invaded first. (Statements by Maj. Gen. Masakazu Amano, Chief, Operations Section, Imperial General Headquarters, Army Section, and Col. Hosoda, previously cited.)

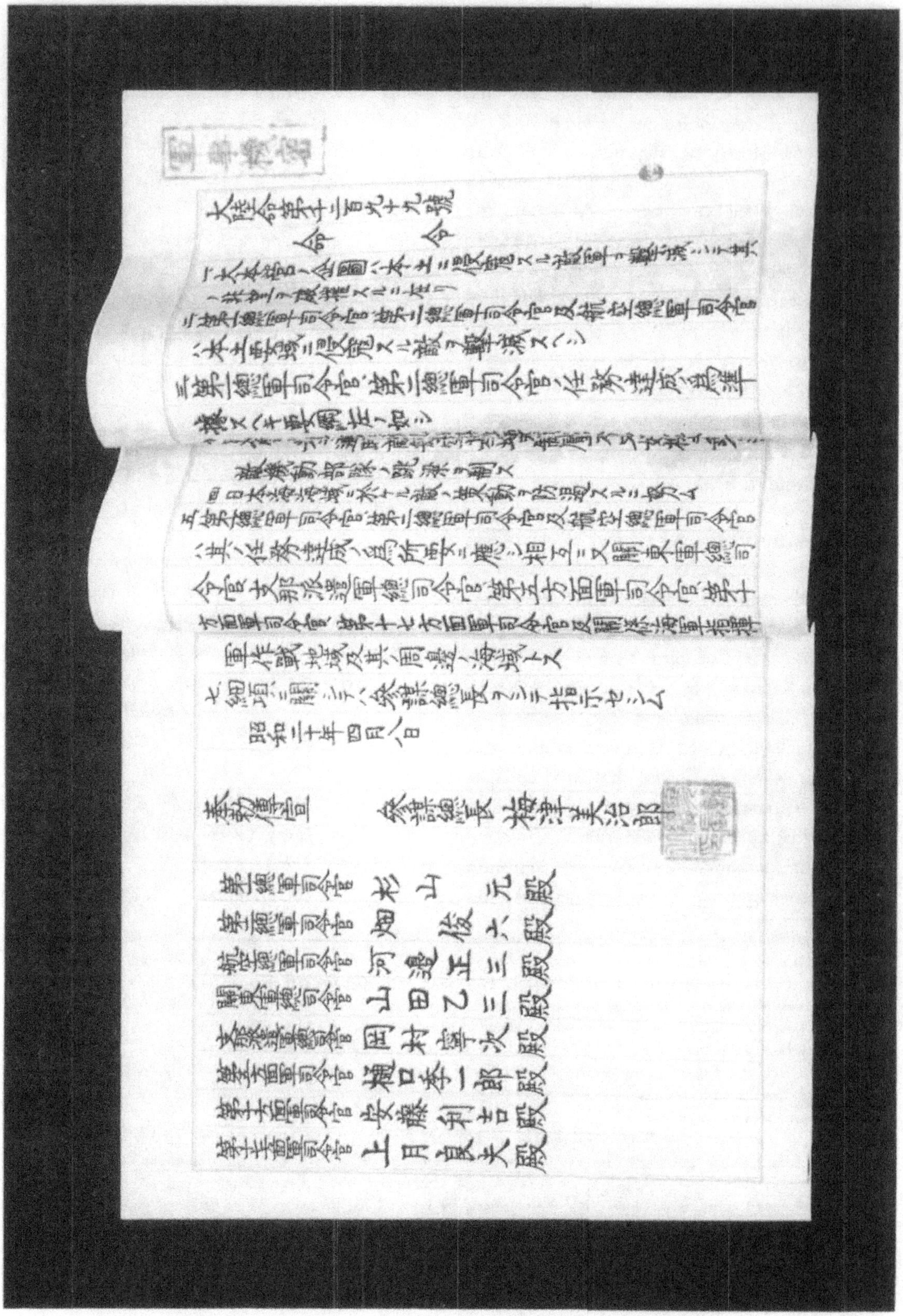

軍事機密

大陸命第千二百九十九號

命令

一、大本營ノ企圖ハ本土ニ侵寇スル敵軍ヲ撃滅シ [illegible]

二、第一總軍司令官、第二總軍司令官及航空總軍司令官ハ本土要域ニ侵寇スル敵ヲ撃滅スヘシ

三、第一總軍司令官、第二總軍司令官ノ任務達成ノ為準據スヘキ要綱左ノ如シ [illegible]

四、[illegible]

五、第一總軍司令官、第二總軍司令官及航空總軍司令官ハ其ノ任務達成ノ為所要ニ應シ相互ニ又關東軍總司令官、支那派遣軍總司令官、第五方面軍司令官、第十方面軍司令官、第十七方面軍司令官及聯合艦隊司令長官ト [illegible]

[illegible] 軍作戰地域及其ノ周邊ノ海域トス

七、細項ニ關シテハ參謀總長ヲシテ指示セシム

昭和二十年四月八日

奉勅傳宣　參謀總長　梅津美治郎

第一總軍司令官　杉山元殿

第二總軍司令官　畑俊六殿

航空總軍司令官　河邊正三殿

關東軍總司令官　山田乙三殿

支那派遣軍總司令官　岡村寧次殿

第五方面軍司令官　樋口季一郎殿

第十方面軍司令官　安藤利吉殿

第十七方面軍司令官　上月良夫殿

PLATE NO. 147

Facsimile of Imperial General Headquarters Army Order No. 1299, 8 April 1945

2. *Operational Policy*

a. *The Imperial Army will hasten preparations to meet and crush the attack of U. S. forces in the above key areas. Emphasis will be on the Kanto area (Ketsu No. 3) and Kyushu (Ketsu No. 6).*

b. *Preparations for operations prescribed in this plan will fall into the following general phases:*

First April through July
Second August through September
Third from 1 October

Emergency preparations in Kyushu will be completed by early June. Dispositions will continue to be strengthened during the second phase, and all tactical plans completed. Final deployment of field units and perfection of the field positions will be completed in the early part of the third phase.

c. *Air Operations*

(1) *A close watch will be maintained over all enemy fleet movements, particularly transport convoys. Air search over the approaches to the Homeland will be continuous and aggressive.*

(2) *Enemy amphibious task forces attempting to invade the Homeland will be destroyed on the water.*

(3) *The primary target of the air offensive will be transports.*

(4) *All air operations will be curtailed rigorously until the enemy main convoy approaches. Fighting strength will be preserved until the moment for the decisive effort.*[73]

(5) *Air support for ground forces will be restricted to liaison missions and tactical reconnaissance in extreme emergencies.*

(6) *Long-range, surprise air raids against such enemy bases as Iwo Jima and Okinawa will be carried out.*

d. *Ground Operations*

(1) *The ground forces will win the final decision by overwhelming and annihilating the enemy landing force in the coastal area before the beachhead is secure.*

(2) *Speedy maneuver of the largest possible force against the enemy landing sector is the key to success in such operations. As many local reserves as possible must be maneuvered into the expected landing sector as soon as it becomes known that the enemy intends to land. After the enemy has landed, additional ground troops from other parts of the Homeland will be deployed to the area in accordance with the plan prescribed in Para. 3 below.*

(3) *In the event of simultaneous invasions of more than one area, the main Japanese counteroffensive will be directed at the main enemy landing. Delaying actions will be fought in other localities. This principle will apply both tactically and strategically.*

(4) *If the location of the enemy's main landing is undetermined, the main Japanese force will be committed in the area which presents the most favorable terrain for offensive operations. Delaying actions will be conducted in other localities. This principle will apply both tactically and strategically.*

(5) *Large-scale and thorough construction of fortifications will be carried out with emphasis on those field positions designed to provide jumping-off, rallying, and support points for local*

73 Air concentration for *Ketsu-Go* was to be over and above the commitment to *Ten-Go*. (Cf. Chapter XIX, n. 45) At this time the utilization of aircraft was running about two months behind production. Thus, new aircraft being used in the *Ten-Go* operation in April and May were largely out of February and March production. Assuming that *Ten-Go* would be completed during May, then all aircraft production after 1 April could be made available for *Ketsu-Go*. (Statement by Maj. Gen. Sadanori Harada, Air Ordnance Department, Munitions Ministry.)

offensives.

(6) *Special security precautions will be taken at vital installations to forestall enemy airborne penetrations.*

e. *Air Defense*

(1) *Air defense of the Homeland will be emphasized. Points of first priority will be Tokyo, cities lying on the principal routes of communication, vital industrial facilities, airfields and munitions dumps.*

(2) *Increasing emphasis will be placed on air raid precautions and perfecting warning facilities.*

(3) *Dispersal of facilities, particularly airfields, will be accomplished wherever possible.*

(4) *When the ground armies initiate their operational movements to counter enemy landings, their assembly must be covered by antiaircraft units. Plans will be made for the large-scale diversion of such units to this mission at the appropriate time.*

f. *Channel Defense*

(1) *In cooperation with the Navy, the defenses of such important straits as Bungo, Kii, and Shimonoseki will be strengthened. Principal harbors will also be strongly defended.*

(2) *Batteries will be stationed so as to prevent penetration by enemy vessels and amphibious landings. Channel defense units will strengthen bomb and shell-proof installations, and will dispose armed boats at strategic points.*

g. *Guerrilla Resistance and Internal Security*

(1) *We will strive to realize our operational objectives through exploitation of the traditional spirit of "Every citizen a soldier."*

(2) *Guerrilla resistance will aim at the obstruction of enemy activities and the attrition of enemy strength through guerrilla warfare, espionage, deception, raids on rear areas, and demolition of enemy installations. Such resistance will be carried out as part of the overall operation to assist line units, to meet enemy airborne operations and small secondary amphibious landings, and to cut off and harass enemy elements which penetrate into the interior.*

(3) *Internal security will aim at protecting military activities, vital communications, transport, power sources, and secret areas. If necessary, internal security will quell public disorder arising as a result of air raids, bombardment, invasion, propaganda, and natural disaster.*

(4) *Forces for guerrilla resistance and internal security will be drawn from the entire body of the citizenry as the situation may dictate. Guard units and civilian defense organizations will provide manpower, organized around small elements of the field forces as a guiding nucleus. Such units will be under the command of the various district army commands.*

3. *Redeployment Plan*[74]

a. *In event of an enemy invasion of Kyushu, the following steps will be taken:*

(1) *Four line combat divisions will be dispatched to Sixteenth Area Army from forces available to Thirteenth and Fifteenth Area Armies in central and western Honshu and Shikoku.*

(2) *Preparations will be made for the advance of a second increment. Three or four additional*

74 This redeployment plan was based on the assumption that sea and rail communications in Japan would be largely destroyed and that most of the troop movements would be made on foot. Maximum possible use was to be made, however, of existing highway, water, and rail routes. The main communications route was to be the Aomori-Sendai-Tokyo-Nagoya-Osaka-Shimonoseki line, backbone of the rail system on Honshu. Maximum use was also to be made of secondary routes and highways. Scheduled water routes from Honshu to Kyushu and Shikoku were to be set up with a total capacity of 9,500 tons daily. Even so, considerable difficulty was anticipated in the execution of this plan. For example, 65 days would have been required to carry out the planned move of two divisions from Kyushu to Nagano Prefecture. Another ten days would be needed to move these divisions into the line in the Kanto area.

divisions will be dispatched from the Eleventh Area Army in northern Honshu and from Twelfth Area Army in the Kanto district to the Osaka-Kobe area, where they will be held in readiness for further advance to Kyushu.

b. *In event of an enemy landing in the Kanto district, without a prior landing in other areas, the following steps will be taken:*

(1) *Three line combat divisions from Eleventh Area Army in northern Honshu, three from Fifteenth Area Army in the Osaka-Kobe area, and two from Sixteenth Area Army in Kyushu will be immediately sent to Twelfth Area Army and held in readiness in Nagano Prefecture as a mass counterattack force.*

(2) *If the situation permits, two line combat divisions will be speedily dispatched by Thirteenth Area Army from the Nagoya area to Twelfth Area Army in the Kanto district.*

(3) *Preparations will be made for the advance of a second increment. Two line combat divisions will be sent by Fifth Area Army in Hokkaido to Eleventh Area Army in northern Honshu, and five divisions from Sixteenth Area Army on Kyushu to Fifteenth Area Army in the Osaka-Kobe district, all these divisions to be held in readiness for further advance to the Kanto district.*

4. *Logistics Plan*

a. *General Aims*

(1) *The Homeland will immediately enter the status of a battle theater communications zone.*

(2) *Emergency logistic preparations called for under this plan will be completed by the end of June. The full program will be perfected by the end of October.*

(3) *Successful completion of this program is dependent upon securing maximum efficiency in production and procurement, use of transportation facilities, security of logistics installations, dispersal operations, transfer of service units and supplies from the Continent,*[75] *and maximum utilization of idle materials.*

(4) *In order to meet the expected gradual increase in pressure upon land and sea transportation services, self-sufficiency in each district and area Army will be emphasized, particularly as regards food, materiel repair, and certain classes of arms and equipment procurement.*

(5) *The logistics base (excluding Korea) will be 2,903,000 men, 292,000 horses, and 27,500 motor vehicles.*

b. *Logistics Build-up*

(1) *As far as items of organizational equipment are concerned, the furnishing of new troop units will have top emergency priority over all other supply activities, emphasis being placed on units in the Kanto area and Kyushu.*

(2) *The over-all build-up targets for expendable items (excluding the normal unit allowance) will be, in the case of ammunition, enough for one campaign by the entire projected field strength (2,000 tons per division), and of fuel, food, and forage, enough for* $1\frac{1}{2}$ *months of combat operations.*[76] *The main stockpiling areas will be the Kanto and Kyushu areas.*

(3) *Only in Kyushu and the Kanto will the full ammunition unit of one campaign be available in operational stockpiles. Other areas will have such fractions of one unit as are prescribed.*

---

75 The plan called for the emergency transportation of 296,100 barrels of fuel and lubricants and 20,000 tons of ammunition from the continent to Japan before the end of August.

76 Build-up targets included ammunition for 48 divisions for one campaign (96,000 tons), of which 22 were to be disposed in the Kanto, and 12 in Kyushu; 223,600 barrels of fuel (excluding aviation gasoline), of which 77,500 barrels were to be disposed in the Kanto and 51,000 in Kyushu; and two and one-quarter months supply of rations and clothing for 2,900,000 troops.

本土方面陸軍部隊指揮系統一覽表
ARMY CHAIN OF COMMAND FOR HOMELAND OPERATION

(一九四五年四月)
APRIL 1945

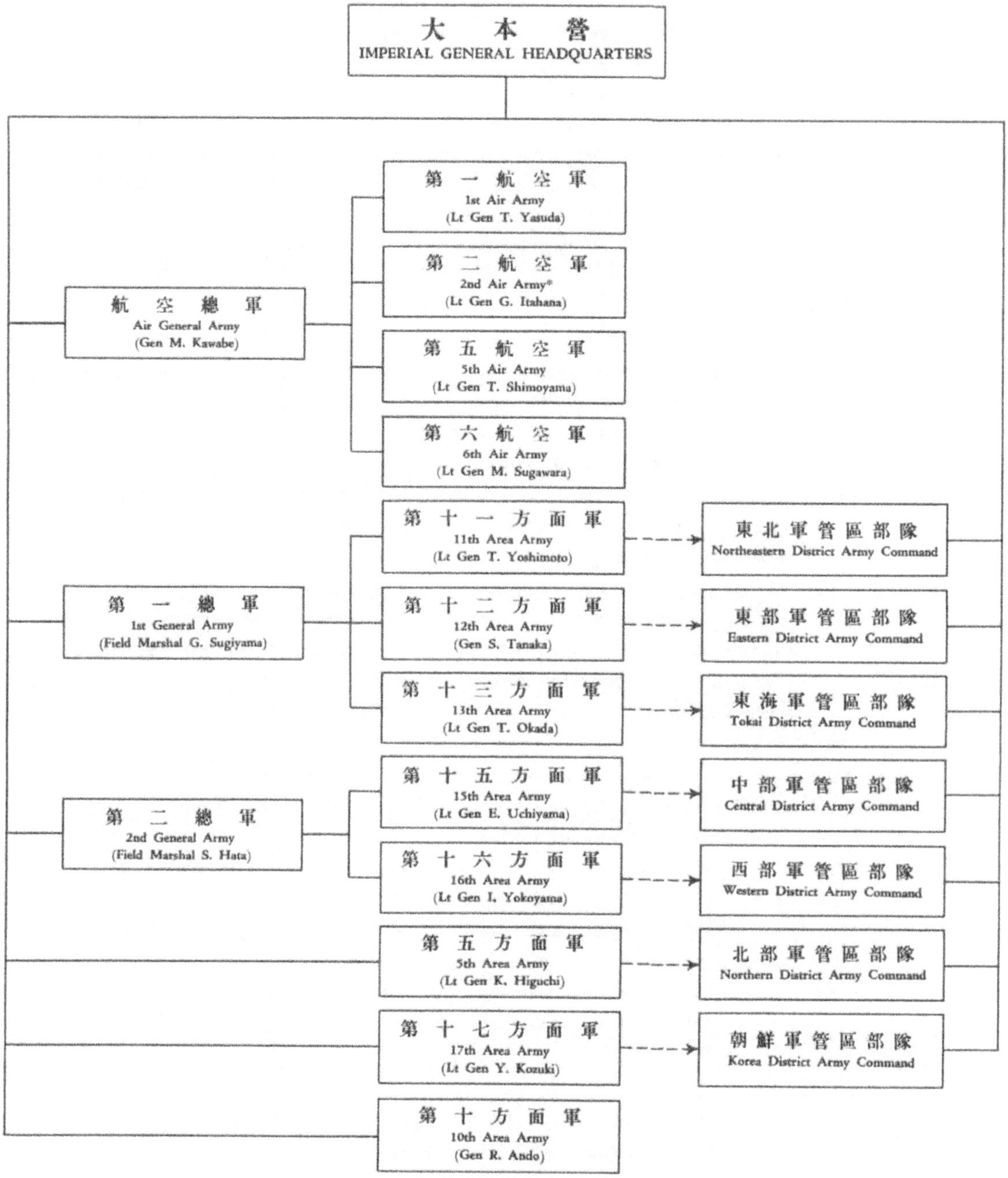

– – – Operational command

* 2nd Air Army in Manchuria was under the operational command of the Kwantung Army.

PLATE NO. 148
Army Chain of Command for Homeland Operation, April 1945

All main operational areas will have one month's supply of items of continuous issue. Certain remote areas may have larger stockpiles as prescribed.[77]

(4) *Certain reserve stockpiles in all categories of supplies will be established by Imperial General Headquarters. Such supplies will be held available for rapid transfer to the active battle theater when it develops.*[78]

c. *Completion Deadlines*

(1) *Operational stockpiling in Kyushu and the Kanto: 31 May*

(2) *Operational stockpiling in other areas: 30 June*

(3) *Reserve stockpiling in Kyushu and the Kanto: 31 August*

(4) *Reserve stockpiling in other areas: 31 October*

d. *Redeployment of Supplies*

(1) *At the commencement of the operation, operational supplies will be released to the engaged units immediately in sufficient quantity to raise the supply level of the active theater up to two campaigns in ammunition and four months supply of fuel, forage, and rations.*

(2) *Supplies will be released from the reserve by Imperial General Headquarters as the situation requires.*

e. *Air Logistics*

(1) *Emphasis will be placed on expanding facilities for aircraft maintenance, preparation of ordnance for special-attack units, and dispersal of fuel dumps.*

(2) *Special-attack units will have first priority in all matters of supply.*

(3) *With two sorties per unit as the buildup target, bombs will be delivered and stockpiling completed by 30 June.*[79]

Concurrently with the publication of the *Ketsu-Go* plan, Imperial General Headquarters activated the headquarters of the First and Second General Armies according to previous plan, simultaneously deactivating the General Defense Command. (Plate No. 148) The First General Army under Field Marshal Sugiyama established headquarters in Tokyo and assumed command of the Eleventh, Twelfth, and Thirteenth Area Armies.[80] The headquarters of the Second General Army under Field Marshal Shunroku Hata was established at Hiroshima and included command of the Fifteenth and Sixteenth Area Armies.[81] To provide a similar command level for all Army air units participating in the campaign, an Air General Army headquarters was established under General Masakazu Kawabe. This headquarters took complete command of all Army air units in Japan and Korea,[82] except the

77 The operational stockpiling program included ammunition for one campaign by 24 divisions (48,000 tons), of which 8 would be in the Kanto and 8 in Kyushu; 25,800 barrels of fuel (excluding aviation gasoline), of which 7,500 would be in the Kanto and 6,900 in Kyushu; and one month's supply of rations and clothing for 820,000 men.

78 The reserve stockpiling program called for ammunition for one campaign by 24 divisions, of which 14 would be in the Kanto and 4 in Kyushu; 204,000 barrels of fuel; and rations and clothing for one month for 575,000 men.

79 Under the arming program, approximately 3,790 metric tons of bombs were scheduled for delivery to operational units.

80 Also provided for in the same order was the transfer of the Thirty-sixth Army, formerly the central mobile reserve directly under the General Defense Command, to the Twelfth Area Army. (1) Imperial General Headquarters Army Order No. 1297, op. cit. (2) Imperial General Headquarters Army High Command Record, op. cit., pp. 332–6.

81 Ibid.

82 The Air General Army assumed command of the First Air Army, the 51st, 52d and 53d Air Divisions, and the 30th Fighter Group. (1) *Dairikumei Dai Sennihyakukyujuhachi-go* 大陸命第千二百九十八號 (Imperial General Headquarters Army Order No. 1298) 8 Apr 45. (2) Imperial General Headquarters Army Order No. 1299, op. cit. (3) Imperial General Headquarters Army High Command Record, op. cit., pp., 332–4.

Sixth Air Army which, until the completion of the *Ten-Go* operation, was to remain under the operational command of Combined Fleet. The organization of all these headquarters was completed by 15 April.

Meanwhile, the first general mobilization called for in the 26 February plan was successfully carried out. Thirteen new divisions, all of the coastal combat type, and one new independent mixed brigade were activated and assigned as follows:[83]

First General Army
- *Eleventh Area Army*
  - *142d Division*
  - *157th Division*
- *Twelfth Area Army*
  - *140th Division*
  - *151st Division*
  - *152d Division*
- *Thirteenth Area Army*
  - *143d Division*
  - *153d Division*

Second General Army
- *Fifteenth Area Army*
  - *144th Division*
  - *155th Division*
- *Sixteenth Area Army*
  - *145th Division*
  - *146th Division*
  - *154th Division*
  - *156th Division*

107th Independent Mixed Brigade

On 8 April, concurrently with the publication of the *Kestu-Go* plan, the High Command also issued the text of an Army-Navy Joint Agreement regarding ground operations. This document contained the following general provisions:[84]

1. It shall be the general rule that the Army will take charge of land operations and the Navy of sea operations, both surface and underwater. Provisions for air operations will be set forth separately.[85]

2. The Commanders of the First General Army, Second General Army, and Fifth Area Army, hereinafter referred to as the senior commander(s) of the Army forces concerned, will be in command of all land operations in their respective zones of responsibility, including those conducted by naval troops.

3. All land operations at naval bases and naval stations will be under the immediate command of the commandant of the appropriate naval station. This officer will in turn come under the senior commander of the Army forces concerned.

4. During the period of operational preparations, the senior commander(s) of the Army forces concerned are empowered to issue general instructions to naval land troops concerning security, defense plans, tactics, and training.

5. Designated Army coastal installations in the vicinity of large naval bases will be placed under local navy command.

83 In addition, two line combat divisions (the 88th and 89th), one coastal combat division (the 167th) and one independent mixed brigade (the 101st) were activated for Fifth Area Army on Hokkaido and two coastal combat divisions (the 150th and 160th) and one independent mixed brigade (the 108th) were activated in Korea. An important redeployment had also been effected with the transfer of the 44th and 84th Divisions from Fifteenth Area Army to Twelfth Area Army to strengthen the defenses of the Kanto. (1) Imperial General Headquarters Army High Command Record, op. cit., pp. 332–4. (2) Homeland Operations Record, op. cit. Vol. II, p. 8 and Map. I; Vol. III, Chart I; and Vol. IV, pp. 28–9; Vol. V, pp. 18–20. (3) Imperial General Headquarters Army Order No. 1297, op. cit. (4) *Dairikumei Dai Sennihyakushichijuroku, Dai Sennihyakuhachijusan-go* 大陸命第千二百七十六、第千二百八十三號 (Imperial General Headquarters Army Orders No. 1176, No. 1283) 10 Mar 45, 26 Mar 45. (5) Unit Organization Tables, op. cit.

84 (1) *Dairikushi Dai Nisenyonhyakuyonju-go* 大陸指第二千四百四十號 (Imperial General Headquarters Army Directive No. 2440) 8 Apr 45. Annex: Army-Navy Joint Agreement. (2) *Dairikumei Dai Sensambyakuni-go* 大陸命第千三百二號 (Imperial General Headquarters Army Order No. 1302) 8 Apr 45.

85 Inability to conclude an air agreement at this time resulted from the Navy's preoccupation with the *Ten-Go* operation. This agreement was actually delayed until July. (Cf. Chapter XIX, p. 595.)

6. Upon activation of the operation, the Navy (Army) ground antiaircraft units in the zone of operations of Army (Navy) ground units will be placed under the appropriate Army (Navy) commander.

7. Sea and air special-attack bases will be defended against land attack by the service operating the base. However, upon activation of the operation, such defense troops will be placed under the command of the Army or Navy commander controlling ground operations in the area concerned.

8. Local agreements necessary to carry out this general agreement will be concluded immediately.

Following the adoption of this joint agreement, the Navy, on 25 April, established a General Navy Command to exercise supreme operational control of all Navy surface and air forces. Admiral Soemu Toyoda, Commander-in-Chief of Combined Fleet, was designated Commander-in-Chief, General Navy Command, holding both positions simultaneously.[86] (Plate No. 149)

Tactics and Techniques

The wholesale mobilization of the national potential required to implement the *Ketsu-Go* plan obliged Imperial General Headquarters to give the nation frank warning that the Japanese home islands, inviolate through centuries, now stood in imminent peril of feeling the tread of an invader's heel. This dread prospect evoked in the Armed Forces and civil population alike a readiness to take the most extraordinary measures to repel the enemy.

The national attitude became the basis of tactical policy This was gradually elaborated in many official directives, guidance manuals and public announcements issued to implement the *Ketsu-Go* plan. The general concept of Homeland defense set forth in these documents was as follows:[87]

1. Phase I—Prior to Sortie of Enemy Convoy

a. The Fleet submarine force will disrupt enemy communications in the Marshalls, the Marianas, the Philippines, and the Ryukyus Islands areas. Submarine-launched aircraft will attack enemy forward bases such as Ulithi.[88]

b. Long-distance surprise air raids will be launched against enemy advance bases such as Ulithi, Iwo Jima, and Guam.

2. Phase II—After Sortie of Enemy Convoy

a. When the advance elements of the enemy invasion task force enter the outer Japanese air perimeter, long-range Navy bombers will attack them vigorously, and short-range submarines will launch torpedo attacks.

b. When the main invasion convoy reaches a point 180–200 miles off Japan, the massed strength of all Army and Navy air tokko will begin to be committed. Koryu (midget submarines) will also join the attack.

c. As soon as the target and time of landing have been determined, airborne raids will be carried out against forward enemy air bases supporting the

86 *Daikairei Dai Sanjukyu-go* 大海令第三十九號 (Imperial General Headquarters Navy Order No. 39) 25 Apr 45.

87 (1) *Hondo Joriku ni taisuru Hangeki Sakusen Jumbi* 本土上陸に對する反撃作戰準備 (Preparations for Counterattack Operations in Defense of the Homeland) 2d Demobilization Bureau, Mar 47, pp. 17–22, 24–5. (2) Imperial General Headquarters Army Directive No. 2438, Annex, op. cit. (3) *Hondo Homen Koku Sakusen Kiroku* 本土方面航空作戰記錄 (Homeland Area Air Operations Record) 1st Demobilization Bureau, Dec 46., pp. 41, 101–2, 113. (4) Homeland Operations, op. cit. Vol. II, pp. 55–6; Vol. III, Supplement: Essentials of Sixteenth Area Army *Ketsu-Go* Operation Plan, pp. 5–10. (5) *Kokumin Kosen Hikkei* 國民抗戰必携 (National Resistance Manual) Army Section, Imperial General Headquarters, 26 Apr 45. (6) *Kokudo Kessen Kyorei* 國土決戰敎令 (Decisive Battle Manual) Army Section, Imperial General Headquarters, 20 Apr 45.

88 Due mainly to the fuel situation, Fleet participation was to be limited to submarine and destroyer operations. The remainder of the surface fleet was to remain in port and function as stationary AA batteries. The Navy still had 13 large submarines for use in enemy rear areas. In addition, four mammoth submarines, each carrying two or three seaplanes, and 21 short-range submarines were available. (1) Japanese Naval Vessels at the End of War, 2d Demobilization Bureau, 25 Apr 47, pp. 35–6. (2) Preparations for Counterattack Operations in Defense of the Homeland, op. cit., p. 22.

本土方面海軍部隊指揮系統一覽表
NAVY CHAIN OF COMMAND FOR HOMELAND OPERATION

(一九四五年四月)
APRIL 1945

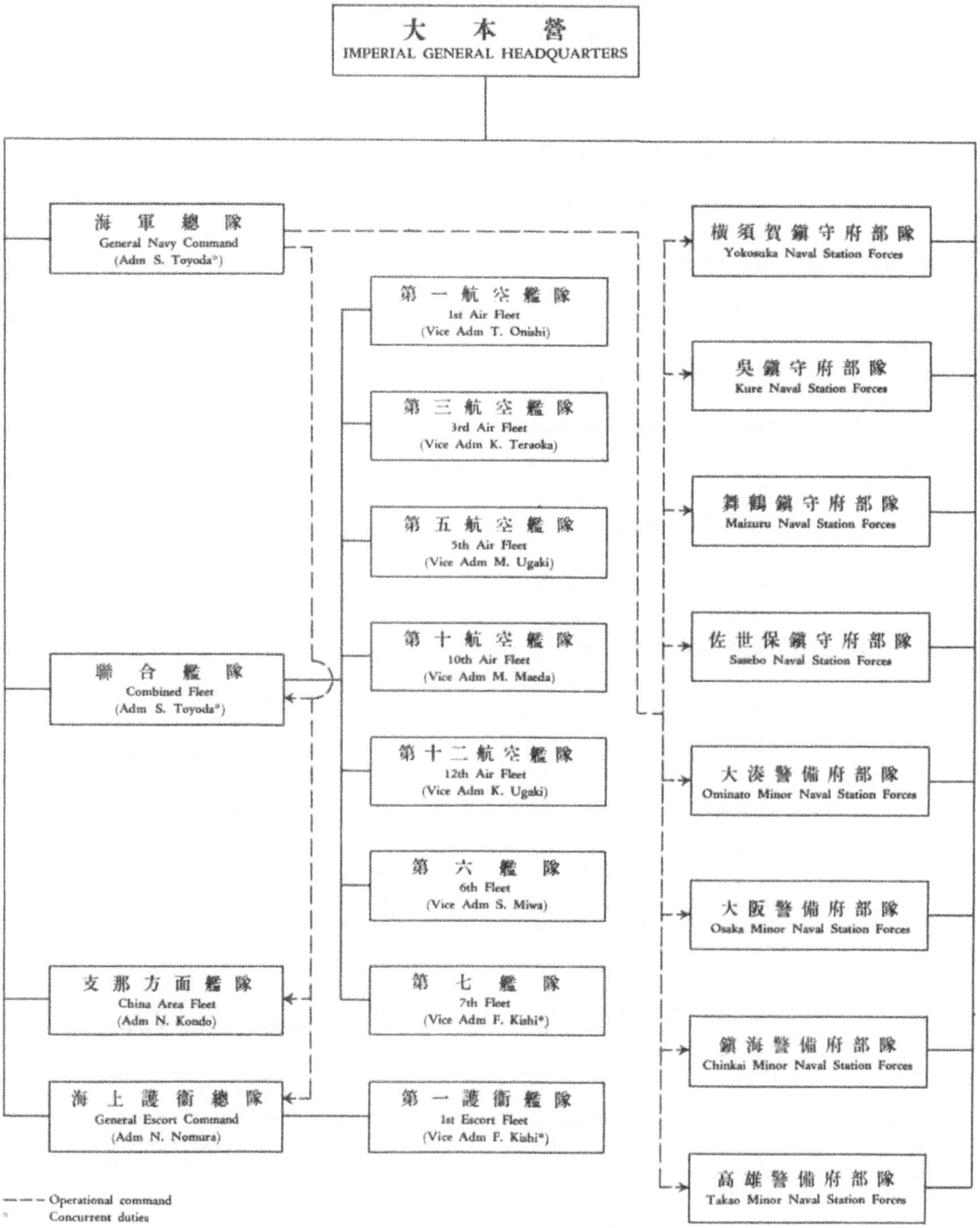

PLATE NO. 149
Navy Chain of Command for Homeland Operation, April 1945

invasion. The purpose of these attacks will be to disrupt enemy air activity and thus facilitate Japanese special-attack operations.

3. Phase III—After Arrival of Enemy Convoy

a. When the convoy is in the process of anchoring, Koryu (midget submarines) will attack in force.

b. Kaiten (human torpedo) will strike in force at gunfire support ships and transports.

c. The anchorage will be kept under constant interdiction, particularly at night, by means of attacks by Shinyo and Renraku-tei (crash-boats).

d. Long-range artillery will shell the anchorage.

e. The air tokko offensive will continue.

f. While the enemy is engaged in ship-to-shore movement, transports and landing craft will be swept with fire by all available artillery strength.

4. Phase IV—After Enemy Troop Landing

a. The artillery will shift its fire from water-borne targets to the landing beaches.

b. Air and sea special-attack forces will continue their attacks.

c. Infantry units of the coastal combat divisions will counterattack the invaders immediately from positions close to the beach. These counter-attacks will be persistent and continuous so as to disrupt the consolidation of the beachhead and merge the lines in a confused struggle in which enemy air, artillery, and naval gunfire will be seriously hampered in choice of targets.

d. As soon as the enemy objective has been determined, the line combat divisions making up the decisive battle reserve will move into the main attack area and will occupy previously prepared jump-off positions. Tanks, heavy artillery, and other elements will be committed upon arrival in the forward area without waiting for the assembly of the entire counterattack force. Units will be fed into the battle as they arrive. They will be committed on a narrow front and in great depth against a thin enemy beachhead, already weakened by the activities of the coastal combat groups. These attacks will be continuous as long as there are units available in the rear areas. If properly executed, they will result in the complete collapse of the beachhead before the enemy has a chance to get ashore his heavy striking elements.[89]

e. Should enemy elements succeed in penetrating into the interior, they will be met with fierce and determined guerrilla resistance.

Under this general scheme of tactics, the chief burden of blunting the American invasion spearhead fell upon the air and sea special-attack forces. However, the application of the *tokko* principle was to be carried much farther than ever before. For the first time in the war, *tokko* methods were to be used in ground combat on a large scale, by both formal military organizations and partisan groups.

In land combat operations, the tank was the enemy's most effective weapon, especially when equipped with a flame-thrower. It was therefore expected that large armored formations would be used in the Homeland invasion. The Japanese forces were ill-equipped to meet such an attack due to marked inferiority in both tanks and antitank guns.[90] This made

89 Under the original *Ketsu-Go* plan, the missions of the coastal combat divisions and line combat divisions were slightly different from those outlined here. The coastal combat divisions were to be charged only with containing the enemy beachhead in a shallow salient until the decisive battle reserve could participate in a general counteroffensive. This was not to be launched until all units were in positions and ready. Largely due to fear that enemy consolidation of a beachhead, no matter how shallow, would make it impossible to dislodge the invaders, this tactical policy was later modified as set forth here. (1) Decisive Battle Manual, op. cit., pp. 14–24. (2) Statements by Lt. Gen. Torashiro Kawabe, Deputy-Chief, Army General Staff, and Lt. Col. Shiro Hara, Staff Officer (Operations), Imperial General Headquarters, Army Section.

90 The heaviest tank available to the Japanese was the 17.2-ton Type-1, having 50 mm armor and mounting a 47 mm gun. The American forces, on the other hand, were employing large numbers of M 4 medium tanks weighing 30 tons, carrying 85 mm armor, and mounting a 75 mm gun. The Japanese Type-4 medium tank, similar in technical characteristics to the American M 4, was barely out of the pilot model stage. The heaviest antitank gun used by the Japanese was 47 mm. This was effective against the U. S. M 4 only at short ranges. Plans were laid to use the 75 mm Type-90 field gun for antitank purposes. A limited amount of an extremely effective hollow-charge shell, patterned after German designs, was manufactured for these weapons before the end of hostilities. (Statements by Lt. Gen. Tomio Hara, Chief, 4th Technical Research Laboratory (Armor) and Sagami Army Arsenal and Lt. Col. Yoshitaka Yoshinaga, Technical Section, Army Ordnance Department.

it essential to develop an aggressive antitank program relying on other weapons.

Experience in the Philippines and on Okinawa had demonstrated that the only effective means of combatting enemy tank superiority lay in a resort to mass special-attack operations. Manuals and directives issued to implement the *Ketsu-Go* plan therefore placed strong emphasis on the utilization of such tactics. All military units, as well as civilians, were to be trained in the use of the many different types of hand-carried mines and charges designed for such attacks.[91]

Second only to the emphasis laid on *tokko* tactics, the most important aspect of the *Ketsu-Go* plan and its implementing directives was the inclusion of the doctrine of an aggressive beach defense. This doctrine, in brief, called for the defending forces to make their decisive stand on the beach and in the coastal plain rather than on inland positions, no matter how favorable in theory the latter might be. The decision to do this stemmed from the following considerations:[92]

1. There was no assurance that Japanese firepower and assault weapons could successfully penetrate an enemy beachhead once it had expanded to the edge of the coastal plain.

2. The coastal zone contained all of the large key air bases.

The third unique feature of the *Ketsu-Go* plan was the extent to which the civilian population was to participate in the actual military defense of the Homeland. The plan called for a "National Resistance Program," the basic concept of which was that all able-bodied Japanese, regardless of sex, would be called upon to engage in battle.[93] Should the enemy overrun any considerable part of the Homeland, his forces were to be beset from all sides by partisan operations. Each citizen was to be prepared to sacrifice his life in suicide attacks on enemy armored forces. In addition, civilians were to be used in large numbers for behind-the-lines duties such as air raid precautions, construction, transport, and evacuation.[94]

Thus, relying for the most part on the suicidal bravery, ardent patriotism, and fierce loyalty of the people, Japan prepared to wage the final decisive battle against an enemy far superior in both technical resources and manpower.

---

91 Since the Homeland battle was considered a decisive and not a delaying action, it was manifestly impossible for the infantry to flee from the enemy tanks and hole up in caves. Each man was to be prepared to attack enemy tanks with special weapons. These included a hand-thrown, shaped-charge bomb similar to the American "beehive" charge, various types of "Molotov cocktail," the blanket mine, a cushion-shaped explosive bag which could be hung on projections outside the tank, and "satchel" charges which were carried on the back, turning the attacker into a "human bomb." (1) *Dairikushi Dai Nisengohyakunijugo-go* 大陸指第二千五百二十五號 (Imperial General Headquarters Army Directive No. 2525) 12 Jul 45. (2) Statement by Lt. Col. Yoshinaga, previously cited.

92 Statements by Lt. Gen. Kawabe, and Lt. Gen. Miyazaki, both previously cited.

93 It was not the intention of the High Command that children or the sick and aged should be drawn into the Homeland battle. There were, however, seemingly insurmountable obstacles to the evacuation of civilians from the coastal areas. For one thing, almost all food production depended upon the labor of these non-combatants. Food could only be grown in the arable land near the coast. The physical problems of evacuation-transportation, food, shelter, and sanitary facilities for millions of people—also militated against this move. The local commanders had still not found effective measures to cope with these difficult problems by the time of the surrender. (Statements by Col. Fuwa and Lt. Gen. Kawabe, both previously cited.)

94 *Bogyo Senryaku oyobi Senjitsu ni kansuru Kaito* 防禦戰略及戰術に關する回答 (Reply to Questionnaire Concerning Defense Tactics and Strategy) 1st Demobilization Bureau, 31 Mar 46. (2) National Resistance Manual, op. cit.

CHAPTER XIX

HOMELAND DEFENSE: STRATEGIC SETBACKS AND FINAL PREPARATIONS

Air Raids and the Industrial Crisis

From the very beginning, preparations for the defense of the Homeland had been seriously hampered by enemy air activity over the home islands.[1] Shortly after the *Ketsu-Go* plan had been issued, these attacks were intensified and rapidly reached a point at which they threatened to disintegrate the entire social and economic structure of the nation before the decisive battle could even be begun.

From the standpoint of physical destruction and suffering, the worst raids of all were the B-29 incendiary attacks against urban areas. During April, May, and June, the fire-bomb campaign against the large urban complexes was mercilessly continued. In June, it was extended to 12 small and medium-sized cities containing important basic materials and subcontracting facilities.[2] In some of the larger raids it was estimated that as many as 400 B-29's participated, while the monthly cumulative total of B-29 sorties rose by June to 3270.[3]

Carrier planes, as well as fighters and bombers from Okinawa and Iwo Jima also added weight to the attacks. Although these were mainly directed at airfields in Kyushu, urban areas as far east as Tokyo were harassed by strafing and bombing. Fighter and bomber sorties from Okinawa reached a monthly total of 777 by June and fighter sweeps from Iwo Jima 265 for the same period.[4]

The damage caused by these operations was almost incalculable. Large areas of the three urban complexes forming the keystones of Japan's war economy were laid waste, 56% of the Tokyo-Kawasaki-Yokohama area, 52% of Nagoya, and 57% of Osaka-Kobe being burned out causing complete paralysis of production, transportation, and communications facilities. Key industrial installations in the smaller cities were also levelled. Harbors and channels all along the Japan Sea and the Inland

1 This chapter was originally prepared in Japanese by Lt. Col. Iwaichi Fujiwara, Imperial Japanese Army. For duty assignments of this officer cf. n. 1, Chapter XVIII. All source materials cited in this chapter are located in G-2 Historical Section Files, GHQ FEC.

2 Minor urban areas fire-bombed during this period were Amagasaki, Fukuoka, Hamamatsu, Moji, Nobeoka, Okayama, Omuta, Sasebo, Shizuoka, Toyohashi, Tsu, and Yokkaichi. Typical of the destruction was the city of Hamamatsu, one of the largest producers of airplane propellers in Japan and an important railhead on the main Honshu network. This city was 72% burned out. Okayama, an important producer of plastics, explosives, and metals, was 62% destroyed. The incendiary raids on the major urban areas, which had begun with the great Tokyo raid of 9 March, ended in June. The total gross percentage for each city was computed by the author on the basis of the total number of buildings destroyed. Due to the extensive decentralization of Japanese industry, the gross damage percentage furnish a fairly accurate index of the total physical damage to industrial floorage. *Tokei Chosa Hokoku* 統計調査報告 (Statistics and Analysis Reports) Japanese Research Group, G-2 Historical Section, GHQ FEC, 30 Nov 50, Report No. 5: Extent of Air Raid Casualties and Property Damage on Small and Medium Cities during the War.

3 In addition to the urban area raids, B-29 formations conducted precision attacks on a variety of industrial targets using heavy loads of high explosives. Statistics and Analysis Reports, op. cit., Report No. 6: Bombing of Japanese Homeland by Allied Aircraft, May-Jul 45.

4 American carrier planes flew 1645 sorties against Japan during May, but did not show up at all in June, giving the Japanese a short respite from at least one type of attack. Ibid.

Sea were tightly sealed with aerial mines for days at a time. It became almost impossible to move material or commodities either from the continent or between ports in Japan itself.[5]

By the end of June, about 118,000 citizens in the three major urban areas had met with violent and painful deaths, about 170,000 were wounded and missing, about 1,300,000 buildings had been destroyed, and a total of 5,503,000 persons, about 42% of the population of Japan's three largest urban areas had been bereft of their homes, furniture, clothing, and personal effects.[6]

These violent assaults on the very foundations of urban society were rapidly reflected in the morale, efficiency, and availability of the labor force. The attacks so aggravated the already precarious food situation and created such a critical shortage of housing that millions were forced to flee to the countryside to seek the very necessities of life. In Tokyo-Kawasaki-Yokohama, for example, 4,210,000 people, about 53% of the population evacuated, leaving the remains of their homes and their means of livelihood behind. Throughout the nation more than 8,000,000 persons did likewise, meaning that almost 10% of the national citizenry became displaced persons.[7]

The mass evacuations greatly compounded the effects of the air raids on productive capacity. Immediately following the raids, an absentee rate of 70–80% was noted, dropping to about 40% after the restoration of local order. Even in those factory districts not damaged, the rate was about 15%. This high rate of worker absenteeism was an important factor in the decline of industrial productivity.[8]

The evacuations were symptomatic of the steady disintegration of organized urban life. Against the overpowering enemy air offensive the people and their local officials were beginning to feel completely helpless. The entire system of air raid precautions simply collapsed under the unexpectedly heavy assault. Shelters proved insufficient in number and vulnerable to enemy bombs, fire-fighting equipment was entirely inadequate to cope with the cascade of incendiaries, and the welfare, medical, and rehabilitation programs were over-saturated by the flood of casualties and

---

5 (1) Statistics and Analysis Reports, op. cit., Report No. 7: Extent of Air Raids and Property Damage and Casualties on Principal Cities during the War. (2) "Shimonoseki Strait was closed from March 1945 up to the end of the war. During each month, approximately 15 days represented complete closure due to the necessity for sweeping operations. In spite of the supposedly safe periods, the danger to navigation was still existent. . ." U.S. Strategic Bombing Survey (Pacific), Naval Analysis Division, *Interrogation of Japanese Officials*, 1946, Vol. I, p. 19. (Interrogation of Capt. Kyuzo Tamura, Mine Division, Naval Technical Department, Navy Ministry. (3) "Beginning with the mine blockade. . .of Shimonoseki Strait on 27 March 1945, all important ports and navigation routes as far north as Funakawa, Akita Prefecture, and Rashin, Korea, were subjected to repeated aerial mining. . .we were forced to abandon the use of. . .the Inland Sea first and later the major ports on the Japan Sea. The most effective and disastrous blockades were those of Shimonoseki Strait, Osaka-Kobe, Hakata, Maizuru, Fushiki, Nanao, and Niigata. . ." Statement by Capt. Atsushi Oi, Staff Officer (Operations), General Escort Command.

6 Typical of the damage suffered in the hardest hit of the smaller urban areas was Hamamatsu where 2,447 were killed, 1,702 wounded, 31,108 buildings destroyed or damaged, and 129,895 persons, or about 80% of the population rendered homeless. Statistics and Analysis Reports, op. cit., Report Nos 6, 7.

7 Evacuation of the other large urban complexes was as follows: Nagoya, about 740,000 (55%), Osaka-Kobe, 2,250,000 (60%). In each case, a small percentage were evacuated as part of the program of industrial dispersal, and, in some cases, a few of the displaced persons commuted back to their jobs. For the most part, however, the figures represent a total, permanent loss to the labor force of the city in question. Statistic and Analysis Reports, op. cit., Report No. 8: Population, Death by Air Raids and Number of People Evacuated during the War (Eight main Japanese Cities).

8 (1) *Gikai Hokokusho* (Reports to the Diet): Labor Situation in the Greater East Asia War, Sep 45. (2) On the basis of questionnaires prepared by leading Japanese industrialists, it had been found that the absentee rate had climbed to 49% by the end of the war. U.S. Strategic Bombing Survey (Pacific), Urban Areas Division, *The Effects of Air Attack on Japanese Urban Economy*, Mar 47, Table 21, p. 28.

homeless refugees.[9]

Although the dislocation of urban society was bad enough, a much more serious long-term result of the air raids was their effect on the industrial plant of the nation. Prior to April, the submarine and mine blockade, combined with the air attacks of November 1944 through March 1945, had carried forward apace the disintegration of Japanese basic industry which had begun early in 1944.[10] During the spring and early summer, this process assumed alarming proportions.

The most critical shortage continued to be that of coal. By June 1945, the imports of heavy coking coal from North China had ceased, crippling heavy industry. Although the domestic production of coal held up fairly well,[11] it proved to be almost impossible to move it in any appreciable amount due to the submarine and mine blockade of the major ports.[12]

The coal shortage fell most heavily on the iron and steel, non-ferrous metals, cement, and chemicals industries. By the end of June, the production of ferrous metals (iron, steel, and carbon steel) had fallen to a mere 35 % of the wartime peak. Cement production was down to 46 % of the peak figure.[13] The situation was no better in critical non-ferrous metals, with copper, aluminum, and magnesium standing at a bare 35 % of the industry's peak, and in one case, that of aluminum, sagging to a disastrous 16 %.[14]

The chemicals industry, like the others, was

9 (1) "After the May 23–25. . .raids on Tokyo, civilian defense measures in that city, as well as other parts of Japan, were considered to be a futile effort." U. S. Strategic Bombing Survey (Pacific), *Interrogations*, No. 118, 1945 (Interrogation of Mr. Genki Abe, Minister of Home Affairs.) (2) "The public began to realize that equipment and general preparations for air raids were futile. . ." Ibid., Interrogation Nó. 132. (Interrogation of Mr. Kingo Machimura, Chief, Police Section, Air Defense General Headquarters.)

10 Cf. Chapter XVIII, pp. 533–4, 553.

11 The production of coal during the period April-June was 10,890,019 metric tons or 71 % of the wartime peak achieved during the first quarter of 1944. Of this total, only about 4,900,000 tons were actually moved to key industrial zones. (1) Statistics and Analysis Reports, op. cit., Report No. 2: Output of Strategic Materials in the Japanese Homeland. (2) *Saiko Senso Shido Kaigi Tsuzuri* 最高戰爭指導會議綴 (Proceedings of the Supreme War Direction Council) Part I: Report of Lt. Gen. Tsukizo Akinaga, Chief, Cabinet Coordination and Planning Bureau, 6 Jun 45.

12 The importance of the coal shortage and its decisive role in the decline of Japanese industry is attested by leading industrial officials. USSBS, *Interrogations*, op. cit., Nos. 10, 15, 30 and 218. (Interrogations of Adm. Teijiro Toyoda, Munitions Minister; Mr. Y. Taguchi, Chief Engineer, Coal Section, Fuel Bureau; Mr. Matsuo Araki, Chief Electric Power Bureau; and Mr. T. Numabe, Coal Mining Department, Greater East Asia Ministry.)

13 Statistics and Analysis Reports, op. cit., Report No. 2. The production of pig iron and ingot steel was peaked in January-March 1943 and that of carbon steel in April-June 1944. Due to raw material shortage, this industry was well off its peak before the first B–29 appeared over the Homeland. USSBS, *Interrogations*, op. cit., Nos. 281 and 290. (Interrogation of Officials of the Yawata Plant, Japan Iron Works.)

14 Statistics and Analysis Reports, op. cit., Report No. 2. A similar situation obtained in the case of tin, lead, and zinc. The gross domestic production and imports of ferro-alloy ores and concentrates (chromium, manganese, molybdenum, nickel, tungsten, titanium, and vanadium) was 176,557 metric tons in the first quarter of 1945 or only about 77 % of the average quarterly production in the peak year, 1944. The chief cause of the decline was in shutting off of imports by the blockade. No manganese or titanium was imported after 1943 and the inflow of chromium and vanadium ceased at the end of 1944. The average quarterly production of ferro-alloys dropped in 1945 to 29 % of the peak in the second quarter of 1944. The shortage of ferroalloys, copper, and light metals had an exceedingly disruptive effect upon the productivity of many industries, particularly electrical equipment, machine tools, precision instruments, and aircraft. USSBS, *Interrogations*, op. cit., Nos. 22, 143, 196, 326 and 363. (Interrogations of Mr. D. Yasukawa, President, Electrical Equipment Control Association; Mr. S. Nakajima, President, Japan Wireless Equipment Co.; Mr. S. Fujii, Managing Director, Metals Distribution Control Co.; Mr. Sozo Kano, President, and Officials of the Onoda Cement Co.; and Mr. Masatsune Okamoto, Plant Manager, Osaka Plant, Mitsubishi Copper Refinery.)

防空壕　栗原 信

Original Painting by S. Kurihara

PLATE NO. 150
Air Raid Shelter

hard hit by shortages of raw materials, a special problem being the scarcity of salt.[15] In the case of five selected chemical products (ammonia, nitric acid, caustic soda, benzol, and toluol) total production in the quarter ending in June 1945 was only 43 % of the same period in 1944.[16] The most important immediate effect of the collapse of the chemicals industry was on the production of explosives. The gross production of propellants, high explosives, and primer materials in June 1945 was only 62 % of the peak established in March 1944.[17] This made it exceedingly difficult for the Japanese to meet the ammunition production program laid down to support decisive battle plans.

From the standpoint of decisive battle planning, the most dangerous basic material shortage was that of oil.[18] By the end of June, the quarterly gross national production and import of crude and refined petroleum had fallen to 24 % of the wartime peak established in the period July—September 1943, and the inventory of 4,751,000 barrels was only about 8 % of what had been on hand at the beginning of the war.[19] Of this, only about 606,000 barrels were aviation gasoline, of which 333,900 barrels were earmarked for decisive battle operations.

Despite a reduction in operations to within 80 % of expectations, including rigid curtailment of training flights and elimination of all operational flights except those connected with the continued prosecution of *Ten-Go*,[20] consumption still ran about 188,600 barrels in June against a production of about 98,000 barrels.[21] Although aviation gasoline was to be supplemented by alcohol and other substitute fuels, the projected margin between requirements and

15 In the pre-war years, Japanese industry had used about 1,000,000 metric tons of salt per year, while a like amount was used for food. About three-fourths of this was imported. By June 1945, quarterly imports had fallen to 260,000 tons and domestic production to a mere 70,600 tons or 39 % of the wartime peak. (1) Proceedings of the Supreme War Direction Council, op. cit., Report of Adm. Teijiro Toyoda, Munitions Minister, 17 Jul 45. (2) U. S. Strategic Bombing Survey (Pacific), Over-all Economic Effects Division, *The Effects of Strategic Bombing on Japan's War Economy*, Dec 46, p. 146.

16 The situation in other critical chemicals was even worse. Naphthalene, acetone, and butanol were producing at about 20 % of wartime peak, while sulfuric acid, tetraethyl lead, transparent plastic, methanol, soda ash and oxygen were down to an average of 37 %. A special case was alcohol where vigorous production efforts were made to supplement the supply of aviation fuel, resulting in a rise in production of 173 % for the period April-June, when compared with the preceding three months. (1) Statistics and Analysis Reports, op. cit. Report No. 2. (2) USSBS, *The Effect of Strategic Bombing on Japan's War Economy*, op. cit., pp. 156–7, 162–9, 171–2,

17 USSBS, *The Effect of Strategic Bombing on Japan's War Economy*, op. cit, p. 145.

18 Throughout the war, oil had been the most critical basic raw material problem faced by the Armed Forces. As a result of the terrific attrition in tankers caused by enemy submarine operations, imports of oil from the southern area had begun to decline in mid-1943. Ibid., p. 135.

19 (1) Statistics and Analysis Reports, op. cit., Reports No. 2; and No. 9: Inventory of Aviation Fuel in the Japanese Homeland after Apr 45. (2) Notes of Maj. Gen. Kikusaburo Okada, Chief, War Plans Section, Economic Mobilization Bureau, War Ministry. (Cf. Chapter IV, n. 4.)

20 (1) "Originally a plane was given a two-hour flight test, but by January 1943, shortage of fuel had reduced this time to one hour. Moreover, if the plane seemed satisfactory after a one-half hour test, it was landed and accepted...... At the end of the war......types of planes which had been in production for some time were merely flown once or twice around the field. If found satisfactory, they were accepted forthwith." USSBS, *Interrogations*, op. cit., No. 300. (Interrogation of Maj. Gen. Sadao Yui, Commanding Officer, Kagamigahara Air Depot.) (2) "....shortage of fuel prevented planes from averaging more than two hours flying time a month. It worked out that a plane was only used on an average of once every three weeks. No attempt was made to keep the engines in condition by running them every day or so..." USSBS, *Interrogations*, op. cit., No. 299. (Interrogation of Lt. Comdr. Ono, Staff Officer, 13th Air Flotilla.)

21 Data regarding aviation gasoline, including consumption, production, stockpile, and decisive battle reserve, was obtained from the following source: Statistics and Analysis Reports, op. cit., Report No. 9.

inventories was so exceedingly narrow that, particularly if the enemy continued to bomb refineries and tank farms, it was doubtful whether the decisive air battle could be fought later than the end of 1945.[22]

By the end of June, the decline of basic industry had not yet had its full impact on the fabricating industries of Japan. Although they were producing substantially less than they had at peak output, it was felt that the majority of decisive battle needs could be met by the estimated time of invasion. Aircraft production held up best of all, a total of 4,856 aircraft of all types being manufactured in the quarter ending in June. This was 65 % of the wartime peak rate.[23]

The munitions industry was, at this time, somewhat less well off than aircraft, and the production goals set to support decisive battle plans seemed beyond reach in most items.[24] This industry had reached its peak in the period October—December 1944 and thereafter fell off mainly as a result of raw material shortages. By the end of June 1945, Army ordnance industries were producing at a rate of only about 44 % of peak and Navy facilities at about 55 %.[25]

The shipping industry was also caught in the vicious circle of raw material shortages and high attrition through losses.[26] By the end of June, 79 % of all Japan's merchant tonnage had been sunk. Shipyards were producing a bare 27 % of the wartime peak set in January—March 1944. Maritime shipping capacity was down to 850,000 gross tons and was dropping at the rate of 2/300,000 tons per month. In the quarter just ended cargo loading amounted to 2,968,400 tons or roughly 27 % of wartime peak maritime activity. The shipping crisis was severe enough to dash any hope that industry could even be partially revived through efforts to improve the raw material situation.[27]

The blockade and the air raids had the same

22 This estimate was predicated upon a further reduction of monthly consumption, an increase in production of gasoline substitutes, and negligible losses due to air raids on stockpiles, the bulk of which by now was dispersed and concealed in underground caves.

23 Statistics and Analysis, Reports, op. cit., Report No. 2.

24 Cf. Chapter XVIII, pp. 549–0.

25 Particularly "soft" items were infantry cannon, mortars, tanks, artillery, and motor transport. The Navy was able to maintain a fairly good rate of production of land combat munitions due, in part, to the abandonment of the production of ship ordnance. Since the production of fighting vessels had been all but given up, the need for ship ordnance was fast disappearing. Statistics and Analysis Reports, op. cit., Report No. 10: Japanese Ordnance Production Apr–Jun 45 and Percentages to Peak Production.

26 During June the first losses incurred by enemy submarine action in the Sea of Japan were confirmed. Twenty-three vessels were sunk or damaged off Hokkaido and northern Honshu while 24 were sunk or damaged by enemy submarines off Korea, western Honshu and in the Korean Strait area. Anticipating the complete severance of supply lines to the continent, Imperial General Headquarters, on 28 June, set up an emergency operation to run food and war materials through the blockade. This was known as the *Nichi-Go* (日號) Operation The routes were between ports in Korea and northern and western Honshu. Orders were issued to the Commanders-in-Chief, General Escort Command, Air General Army, and First and Second General Armies to give all possible sea and air protection to the convoys. The operation was to begin immediately. (1) *Nihon Kaigun Hensei Suii oyobi Heiryoku Soshitsu Hyo* 日本海軍編制推移及兵力喪失表 (Table Showing Organizational Changes and Losses of Japanese Naval Forces) 2d Demobilization Bureau, Oct 49, pp. I–62–4. (2) *Dairikushi Dai Nisengohyakujugo-go* 大陸指第二千五百十五號 (Imperial General Headquarters Army Directive No. 2515) 28 Jun 45.

27 (1) Statistics and Analysis Reports, op. cit., Report. No. 1: Gross Tonnage of Japanese Merchant Shipping Losses through Allied Action Dec 41–Aug 45 and Report No. 11: Amount of Cargo Transported by Japanese Ships during the War. (2) Statement by Lt. Col. Michinori Ureshino, Staff Officer (Shipping), Imperial General Headquarters, Army Section.

impact on the service industries as elsewhere. Of these, by far the most important was food. Before the war, Japan imported about 20% of her most important staple, rice, from Korea, Formosa, and the southern area. The shipping shortage had reduced this to a trickle, no rice at all being imported from the southern area after March 1944. The ambitious program of subsisting the nation on continental imports, which had been laid down in January 1945,[28] fell far short, only 60% of the quota being met in the quarter January—March and even less by June. Moreover, rice for the Armed Forces had been withdrawn in ever increasing quantities so that, by 1945, there was on the average 40% less rice for civilian consumers than had been the case in 1941. An even worse situation obtained in the case of soy beans, meat, fish, vegetables and salt.[29]

As a result of the dwindling inventories of basic foods, the daily ration amounted to fewer than 1,500 calories, about 65% of the minimum Japanese standard for the maintenance of health and work efficiency.[30] While this was not actually a starvation diet, the prospect of the tightening of the blockade plus attacks on Japan's land transportation system, and the possibility of attacks on the unharvested fields themselves all gave rise to a distinct danger of famine, at least in the more densely populated areas.

Even more frightening was the possibility that the U. S. invasion might be so long postponed that the troops disposed for decisive battle would run through their reserve rations. This was one more reason why an invasion in the fall was to be preferred if Japan were to strike the heavy counter-blow which she was so carefully planning.[31]

Progress of Mobilization and Deployment

As enemy air raids hammered away at the threadbare fabric of Japan's industrial economy, preparations for the decisive battle were carried resolutely forward. In a series of orders issued during the first week in April, Imperial General Headquarters set in motion the second mobilization of ground troops according to plan.[32] Six new Army headquarters,[33] eight line combat

28 Cf. Chapter XVIII, n. 4.

29 Soy beans, largely imported from the Continent, were 31% off the 1941 level as a result of the shipping crisis; meat was 70% off, due to lack of imported forage; fish products, the chief source of animal protein in the Japanese diet, were down a disastrous 78% as a result of lack of boats and fuel and enemy air and naval interdiction of the fishing grounds; vegetables were down 81% due to shortage of manpower and fertilizer, and salt was off 52% as a result of the conversion of domestic table salt to industrial uses. Statistics and Analysis Reports, op. cit., Report No. 12: Japan's Food Supply Situation during the War.

30 The food market was actually in the grip of a ruinous inflation which the Government seemed powerless to control. By the end of June, for example, the black market (i.e. free market) price of rice was 46 times the legal ceiling, of flour 76 times, of potatoes 24 times and of sugar 200 times. However, through the use of home gardens, foraging trips into the countryside, and purchases on the black market, the average daily food consumption was actualy raised to around 1,800 calories.

31 (1) Proceedings of the Supreme War Direction Council, op. cit., Part I. (2) Statement by Maj. Tosaku Hirano, Staff Officer (Logistics), Second General Army.

32 Two Army headquarters (Fiftieth and Fifty-fourth) scheduled for activation during the second mobilization were postponed until the third. (Cf. n. 69) *Hondo Sakusen Kiroku* 本土作戰記錄 (Homeland Operations Record) Vol. II: First General Army, 1st Demobilization Bureau, Oct 46. Attached Map.

33 In addition, one Army headquarters (Fifty-eighth) was activated in Korea. (1) Homeland Operations Record, Volume V: Seventeenth Area Army, p. 19. (2) *Dairikumei Dai Sennihyakukyujushichi-go* 大陸命第千二百九十七號 (Imperial General Headquarters Army Order No. 1297) 8 Apr 45.

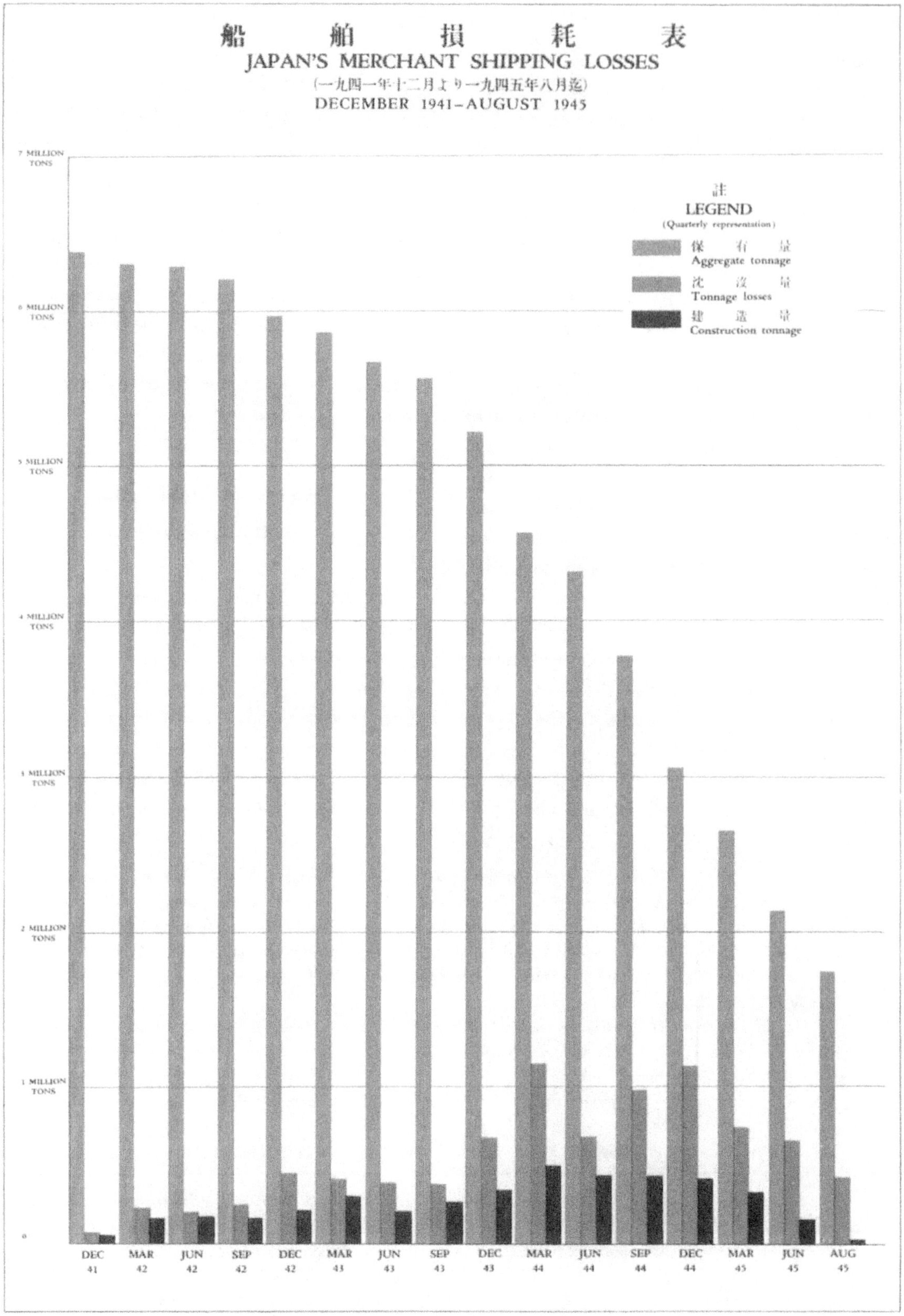

PLATE NO. 151
Japan's Merchant Shipping Losses

divisions,[34] and six armored brigades were activated in the Homeland and subsequently assigned as follows:[35]

- *First General Army*
  - *Twelfth Area Army*
    - *Thirty-sixth Army*[36]
      - *201st Division*
      - *202d Division*
      - *214th Division*
    - *Fifty-first Army*
      - *7th Tank Brigade*
    - *Fifty-second Army*
      - *3d Tank Brigade*
    - *Fifty-third Army*
      - *2d Tank Brigade*
  - *Thirteenth Area Army*
    - *209th Division*
- *Second General Army*
  - *Fifteenth Area Army*
    - *Fifty-fifth Army*
  - *Sixteenth Area Army*
    - *Fifty-sixth Army*
      - *4th Tank Brigade*
    - *Fifty-seventh Army*
      - *5th Tank Brigade*
      - *6th Tank Brigade*
    - *205th Division*[37]
    - *206th Division*
    - *212th Division*
    - *216th Division*

Concurrently with this second phase mobilization, the planned diversion of troop units from Manchuria was executed.[38] The 1st Armored Division began to arrive in the Thirty-sixth Army zone in early April, closing into the new position before the end of the month. The 25th and 57th Divisions, scheduled for assignment to Sixteenth Area Army and Thirty-sixth Army, respectively, arrived in Kyushu by mid-May, while the 11th Division displaced into the Fifteenth Area Army zone in southern Shikoku in early May.[39] Although not called for in the original plan, the 8th Tank Brigade was also included in the redeployment and joined the Thirteenth Area Army in the Tokai district of central Honshu.[40]

The High Command had meanwhile undertaken a reexamination of the strategic situation in light of the increasingly unfavorable developments on Okinawa. By early May Imperial General Headquarters concluded that the enemy might soon be free to take the initiative

34 Cf. Chapter XVIII, n. 38.

35 (1) *Daihonyei Rikugun Tosui Kiroku* 大本營陸軍統帥記錄 (Imperial General Headquarters Army High Command Record) 1st Demobilization Bureau, Nov 46, pp. 348–9. (2) *Hengo Butai Gaikenhyo* 編合部隊概見表 (Unit Organization Tables) Army General Staff. (3) *Dairikumei Dai Sensambyakujugo, Sensambyakunijuhachi-go* 大陸命第千三百十五・千三百二十八號 (Imperial General Headquarters Army Orders No. 1315 and 1328) 21 Apr and 10 May 45. (4) Imperial General Headquarters Army Order No. 1297, op. cit.

36 This Army headquarters (Thirty-sixth) had been activated in 1944 as a part of the *Sho* No. 3 plan. All other Army headquarters in the list were activated during the second mobilization.

37 The 205th Division was assigned directly to the Sixteenth Area Army upon activation. Late in May, however, the Second General Army transferred the division to the Fifty-fifth Army to strengthen the defenses of Shikoku. Homeland Operations Record, op. cit. Vol. III: Second General Army, p. 20.

38 Concurrently with this redeployment from Manchuria to the Homeland three coastal combat divisions, (111th, 120th, and 121st) were diverted from Manchuria to Korea. Homeland Operations Record, op. cit., Vol. V, p. 20.

39 The transfer of these units from the Continent was very successful. In spite of the shipping shortage and enemy air interdiction of the Japan Sea, it was still possible to operate at night across the Korea Strait. Units transferring from Manchuria were therefore able to arrive in Kyushu and Japan Sea coast of central Honshu with even more supplies and equipment than had been called for in the *Ketsu-Go* plan. (1) Imperial General Headquarters Army High Command Record, op. cit., pp. 328–9. (2) Homeland Operations Record, op. oit., Vol. II, Attached Map; Vol. III, Attached Map. (3) Imperial General Headquarters Army Order No. 1297, op. cit.

40 *Dairikumei Dai Sensambyakujukyu-go* 大陸命第千三百十九號 (Imperial General Headquarters Army Order No. 1319) 28 Apr 45.

in the next major move and that Kyushu now appeared the more probable target, the earliest possible invasion date falling after late June.[41]

The possibility of an invasion two months earlier than previously estimated resulted in an immediate acceleration of the defensive preparations. The first move was the transfer, 1 May, of units from the less threatened northeastern flank of the Homeland, the 147th Division being diverted to Kanto from Hokkaido and the 3d Amphibious Brigade to southern Kyushu from the Kuriles.[42]

In view of the extremely remote probability of an enemy invasion of the northeastern area and of further contemplated withdrawals from this area, Imperial General Headquarters ordered the Fifth Area Army on 9 May to concentrate the bulk of its troops on Hokkaido, leaving garrisons to defend only the more important islands in the Kurile chain.[43] This new mission, in effect, ruled out possible activation of *Kestu* No. 1, and rendered any subsequent action in this area of a delaying nature rather than decisive.[44]

In the week which followed, the High Command further strengthened the defenses of Kyushu and its approaches with the assignment of two additional divisions and the activation of a new independent mixed brigade. The 57th Division just assembling at Hakata, Kyushu, en route from Manchuria to join the Thirty-sixth Army, was reassigned to the Sixteenth Area Army on 10 May.[45] Four days later the 77th Division (Hokkaido) was transferred from the Fifth Area Army to Sixteenth Area Army while on the same day the lone infantry regiment securing the island of Tanega-shima lying just South of Kyushu was reorganized as the 109th Independent Mixed Brigade.[46]

Concurrently with these steps taken by the High Command, the Second General Army Commander Field Marshal Shunroku Hata, drafted an operational plan to meet the situation. This called for forces in Kyushu to wage a delaying action if the enemy should land during July, trading space for time until the assembly, equipping, and training of the decisive battle reserve could be completed and a general counteroffensive launched. Preparations for the delaying campaign in the critical invasion areas were to be completed by the end

41 (1) Statements by Lt. Gen. Torashiro Kawabe, Deputy-Chief, Army General Staff, and Lt. Gen. Seizo Arisue, Chief, Second Bureau (Intelligence), Imperial General Headquarters, Army Section. (2) Homeland Operations Record, op. cit., Vol. III, pp. 9–11.

42 The 3d Amphibious Brigade suffered heavy losses while displacing from the Kuriles to Hokkaido en route to Kyushu. Although the remnants immediately took up positions upon arrival in their newly assigned sector, the southeastern tip of Satsuma Peninsula, they were subsequently absorbed by the 125th Independent Mixed Brigade upon activation of the latter in the third mobilization. (Cf. p. 584) (1) *Dairikumei Dai Sensambyakuniju-go* 大陸命第千三百二十號 (Imperial General Headquarters Army Order No. 1320) 1 May 45. (2) *Rikugun Butai Chosa Hyo* 陸軍部隊調査表 (Table of Army Units) War Ministry, 28 Oct 45, Part I, p. 95. (3) Statement by Maj. Yasunobu Haba, Staff Officer (Operations), Sixteenth Area Army. (4) Homeland Operations Record, op. cit. Vol. IV: Fifth Area Army, pp. 30–1.

43 The Area Army was also charged with preventing the entry of enemy submarines into the Japan Sea. *Dairikumei Dai Sensambyakunijuroku-go* 大陸命第千三百二十六號 (Imperial General Headquarters Army Order No. 1326) 9 May 45.

44 It was the intention of Imperial General Headquarters not to reinforce Fifth Area Army in the event of an enemy landing in the northeastern area. The forces then deployed there would exert the maximum possible delay on the enemy. (Statement by Lt. Col. Shiro Hara, Staff Officer (Operations), Imperial General Headquarters, Army Section.)

45 Imperial General Headqurters Army Order No. 1328, op. cit.

46 *Dairikumei Dai Sensambyakunijushichi, Sensambyakunijukyu-go* 大陸命第千三百二十七・千三百二十九號 (Imperial General Headquarters Army Orders No. 1327 and 1329) both 14 May 45.

of June.[47]

While the mobilization and redeployment of ground troops was on schedule the preparation of battle positions in the critical invasion areas was falling seriously behind. Although the troops were assisted by thousands of civilian volunteers, by early May the construction of coastal positions in Kyushu had barely begun. The 86th Division and the 98th Independent Mixed Brigade, garrisoning the Ariake Bay area, had brought their field fortifications to about 50% completion, while the 156th Division on the Miyazaki coast, the 146th Division on the Satsuma peninsula, and the 145th Division in the Fukuoka area were only 10–20% finished. Troops in all these units were in a low state of training,[48] headquarters command arrangements were inadequate, and weapons, ammunition, equipment, and horses were still in short supply.[49] In the Kanto district, an even worse situation obtained.[50] There were permanent coast defense batteries in the entrance to Tokyo Bay and a few scattered semi-permanent heavy artillery positions forming the nuclei of projected strongpoints, but an effective field fortification system did not as yet exist.[51]

More discouraging than the lag in field construction was the extent to which local commanders misunderstood the policy of aggressive beach defense and failed properly to indoctrinate their subordinates.[52] Based largely on precedents set in the southern area, many coastal combat units located their rallying positions much too far from the beach. Although this was an understandable attempt to avoid the devastating effects of enemy naval bombardment, it had the highly undesirable

47 To implement these emergency preparations, plans were made to transport all the remaining supplies scheduled for *Ketsu* No. 3 to Kyushu during the last week in May. Due to transportation difficulties only about 80% of the amount had actually been concentrated on the island by early June. (1) Homeland Operations, Record, op. cit. Vol. III, pp. 9–15, 47–8. (2) Statement by Maj. Hirano, previously cited.

48 Units of the second mobilization scheduled for Kyushu had not yet even begun their training in early May. There was a 6–8 week time lag between the activation of new units and completion of the assembly of the cadre and filler replacements.

49 Fifty-seventh Army headquarters, responsible for southeastern Kyushu, had just completed organization. Fifty-sixth Army, responsible for northern Kyushu had not yet completed organization. Fortieth Army, designated for transfer from Formosa to assume responsibility for southwestern Kyushu, was, in mid-May, still awaiting transportation. (1) Homeland Operations Record, op. cit. Vol. III, pp. 5–6, 23, 26, 46–7, Chart I. (2) Imperial General Headquarters Army Order No. 1327, op. cit.

50 The lone exception was the Thirty-sixth Army. This decisive battle reserve continued to be in a high state of training and preparedness. At the beginning of May it was composed of the 81st, 93d, 201st, 202d, and 214th Divisions and the 1st and 4th Armored Divisions.

51 Major units disposed in the Kanto district at this time were as follows:

Kujukuri Beach area	Kashima Sea area	Sagami Bay area
Fifty-second Army	Fifty-first Army	Fifty-third Army
3d Imperial Guards Division	44th Division	84th Division
152d Division	151st Division	140th Division

In addition, the 147th Division was beginning to arrive in the Kujukuri Beach area. This transfer was expected to be completed about the end of May. (1) Homeland Operations Record, op. cit. Vol. II, pp. 8–10, Map. No. 1. (2) Statement by Col. Hiroshi Fuwa, Staff Officer (Operations), First General Army.

52 This misunderstanding was attributable to the haste with which the *Ketsu-Go* plan was evolved in Imperial General Headquarters. The implementing directives failed to describe clearly and in sufficient detail the conception of the policy for an aggressive beach defense. The maximum effort of the participating staff officers had been devoted to planning the broader aspects of the decisive battle, thus allowing little time for a reexamination of the doctrines of tactics and techniques or inspection of progress registered. (Statement by Col. Ichiji Sugita, Staff Officer (Operations), Imperial General Headquarters, Army Section.)

相模灣を俯瞰せる側防砲陣地　向井潤吉　昭和二十三年三月)

Original Painting by Junkichi Mukai

PLATE NO. 152
Flanking Gun Emplacement Overlooking Sagami Bay

effect of allowing the enemy to seize and consolidate a sizeable beachhead, which was entirely contrary to *Ketsu-Go* doctrine. Line combat divisions of the decisive battle reserve were also establishing their concentration points too far inland.

These discrepancies were uncovered during May by staff officers of Imperial General Headquarters while inspecting the critical invasion areas. Immediate remedial action was clearly indicated. On 6 June, therefore, the Army Section of Imperial General Headquarters issued a new manual of decisive battle tactics. This document contained an even stronger statement of the aggressive beach defense doctrine than had been laid down in the *Ketsu-Go* plan and its implementing directives. At the same time, the Chief of the Army General Staff, General Yoshijiro Umezu, drew up a detailed explanation of the desires of the High Command and disseminated it to all units. Pursuant to these instructions, the Commanders-in-Chief of the First and Second General Army immediately initiated strong orientation and indoctrination programs in their respective areas. However, much valuable time had already been lost.[53]

The lag in the construction of coastal positions, the costly misunderstandings concerning tactical doctrine, and the difficulties of redeployment and training were especially dangerous in Kyushu since it was earliest on the estimated invasion schedule. To help meet the situation, Marshal Hata, in late May forwarded a strong recommendation to Imperial General Headquarters that the forces scheduled for redeployment to Kyushu under the *Ketsu-Go* plan be moved immediately without waiting for activation of the operation.[54] The High Command, however, was so worried about the dangers inherent in what they felt to be a premature weakening of the Kanto district that no action was taken on this recommendation.[55]

While the major ground commands were engaged in these preparations, Imperial General Headquarters took steps to ready the Army air establishment for the decisive battle. On 8 May, orders were issued releasing the Second Air Army in Manchuria from the order of battle of the Kwantung Army, the Fifth Air Army in China from the China Expeditionary Army, and the 1st Air Division in the Northeast area from Fifth Area Army. All these forces were assigned to the Air General Army.[56]

On the same day, the High Command issued a directive outlining the air redeployment plan for *Ketsu-Go*. The essentials of this document were as follows:[57]

1. *Policy*

Although participation in the Ten-Go operation will continue, primary emphasis will be on completing preparations for Ketsu-Go by the end of June.

---

53 (1) *Kokudo Kessen Sempo Hayawakari* 國土決戰戰法早わかり (Key to Homeland Decisive Battle Tactics) Imperial General Headquarters, Army Section, 6 Jun 45. (2) Homeland Operations, Record, op. cit. Vol. II, pp. 7–8, 20–3, 29, 51–8.; Vol. III, pp. 6–7, 29–33.

54 Cf. Chapter XVIII, pp. 559–60.

55 Statements by Maj. Gen. Joichiro Sanada, Deputy-Chief of Staff, Second General Army; Lt. Col. Masakatsu Hashimoto, Staff Officer (Operations), Second General Army; and Col. Ichiji Sugita, previously cited.

56 Pending alert for the *Ketsu-Go* operation, the 1st Air Division still remained under the operational control of Fifth Area Army, the Second Air Army under Kwantung Army, and an element of Fifth Air Army was retained under the Commander-in-Chief, China Expeditionary Army. (1) *Dairikumei Dai Sensambyakunijugo-go* 大陸命第千三百二十五號 (Imperial General Headquarters Army Order No. 1325) 8 May 45. (2) *Hondo Koku Sakusen Kiroku* 本土航空作戰記錄 (Homeland Air Operations Record) 1st Demobilization Bureau, Dec 46, pp. 19–20.

57 *Dairikushi Dai Nisenyonhyakushichijugo-go* 大陸指第二千四百七十五號 (Imperial General Headquaters Army Directive No. 2475) 8 May 45.

2. *Redeployment Outline*

(*a*) *The main strength of Fifth Air Army will immediately proceed to Korea. An element will remain in China under the command of the Commander-in-Chief, China Expeditionary Army.*

(*b*) *When an alert for the Ketsu-Go operation is issued, the main strength of the 1st Air Division will be immediately transferred to the anticipated operational area.*

(*c*) *Upon activation of the Ketsu-Go operation, the main strength of the Second Air Army and that part of the Fifth Air Army remaining in China will be immediately transferred to the active battle theater.*

(*d*) *Should the enemy land in the China coastal sector prior to the Ketsu-Go operation, Fifth Air Army, reinforced by designated units of Second Air Army will proceed to the enemy landing area and attack. These operations will be supervised by the Commander-in-Chief, China Expeditionary Army.*

By late May, it had become apparent to the Army High Command that further pursuance of the *Ten-Go* operation would result in a needless expenditure of air strength without appreciably delaying the enemy in his final approach to the Homeland.[58] On 26 May, therefore, orders were issued releasing the Sixth Air Army from attachment to Combined Fleet and reverting it to control of the Air General Army. At the same time, General Masakazu Kawabe, Commander-in-Chief of Air General Army, was given a directive concerning his immediate mission, the gist of which was as follows:[59]

The Commander-in-Chief, Air General Army, will accelerate operational preparations with the main objective of destroying enemy forces invading the Homeland. Emphasis will be on the Kyushu and Korea Strait area.

The most serious operational problem facing the Japanese remained that of air defense. In the struggle to provide an antidote to the violent enemy air offensive, the nation was handicapped by many adverse circumstances. In the first place, having lost the Marianas and Iwo Jima, and with Okinawa already being used as a forward base, the Japanese were deprived of patrol and reconnaissance bases so necessary to the maintenance of an adequate warning network. Moreover, the enemy, using these same bases was now able to mount fighter-escorted bombing attacks over almost all of the Homeland. Another serious handicap was the chronic shortage of antiaircraft guns and ammunition, brought about mainly through the decline in production of these items. Finally, it was found that even those guns that were available were ineffective against night attacks by high-altitude planes.[60]

In the face of these obstacles, both the Army and the Navy undertook to reorganize, strengthen, and redeploy air defense forces. In

---

58 Even after the Army began to shift the emphasis from *Ten-Go* to the forthcoming *Ketsu-Go* operations, the Navy clung tenaciously to the idea of preventing enemy consolidation of his Okinawa base. The participating units had reported that, in operations between 18 March and 28 May, 358–360 enemy vessels had been sunk or heavily damaged. The Navy was confident that, if this high attrition rate could be maintained, the Homeland invasion would be indefinitely postponed. Navy pursuance of the *Ten-Go* objectives thus continued to delay formulation of joint arrangements for *Ketu-Go*. (1) *Jokyo Handan* 情況判斷 (Estimate of the Situation) 1st Section, Navy General Staff, 1 Jun 45. (2) Statements by Maj. Gen. Kazuo Tanikawa, Staff Officer, and Col. Hiromu Hosoda, Staff Officer (Operations), both of Imperial General Headquarters, Army Section.

59 (1) *Dairikumei Dai Sensambyakusanjuroku-go* 大陸命第千三百三十六號 (Imperial General Headquarters Army Order No. 1336) 26 May 45. (2) *Dairikushi Dai Nisenyonhyakukyujuichi-go* 大陸指第二千四百九十一號 (Imperial General Headquarters Army Directive No. 2491) 26 May 45.

60 Statement by Lt. Col. Takeshi Murata, Staff Officer (Air Defense), Imperial General Headquarters, Army Section, and later, Staff Officer (Air Defense), Second General Army.

the case of ground units, this involved the conversion of several antiaircraft groups and units into antiaircraft divisions, at the same time strengthening each of the newly organized units.[61] Some of the batteries were redeployed from the large urban areas to smaller cities and to key positions on railroads and harbors.[62] The organization of interceptor units remained as before although the shortage of aircraft continued due to the higher priority enjoyed by the offensive air establishment.[63]

The trend of Navy action in the matter of air defense had been towards achieving independence from Army control. As enemy carrier task forces maneuvered closer and closer to the Homeland, the distinction between interception and offensive operations tended to break down. The Navy accordingly decided to organize its own interceptor pools in the various Homeland areas. On the 23d, the 71st and 72d Air Flotillas were activated under Third and Fifth Air Fleets, respectively, to take charge of interception in eastern and western Honshu, Shikoku, and Kyushu.[64] Together, the Army and Navy were able to muster about 970 aircraft for air defense purposes.[65]

By the beginning of June, the Homeland defense situation, as well as developments in other theaters, had reached a point where even a greater mobilization of national effort was considered an essential prerequisite to waging a successful decisive battle. With military preparations rapidly passing from the planning stage, the High Command concluded that it was now necessary to strengthen still further the all-out effort by drawing in the Government as well as the military. On 8 June, therefore, a conference attended by members of the Supreme War Direction Council and other high officials of the Government was held in the Imperial presence. At this gathering a basic war policy was adopted which called for the full participation of the entire nation in prosecuting the war to the end.[66]

---

61 On 6 May, the Nagoya AAA Unit was reorganized into the 2d AAA Division, the Chubu (Osaka) AAA Group into the 3d AAA Division, and the Seibu (northern Kyushu) AAA Group into the 4th AAA Division. *Dairikumei Dai Sensambyakunijuyon-go* 大陸命第千三百二十四號 (Imperial General Headquarters Army Order No. 1324) 6 May 45.

62 Homeland Operations Record, op. cit. Vol. II, p. 38, Vol. III, pp. 35–6.

63 The fighters which the Navy had attached to the air defense forces (Cf. Chapter XVIII, p. 541) had long since been withdrawn. *Daikaishi Dai Gohyakukyu-go* 大海指第五百九號 (Imperial General Headquarters Navy Directive No. 509) 19 Feb 45.

64 *Hondo Boei Sakusen Sono San* 本土防衛作戰其の三 (Homeland Defense Naval Operations, Part III) 2d Demobilization Bureau, May 49, p. 33.

65 (1) *Rikugun Chosabu Shitsumon Sho (Sono Juroku) Kaito* 陸軍調査部質問書(其十六)回答 (War Ministry Reply to USSBS Questionnaire No. 16) 20 Nov. 45. (2) Homeland Defense Naval Operations Part III, op. cit., p. 35. (3) One consequence of the growing ineffectualness of Japan's air defenses was a fear that various sectors in the Homeland might be isolated from the remainder through raids on the transportation network. In anticipation of such a situation and to strengthen local passive air defense measures, the Government, on 10 June, issued instructions setting up local authorities to assume complete responsibility for civil administration in event of emergency. These officers, designated governors-general, were located at or near the headquarters of each district Army, and their zone of authority was coextensive with that of the district Army. They were to render such cooperation and assistance to the district Army and naval station as needed to expedite operations, to maintain law and order, and, in general, to function *in loco regis* in event of severance of communications with the civil government. (a) *Chokurei Dai Sambyakugoju-go* 勅令第三百五十號 (Imperial Ordinance No. 350) 10 Jun 45. (b) *Naikaku Kunrei Dai San-go* 內閣訓令第三號 (Cabinet Instructions No. 3) 10 Jun 45.

66 Proceedings of the Supreme War Direction Council, op. cit. Part I. Cf. Chapter XX, pp. 643–4.

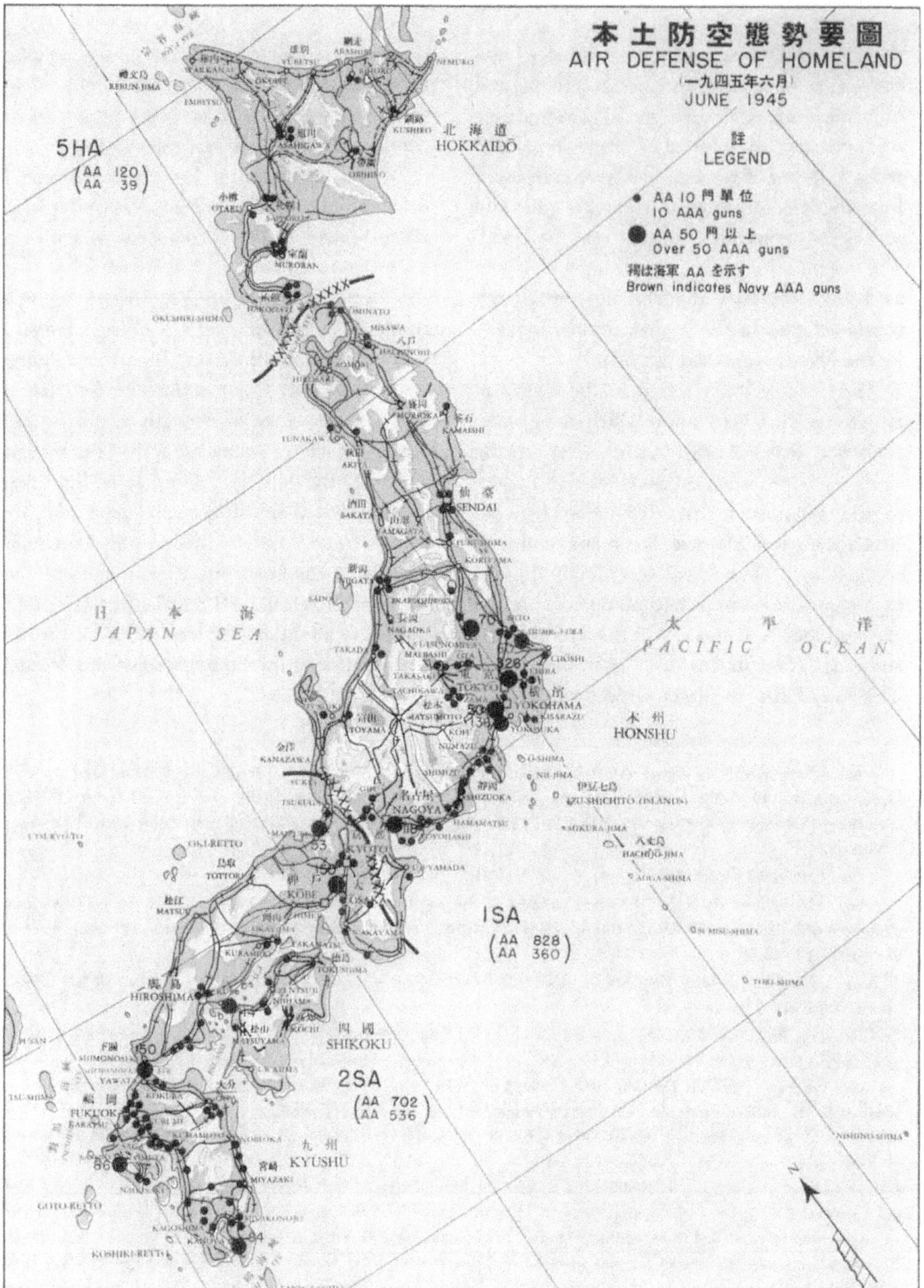

PLATE NO. 153
Air Defense of Homeland, June 1945

Pursuant to the decisions of the Imperial conference, the Government on 23 June promulgated the National Volunteer Service Act, which, together with its implementing directives, contained the following general provisions:[67]

1. National volunteer units will be immediately organized in each prefecture, each railway and communication division control district, and in designated major industrial installations, the prefecture constituting a regimental district. All men 15–60 and all women 17–40 will be subject to such service.

2. General supervision of the organization of these units will be exercised by the District Army (Area Army) commander assisted by the local governor-general. Existing Civilian Defense units, Neighborhood Associations, etc., will be absorbed into the new organization.

3. Volunteer unit missions will include transportation, communications, construction, repair, and other logistics duties, security of vital installations, intelligence work, and, should the need arise, fighting beside the regular line units.

4. In general, arms and equipment will be limited to small arms, grenades, and grenade dischargers.

In the meantime, Imperial General Headquarters had taken steps to effect the third mobilization of ground combat units almost two months earlier than originally planned.[68] Pursuant to a series of organization orders, the first of which was issued on 23 May, four new Army headquarters,[69] 10 coastal combat divisions,[70] eight line combat divisions,[71] and 14 independent mixed brigades were activated in the Homeland[72] and on 19 June, assigned as follows:[73]

First General Army
Eleventh Area Army
Fiftieth Army
308th Division

---

67 (1) *Bogyo Senryaku oyobi Senjutsu ni kansuru Kaito* 防禦戰略及戰術に關する回答 (Reply to a Questionnaire Concerning Defense Tactics and Strategy) 1st Demobilization Bureau, 31 Mar 46. (2) *Giyu Heieki Ho (Horitsu Dai Sanjuku-go)* 義勇兵役法(法律第三十九號) (Volunteer Service Act Law No. 39) 87th Diet, 23 Jun 45. (3) *Kokumin Giyu Sentotai Tosotsu Rei* 國民義勇戰鬪隊統率令)Regulations for Command and Organization of National Volunteer Units) 23 Jun 45. (4) *Kampo* 官報 (Official Gazette) Special Issue, 9 Jun 45, containing Stenographic Record Proceedings of the House of Representatives, 87th Diet.

68 The third mobilization was advanced to enable the units slated for Kyushu to gain some time in their emergency preparations. Because of this as well as the delay in fulfilling production goals, units of the third mobilization were slightly weaker in artillery than the early mobilizations. (Cf. Chapter XVIII, p. 548) (1) Statements by Maj. Hirano, previously cited and Maj, Yasuji Komuratani, Staff Officer (Operations), Imperial General Headquarters, Army Section, (2) Homeland Operations Record, op. cit. Vol. III, p. 19.

69 These were the Fiftieth and Fifty-fourth Army headquarters activation of which had been postponed from the previous mobilization and the Fifty-ninth and Tokyo Defense Army headquarters which were not provided for in the original plans drafted in February. The Fifty-ninth Army headquarters was activated to command the defense of western Honshu, and the Tokyo Defense Army headquarters to command the units in the capital city. (1) Table of Army Units, op. cit. (2) Homeland Operations Record, op. cit. Vol. II, pp. 10–1, 26; Vol. III, pp. 24–5.

70 The 355th Division, not included in the original plans, was organized to strengthen the defenses of Nagoya area about which concern was being increasingly felt. (1) Unit Organization Tables, op. cit. (2) Statement by Lt. Col. Hara, previously cited.

71 The 234th Division was not included in the original mobilization plans. Unit Organization Tables, op. cit.

72 In addition, one coastal combat division (320th) and one independent mixed brigade (127th) were activated in Korea. The 320th Division, not provided for in the original plans, was organized to strengthen the defenses of southern Korea. (1) Unit Organization Tables, op. cit. (2) Homeland Operations Record, op. cit. Vol. V, pp. 28–30 32.

73 (1) *Dairikumei Dai Sensambyakugojuichi, Sensambyakugojugo-go* 大陸命第千三百五十一、千三百五十五號 (Imperial General Headquarters Army Orders No. 1351 and 1355) 19 and 23 Jun 45. (2) Unit Organization Tables, op. cit. (3) Homeland Operations Record, op. cit. Vol. II, pp. 10–1, Map 1; Vol. III, p. 19, Chart 3, Attached Map. (4) Imperial General Headquarters Army High Command Record, op. cit., pp. 349–50, 402–3.

222d Division
322d Division
113th Independent Mixed Brigade
Twelfth Area Army
Tokyo Bay Group[74]
354th Division
114th Independent Mixed Brigade
Tokyo Defense Army
Fifty-first Army
221st Division
115th Independent Mixed Brigade
116th Independent Mixed Brigade
Fifty-second Army
234th Division
Fifty-third Army
316th Division
117th Independent Mixed Brigade
321st Division
Thirteenth Area Army
Fifty-fourth Army
224th Division
355th Division
119th Independent Mixed Brigade
120th Independent Mixed Brigade
229th Division
Second General Army
Fifteenth Area Army
Fifty-fifth Army
344th Division
121st Independent Mixed Brigade
Fifty-ninth Army
230th Division
231st Division
124th Independent Mixed Brigade
225th Division
123d Independent Mixed Brigade
Sixteenth Area Army
Fortieth Army[75]
303d Division
125th Independent Mixed Brigade
Fifty-sixth Army
312th Division
351st Division
118th Independent Mixed Brigade
122d Independent Mixed Brigade
126th Independent Mixed Brigade

With these activations, the line strength of the Homeland defense armies (including Hokkaido) was brought to 30 line combat divisions,[76] 24 coastal combat divisions, two armored divisions, seven a tank brigades, 23 independent mixed brigades, and three infantry brigades.[77]

Concurrently with the implementation of the third mobilization, a further redeployment of major combat units was effected. The Fifth Area Army transferred the 42d Division and Chishima 1st Brigade from the Kurile Islands to Hokkaido while at the same time the High Command ordered the transfer of the 4th Amphibious Brigade from the Kuriles to the Twelfth Area Army.[78] In addition, the 209th Division was transferred from Thirteenth Area

74 Tokyo Bay Group was the designation assigned to the Tokyo Bay Fortress Unit on 8 April by Imperial General Headquarters. This step authorized the commander, Tokyo Bay Fortress Unit, to command other combat units assigned for the defense of Tokyo Bay as well as the fortress units. Imperial General Headquarters Army Order No. 1297, op. cit.

75 The Fortieth Army headquarters completed its displacement from Formosa by mid-June at which time it took over control of operations in southwestern Kyushu. Homeland Operations Record, op. cit. Vol. III, pp. 25–6.

76 This total of line combat divisions did not include the 1st Imperial Guards Division which was assigned as security force for the Imperial Household. (Cf. Chapter XVIII, n. 17.)

77 Of this total line strength the 7th and 42d Divisions and 101st Independent Mixed Brigade were stationed on Hokkaido. The total does not include the 89th and 91st Divisions and the 129th Independent Mixed Brigade (organized 16 July 1945) stationed in the Kurile Islands, the 88th Division on Karafuto, or the seven coastal combat divisions, three line combat divisions, and two independent mixed brigades in Korea. (1) Homeland Operations Record, op. cit. Vol. IV, pp. 37–8, Attached Map No. 2; Vol. V, pp. 4–5, 17–21, 28–30, 32, 38. (2) Unit Organization Tables, op. cit.

78 The 4th Amphibious Brigade was absorbed by the 1st Armored Division. Unit Organization Tables, op. cit.

Army to Thirty-sixth Army, further strengthening this large reserve Army to a total of six line combat and two armored divisions.[79]

By this time, production trends clearly indicated that industry was incapable of adequately supporting the immense structure of the Homeland defense forces and the tactical and strategic plans devised for their use. The priority enjoyed by the units in Kyushu was draining off almost all of the material strength of the military establishment. While this meant that preparations for the decisive battle in Kyushu would be almost certainly completed, it also meant that Kyushu would very likely be the only area in which a decisive battle could be supported at all.

The brightest spot in the otherwise clouded production picture was in aircraft. By the end of June, almost 8,000 planes, mostly *tokko* types, had been hoarded for the decisive battle, while it was fairly certain that an additional 2,500 could be produced by the end of September.[80] Although this program was running slightly behind the schedule laid down in February,[81] it remained a single encouraging item in an atmosphere of disaster. Facilities for basing this vast special-attack armada were also being rushed to completion. A total of 325 airstrips, some of which were simple one-way strips, were under construction throughout the Homeland, 95 of them on secret sites far in the interior.[82]

Although doing fairly well in aircraft, the Japanese were far behind schedule in ground combat ordnance. At the end of June, output in every item was short of the schedule that had been set earlier in the year. Percentages of scheduled production actually achieved by this time were as follows:[83]

Rifles	78 %
Light machine guns	62 %
Machine guns	89 %
Infantry cannon	13 %
Antiaircraft guns	62 %
Mortars	8 %
Self-propelled cannon	25 %
Light artillery	17 %
Heavy artillery	60 %

79 Imperial General Headquarters Army Order No. 1351, op. cit.

80 These figures include Fifth Air Army (about 700 planes) but do not include 1st Air Division and Second Air Army which were to contribute 65 and 120 aircraft, respectively. Of the total projected for September, about 5,400 were scheduled for the Army and the balance for the Navy. Fighters, reconnaissance planes, bombers, and trainers were all being converted for *tokko* purposes. In addition, the Navy was concentrating on the production of types engineered for *tokko* use exclusively. In addition to *Oka* 櫻花 (Cherry Blossom) which participated in the *Ten-Go* operation (Cf. Chapter XVIII, pp. 555–7), the Navy had developed *Kikka* 橘花 (Orange Blossom) and the *Toka* 藤花 (Wisteria Blossom). The former was a ground-launched missile, propelled by rocket turbines, with a range of about 350 miles at 400 m. p. h. carrying a 1,100 pound bomb. The *Toka* was a simply-built, generally conventional type plane with a 700 mile range at 300 m. p. h. carrying a 1,100 pound bomb. The Navy planned to make 1,000 of these special types before September. (1) *Kokuki Seisan Jisseki oyobi Kushu To ni yoru Seisan Noryoku Soshitsu Gaikyo* 航空機生產實績及空襲等による生產能力喪失概況 (Summary of Production of Aircraft, Loss of Capacity through Air Raids and Allied Matters) Ministry of Commerce and Industry, 1 Sep 45. (2) *Koku Tokko Sembi* 航空特攻戰備 (Battle Preparations of Naval Air *Tokko*) 2d Demobilization Bureau, Mar 46, pp. 2, 4–5, and Attached Table (Data of Special Planes). (3) Homeland Air Operations Record, op. cit., p. 39. (4) Homeland Operations Record, op. cit. Vol. IV, pp. 48–9. (5) Statements by Rear Adm. Katsuhei Nakamura, Chief, General Affairs Bureau, Naval Aeronautical Department; Lt. Col. Katsuo Sato, Staff Officer, Air General Army; and Lt. Col. Koji Tanaka, Staff Officer (Air Operations), Imperial General Headquarters, Army Section.

81 Cf. Chapter XVIII, p. 534.

82 Dispersion to a large number of small strips was emphasized as was the use of caves for hangers. (1) Statement by Lt. Col. Katsuki Mizumachi, Staff Officer (Operations), Sixth Air Army. (2) Battle Preparations of Naval Air *Tokko*, op. cit., p. 7 and Attached Chart (Air Installations).

83 Statistics and Analysis Reports, op. cit., Report No. 17: Production of Japanese Army Ordnance.

This discouraging situation in weapons production naturally made it very difficult to equip new units and almost impossible to achieve adequate levels in reserve dumps. The bulk of production was channeled to units in Kyushu and it was felt that at least sufficient weapons would be available to fight the decisive battle in that area by September.

Another bad situation facing the Japanese at this time was the extent to which preparations for naval surface participation in the *Ketsu-Go* were falling behind. In the quarter ending 30 June, only 1,235 surface special-attack boats had been produced and only 324 underwater types.[84] These were 18% and 15% respectively of the targets set for September. While it seemed hardly likely that these goals could be met, as in the case of ordnance, it was felt that sufficient quantities would be available to fight the Kyushu decisive battle at least.

The failure of production efforts, coupled with transportation and communications difficulties, shortages of fuel and rations, and continued enemy air attacks had, by the end of June, dealt serious blows to the national preparations for decisive battle. As far as organization of units was concerned, there were few difficulties. In both Kyushu and Kanto, units of the third mobilization were expected to be fully organized, although untrained, by mid-July. In Kyushu, 60% of these had already begun their training.[85]

Equipment, however, was another story. In Kanto, units of the third mobilization were short in every item, with no prospect of catching up as long as Kyushu held a higher priority. This was particularly true of small arms, antitank guns, mortars, and self-propelled cannon. Even in Kyushu, stocks of equipment on hand were only about 50% of third mobilization requirements with 31 August as the tentative completion date.[86]

Stockpiling of expendables was also in a confused state. In Kanto, no munitions and ordnance stockpiles had been established at all, while rations stocking was about 50% complete. Units in Kyushu, however, had built ammunition stockpiles to 100% of *Ketsu-Go* requirements, fuel to 94%, and rations to 164%.[87] This was encouraging as regards the conduct of the Kyushu battle, but pointed up the fact that the national war potential was

84 The total numbers of each type available by the end of June were as follows:

Underwater Types		Surface Types	
Koryu	73	Shinyo	2150
Kairyu	252	Renraku-tei	700 (approx)
Kaiten	119		

(1) *Hondo Joriku ni taisuru Hangeki Sakusen Jumbi* 本土上陸に對する反擊作戰準備 (Preparations for Counterattack Operations in Defense of the Homeland) 2d Demobilization Bureau, May 47, Attached Chart No. 3. (2) *Showa Nijunendo Zenki no Kokuki Tokko Heiki no Seisan Keikaku to Seisan Jisseki* 昭和二十年度前期の航空機特攻兵器の生產計画と生產實績 (Planned and Actual Production of Aircraft and *Tokko* Weapons in the First Half of 1945) 2d Demobilization Bureau, 15 Jul 49. (3) Statement by Maj. Kanetoshi Mashita, Staff Officer (Operations), Imperial General Headquarters, Army Section.

85 (1) Statement by Col. Fuwa, previously cited. (2) Author's memory. (Lt. Col. Fujiwara was a staff officer of Second General Army at the time and later of the Fifty-seventh Army.)

86 Tables of equipment for artillery units of the third mobilization had already been cut almost to nothing. Artillery weapons on hand were therefore sufficient to immediately equip these units. Units of the second mobilization had completed their equipment in Kanto except for antitank guns and 120 mm mortars. In Kyushu, the second mobilization units were about 70% equipped with 31 July as a completion target. First mobilization units were already equipped. (Statements by Col. Fuwa and Maj. Hirano, previously cited.)

87 (1) Homeland Operations Record, op. cit. Vol. II, p. 44. (2) Statements by Maj. Hirano, previously cited and Lt. Col. Shinroku Iwakoshi, Staff Officer (Supply), Imperial General Headquarters, Army Section.

PLATE NO. 154

Dispositions of Naval *Tokkotai* in Western Japan, July 1945

probably incapable of supporting more than one such decisive operation.

While preparations in the Homeland went slowly forward, the Okinawa campaign was brought to its unhappy end. The superhuman courage of the vastly outnumbered ground and naval forces and the expenditure of 2258 aircraft had completely failed to halt enemy seizure and consolidation of this most vital base.[88] Organized resistance by Thirty-second Army ceased on 23 June[89] and on the 25th the painful fact of Okinawa's loss was made known to the general public.[90]

Okinawa was but one strand in the web of disaster in which the Japanese were now caught. The pattern was repeated in the Philippines with the fall of Baguio and the virtual end of strategic delaying operations in Luzon, in the southern area where Allied landings at Brunei Bay, North Borneo and other points cut Japanese remnants to ribbons, in southeast Asia with the loss of Rangoon, and finally in Europe where the unconditional surrender of Germany had released millions of troops for redeployment against Japan Against this background, President Truman on 2 June made public the basic strategy to be used in the invasion of Japan, an operation that could not now be far off.[91]

A New Estimate of the Situation

Being inferior in the technical equipment of war and having been forced into an extremely disadvantageous strategic position, Japan's principal hope of success in the forthcoming Homeland battle lay in outguessing the enemy in accurately predicting the time and place of the invasion and meeting it with overwhelming force.

Quick formulation of a strategic estimate was, however, rendered impossible by the existence within Imperial General Headquarters of a wide range of opinion on every aspect of the problem. The only point on which there

88 After the abortive general offensive of 4–5 May, the Japanese Army and Navy air forces flew five more general attacks interspersed with minor operations. A total of 2,784 sorties were flown between 5 May and 22 June. Reports of damage to enemy vessels continued to be high. Although the official end of the Okinawa campaign came on 25 June and the Army had long since withdrawn its units from *Ten-Go,* the Navy refused to give up the idea of preventing enemy use of Okinawa and continued the *Ten-Go* operation using hit and run tactics until early in July. In the meantime, the 7th and 98th Air Regiments (Torpedo bombers), which had been under the operational command of the Commander-in-Chief, Combined Fleet, since 1 February 1944, reverted to the Sixth Air Army on 28 June. (Cf. Chapter XVIII, n. 10.) (1) Statistics and Analysis Reports, op. cit., Report No 4: Commitment and Losses of Japanese Aircraft and Damages Inflicted on Allied Forces in the Okinawa Air Operations, 25 Mar–22 Jun 45. (2) *Dairikumei Dai Sensambyakugojushichi-go* 大陸命第千三百五十七號 (Imperial General Headquarters Army Order No. 1357) 28 Jun 45. (3) *Daikaishi Dai Sambyakunijuhachi-go* 大海指第三百二十八號 (Imperial General Headquarters Navy Directive No. 328) 1 Feb 44.

89 (1) The end of the Okinawa campaign was dramatized by the ceremonial *harakiri* of Lt. Gen. Mitsuru Ushijima, Thirty-second Army commander and members of his immediate staff. This gesture followed closely the *samurai* tradition constituting an apology for failure. *Dai Sanjuni-Gun Shijitsu Shiryo* 第三十二軍史實資料 (Historical Data, Thirty-second Army) Home Depot Division, Mar 47, pp. 73–4.

90 The end of the Okinawa campaign brought home to the people the desperate plight of the nation. The beginning of the battle had been accompanied by a great deal of public fanfare hailing the "decisive stand" at Okinawa. Failure there was a great blow to national morale.

91 According to the President, the Americans would, (a) nail down and isolate Japanese forces and annihilate them one by one, (b) concentrate overwhelming strength against attack objectives. (c) attain victory through massive concentrations of weapons, keeping personnel losses to a minimum, and (d) mobilize maximum strength and bring unremitting pressure giving the Japanese no chance to rally. *Asahi Shimbun* 朝日新聞 (Asahi Newspaper) Tokyo, 4 Jun 45.

was general agreement was that the enemy would immediately intensify the sea and air blockade of Japan and would sooner or later attempt an invasion.

The most important of the controversial issues was whether the United States would seek an early end to the war by moving immediately, or, on the other hand, initiate a long blockade designed to reduce Japan to the point of complete helplessness.[92] A dominant majority adhered to the former alternative, and it became the official position of Imperial General Headquarters.[93]

Proceeding on the assumption that United States policy would call for a quick decisive battle, there were two possibilities regarding the direction and objective of enemy operations. First of all, the United States might seek an immediate decision by moving directly to the main Japanese islands. In this case, the enemy might first seize air and sea bases in such areas as the northern Ryukyus and the Izu Island chain lying between Tokyo and the northern Bonins. He would most certainly drop such plans, however, and move at once against the main Japanese islands if he ever became convinced that Japanese air power had completely collapsed.

On the other hand, with Japan Proper as the ultimate goal, the enemy might first seize additional major advance bases.[94] A part of the Army intelligence staff in Imperial General Headquarters, particularly those officers connected with Chinese intelligence, was convinced that the United States forces would land in central and/or north China in order to give military support to the Chungking regime. Other groups within the headquarters, particularly in the Army operations group, held that two strong additional possibilities were an invasion of southern Korea or of Saishu (Quelpart) Island lying in the key Korea Strait area. Any of these three would gain important advance bases, would cut Japan's continental supply lines, and, in the first two cases, would check Soviet influence in north China and Korea.

92 A majority of the officers in Imperial General Headquarters actually desired that the enemy attempt to invade the Homeland before the end of the year. Not only would this give the nation a chance to strike a heavy blow, but also the over-all logistics situation was deteriorating so rapidly that the longer the invasion was postponed, the weaker the nation became. Actually, a tightening of the blockade with no attempt to invade the home islands was feared by the leaders and, for Japan, was the worst of all the alternate courses open to the enemy. (1) Statements by Lt. Gen. Kawabe and Lt. Gen. Arisue, both previously cited. (2) *Showa Nijuichi Nen Haru goro o Mokuto to Suru Josei Handan* 昭和二十一年春頃を目途とする情勢判断 (Situation Estimate for the Latter Half of 1945 and the Spring of 1946) Imperial General Headquarters, Army Section, 1 Jul 45, p. 5–7.

93 Arguments advanced by the "quick decision" school of thought were: (a) If termination of the Pacific war were delayed, the United States would be harassed by many foreign and domestic political problems; (b) American public opinion will demand an early decisive battle in keeping with the fast-paced advance across the Pacific; and, (c) the United States, believing that defeat of the nucleus of the Homeland Army is an essential prerequisite to national surrender, and having the power and the confidence necessary for opening the campaign, will not allow their powerful military establishment to stand idly by when the job remains to be done. (Statements by Lt. Gen. Arisue, previously cited, and Capt. Toshikazu Ohmae, Staff Officer (Operations), Imperial General Headquarters, Navy Section.)

94 Adherents to the "advance bases first" school of thought argued that the United States, with the lessons of the Okinawa campaign in mind, would use the full strength of their air and sea forces to support the invasion of Japan, attacking only when assured of an absolutely safe margin of strength. But since bases in the Marianas, on Okinawa, and in the Bonins would not be sufficient for the necessary forward air deployment and were too far from Japan to offer advantageous turn around time for shipping, the enemy would undoubtedly endeavor to secure bases farther forward. Ibid.

These possibilities were the subject of discussion in the High Command, with the majority opinion adhering to the view that the northern Ryukyus were the most probable target with Quelpart as a possible secondary invasion area.[95]

Following these preliminary operations, the enemy was expected to move as soon as possible against Japan Proper. In this connection, there were two possibilities considered, the first being a direct move to the Kanto district, and the other a campaign to gain air and sea bases in Kyushu and Shikoku first, followed by an advance on Kanto. While there was considerable anxiety over the possibility of a direct attack on Kanto,[96] a great majority of the officers in Imperial General Headquarters agreed that the invasion of Kyushu seemed most probable.[97] In any case, the consensus was that, so long as Japan was resolved to resist to the end, the final battle would be on the plains of Kanto, since it was not only the political and strategic center of the nation, but also the most favorable area for the deployment of enemy armored and mechanized equipment.

The various opinions concerning enemy capabilities and intentions were contained in formal estimates, memoranda, and other communications that circulated in High Command circles from May to July. Each idea was considered for acceptance or rejection by the Chiefs of the Army and Navy General Staff. No written estimate, however, was ever produced representing the combined opinion of the highest command levels. The study and circulation of the basic documents resulted, nevertheless, in the formulation of an official position that represented a blend of all the accepted recommendations and provided a firm basis for future action. A detailed summary of the final estimate is as follows:[98]

1. Over-all Strategy

*a. **Strategic Objective**: The enemy's objective is to bring the war to an early end by securing the unconditional surrender of Japan.*

*b. **General Method**:*

---

95 (1) Situation Estimate for the Latter Half of 1945 and the Spring of 1946, op. cit., pp. 6, 34. (2) Statements by Rear Adm. Tomioka, previously cited, and Maj. Gen. Masakazu Amano, Chief, Operations Section, Imperial General Headquarters, Army Section.

96 Several times during June, Second General Army renewed its recommendation that the decisive battle reserve scheduled for Kyushu be shifted immediately without waiting for activation of the *Ketsu-Go* operation. Due to the latent fear of a direct attack on Kanto, Imperial General Headquarters had not, by the end of the war, reached a final decision to transfer the planned strength. (Statements by Maj. Gen. Sanada, Col. Sugita, and Maj. Mashita, all previously cited.)

97 There existed a divided opinion in the High Command as to whether enemy operations against Kyushu would take the form of a limited objective invasion designed only to seize advance bases or of an annihilation battle to crush the Japanese Army. The majority opinion held that, although the invasion of Kyushu would be large scale, it would be only a limited objective operation preliminary to the even bigger landing at Kanto. (Statements by Maj. Gen. Amano and Capt. Ohmae, both previously cited, and Capt. Mitsuo Fuchida, Staff Officer (Air Operations), Combined Fleet.

98 (1) Situation Estimate for the Latter Half of 1945 and the Spring of 1946 op. cit., pp. 5–13, 3–49, 46–60, 78–80. (2) Estimate of the Situation, op. cit., pp. 6–8. (3) Preparations for Counterattack Operations in the Defense of the Homeland, op. cit., pp. 3–9. (4) Proceedings of the Supreme War Direction Council, op. cit., Part I. (5) Homeland Operations Record, op. cit. Vol. III, pp. 9–12, 28–9, (6) Statements by Maj. Gen. Amano and Rear Adm. Tomioka, both previously cited, and Lt. Gen. Shuichi Miyazaki, Chief, First Bureau (Operations), Imperial General Headquarters, Army Section.

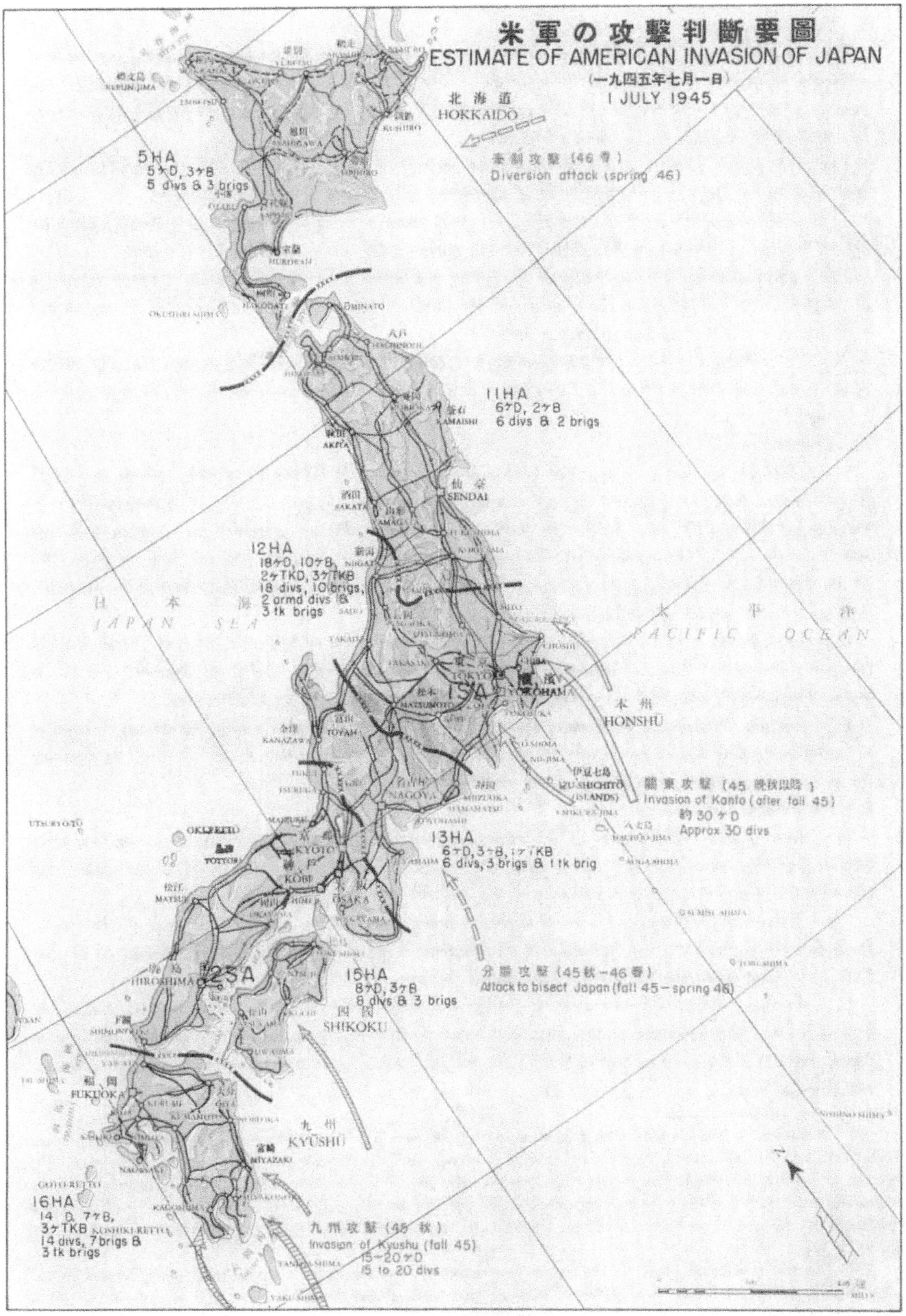

PLATE NO. 155
Estimate of Allied Invasion, 1 July 1945

(1) In pursuance of his fundamental objective, the enemy will first isolate and neutralize Japan through continued application of the air and sea blockade. New forward bases will be acquired and utilized for this purpose. Finally, after thorough preparation, the enemy will attempt to destroy the main strength of our Army by one or more invasions of the Home Islands.

(2) The general direction of the enemy advance against the Homeland will be from the south (i.e. Philippines and central Pacific). The possibility of an invasion from the north is extremely remote.

(3) There is a remote possibility that the enemy may endeavor to bring about the surrender of Japan without an invasion by intensifying the blockade so as to destroy totally the national combat potential.

(4) The probability is very small that the enemy will invade the Homeland without the prior acquisition of advance bases. However, at the first sign of the total collapse of Japanese air strength, the enemy will probably move against the Home Islands at once.

(5) In conjunction with the campaign against the Homeland, the enemy will also carry out a large number of peripheral campaigns in the southern area and in China, designed to weaken further Japan's strategic position.

2. *Strength*

a. ***Ground***[99]

(1) Over-all National Strength—The United States plans to fight Japan with an Army reduced in strength from 8,000,000 men (120 divisions) to 7,000,000 men (103 divisions) by next March. The Army will be reduced to 100 divisions by June. Subtracting from this figure the 10 divisions scheduled for retention in Europe and 10 for the zone of the interior garrison, it appears that about 80 Army divisions will form the troop basis for the continued prosecution of the war against Japan. In addition, 10 Marine divisions will be available, bringing the total ground strength to 90 divisions as of June 1946.

(2) Strength in the Pacific—It is estimated that there are at present in the Pacific 40–45 American divisions (including Marines). Reinforcements expected to arrive by September 1945 total 10–15 divisions, and by the end of December an additional 15–20 divisions. Total: 65–80 divisions by 31 December.

(3) Invasion Strength—It is estimated that the United States will employ a large number of divisions in peripheral campaigns and in rear area garrisons. Consequently, not more than 60 divisions will be available for operations against the approaches to Japan and the Homeland itself.

b. ***Land-based Air***[100]

(1) Over-all National Strength—The United States is estimated to have approximately 19,700 land-based aircraft (including naval planes). About 3,600 are stationed in Europe and 4,800 in the United States and other areas, leaving a total of approximately 11,300 available for use against Japan.

(2) Strength in the Pacific—There are at present approximately 8,500 land-based aircraft in the Pacific. By September of this year, it is believed that an additional 2,200 will be redeployed from Europe to the Far East. An additional 1,000 will arrive by the end of the year.

(3) Invasion Strength—Some enemy air strength will be diverted to peripheral campaigns and some will be kept in reserve. The estimated balance available for use against the approaches to the Homeland and against Japan proper is about 6,000 aircraft of which 1,500 will be B-29s, 1,400 heavy bombers, 1,100 medium bombers, and 2,000 fighters.

---

99 In addition to the United States ground strength there were 67 divisions of the British Commonwealth Forces in East Asia and the Pacific and it was expected that this number would be increased to 82 by the end of the year. The Japanese did not expect the British to contribute ground strength to the invasion of Japan. At this time, too, it was estimated that the Chinese had 10 fully American-equipped divisions and 25 partially equipped, the number of the former to reach 20 by the end of the year. Situation Estimate for the Latter Half of 1945 and the Spring of 1946, op. cit., pp. 7, 10–11.

100 British land-based planes in the Orient were estimated at 2,200 to be increased to 2,700 by September and 3,400 by the end of the year. The Japanese believed that these aircraft would be used only to support peripheral operations by British Commonwealth Forces. Ibid., pp. 10–1.

c. ***Carrier Air***[101]

(1) Over-all National Strength—Present U. S. carrier air strength is estimated at 2,100 aircraft. This will be expanded to 4,100 by September, 4,800 by the end of the year, and 5,100 by next spring.

(2) Strength in the Pacific—Present fleet carrier air strength in the Pacific is about 1,600.

(3) Invasion Strength—A small number of carrier aircraft will remain stationed in the Atlantic. A large number will be used for miscellaneous duties in the Pacific such as convoy escort, training, and supporting peripheral operations. Deducting all these activities and those held for repair and replacement, it is estimated that 2,400 planes will be available by September, 2,600 by the end of the year, and 3,100 by next spring for use over the approaches to the Homeland and against Japan itself.

d. ***Naval Strength***[102]

(1) Over-all National Strength—The estimated over-all composition of the United States Fleet by the end of December is 32 aircraft carriers, 85 escort carriers, 21 battleships, 58 cruisers, and 450 destroyers.

(2) Invasion Strength—It is estimated that 25 carriers, 25 escort carriers, 21 battleships, 54 cruisers, and 330 destroyers will be available for operations on the approaches to the Homeland and against Japan itself.

3. Targets

a. ***Preliminary Operations***

(1) In order to intensify the sea and air blockade and to prepare for future operations against Japan proper, the enemy will seize advance bases during July and August.

(2) An advance into the northern Ryukyus and the Izu Islands is most probable. Strengths committed will be: (a) To the northern Ryukyus (Toku-no-Shima, Kikaiga-Jima, and Amami Oshima), 1–2 divisions, and (b) to the Izu Islands, 1–2 divisions.

(3) As an additional possibility, the enemy may advance into the coastal sector of central and/or north China. Strategic areas to be attacked are, in the order of probability, the Ning-po-Shanghai area and southern Shantung Province. Commitment to this enterprise will be about 10 divisions. If scheduled, this operation will have the additional purpose of giving political and military support to Chungking and checking the advance of Soviet influence in China.

(4) Two additional possibilities are an invasion of Quelpart Island or southern Korea. One of these operations is less probable than either of the above, although Quelpart may be invaded as an extension of the enemy's program in the northern Ryukyus outlined in (2) above. Commitment to this operation will be about 6 divisions.

b. ***First Mainland Attack***

(1) Due to the fact that it is the political and economic center of the Japanese Empire as well as the best tactical terrain in the Home Islands, the enemy will fight the final decisive battle with the Japanese Army in the Kanto area.

(2) Contingent upon the availability of shipping, the enemy can mount a 30 division operation, or any series of operations involving a cumulative total of 30 divisions, by late fall of this year. He can mount a 50 division operation, or any series of operations involving a cumulative total of 50 divisions, by next spring.

(3) In order to prepare for the final decisive battle in the Homeland, the enemy will desire to secure additional forward sea and air bases to cover his approach to Kanto.

(4) Following the preliminary operations, the enemy will attempt to secure advance bases by a large-scale

---

101 By September the strength of British carrier air in the Far East was expected to reach 700 aircraft. The Japanese definitely expected that this strength would be used aganist the Homeland. Ibid., p. 11.

102 In addition to the U. S. strength, the British Pacific Fleet was estimated to have four battleships, five carriers, eight escort carriers, three light cruisers, five heavy cruisers, and about 40 destroyers. Also available if needed were an estimated 10 batteships, seven carriers, 23 escort carriers, 35 cruisers, 106 destroyers, and 50 submarines of the British fleets in the South China Sea and Indian Ocean. Ibid., pp. 7, 79–80.

amphibious lodgement in the southern part of Japan proper. Targets will be Tanega-Shima, southern Kyushu, and the south coast of Shikoku. A commitment of 15–20 divisions is anticipated. This operation can be mounted any time after September contingent upon no commitment to the China mainland.[103] *If the China operation is executed, the Kyushu operation cannot be mounted until after late fall.*

(5) It is possible that the enemy may seek to open a decisive battle in Kyushu. If so, the operation will be executed in late fall with about 30 divisions. Targets will be southern Kyushu and the Hakata Bay-Shimonoseki-Moji area of northern Kyushu. This operation has a very low degree of probability. It cannot occur at all this year, if the enemy commits any strength to the China mainland.

c. ***Main Invasion***

(1) Upon completion of his operational objectives in southern Japan and assembly of the necessary strength, the enemy will invade Kanto. This operation may be covered by a diversionary feint at Hokkaido. Commitment to the operation will be about 30 divisions and it will be mounted next spring.

(2) If it becomes apparent to the enemy that Japanese air power has completely collapsed, he will launch a direct invasion of Kanto late this fall, using about 30 divisions.

(3) If the enemy tries a blockade operation (See 1 (b) above), which fails to force Japan to surrender, a direct invasion of Kanto with 50 divisions next spring is a very remote possibility.

4. Allied Participation

a. Participation in the Homeland invasion by Allies of the United States is expected to be little more than token in nature, with the single exception of the British Pacific Fleet. However, the Allies are expected to engage in a variety of peripheral operations designed to weaken further the Japanese strategic position. It is believed that the enemy plans a consistent policy of coordinating his operations in China and the southern area with those against the Homeland.

b. ***Southern Area***

(1) Great Britain undoubtedly plans to recapture Malaya and Hongkong, to complete the reoccupation of Borneo, and to bring Thailand within her sphere of influence. The British will also seek to occupy areas of military importance in the Netherlands East Indies and French Indo-China.

(2) Burma-Malaya Area—The enemy will probably open a general offensive in Burma with the front line strength now available there (about 11 divisions). If this operation meets with success, amphibious operations will be launched against the southwest coast of the Malay Peninsula and thence south by shore-to-shore bounds, the final target being Singapore. The initial lodgement will be secured during July or August with 3–5 divisions, and the total commitment to the operation will be 9–12 divisions. A limited objective operation against the northwestern tip of Sumatra is probable in order to secure air bases and cover the flank of the drive toward Singapore. Simultaneous overland invasion of central and southern Thailand with 4–5 divisions is expected.

(3) Indo-China—A composite force of approximately five American, French, and Australian divisions will probably enter southern French Indo-China in the summer or early fall.

(4) Netherlands Indies—The reconquest of Borneo will be completed using 3–5 divisions. The probability of an invasion of Sumatra, Java, and the Lesser Sundas is very small. If the redeployment of Dutch troops proceeds satisfactorily, the enemy may launch a 2–3 division operation in that area next year. Meantime, the area will be neutralized by air and naval bombardment.

---

103 The estimated earliest possible date of the Kyushu invasion was moved up at least a full month between the the time of the Imperial conference on 8 June and the promulgation of the Army Section estimate on 1 July. This change was made for the following reasons: (a) The enemy had not yet begun air raids against tactical objectives; (b) The Okinawa campaign had lasted until late in June, much later than anticipated; (c) Signal intelligence did not report any unusual increase in traffic and no sign of immediate preparations for a task force sortie; and (d) It was felt that the enemy would prefer to postpone the invasion until after the typhoon season. (Statements by Lt. Gen. Kawabe and Lt. Gen. Arisue, both previously cited.)

c. China Area

(1) The enemy will employ 3–5 divisions in landings on the south China coast, most probably in the Canton-Hongkong sector but possibly in the Hainan Island-Luichow Peninsula area. Upon completion of operations in south China, these forces may enter northeastern French Indo-China.

(2) Chungking will open a general offensive in central China in July or August, utilizing mainly American-equipped and trained troops.

5. The Soviet Union

a. Since the Soviet Union announced its intention to abrogate the Russo-Japanese neutrality pact, it has been heavily reinforcing its military strength in the Far East. Since mid-February the estimated reinforcement has amounted to about 550,000 troops, 3,700 aircraft, 2,000 tanks, 6,700 field pieces, and 13,400 motor vehicles. Present strength is about 1,300,000 troops, 5,400 planes, and 3,000 tanks.

b. The Soviet Union, in order to further its political objectives in the Far East, will probably enter the Greater East Asia War at the earliest opportunity.

c. If the present reinforcement rate continues, the Soviet Union will be capable of commencing military action against Japan in August or September.

6. Evaluation

a. The enemy is mustering enormous and overwhelming military strength for use against Japan, and the issue will be joined between now and next spring.

b. Although Japan is faced with an exceedingly precarious strategic situation, there are certain circumstances that are working to her advantage.

(1) While the end of the war in Europe has given the United States a comfortable reserve of national war potential, industrial mobilization and reconversion have already begun due to the desire to grab quick post-war profits.

(2) The fighting morale of the United States is being weakened by the fear of large casualties.

(3) There has been an increase in labor strife, criticism of the military, and agitation from the ranks to engage in a precipitous demobilization.

c. Should the United States be defeated in the battle for Japan itself, public confidence in the President and the military leaders will decline abruptly, fighting morale will deteriorate in the flurry of recriminations, and Japan will be placed in a much more favorable strategic position.

July Developments

On 5 July, the Navy Section of the High Command formally abandoned its concept of the decisive air battle over the East China Sea, the *Ten-Go* air operation having long since deteriorated into a series of small scale hit-and-run raids.[104] The way was now clear to the formulation of a joint Army-Navy Air Agreement for the *Ketsu-Go* operation, an annex that had long been missing from the basic plan. This agreement was issued to the field on 13 July, its basic features being as follows:[105]

1. Missions

a. The primary mission of the Army and Navy air forces will be to locate and destroy the American expeditionary force while it is on the water. Emphasis will be placed on special-attack operations.

b. Secondary missions will include:

(1) Air defense

(2) Anti-submarine operations

(3) Air attacks to delay the enemy preliminary operations against the approaches to the Homeland.

---

104 *Daikaishi Dai-Gohyakunijugo-go* 大海指第五百二十五號 (Imperial General Headquarters Navy Directive No. 525) 5 Jul 45.

105 (1) *Daikaishi Dai-Gohyakunijuroku-go* 大海指第五百二十六號 (Imperial General Headquarters Navy Directive No. 526) 13 Jul 45. (2) Homeland Air Operations Record, op. cit., pp. 33–40.

PLATE NO. 156

Plan for Employment of Army Air Forces in *Ketsu* Operation, July 1945

決號作戰に於ける陸軍航空運用計畫

PLAN FOR EMPLOYMENT OF ARMY AIR FORCES IN "KETSU" OPERATION

(一九四五年七月)
JULY 1945

部隊 Unit	展開地域 Deployment Area	種別 機數[a] Type and Number of Planes	用法 Mission
第一航空軍 1st Air Army	鈴鹿以東本州 East of Suzuka Mountain Range, Honshu	特攻 六百 Special-attack 600 一般 五百 Standard 500	一、九州四國方面の作戰の場合は第六航空軍の後詰兵力 二、關東地方の作戰の場合の骨幹兵力 三、決戰生起に至る間、敵大型機に對する邀擊 1. Support 6th Air Army in event of operations in Kyushu and Shikoku. 2. Main striking force in event of operation in Kanto area. 3. Intercept bombers during preliminary enemy operations.
第六航空軍 6th Air Army	鈴鹿以西本州、九州、四國 West of Suzuka Mountain Range, Honshu; Kyushu; and Shikoku	特攻 一千 Special-attack 1,000 一般 四百 Standard 400	一、九州四國方面の作戰の場合の骨幹兵力 二、關東地方の作戰の場合は第一航空軍の後詰兵力 三、決戰生起に至る間、敵大型機に對する邀擊 1. Main striking force in event of operations in Kyushu and Shikoku. 2. Support 1st Air Army in event of operation in Kanto area. 3. Intercept bombers during preliminary enemy operations.
第五航空軍 5th Air Army	主力朝鮮、一部滿州北支 Main force in Korea; elements in Manchuria and North China	特攻 五百 Special-attack 500 一般 二百 Standard 200	一、九州四國方面の場合は第六航空軍の助攻又は後詰兵力 二、關東方面の作戰の場合は第一航空軍の後詰兵力 三、南鮮方面の作戰の場合の骨幹兵力 1. Attack jointly with or support 6th Air Army in event of operations in Kyushu and Shikoku. 2. Support 1st Air Army in event of operation in Kanto area. 3. Main striking force in event of operation in southern Korea.

[a] 本表の外、決號作戰の爲、次の飛行機の使用を豫期す　一、五百乃至一千機（七、八月頃迄に特攻機に改修整備）　二、在滿第二航空軍の約百二十機　三、在北海道第一飛行師團の約六十五機

[a] In addition were following reserves for KETSU Operation: 500-1,000 planes (to be converted for special-attack by July and August); approx 120 planes of 2nd Air Army in Manchuria; and approx 65 planes of 1st Air Division in Hokkaido.

(4) *Air attacks on enemy carrier task forces.*

(5) *Tactical support of defending ground troops.*

2. *Outline of Operations*

a. *Theater*

(1) *Operational preparations will be completed in full in Kyushu, Shikoku, and southern Korea.*

(2) *Preliminary preparations will be completed for possible action in other Ketsu-Go operational areas especially Kanto.*

b. *Operational Concept—Primary Mission*

(1) *Reconnaissance of enemy advance bases is essential in order to obtain warning of the enemy approach.*

(2) *First priority target will be enemy transports.*[106]

(3) *The landing convoys will be attacked and destroyed at the first stage of the landing (i. e. during about the first ten days).*

c. *Operational Concept—Secondary Missions*

(1) *The Army will operate its fighting strength as economically as possible, engaging in short, opportune interception operations against enemy heavy bombers. Both services will cooperate in staging long range hit-and-run raids against enemy heavy bomber bases in the Marianas, Bonins, and on Okinawa.*

(2) *The Navy will intensify anti-submarine operations in the Japan Sea area by clearing the waters and preventing further penetration.*

(3) *A part of the fighting strength will be utilized to inflict casualties and delay enemy preliminary operations against the China coast, the northern Ryukyus and/or Bonins, the Izu Islands, Tanega-Shima, Goto, Quelpart, and/or southern Korea. Such operations will be mounted only by locally available forces, and they will not be allowed to compromise the conduct of the Homeland decisive battle.*

(4) *Enemy carrier task forces will be attacked with a part of our strength for the purpose of interdicting tactical air support of the enemy landings.*

(5) *In rendering support to the ground effort, the main emphasis will be on attacking enemy gunnery ships engaged in shore fire support. Such attacks will be synchronized with the local offensive activities of the ground units.*

3. *General Development and Strength*

a. *Initial dispositions (Plates No. 156–157)*

(1) *In western Honshu, Shikoku, Kyushu, and Korea will be stationed the Fifth and Sixth Air Armies, the Fifth Air Fleet, and a part of the Tenth Air Fleet.*

b. *A total commitment of all available air forces will be made in the area in which the enemy first invades Japan, most probably Kyushu. Over-all strength committed will be as follows:*[107]

(1) *Army: 3,200 aircraft*

(2) *Navy: 5,225*

4. *Command Relationships*

a. *The general basic command relationship will be one of inter-service cooperation.*

b. *Commanders of major air commands will be stationed at the same locations or as close as possible to their respective opposite numbers. Command posts for the decisive battle will be chosen immediately.*[108]

---

106 In the past there had been considerable controversy between the Army and the Navy over the mission of land-based aviation. In the Philippines the Army, by joint agreement, had concentrated on transport convoys, and the Navy on enemy carriers. (Cf. Chapter XI, p. 296.) At the outset of the *Ten-Go* operation the Navy kept this arrangement until late April when it was agreed that henceforth all air forces would concentrate on transport convoy. This agreement was extended to include *Ketsu-Go.*

107 Strength figures given in this agreement were based on the 8,500 operational aircraft available at the time the agreement was formulated. It was hoped that an additional 2,000 planes could be produced by the time of the invasion. The Japanese, however, were not counting on their availability. (Statement by Lt. Col. Tanaka, previously cited.)

108 In compliance with this provision, Sixth Air Army was to move its command post to Oita to join Fifth Air Fleet headquarters. First Air Army and Tenth Air Fleet were both to be located in Takasaki, Gumma Prefecture, Air General Army was at Taisho airfield on the outskirts of Osaka, and Third Air Fleet and General Navy Command were to be at Yamato airfield in Nara Prefecture. (1) Preparations for Counterattack Operations in Defense of the Homeland, op. cit., p. 20. (2) Statement by Capt. Ohmae, previously cited.

Tables appended to this agreement indicated in detail the desired deployment of all available air strength.

As the summer wore on, many complex planning difficulties arose in high-level Japanese headquarters, the most important of which was brought about by the shift of operational emphasis to Kyushu. It was feared that the high priority accorded to preparations in the south would so drain off field engineering supplies, ammunition, food, fuel and lubricants, and all other classes of military supplies that it would be impossible to conduct properly the defense of Kanto should that area be attacked directly before the end of the year.

A second source of anxiety to the High Command were the shortcomings of Japanese strategic intelligence. This was largely a result of the drying up of the major sources of information. With both air and sea superiority lost right up to the very shores of Japan itself, direct and frequent observation of enemy invasion bases by Japanese submarines and planes was impossible. The High Command was forced to rely almost entirely on radio intelligence, although this means was not completely reliable, and certainly not so in determining absolutely the direction, time, and strength of the attack. This inability to get sufficient warning naturally compounded the anxiety that was being felt about Kanto.[109]

In addition to these worries, the month of July found Imperial General Headquarters increasingly concerned about the possibility of an enemy landing in the Tokai (Nagoya) district of central Honshu. From the very beginning, the High Command had believed in the remote possibility that this area would be selected as a target.[110] This feeling now became particularly strong in view of the fact that the Tokai district was the narrowest part of Honshu, that its defenses were relatively weak, and that with both Kyushu and Kanto becoming stronger day by day, it might tempt the enemy to cut Japan in two by landing in the vicinity of Ise Bay and occupy the areas around Nagoya, Kyoto, and Osaka. On 20 July, a staff group from Imperial General Headquarters held a special meeting in Kyoto to discuss countermeasures.[111]

A fourth troublesome problem faced by the High Command was the weakness of the defenses of Manchuria, northern Korea and Karafuto. Despite indications that the Soviet Union was on the verge of commencing hostilities against Japan, the Kwantung Army, already weakened by the withdrawal of troops and munitions for the Philippines, Formosa, and Okinawa, had been further reduced by the diversion of four crack divisions to the Homeland.[112] Having decided on all-out commitment against the forthcoming American invasion attempt, there was little that Imperial General Headquarters could do about this particular problem except hope that the Japanese units could hold off the Soviet tide while a decision was being sought in the Homeland.[113]

109 Statement by Lt. Gen. Arisue, previously cited.

110 Cf. Chapter XVIII, n. 23.

111 Another problem concerning the Tokai district was that it lay on the boundary between general armies. This presented special command and communication difficulties. Homeland Operations Record, op. cit. Vol. III, pp. 26–7.

112 The Japanese High Command, beginning in September 1942, had effected the transfer from the Kwantung Army of many front-line combat units to reinforce other fronts. In all, 17 infantry and two armored divisions, one independent mixed brigade, one tank brigade, two amphibious brigades, numerous smaller combat units and many miscellaneous supporting units were withdrawn. Although later mobilizations helped fill this gap insofar as personnel requirements were concerned, the heavy drain of equipment could not be replaced. Moreover, the over-all level of experience and training of the troops was sharply reduced. Statistics and Analysis Reports, op. cit., Report No. 14: Number of Japanese Units Transfered from Manchuria to the Pacific Front, Jan 42-Aug 45.

113 Statement by Maj. Gen. Amano, previously cited.

PLATE NO. 157

Plan for Employment of Navy Air Forces in *Ketsu* Operation

決號作戰に於ける海軍航空運用計畫

PLAN FOR EMPLOYMENT OF NAVY AIR FORCES IN "KETSU" OPERATION

(一九四五年七月)
JULY 1945

部隊 Unit	展開地域 Deployment Area	機種、機數 Type and Number of Planes	用法 Mission
第三航空艦隊 3rd Air Fleet	鈴鹿以東本州 East of Suzuka Mountain Range, Honshu	戰鬪及偵察 五百十 特攻、爆擊、雷擊 千百四十 Fighter and reconnaissance 510 Special-attack, bomber, and torpedo 1,140	一、九州四國方面の作戰の場合は第五航空艦隊の後詰兵力 二、關東地方の作戰の場合の骨幹兵力 1. Support 5th Air Fleet in event of operations in Kyushu and Shikoku. 2. Main striking force in event of operation in Kanto.
第五航空艦隊 5th Air Fleet	鈴鹿以西本州、九州、四國 West of Suzuka Mountain Range, Honshu; Kyushu and Shikoku	戰鬪及偵察 六百三十 特攻、爆擊、雷擊 千七百六十五 Fighter and reconnaissance 630 Special-attack, bomber, and torpedo 1,765	一、九州四國方面の作戰の場合の骨幹兵力 二、關東地方の作戰の場合は第三航空艦隊の後詰兵力 1. Main striking force in event of operations in Kyushu and Shikoku. 2. Support 3rd Air Fleet in event of operation in Kanto.
第十航空艦隊 10th Air Fleet	鈴鹿以東本州 朝鮮 East of Suzuka Mountain Range, Honshu and Korea	戰鬪 三十 特攻、爆擊、雷擊 千七十 Fighter 30 Special-attack, bomber, and torpedo 1,070	第三、第五航空艦隊の夫々の地域の作戰の場合各艦隊に増加 1. Reinforce 3rd or 5th Air Fleets, as required.

註 1. 本表の外在東カロリン第四艦隊の偵察機八十は敵進攻部隊の偵察に任ず　2. 本土作戰生起の場合は第五航空艦隊司令長官は全海軍航空部隊を指揮す

Note: 1. In addition, 4th Fleet in East Carolines (80 reconnaissance planes) was assigned the mission of reconnaissance of enemy invasion forces. 2. In event of Homeland operations, commander of 5th Air Fleet was to command all naval air forces.

Still another problem which caused considerable concern within the High Command was the critically weak coastal defenses along the Japan Sea side of the Homeland. Recognizing that Japanese sea and air forces could not deny the enemy use of the vital Korea Strait, Imperial General Headquarters and Second General Army took some steps during July to strengthen the defenses along the coast of western Honshu.[114] This flank, nevertheless, remained extremely vulnerable.

Far more serious than even these strategic planning problems was the extent of the damage inflicted on Japan by enemy aircraft during the month of July. After the devastation of March through June it seemed almost impossible that the American air offensive could gain in intensity. Such was far from the case. During the month, the Japanese counted a total of 20,859 sorties flown over the Homeland by enemy aircraft, more than four times as many as in any previous month. Only six days in the month were free of air raids of any kind, and on one day (28 July) more than 3,400 sorties were flown.[115]

During July, B-29 operations against the Homeland were expanded about 25%. Abandoning the offensive against large urban areas, the enemy concentrated his July attacks against 35 small and medium-sized cities, each of which received more than 100 tons of bombs, mostly incendiaries, and fourteen of which received more than 1,000 tons.[116] At the same time, the aerial mining campaign against Japan's coastal waterways was continued.

During the same period, other enemy air forces were far from idle. Okinawa-based bombers and fighters swarmed all over southern Japan piling up a total of 3,193 sorties, almost four times as many as in June. Fighters based on Iwo Jima, operating mainly over the Kanto, the Osaka-Kobe, and Nagoya areas, flew 1,787 sorties, almost seven times as many as in the previous month. Both these short-range air forces hammered air bases, shipping factories, railroads, tactical positions, and even fishing villages throughout all of Japan west of Tokyo. Air reconnaissance activities over Kyushu, Shikoku, and Kanto were even more pronounced than in June.

About 10 July, the American carrier task force appeared again, hitting targets throughout Japan, but concentrating on Hokkaido and northeastern Honshu. These atttacks, directed mainly against airfields, shipping, industrial targets, and transportation, added up to a total for the month of 12,213 sorties.[117]

Urban area incendiary raids continued to cause the deepest and most lasting damage to the Japanese war potential. Large sections of the cities which contained some of Japan's most vital fabricating and sub-contracting facilities lay in smouldering ruins[118] Next in importance were the enemy carrier task force

---

114 The 230th Division and 124th Independent Mixed Brigade, both of which were organized during the third mobilization, were deployed in the beach sector of Yamaguchi Prefecture during July. Moreover, a strong element of the 231st Division, also organized in the third mobilization, was disposed in the vicinity of Tottori. At the same time, construction of coastal positions near Maizuru and Tsuruga was begun. Homeland Operations Record, op. cit. Vol. III, pp. 21–2, Attached Map.

115 Statistics and Analysis Reports, op. cit., Report No. 6.

116 Cities receiving more than 1,000 tons were Hiratsuka, Ichinomiya, Kochi, Kumamoto, Kure, Kuwana, Numazu, Sendai, Shimizu, Shimotsu, Tokushima, Tsu, Ube, and Uwajima. Ibid., Report No. 13: Extent of Air Raid Casualties and Property Damage on Small and Medium Cities in Japan, July 45.

117 Added to the weight of these air attacks was the bombardment, for the first time, of selected industrial targets along the coast of Honshu by enemy surface units. Ibid., Report No. 6.

118 Typical of the damage to these cities was Hiratsuka, containing important aircraft parts factories and one of

raids. Vital steel production facilities at Ishinomaki, Kamaishi, and Wanishi were heavily damaged in these operations. Widespread havoc was wrought in the ports of Aomori and Hakodate, the most serious loss being the putting out of action of every single one of the rail ferries plying between Hokkaido and Honshu. This cut rail haulage capacity between the two islands from 300,000 tons per month to almost zero. The carrier plane campaign against Japanese airfields throughout the Home Islands continued, interdicting many important installations and heavily damaging several.[119]

Even before the July raids, it had become obvious that an unchallenged continuation of such mass enemy air attacks would render Japan physically incapable of further resistance, particularly if the enemy shifted, as appeared likely, to attacks on the nation's land transport system and other tactical targets. To meet this situation, the High Command took a drastic step. On 30 June, Imperial General Headquarters ordered Air General Army to assume complete responsibility for carrying out a systematic air defense of the Homeland, simultaneously transferring the 10th, 11th and 12th Air Divisions, which had been engaged in air defense operations under the control of respective area armies, to the command of the Air General Army. This new mission represented a switch from the former policy of strict conservation of aircraft.

Operations were carried out during July but, in spite of the addition of eleven fighter regiments to the air defense forces,[120] units were still so thinly spread and quick concentration was so difficult, that enemy air units continued to break through almost at will.[121]

In the midst of these discouraging planning

(118 cont'd) Japan's largest naval arsenals, which was 63% burned out. Numazu, containing both Army and Navy arsenals as well as being one of the chief producers of automatic weapons in the nation, was 58% destroyed. Ibid., Report No. 13.

119 Among the heaviest hit of the air bases was Misawa near Hachinohe in northeast Honshu. The most important single effect of the carrier raids was the virtual severance of Hokkaido from Honshu. The choking off of the flow of coal, iron ore, lumber, and marine products was a death blow to the already crippled Japanese industrial machine. The attacks against Kyushu and the Inland Sea area were particularly severe. On one day (28 July) the amazing total of 2,772 sorties were flown against air bases, railways, factories, and power stations on Kyushu. (1) *Senkyo Shubo* 戰況手簿 (Daily Record of the War Situations) Operations Section, Army General Staff. (2) Homeland Defense Naval Operations, Part III, op. cit., 53–69.

120 The High Command released from the decisive battle reserve 11 fighter regiments which were distributed as follows:

Kanto district	four regiments (80/100 planes)
Tokai district	three regiments (60/75 planes)
Shikoku	two regiments (40/50 planes)
Kyushu	two regiments (40/50 planes)

Plans were made to concentrate the bulk of these forces, together with the fixed air defense units (10th, 11th, and 12th Air Divisions) in the threatened sector during each big raid. Due to lack of warning, interdiction of bases, and other hindrances these plans were seldom successfully executed. (1) *Dairikumei Dai Sensambyakugojukyu-go* 大陸命第千三百五十九號 (Imperial General Headquarters Army Order No. 1359) 30 Jun 45. (2) *Dairikushi Dai Nisengohyakujuhachi-go* 大陸指第二千五百十八號 (Imperial General Headquarters Army Directive No. 2518) 30 Jun 45. (3) Homeland Air Operations Record, op. cit., pp. 27–32.

121 One result of these continued attacks was to interdict Japanese efforts to assemble a long-range bomber attack force for operations against enemy rear areas. The carrier plane raids of 14 July against Misawa airfield on northeast Honshu, for example, completely destroyed 25 naval medium bombers that had been assembled to take an airborne raiding force to enemy bases in the Marianas in late July. The Japanese were never able to carry out any part of the plans for such operations called for in the *Ketsu-Go* policy. (Cf. Chapter XVIII, p. 558–9) Homeland Defense Naval Operations, Part III, op. cit., pp. 55, 57.

and operational setbacks, the only encouraging sign remained the fact that the forces on Kyushu were rapidly reaching a state of full combat readiness. Regardless of what appeared to be disaster for the Japanese nation, the High Command held firm in its resolve to fight the Kyushu battle as scheduled.[122]

Preparations for the Defense of Kyushu

The responsibility for preparing the ground defenses of Kyushu belonged to Field Marshal Shunroku Hata, Commander-in-Chief, Second General Army and, under him, to Lt. Gen. Isamu Yokoyama, Commander of Sixteenth Area Army. Pursuant to the basic *Ketsu-Go* plan, the headquarters of both these officers as well as the staffs of the local army commanders had undertaken to estimate the trend of future enemy tactics to be used against Kyushu. By early July, these efforts had resulted in a consensus of opinion which was substantially as follows:[123]

1. ***General Objective***

The general aim of the enemy invasion of Kyushu will be to destroy the Japanese forces on the southern Kyushu front and immediately occupy important sea and air bases in Miyazaki and Kagoshima Prefectures.[124]

2. ***Preliminary Operations***

a. The enemy will continue all-out attacks with strategic air and naval forces in order to maintain the blockade, annihilate Japanese air strength, and destroy the nation's domestic communications.[125]

b. The enemy will invade Tanega Shima and install a forward fighter base there to support the main landing.

c. Just prior to the main landing on Kyushu proper, the enemy may make a short feint in another area to conceal the real invasion goal.

d. The main landing will be preceded by several days of concentrated shelling and bombing attacks against coastal positions and installations in order to neutralize air bases, knock out fortifications, destroy communications, and isolate the proposed battle area.

3. ***Main Landing***

a. It is most probable that the main attack will be directed against the Ariake Bay and Miyazaki coast areas with a secondary attack against the west side of the Satsuma Peninsula.

b. In coordination with the main amphibious effort, it is highly likely that the enemy will drop strong airborne forces on the group of airfields around Kanoya, Miyakonojo, and Nyutabaru.

c. Simultaneously with the operations against Kyushu, it is highly probable that the enemy will attack southern Shikoku.[126]

---

122 The strategists at Imperial General Headquarters believed that if they could succeed in inflicting unexpectedly heavy damage on the enemy in the Homeland battle, convincing him of the huge sacrifice involved in pursuing the campaign and making him aware of the determined fighting spirit of the entire Japanese nation, it would lead to the termination of hostilities on comparatively advantageous terms for the Japanese. (Statements by Lt. Gen. Arisue and Lt. Gen. Kawabe, both previously cited.)

123 (1) Homeland Operations Record, op. cit. Vol. III, pp. 9–12, 28–9. (2) Statements by Maj. Gen. Yasumasa Yoshitake, Chief of Staff, Fifty-seventh Army, Maj. Ryoichi Tabata, Staff Officer (Operations), Fifty-seventh Army, Lt. Col. Hashimoto, and Maj. Haba, both previously cited.

124 In addition to large airfield groups, such as Chiran, Kanoya, Miyakonojo and Nyutabaru, southern Kyushu embraced large ship bases such as Kagoshima Bay and Ariake Bay. Furthermore, while its location at the tip of Kyushu made the movement of troops difficult for the Japanese, the Americans could easily use small vessels and fighter planes from Okinawa.

125 It seemed inevitable that all Japanese air bases, with the exception of those skillfully concealed, would be destroyed in this strategic and tactical bombing. Furthermore, it was expected that not only would Kyushu, Shikoku, Honshu and Hokkaido be isolated from each other but that the interior of each island would be subdivided, that rail transportation on the battle fronts would be disrupted, and that daytime travel on main highways would become difficult.

126 The specific enemy landing area on Shikoku was estimated as Tosa Bay. Enemy objectives in this operation were believed to be (a) to extend his air perimeter over all of the Inland Sea by establishing forward fighter bases on Shikoku and (b) to neutralize *tokkotai* bases on Shikoku.

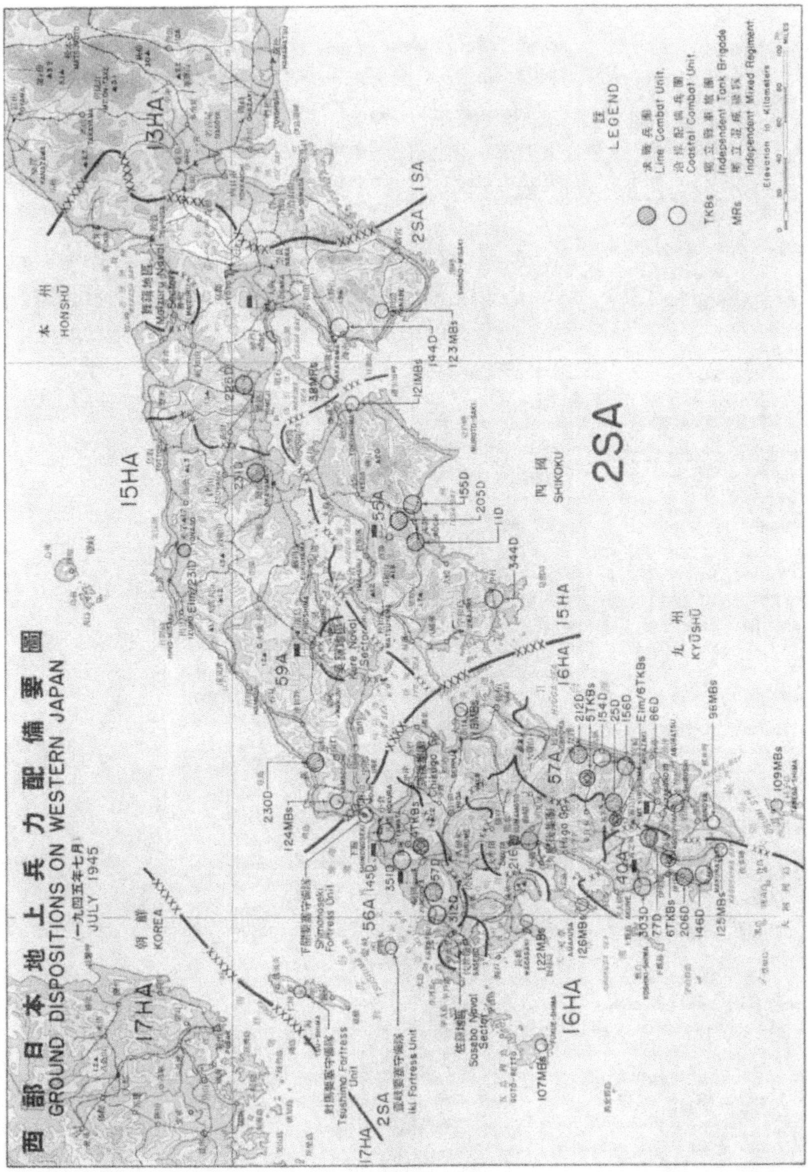

PLATE NO. 158
Ground Dispositions on Western Japan, July 1945

4. ***Strength***

a. The enemy will use approximately 15–18 divisions in this operation.

b. Enemy strength will be distributed as follows:

Ariake Bay—5–6 divisions
Miyazaki coast—3–4 divisions
Satsuma Peninsula—2 divisions
Shikoku—2 divisions
Airborne—1–2 divisions
Army reserve—1–2 divisions

5. ***Post-Landing Tactics***

a. Forces landing in the Ariake Bay area will occupy the Kanoya sector, the coast of Kagoshima Bay, and later the Miyakonojo area. Elements of this force will seize the Mt. Kirishima area.

b. The force landing on the Miyazaki coast will occupy the Miyazaki plain, the Karasebaru, and the Nyutabaru airfield sector. Part of this force will be sent rapidly forward to the Miyakonojo sector to aid the advance of the Ariake Bay landing force.

c. The force landing on the Satsuma Peninsula will occupy Chiran and Kagoshima city. Some elements will secure a line roughly along the boundary between Kumamoto and Kagoshima Prefectures.

6. ***Northern Kyushu***

a. There is a remote possibility that the enemy may seek an all-out decisive battle in Kyushu, including, in addition to the operations in the south, an amphibious invasion of the northern part of the island.[127]

b. Probable landing areas include the Fukuma sector (main landing) and the Hakata and Kokura areas (secondary landings). In order to cover this operation the enemy will also seize Quelpart or the Goto Islands.

c. The objectives of this operation will be to occupy the Shimonoseki-Moji area covering the vital Tsushima and Shimonoseki Straits. Airfields on the Hakata and Kurume Plains will be overrun, while elements will drive deep into the island to block reinforcement from the interior.

It was the Japanese intention to blunt the enemy invasion spearhead off Kyushu chiefly by an all-out attack of the air forces. Immediately available in western Japan for this purpose were units of the Sixth Air Army under Lt. Gen. Michio Sugawara and of Fifth Air Fleet under Vice Adm. Matome Ugaki. These were as follows:[128]

Sixth Air Army
- *11th Air Division*
- *12th Air Division*
- *30th Composite Air Group*
- *51st Air Division*
- *6th Fighter Brigade*
- *100th Fighter Brigade*
- *27th Torpedo Bomber Brigade*

Fifth Air Fleet
- *72d Air Flotilla*
  - *203d Fighter Group*
  - *332d Fighter Group*
  - *343d Fighter Group*
  - *352d Fighter Group*
- *12th Air Flotilla*
  - *12 unnumbered tokko groups*
- *634th Reconnaissance Group*
- *701st Bomber Group*
- *762d Bomber Group*
- *801st Torpedo Bomber Group*
- *931st Torpedo Bomber Group*
- *721st Composite Group*
- *171st Composite Group*

---

127 Although northern Kyushu was the industrial center of western Japan and constituted an area vital to the security of the Korea Strait and the Inland Sea, with ship bases in Hakata Bay and at Shimonoseki-Moji and several airfield groups near Fukuoka, an invasion in this sector would have been complicated by the protection afforded by Japan's strategic naval and air bases on Quelpart, Goto, Tsushima, and Iki Islands.

128 Only those units in which fighting elements were actually present at the end of July are listed here. In addition, in order to lend flexibility to the organization for combat and provide special air task force headquarters, the Sixth Air Army had available the Headquarters, 7th and 21st Air Brigades. (1) Homeland Air Operations, Record, op. cit., pp. 117–8, Attached Chart 1 (2) Table Showing Organizational Changes and Losses of Japanese Naval Forces, op. cit., pp. L–49–50.

In addition, at the outset of the emergency in Kyushu, it was planned to deploy the following units into the area:[129]

First Air Army
- *10th Air Division*
- *52d Air Division*
- *12th Fighter Brigade*
- *26th Fighter-Bomber Brigade*

Fifth Air Army
- *53d Air Division*
- *1st Fighter Brigade*
- *2d Fighter-Bomber Brigade*
- *8th Light Bomber Brigade*

20th Fighter Group

Third Air Fleet
- *13th Air Flotilla*
  - *8 unnumbered tokko groups*
- *53d Air Flotilla*
  - *210th Fighter Group*
  - *2 unnumbered tokko groups*
- *71st Air Flotilla*
  - *302d Fighter Groups*
  - *1 unnumbered tokko group*
- *131st Composite Group*
- *252d Composite Group*
- *601st Composite Group*
- *752d Composite Group*
- *706th Composite Group*

Tenth Air Fleet
- *3 unnumbered tokko groups*
- *2 unnumbered fighter groups*
- *2 unnumbered composite groups*
- *1 unnumbered torpedo-bomber group*

In mid-July the staff of Sixth Air Army and Fifth Air Fleet conducted a joint operational study in Fukuoka, and during the remainder of the month, both these headquarters perfected their operational plans. Combined and summarized, these plans provided as follows:[130]

1. ***Major Objectives***

The U. S. convoy will be destroyed by our air forces at approximately the time of its entry into the anchorage. Preparations for this operation will be completed by the end of September.

2. ***Reconnaissance***

a. Naval air forces will be responsible for all long-range, some short-range, and all night reconnaissance.

b. Army air forces will engage in short-range reconnaissance only.

c. Complete reconnaissance coverage of the offshore waters of Japan will be maintained to a distance of 600 miles.

3. ***Deployment***

a. Deployment of all units will be carried out under conditions of extreme secrecy.

b. Initial and reinforcing deployment will be in accordance with the Army-Navy Central Air Agreement of 13 July.

4. ***Attacks Against Enemy Carriers***

a. The enemy carrier task force will not be attacked until it becomes apparent that a full-scale landing is underway.

b. When the above condition is fulfilled, a crack naval air force (approximately 330 aircraft) reinforced by designated army units will attack the carriers and rob them of their ability to support the landings.

5. ***Attacks Against Enemy Transports***

a. When the convoy enters the attack zone, large type transports will be made the targets of a determined, round-the-clock, series of special-attacks.

b. The period of attack will be ten days, during which all available air strength will be used against the enemy.

c. All available fighter strength will be utilized to seize and maintain air superiority over the anchorage area.

6. ***Attacks Against Enemy Gunnery Ships***

Specially trained army and navy air elements (approximately 250 aircraft) will be assigned the full-time duty of attacking ships engaged in naval gunfire support.

129 As an extra available command organization, First Air Army had the Headquarters, 5th Air Brigade. Ibid.

130 (1) Homeland Air Operations Record, op. cit., pp. 108–19. (2) Preparations for Counterattack Operations in Defense of the Homeland, op. cit., pp. 16–12

7. ***Attacks Against Rear Bases***

Approximately 1,200 airborne troops will be readied for landings at enemy air bases in Okinawa.

8. ***Direct Support of Ground Units.***

There will be no direct support of friendly ground units on Kyushu.

Operating simultaneously with the air special-attack forces against the enemy were to be the sea *tokko* units. In the Kyushu and southern Shikoku area, these were as follows:[131] (Plate No. 154)

Combined Fleet
10th Special-Attack Squadron
101st Assault Unit[132]
102d Assault Unit
Sasebo Naval Station
5th Special-Attack Squadron
32d Assault Unit
33d Assault Unit
35th Assault Unit
3d Special-Attack Squadron
31st Assault Unit
34th Assault Unit
Kawatana Assault Unit
Kure Naval Station
8th Special-Attack Squadron
21st Assault Unit
23d Assault Unit
24th Assault Unit

Apart from the sea *tokko* effort, the naval surface forces were scheduled to play little part in the defense of Kyushu. Most of the vessels of 31st Destroyer Squadron (19 operational vessels in all) were to be used for lifting *Kaiten* (midget submarine) to the scene of action and for subsequent night action against transports. In consideration of the fuel problem, other gunnery ships were barred altogether from participation in the operations.[133]

Japan's hope of success in the battle for Kyushu depended almost entirely on the results expected from the air and surface special-attack operations. Even as early as June, the Navy had estimated that about 30–40% of the invading convoy could be sunk by the *tokko* attacks.[134] As reports of the damage inflicted on the enemy off Okinawa in the *Ten-Go* operation became available for study,[135] this estimate was raised to 30–50%. It was estimated that this would cost the enemy at least five assault divisions before the landing even began.[136]

---

131 In addition, the Army had eight surface raiding regiments (the 31st through the 38th), of about 90 boats each, deployed on Kyushu. The control of these units was decentralized to local commanders, and it was expected that the Army boats would attack in cooperation with the Naval surface *tokko* forces' mass attack. The deployment of all these special-attack forces was completed by 1 August. Tactics to be employed by them have already been discussed in Chapter XVIII, pp. 564, 566. (1) *Dairikumei Dai Sensambyakurokujuni-go* 大陸命第千三百六十二號 (Imperial General Headquarters Army Order No. 1362) 11 Jul 45. (2) Preparations for Counterattack Operations in Defense of the Homeland, op. cit., Attached Charts 3, 4.

132 The composition of the assault unit was extremely flexible, depending on the mission. Some units had all three general types of sea special-attack weapons (suicide boats, midget submarines, and human torpedoes), while others had only boats or only undersea craft. reparations for Counterattack Operations in Defense of the Homeland, op. cit., Attached Chart 3.

133 (1) Ibid., p. 21. (2) Another Navy contribution to the battle was in the realm of mine warfare. All key harbor and bay entrances were sewn with large mines, while plans were being layed for the "seeding" of all the principal landing beaches with small anti-boat mines which would be layed along the water's edge. *Bakugeki Chosa Dan ni taisuru Dai Ni Fukuinsho Kaito Sakusen Tosui Kankei Joho Yokyu no Ken Kaito Sono Ni* 爆擊調査團に對する第二復員省回答, 作戰統帥關係情報要求の件回答其二 (Information Concerning Task Organization and Operations, Submitted by the 2d Demobilization Bureau in Reply to an USSBS Questionnaire, Part II) Nov 45.

134 Proceedings of the Supreme War Direction Council, op. cit., Part I.

135 The final tally of damages inflicted upon the enemy forces in the *Ten-Go* operations included 196 ships reportedly sunk. It was not until after the war that access to Allied records revealed to the Japanese that only 35 ships were actually sunk though about 31 other vessels were heavily damaged and a much larger number less severely damaged.

136 This estimate was arrived at during a joint study conducted at Fukuoka, Kyushu, on 4–5 July at which time Army and Navy air and sea *tokko* operations were discussed in detail. The conference discussions, presided over by Maj.

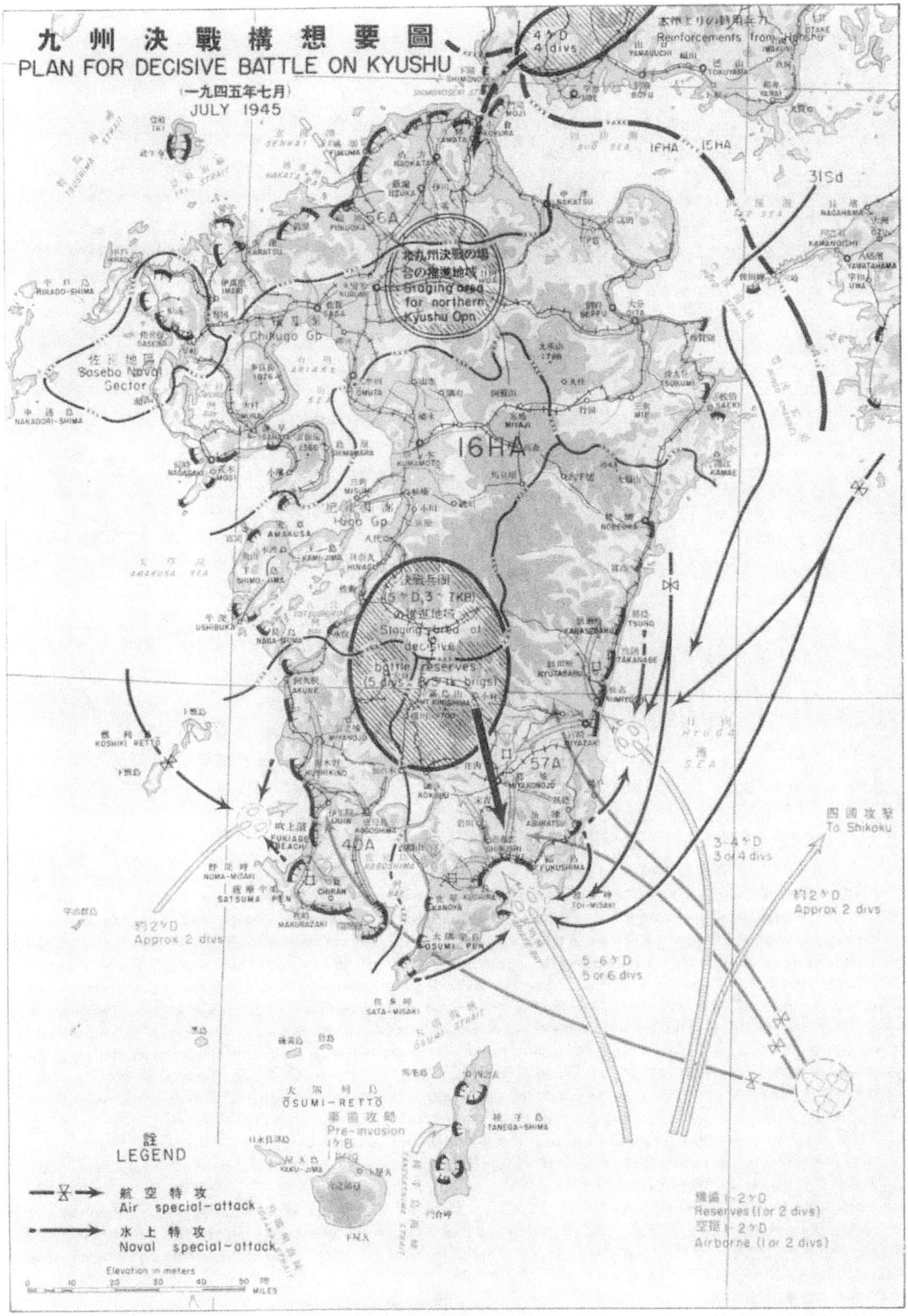

PLATE NO. 159
Plan for Decisive Battle on Kyushu, July 1945

Although this estimate of enemy losses was officially accepted by Imperial General Headquarters, the Army commanders responsible for the conduct of ground operations within the scope of *Ketsu* No. 6 recognized that these estimates were excessively high. It was felt that, in view of the difficult conditions under which the *tokko* operations would necessarily have to be conducted, a more realistic approach to the problem indicated a possible maximum loss to the enemy of 20% of the invasion fleet of transports or about three divisions.[137]

After receiving this initial setback at the hand of the *tokko* forces, the enemy was scheduled to meet with strong resistance on the ground from the very beginning of the landings. By the end of July, the tactical commander of ground forces on Kyushu, Lt. Gen. Yokoyama, had at his disposal 14 divisions, six independent mixed brigades, and three tank brigades.[138] Although some of these units were still deficient in training, the equipping and deployment of troops were generally completed as was the operational stockpiling of munitions.[139]

The deployment of major units of Sixteenth Area Army was as follows:[140] (Plate No. 159)

Miyazaki-Ariake Bay Area

Fifty-seventh Army—Lt. Gen. Kanji Nishihara

Coastal Forces

Miyazaki Coast—154th and 156th Divisions

Ariake Bay—86th Division

Osumi Peninsula—98th Ind. Mixed Brigade

Mobile Reserve

Kirishima Mt. (Staging area)—25th Division, 5th Tank Brigade, one regt., 6th Tank Brigade

Northern Miyazaki Plain—212th Division

Tanegashima Detachment

Tanegashima—109th Independent Mixed Brigade

Satsuma Peninsula

Fortieth Army—Lt. Gen. Mitsuo Nakazawa

Coastal Forces

Makurazaki Coast—146th Division

Kaimon Mt. District—125th Independent Mixed Brigade

Fukiage Coast—206th Division

Kushikino Dist.—303d Division

Mobile Reserve

Kagoshima area—6th Tank Brigade (less one regt.)

Kirishima Mt. area—77th Division

Northern Kyushu

Fifty-sixth Army—Lt. Gen. Ichiro Shichida

Coastal Forces

(136 cont'd) Gen. Kazuo Tanikawa, Imperial General Headquarters, Army Section, and attended by staff officers from all Army and Navy headquarters connected with *Ketsu* No. 6 operations, proceeded from the following basic assumptions: (a) Enemy invasion strength would number 16 divisions: (b) Invasion armada would include many small troop carrying craft in addition to large transports; (c) At the time of the invasion, the strength of the Japanese special-attack forces would be equal to the strength at the end of July. It was concluded that about 500 enemy transport craft, lifting about four divisions, would be destroyed by air *tokko*, and an additional 100 craft by surface special-attack units, making the total estimated losses about 600 craft carrying about five divisions or nearly 30% of the invasion force. On the other hand, if the Japanese *tokko* strength could be materially increased over the July figure or if the enemy invasion fleet included more large transports than the Japanese estimated, the staff officers participating in the conference felt that destruction of 50% of the enemy convoy would not be impossible. (Statements by Maj. Gen. Tanikawa, Lt. Col. Hara, Lt. Col. Sato, and Lt. Col. Mizumachi, all previously cited; and Comdr. Yoshimori Terai, Staff Officer (Air Operations), Imperial General Headquarters, Navy Section.

137 Statements by Lt. Col. Hashimoto and Maj. Haba, both previously cited.

138 Homeland Operations Record, op. cit. Vol. III, Supplement, Essentials of Sixteenth Area Army *Ketsu* Operation Plan, Attached Map. No. 1.

139 (1) Statements by Maj. Gen. Yoshitake, Col. Fuwa, and Maj. Hirano, all previously cited. (2) Homeland Operations Record, op. cit. Vol. III, p. 49. (3) *Heiseiso Mitsu Dai Hachihyakushichijukyu-go: Heiki Jusoku Jokyo* 兵政總密第八百七十九號兵器充足狀況 (Dispatch No. 879 Secret, Report on the Condition of Repletion of Ordnance) Army Ordnance Department, 10 Aug 45.

140 Homeland Operations Record, op. cit. Vol. III, Suppl., Attached Map No. 1.

Shimonoseki-Moji Area—Shimonoseki Fortress Unit
Fukuma Area—145th and 351st Divisions
Karatsu Area—312th Division
Iki Island—Iki Fortress Unit
Mobile Reserve
Hakata Plain—57th Division and 4th Tank Brigade

Central Kyushu

Chikugo Group—Lt. Gen. Waichiro Sonoda
Coastal Forces
Nagasaki Area—122d Ind. Mixed Brigade
Oita Area—118th Independent Mixed Brigade
Mobile Reserve—None
Higo Group—Lt. Gen. Ichiji Tsuchihashi
Coastal Forces
Amakusa—126th Independent Mixed Brigade
Mobile Reserve
Kumamoto Dist.—216th Division

Sasebo

Sasebo Naval Station—10 battalions

Tsushima

Tsushima Fortress Unit

Goto Islands

107th Independent Mixed Brigade

In general, the tactics to be used by the ground formations on Kyushu were based upon the concepts which had given rise to the coastal combat and line combat divisional organization.[141] These principles applied both to local areas such as Ariake Bay (Plate No. 160), where strong coastal combat formations were backed up by local mobile reserves, and to Kyushu as a whole which was to receive reinforcements from Honshu under the *Ketsu-Go* plan.[142]

Salient features of the ground operational plans were as follows:[143]

1. Theater

a. In southern Kyushu, the principal enemy landing points will be the Sumiyoshi Beach (Miyazaki Prefecture), on the right bank of the mouth of the Hishida River (Ariake Bay), and Fukiage Beach on the Satsuma Peninsula.

b. In northern Kyushu, the principal enemy landing areas will be the Fukuma and Hakata Bay areas.

c. The decisive battle will be sought in the area in which the enemy launches his main effort. If the location of the enemy main effort cannot be determined, the decision will be sought in the Ariake Bay front in southern Kyushu and Fukuma Bay in the case of northern Kyushu.

2. Initial Engagement

In accordance with previously announced tactical doctrine, the coastal combat units, firmly entrenched in large deep cave and tunnel positions and supported by fortress and siege artillery, will immediately engage the enemy close to the beach.

3. Employment of Reserves

a. Success in the battle for Kyushu depends on speed and flexibility in the employment of the decisive battle reserve.

b. As soon as the location of the enemy main effort has been determined, all reserves available in the Area Army will be redeployed to the decisive battle front, leaving the coastal combat units to wage holding actions on the secondary fronts. These decisive battle reserve units will be redeployed through centrally located staging areas from which they can be committed to the battle as preparations are

---

141 Cf. Chapter XVIII, p. 549.

142 The *Ketsu-Go* plan called for the rapid advance to Kyushu of four line combat divisions from Thirteenth and Fifteenth Area Army forces as so on as the operation was activated. (Cf. Chapter XVIII, pp. 559–60). Three or four additional line combat divisions (Thirty-sixth Army) were to be sent from Kanto to western Honshu, there to wait in readiness for commitment to Kyushu. Although serious consideration had been given throughout the summer to advancing the Thirty-sixth Army directly to Kyushu without waiting for activation of *Ketsu-Go*, Imperial General Headquarters remained reluctant until the end of the war to weaken the defenses of Kanto for such a move.

143 (1) Homeland Operations Record, op. cit. Vol. III, pp. 13–6, 30–3 and Suppl. (2) Statement by Lt. Gen. Kanji Nishihara, Commanding General, Fifty-seventh Army; Maj. Haba and Maj. Tabata, both previously cited.

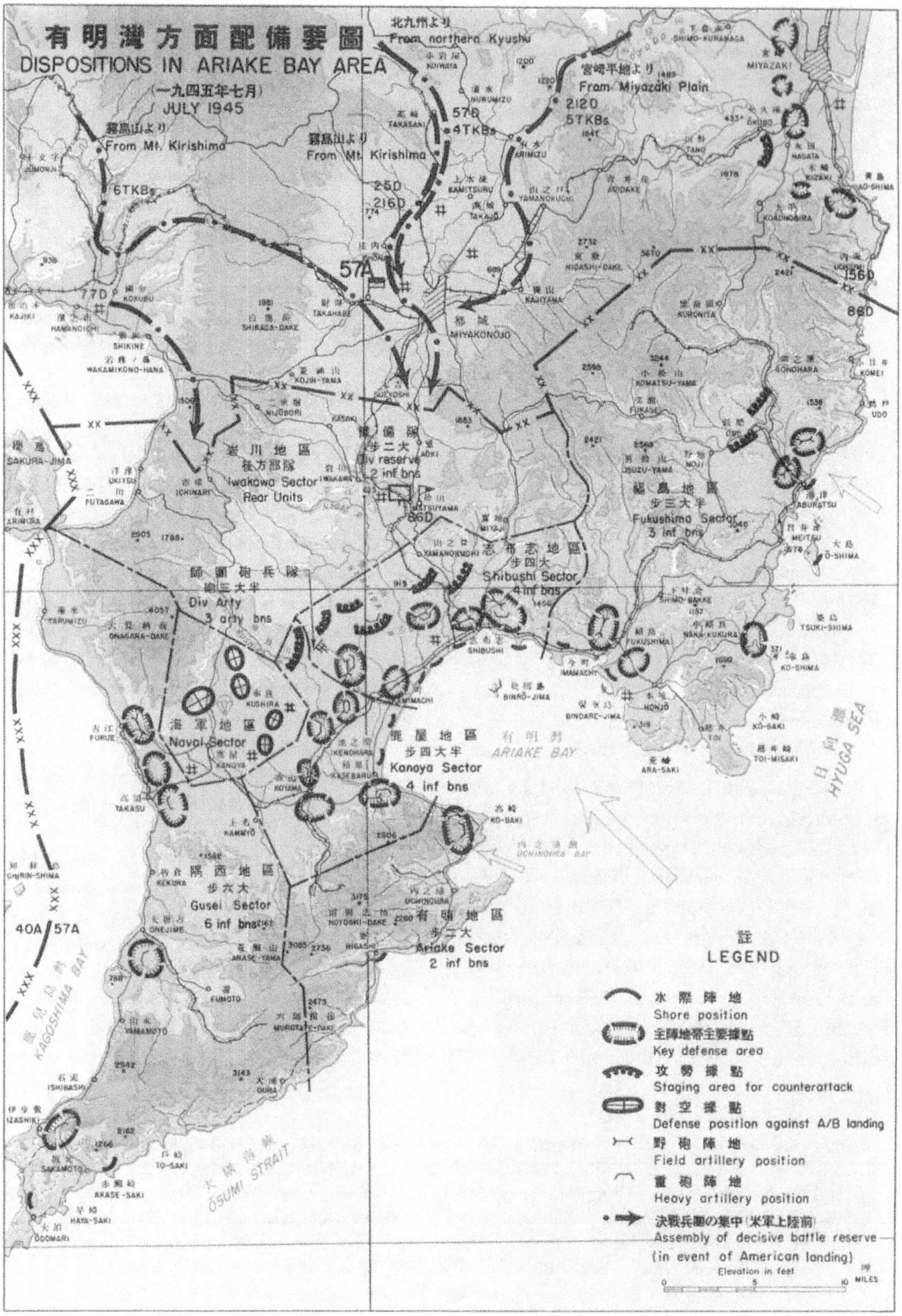

PLATE NO. 160
Dispositions in Ariake Bay Area, July 1945

completed. In the case of southern Kyushu, this will be the Kirishima Mt. area, in the south-central part of the island. In northern Kyushu it will be the Kurume-Iizuka area, in the north-central part of the island.

c. The Area Army mobile reserve, thus constituted at a maximum strength of five divisions and three armored brigades will be later reinforced by the quick arrival of four combat divisions from Thirteenth and Fifteenth Area Armies on Honshu.

4. Missions

a. The missions of Fifty-seventh, Fortieth, and Fifty-sixth Armies are self-evident.

b. Air base troops at Kanoya, Chiran, Miyakonojo, and Nyutabaru have the special mission of strengthening anti-ground and anti-airborne positions at their respective installations.

c. Chikugo and Higo Groups located in central Kyushu will keep open the northern and southern approach routes, secure harbors and channels, and wage vigorous holding actions if attacked.

d. The missions of the Sasebo Naval Station Force, Tsushima Fortress Unit, and Goto Detachment are to secure their respective areas and prevent either enemy establishment of forward bases in the vital Korea and Tsushima Straits or break-through in the same sector.

Ground operations were expected to complete the destruction of enemy forces begun by the *tokko* units. It was estimated that the artillery and automatic weapons fire on landing craft, coupled with the fierce attacks of the coastal combat formations would reduce the remaining enemy force by 30–50 %.[144]

If the losses sustained in the initial landing were as high as expected, the Japanese ground forces would be faced with no more than nine enemy divisions established ashore, of which only four or five would be located on the principal attack front. Available to cope with this remaining force was the Area Army mobile reserve of five divisions and three tank brigades soon to be reinforced by the arrival of four divisions from Honshu.[145]

While these estimates were being made and the summer wore on, combat preparations on Kyushu reached fruition. To Imperial General Headquarters it seemed that chances were steadily improving for dealing the enemy staggering losses if he chose to invade Kyushu.

Preparations for the Defense of Kanto

In marked contrast to the favorable progress

144 This hope was based on the fact that the Kyushu coast was well suited to the use of enfilade fire. Heavy artillery installed in caves was so arranged as to cover the entire approaches to the principal beaches. In Ariake Bay, for example, four 280mm howitzers, four 240mm howitzers, seven 150mm guns, two 120mm guns, eight 100mm guns as well as the light artillery and infantry cannon of the 86th Division were emplaced in coastal cave positions east of Shibushi, south of Kasebaru, north of Uchinoura and on Bindare and Binro Islands so as to cover the entire bay. In addition, preparations had been made so that heavy artillery units under Army control and the divisional artillery units of the mobile reserve units could be moved into previously prepared positions wherever the decisive battle developed. Statements by Lt. Col. Fujiwara, author (cf. n. 1, Chapter XVIII) and Lt. Gen. Wataro Yoshinaka, Commanding General, and Col. Takashi Okuyama, Chief of Staff, 86th Division.

145 The ground strength which it was expected could be massed by the Japanese within a week in the decisive battle area, including coastal combat divisions already in the areas, was about six divisions at Ariake Bay, seven divisions on the Miyazaki plain, and seven on the Satsuma Peninsula, each case being exclusive of the other two. Hope of victory was strengthened by the remembrance of the Okinawa campaign, in which Thirty-second Army, though ringed by enemy naval fire and possessing a strength of only 2½ divisions, held out for 100 days against an American force five or six times as large. The decisive battle on Kyushu was to be fought under conditions incomparably more advantageous to the Japanese.

being registered by Sixteenth Area Army, preparations in the Kanto area were moving ahead at a snail's pace.[146] Responsibility for this sector fell to the First General Army, the top tactical command in northern Honshu, (Plate No. 161) and to its Twelfth Area Army, commanded by General Shizuichi Tanaka.

By the end of July, opinions advanced in various staff studies reflected a fairly well defined appraisal of the enemy's intentions in the event of a direct invasion of Kanto.[147] It was believed that the preliminary operations, insofar as they concerned the tightening of the blockade around the Homeland and the prelanding shelling and bombardment, would follow a pattern similar to that expected if the initial target proved to be Kyushu. Opinion as to other specific enemy intentions revolved around the following major points:[148]

1. ***General Objectives***

It is expected that the major objectives of an enemy invasion of the Kanto area will be the annihilation of the Japanese forces deployed in this sector and the occupation of Tokyo, the political and economic center of the nation.

2. ***Preliminary Operations***

Prior to the main attack, the enemy will occupy the Izu Islands and will launch invasions of the Omaezaki and/or the Tateyama sectors to establish forward fighter bases to support the main landing.

3. ***Main Landing***

a. It is most probable that the main attack will be directed against the Kujukuri Beach sector with a secondary attack being carried out either in Sagami Bay or the Kashima Sea sector. Slightly less probability is seen in the main landing taking place in the Sagami Bay area with a secondary attack in the Kujukuri Beach sector.[149]

b. In coordination with the main amphibious landing, it is probable that enemy airborne elements will be dropped on the airfields in the vicinity of Yachimata and Atsugi to support the attacks of the landing forces or permit use of the fields by fighter planes.

4. ***Strength***

a. The enemy will use approximately 27–30 divisions in this operation.

b. If the main enemy attack falls along Kujukuri Beach strength will be distributed as follows:

Preliminary operations—3 divisions
Kujukuri Beach—15 divisions
Sagami Bay or Kashima Sea—5 divisions
Airborne—1–2 divisions
Army reserve—3–5 divisions.

c. In the event the main landing occurs in the Sagami Bay sector 15 divisions will land there and 5 divisions on Kujukuri Beach. Distribution of other forces will be the same regardless of the target of the main force.

5. ***Post-Landing Tactics***

a. Forces landing in the Kujukuri Beach sector will destroy the Japanese forces in that area, advance rapidly to the east of Tokyo, and from there, in coordination with the enemy forces landing in Sagami Bay, attack the capital in strength. Simultaneously,

---

146 Cf. Chapter XVIII, n. 72.

147 As was the case with regard to Kyushu, these various opinions were not assembled in a single publication representing the view of one staff or commander. (Statement by the author.)

148 (1) Situation Estimate for the Latter Half of 1945 and the Spring of 1946, op. cit., pp. 36–9. (2) Estimate of the Situation, op., cit. pp. 6–7. (3) Preparations for Counterattack Operations in Defense of the Homeland, op. cit., pp. 10–11. (4) Homeland Operations Record, op. cit. Vol. II, p. 39, Attached Map No. 2. (5) Statements by Col. Fuwa, Lt. Col. Hara and Rear Adm. Tomioka, all previously cited.

149 The specific sectors estimated to be the probable targets in each of the three areas were the Togane sector of Kujukuri Beach, left bank of the Sagami River, and the Hokota sector on Kashima Sea. Homeland Operations Record, op. cit. Vol. II, Attached Map No. 2.

other elements of this force will mop up the Japanese forces in the northeastern portion of Kanto. If the secondary attack falls along the Kashima Sea front rather than Sagami Bay, these forces will destroy the Japanese troops directly opposing them and then move into the district on the north bank of the Tone River to prevent a retreat from the Kanto district. Other elements of this force will mop up in the northern Kanto district and check reinforcements moving to Kanto from northern Honshu.

b. The forces landing in the Sagami Bay area will first destroy the Japanese forces in the immediate vicinity. Strong elements will then move against Yokohama and Miura Peninsula, while the main strength advances to the area west of Tokyo, from where, in coordination with other enemy forces attacking from the east, strong attacks will be launched against the city. Other elements of this force will control the Tokaido and Koshu highways, check Japanese reinforcements moving to Kanto from the Tokai and Kofu sectors, and mop up in the western Kanto district.

Pursuant to the joint Army-Navy Air Agreement concluded during July the combined weight of all Army and Navy air forces was to be thrown into a gigantic, all-out aerial offensive over the Kanto area if *Ketsu* No. 3 were activated. Immediately available for this purpose were units of the First Air Army under Lt. Gen. Takeo Yasuda and of the Third Air Fleet under the command of Vice Adm. Kimpei Teraoka.[150] In addition, upon issuing an alert for the implementation of *Ketsu* No. 3, it was planned to deploy the remainder of the air units to fields from which they could support the decisive attacks.

By the end of July no operational plans for the coordinated commitment of these forces over Kanto were yet under consideration. The full energies of the staffs were still directed towards perfecting the plans to support *Ketsu* No. 6 operation which occupied top priority.[151]

In accordance with this same fundamental assumption, the bulk of the surface special-attack units were deployed in the Kyushu area. Only a small portion of these forces could be redeployed to the Kanto area if that proved to be the target of the initial invasion.[152] Already deployed around Tokyo Bay, however, were the following naval surface special-attack forces under the command of Vice Adm. Michitaro Totsuka, Commander, Yokosuka Naval Station:[153]

7th Special-Attack Squadron
12th Assault Unit
14th Assault Unit
17th Assault Unit
1st Special-Attack Squadron
11th Assault Unit
15th Assault Unit
16th Assault Unit
18th Assault Unit
Yokosuka Assault Unit[154]

With these relatively weak surface special-attack forces it was recognized that air and surface *tokko* operations against an initial enemy invasion in the Kanto area would not achieve the success hoped for in *Ketsu* No. 6 operations.

150 The order of battle of these air forces was the same as outlined for employment in *Ketsu* No. 6 operation. (Cf. pp. 608–9.)

151 Statement by Lt. Col. Koji Tanaka, previously cited.

152 Only the small type submarines and the navy *Koryu* special-attack boats were to move up from bases near Kyushu and participate in *Ketsu* No. 3. 31st Destroyer Squadron could not be employed because of fuel restrictions. The large type submarines would carry out the same mission regardless of the target of the initial invasion. Preparation for Counterattack Operations in Defense of the Homeland, op. cit., pp.21–4.

153 Ibid., Attached Chart 3–4.

154 The Yokosuka Assault Unit, responsible for training *Kairyu* pilots until the enemy invasion commenced, was composed of the 101st, 102d and 103d *Kairyu* Units and included a total of 36 *Kairyus*. Ibid.

PLATE NO. 161
Ground Dispositions on Eastern Japan, July 1945

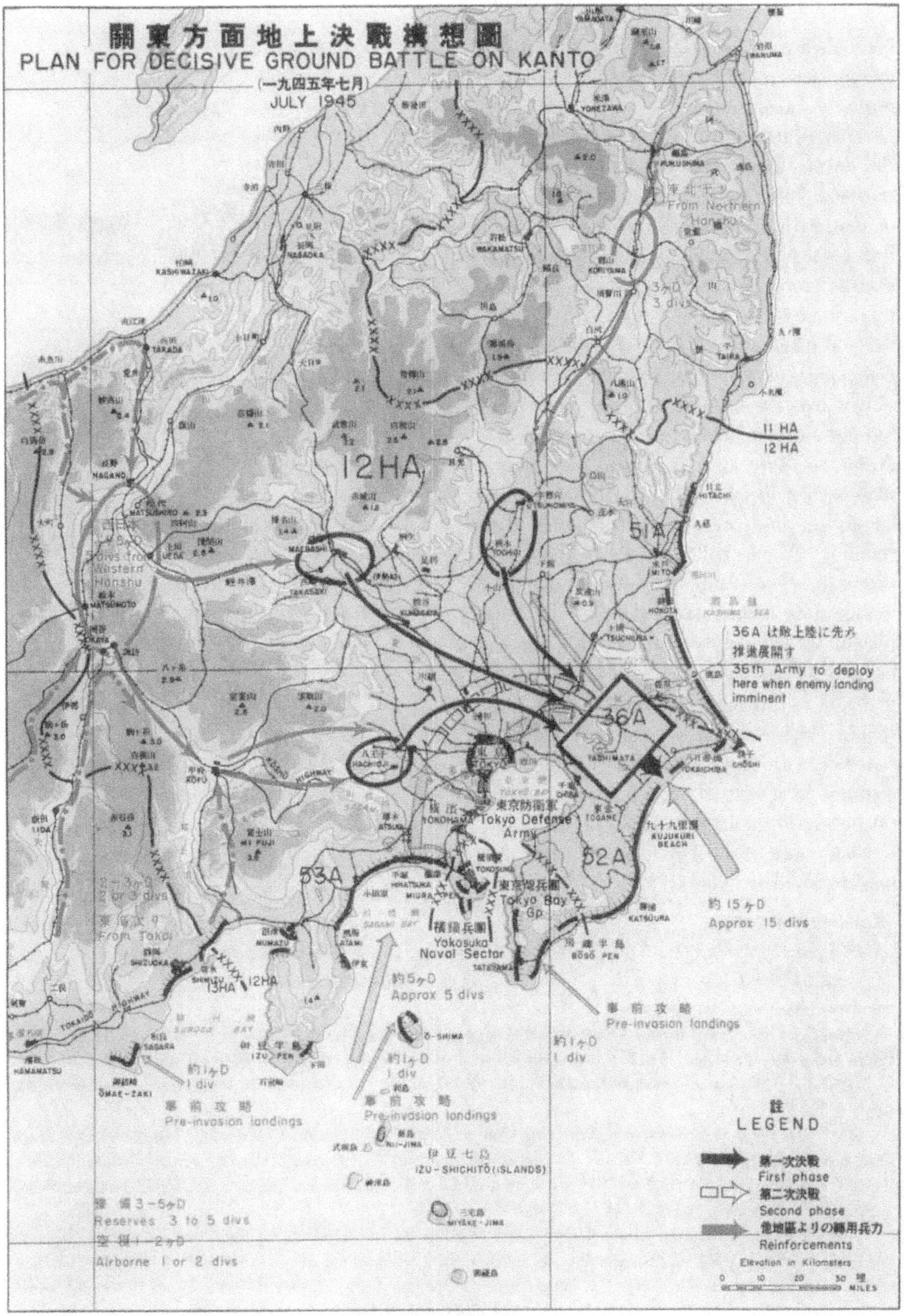

PLATE NO. 162

Plan for Decisive Ground Battle on Kanto, July 1945

Nevertheless, it was felt that a sufficiently strong blow could be delivered to blunt seriously the enemy attack.

After penetrating this special-attack screen, the enemy would next encounter determined resistance from ground units at the very edge of the beach. By the end of the month, 11 line combat and seven coastal combat divisions,[155] seven independent mixed brigades, two armored divisions and three tank brigades were available to General Tanaka, tactical commander of ground forces in the Kanto area.

Combat effectiveness of this vast force, however, was far below that of Sixteenth Area Army on Kyushu. Troops of the second mobilization were being issued the first of their equipment and were only beginning their training. Worse still, it would be the end of the year before equipping units of the third mobilization would be even generally completed. Despite these shortages, all units were being deployed to their assigned areas in order to begin defensive preparations. Accumulation of reserve supplies was even further behind schedule, with the prospect that Kyushu would continue to siphon off the total output of production until the latter part of August.[156]

Major units of the Twelfth Area Army were now deployed as follows:[157]

Kashima Sea Area

Fifty-first Army—Lt. Gen. Kengo Noda

44th Division

151st Division

221st Division

115th Independent Mixed Brigade

116th Independent Mixed Brigade

7th Tank Brigade

Kujukuri Beach Area

Fifty-second Army—Lt. Gen. Tokumatsu Shigeta

3d Imperial Guards Division

147th Division

152d Division

234th Division

3d Tank Brigade

Sagami Bay Area

Fifty-third Army—Lt. Gen. Yaezo Akashiba

84th Division

140th Division

316th Division

117th Independent Mixed Brigade

2d Tank Brigade

Central Kanto *(Central Mobile Reserve)*

Thirty-sixth Army—Lt. Gen. Toshimichi Uemura

81st Division

93d Division

201st Division

202d Division

209th Division[158]

214th Division

1st Armored Division

4th Armored Division

Southern Chiba Prefecture

Tokyo Bay Group—Lt. Gen. Shihei Oba

354th Division

96th Independent Mixed Brigade

Miura Peninsula

---

155 The line combat divisions were the 44th, 221st, 3d Imperial Guards, 234th, 84th, 81st, 93d, 201st, 202d, 209th, and 214th Divisions. The first five of these were deployed in positions near the beaches. (Cf. Plate No. 161.)

156 (1) Homeland Operations Record, op. cit. Vol. II, p. 44. (2) Statement by Col. Hiroshi Fuwa, previously cited.

157 (1) Reply to Questionnaire Regarding Defense Tactics and Strategy, op. cit. (2) Homeland Operations Record, op. cit. Vol. II, Attached Map 3. In addition to the forces outlined above, the 1st Imperial Guards Division, charged with security of the Imperial Palace, was located in Tokyo under the command of the Eastern District Army Commander. Statement by Col. Fuwa previously cited.

158 The 209th Division, although assigned to the Thirth-sixth Army on 19 June, was actually stationed in Kanazawa because of more adequate training and billeting facilities at its old station, movement of which had not been started as of the end of July. (1) *Dairikumei Dai Sensambyakugojuichi-go* 大陸命第千三百五十一號 (Imperial General Headquarters Army Order No. 1351) 19 Jun 45. (2) Statement by Col. Fuwa, previously cited.

Yokosuka Naval Station Force—Vice Adm. Michitaro Totsuka

Yokosuka Combined Special Landing Force

11th to 16th Naval Landing Parties[159]

114th Independent Mixed Brigade

Special-Attack forces

Tokyo

Tokyo Defense Army—Lt. Gen. Jo Iimura

1st Garrison Brigade

2d Garrison Brigade

3d Garrison Brigade

Izu Islands

O-shima

321st Division

Nii-Jima

66th Independent Mixed Brigade

Hachijo-Jima

67th Independent Mixed Brigade

With regard to the deployment of these forces a somewhat different situation prevailed than was the case in Kyushu. The Twelfth Area Army's large central mobile reserve, the Thirty-sixth Army, permitted the Area Army to assign a greater portion of its strength farther forward. In each of the Fifty-first, Fifty-second and Fifty-third Armies, therefore, all of the line combat and coastal combat divisions were made responsible for a sector along the beaches, leaving in each Army only a tank brigade as mobile reserve.

Meanwhile, on the basis of the estimate that an initial invasion in the Kanto area would not begin until after fall, the same urgency for speedy preparation of operational plans did not obtain as was the case with the Second General Army. The First General Army, early in May, had therefore ordered the Twelfth Area Army to continue the operational policy then in force while the General Army undertook a more thorough and deliberate study of preliminary drafts covering plans for the employment of these ground forces.[160]

This prolonged study culminated in the publication, 17 July, of the First General Army operational plan.[161] On the same day a conference attended by the chiefs-of-staff of the component Area Armies was convened at the General Army headquarters in Tokyo to discuss the more important details. Shortly thereafter, Twelfth Area Army published its own operational plan based upon the General Army plan. Summarized and combined, the essential features of these two plans included the following:[162]

1. Theater

a. In the Kanto plain, the principal enemy landing points will be the Togane-Yokaichiba sector (Kujukuri Beach), left bank of the Sagami River and the Hokota sector (Kashima Sea).

---

159 Each of these landing parties was composed of from two to four battalions. Supporting units, such as artillery and engineers, were included in the battalion organization. Total strength of these landing parties was about 18 battalions. (Statements by Vice Adm. Kyuichi Kudo, Commander, Yokosuka Combined Special Naval Landing Force and Lt. Comdr. Satoru Yunoki, Staff Officer (Operations), Yokosuka Combined Special Naval Landing Force.

160 Homeland Operations Record, op. cit. Vol. II, pp. 15, 18–20; Vol. III, pp. 12–3.

161 First General Army simultaneously released a Decisive Battle Outline which set forth major points emphasized by the High Command with regard to the conduct of the decisive battle. The following points were discussed in detail: (a) The Homeland Operation must be a decisive one in which the invasion forces will be quickly sought out and annihilated. (b) The key to ultimate victory rests in the annihilation of the enemy at the water's edge in the period when his landings are still in progress. The assault must be undertaken with the resolve that each man will take an enemy to death with him at the water's edge. (c) The decisive battle area must be determined irrespective of the enemy's plans, and thee nemy must be made to fight at that point. (d) All available manpower must be concentrated for a swift offensive in depth. (e) Fortifications for offensive purposes must be emphasized. Homeland Operations Record, op. cit. Vol. II, pp. 21–2, 51–60.

162 (1) Ibid., pp. 18–26, 51–63 and Attached Map No. 2. (2) Statements by Col. Fuwa and Lt. Col. Hara, both previously cited.

b. The Kujukuri Beach is designated as the sector in which the decisive battle will be fought. The second decisive battle will be fought in the Sagami front.

2. *Initial Engagement*

In the sectors in which the enemy lands, the coastal Armies (Fifty-first, Fifty-second and Fifty-third), with their line and coastal combat divisions firmly entrenched in beach fortifications and supported by fortress and siege artillery, will immediately engage the enemy close to the beach.

3. *Employment of Reserves*

a. As soon as it becomes apparent that an enemy landing is imminent in the Kanto area, the Thirty-sixth Army will be moved up and deployed directly behind the Fifty-second Army.

b. The Armies in whose area the enemy attacks fall will secure the beach positions and prevent establishment of an enemy beachhead.

c. If the enemy fails to launch an attack against one of the coastal armies, as many troops as possible will be withdrawn from that Army and redeployed to the decisive battle front.

d. Taking advantage of the confusion resulting from the determined local offensives against the enemy landing point, the main force of the Thirty-sixth Army will attack on the third or fourth day after the initial enemy landing.

e. Area Army reserves will be later reinforced by the addition of three line combat divisions from Eleventh Area Army, two or three divisions from Thirteenth Area Army, three from Fifteenth Area Army and two from the Sixteenth Area Army. These divisions will assemble in Nagano prefecture and later will be employed as a mass counterattack force.

4. *Missions*

a. Missions of the Fifty-first, Fifty-second, Fifty-third and Thirty-sixth Armies as outlined above.

b. The Tokyo Defense Army will defend Tokyo, particularly the important section of the city centering around the Imperial Palace.[163]

c. The Tokyo Bay Group and the Yokosuka Naval Station Force will defend the Boso and the Miura Peninsulas, respectively, prevent a breakthrough into Tokyo Bay and serve as the pivotal center of the Japanese offensives in the Kujukuri Beach and Sagami Bay sectors.

d. The forces stationed on the Izu Islands will prevent enemy establishment of forward bases.

As of the end of July no conclusions had yet been reached regarding the expected results of each phase of operations against an enemy landing in the Kanto area. It was still too early to accurately predict the defensive poten-

163 Preparations for the defense of Tokyo in case a powerful enemy force should break through the coastal positions during the decisive battle received individual emphasis from the High Command due to the place the capital occupied in the minds of the Japanese. It was recognized that maintenance of morale required adoption of special measures to defend the city to the last. Imperial General Headquarters accordingly issued the Tokyo Defense Operations Outline on 23 June 1945, simultaneously with the activation of the Tokyo Defense Army headquarters. The plan provided for the construction of underground fortifications capable of withstanding an enemy assault for one year. By the end of July, however, positions were still being reconnoitered and surveyed. It was estimated that construction would be started about 20 August. Ground forces to strengthen these defenses were not yet designated by the end of the war although it was planned that two or three divisions, one tank brigade and other supporting units would be added to the three brigades garrisoning the city at this time. Greatest emphasis was to be placed on preparation of positions west of the city and around the Imperial Palace on the basis that the Sagami Bay sector would be the most seriously threatened while the Thirty-sixth Army was being committed on the Kujukuri Beach sector. Almost a year previously, however, as a precautionary measure to prepare for the worst eventuality, the High Command, in complete secrecy, had begun the construction of facilities in Matsushiro, Nagano Prefecture, capable of housing the Government and the Imperial General Headquarters. All installations, set underground and shielded by hard cement capable of withstanding powerful and sustained bombing attacks, were practically completed by the end of July 1945. (1) *Dairikushi Dai Nisengohyakujusan-go* 大陸指第二千五百十三號 (Imperial General Headquarters Army Directive No. 2513) 23 Jun 45. (2) Imperial General Headquarters Army Order No. 1355, op. cit. (3) Homeland Operations Record, op. cit. Vol. II, pp. 26, 61–3. (4) Statements by Lt. Gen. Jo Iimura, Commander, Tokyo Defense Army, and Maj. Gen. Sanada, previously cited.

PLATE NO. 163
General Situation of Japanese Forces, 1 August 1945

tial of the Twelfth Area Army as it would exist by late fall. Much remained to be accomplished before the defenses would give cause for the same optimism which prevailed in the case of Kyushu.

Strategic Situation—1 August 1945

While the High Command was thus pushing final preparations for the defense of Kyushu and the First General Army, anticipating scarcely a trickle of supplies until the requirements of the Kyushu area were completely satisfied, was placing primary emphasis on the construction of coastal fortifications and training of newly mobilized units, the progressively worsening strategic situation made it appear doubtful whether any of these preparations would be tested in battle.

In the first place, the terrific aerial bombardment to which the Homeland had been subjected during July showed no evidence of abating.[164] Although the casualties were now less than in the earlier, giant fire bomb raids on the larger complexes,[165] the disruptive effects were being spread over a wider segment of urban society. The cumulative total of bombs which had been dropped on the four main islands by the end of July had risen to over 130,200 tons of which more than 98% had been directed at 98 urban areas or precision targets within these areas.[166]

Of greatest concern within the High Command was the paralysis spreading at an accelerated rate throughout the Japanese industrial plant. Output of the basic industries continued to decline during July, primarily because of the rapid drying-up of raw material stockpiles although enemy air attacks were contributing to the lowered production to an increasing extent because of plant destruction, transportation dislocations and the increase in absenteeism among the labor force.[167]

Concurrently, the impact of this decline,

---

164 During the first six days of August, in spite of a brief respite on 3–4 August because of a typhoon, the Japanese counted a total of 1,703 sorties flown over the Homeland by enemy land based aircraft. Of these 1,015 were flown by B-29's attacking cities in the Kanto, Tokai, Osaka-Kobe and Japan Sea coast areas and sewing mines in the Korea Strait, Shimonoseki-Moji channel and in the Inland Sea. The remaining sorties were carried out by fighters based on Okinawa and Iwo Jima. Enemy carrier planes did not appear over the Homeland during this period. Daily Record of the War Situation, op. cit.

165 By 31 July about 188,300 Japanese had been killed and 257,000 wounded in the air attacks on the four main islands. Statistics and Analysis Reports, op. cit., Report No. 16: Number of Air Raid Casualties in the Japanese Homeland through Jul 45.

166 Twenty-six of these urban area were towns in the immediate vicinity of important military installations, such as airfields or coastal defense positions. The remaining 72 urban areas were classed as cities, the population of which exceeded 30,000. (1) U. S. Strategic Bombing Survey (Pacific), Civilian Defense Division, *Final Report Covering Air Raid Protection and Allied Subjects in Japan*, 1947, pp. 200–3. (2) *Jiji Nenkan* 時事年鑑 (Jiji Yearbook) Jiji Press Co., Tokyo, 5 Jan 47, pp. 390–1.

167 Typical examples of the decline in production within basic industries by the end of July are to be found among the metal and chemical plants. The Wanishi and Kamaishi plants of the Nihon Iron Mfg. Co., two of the largest steel producers, suffered production drops of 21% and 65%, respectively as a direct result of air raids. In addition thereto, production drops attributable to other causes, such as raw material shortages, amounted to 62% and 35% in the same plants. The Hodogaya Chemical Industry Co., Koriyama, suffered a production drop of 50–70% due to air raids and a total production drop of 80% compared to the peak out-put during the war. The Nihon Synthetic Chemical Industry Co., Ogaki, and the Mitsui Chemical Industry Co., Omuta, suffered production drops of 60% and 100%, respectively, from all causes. Domestic coal production was running about 50% of the peak established in January—March 1944. Statistics and Analysis Reports, op. cit., Report No. 15: Effect of Air Raids on Production Facilities of Important Japanese Industries.

added to the cumulative effects of the widespread urban air raids, began to be felt more keenly in the fabricating industries where production levels had previously remained relatively high. Aircraft production, for example, which during the early summer months had held up well, fell off to 62% of the average monthly rate of the preceding quarter.[168] Only 1,003 planes of all types were completed.[169] The planned production goal of 16,000 by September 30 was now clearly beyond the capabilities of the industry. Even worse, to forestall the possibility of further attrition, flight operations, which already were proving ineffective, would have to be further restricted, leaving the enemy practically unopposed in the air.[170]

The munitions industry fell even further behind the established quotas, although production of light artillery had spurted during the month.[171] Percentages of scheduled production actually achieved by the end of July were as follows:[172]

Rifles	*68%*
Light machine guns	*63%*
Machine guns	*90%*
Infantry cannon	*14%*
Antiaircraft guns	*53%*
Mortars	*9%*
Self-propelled cannon	*21%*
Light artillery	*37%*
Heavy artillery	*54%*

In both construction and hauling capacity, the shipping industry was by now approaching a state of complete collapse. Monthly production of new bottoms had established a new low, falling off to about 15,500 tons or less than 10% of the peak monthly average established in January—March 1944. An additional 4% of merchant tonnage had been sunk, bringing the total losses to 83%.[173] Maritime shipping capacity at the end of the month was down to 780,000 tons, most of which continued to operrate in the Japan Sea between Korea and the Homeland.[174] One net effect of this attrition

168 July production amounted to about 40% of the average monthly rate of 2,482 planes established during the peak quarter of April—June 1944. The Mitsubishi Aircraft Co., Oemachi Plant, Nagoya, where airframes were manufactured, and the Mitsubishi Aircraft Co., No. 4 plant, maker of engines, were typical producers within the aircraft industry suffering from both the air raids and raw material shortages. The damage sustained within each plant resulted in a production drop of approximately 50%. An additional 30% drop was attributed to other causes. Ibid., Report Nos. 2 and 15.

169 Of this total only 832 were considered combat operational planes. An additional 768 planes were either repaired or converted for combat purposes, making a total of 1,600 compared with losses during the same month of 1,146. Ibid., Report No. 18: Statistical Situation of Production and Losses of Japanese Aircraft, Jun–Jul 45.

170 Total number of operational combat planes available to the Army and Navy at the end of July numbered 10,308. Ibid.

171 Within the munitions industry, the Aichi Clock and Electric Funakata Plant, producers of guns, torpedoes and mines, and the Kure Naval Arsenal, where guns, torpedoes, searchlights and hydrophones were manufactured, offer examples of this drop in production. Air raid damage caused a drop of from 35–90% and 20–70%, respectively, while the overall production loss due to all causes amounted to 80% and 60%, respectively. Ibid., Report No. 15.

172 Ibid., Report No. 17.

173 Ibid., Report No. 1.

174 *Nihon Kaiun no Hembo* 日本海運の變貌 (Situation of Japanese Sea Transportation) Mr. Genko Tsuboi, Sea Transport Board, Ministry of Transportation, 20 Aug 46, Chart No. 15. (Writer's note: The *Nichi-Go* operation, established for the emergency transportation of troops and supplies from the continent to the Homeland, (Cf. n. 29.) was still being carried out at the end of July though on a considerably reduced scale. Highest priority was now given the transportation of food, all designated units having completed the transfer across the Japan Sea.)

was the virtual suspension of the vitally important inter-island traffic except for small vessels of less than 200 tons.[175]

In the case o oil, one of the most critical items in the decisive battle plans, the outlook appeared even more ominous. Consumption during the month of about 155,390 barrels, had further reduced the stockpile of aviation gasoline to 508,160 barrels, of which 333,900 barrels were now earmarked for decisive battle operations. Production from all sources had declined sharply during July to 60,890 barrels.[176]

Even rail movements were falling off sharply, although the railroad network had not yet been singled out for intensive attacks except in the case of the Hokkaido-Honshu ferry system and a few isolated bridges.[177] The carrying capacity was now reduced to 8,724,165 tons as compared with the peak monthly average of 14,849,749 tons during 1943.[178] Moreover, temporary delays due to the heavy urban raids were becoming more general and shipment of many critical items was falling far behind schedule.

Nor was there any cause for optimism in the High Command regarding the over-all situation now existing in the Allied camp. Almost three months had elapsed since the collapse of Germany—three months for the enemy to lay the ground-work and begin the stupendous task of transfering overwhelming ground, air and sea forces to the Pacific theater. It was estimated that if the U. S. S. R. joined the attack against Japan in August or September, there would be available in the eastern provinces 40–50 divisions, 6–7,000 aircraft and 4,000 tanks.[179]

Such was the situation facing the nation as the hot and humid month of August ushered in the most critical period in Japan's history. In spite of the odds building up against them, the Japanese people well knew that if their leaders were determined to carry out the decisive battle on the sacred soil of the Homeland there was no alternative but to fight to the bitter end.

---

175 (1) *Showa Nijunendo Busshi Doin Keikaku* 昭和二十年度物資動員計画 (Plans of Mobilization of War Materials Based on the Yearly Plans of 1945) Munitions Ministry, Vol. I, 24 Apr 45, Chart Nos. 1, 4; and Vol. II, 30 Jun 45, Chart Nos. 1, 4.

176 Statistics and Analysis Reports, op. cit., Report No. 9.

177 The vital underwater railroad tunnel connecting Shimonoseki and Moji and other important railroad tunnels had not yet been attacked. Enemy strafing operations over Kyushu had now reached such proportions, however, that daylight travel over the railroad network was becoming increasingly difficult. Moreover, the longest bridge on the island, crossing the Chikugo River, was destroyed in an air raid on July. (Statement by Lt. Col. Kiyoshi Ohta, Staff Officer (Supply), Sixteenth Area Army.

178 (1) *Kokuyu Tetsudo Rikuun Tokei* 國有鐵道陸運統計 (Statistics of Transportation in the National Railway) Railway Department, Ministry of Transportation, 1943, 1944 and 1945 editions. (2) *Unyu Seiseki Tokei Nempo* 運輸成績統計年報 (Yearly Statistics of Land Transportation) 1944–45, Railway Department, Ministry of Transportation.

179 Situation Estimate for the Latter Half of 1945 and the Spring of 1946, op. cit., pp. 7–8.

CHAPTER XX
DECISION TO SURRENDER

Background of the Peace Movement

In early August of 1945, Japan strove to complete preparations for a final supreme effort to repel an invasion of the home islands. At the same time a crucial internal struggle on the highest plane of the nation's political leadership was about to culminate in a decision to avert national destruction through the acceptance of Allied surrender terms.[1]

The earliest tangible development in the political evolution toward peace occurred in September 1943. At that time, a small group of the so-called *jushin*,[2] or "senior statesmen" acting on the initiative of ex-Premier Admiral (ret.) Keisuke Okada, requested Premier Hideki Tojo to sit down with them in a frank discussion of the war situation.[3] The veiled purpose of this move was to discredit the Tojo regime with the ultimate objective of forcing its retirement. Admiral Okada, the prime instigator, believed that a general peace would be impossible as long as Tojo remained in office.

Already, by this date, the successive defeats suffered by the Japanese forces in the Southwest Pacific were beginning to have repercussions at home, despite attempts by the Government and High Command to conceal them. In Europe, Italy had dropped out of the war, and the military fortunes of Germany, Japan's sole remaining ally, were already on the wane. These factors, coupled with deteriorating economic conditions at home, steadily undermined Tojo's popularity. An undercurrent of dissatisfaction with existing leadership began to take shape in February 1944 when Premier Tojo and Navy Minister Shigetaro Shimada moved to tighten their control of the fighting services by assuming concurrently the posts of chiefs, respectively, of the Army and Navy General Staffs.[4]

The senior statesmen not only were influenced by these trends but secretly exploited them. By late spring of 1944, they had gained sufficient strength and confidence to urge their views more strongly. In mid-July the disastrous loss of Saipan gave them an opportunity to

1 This chapter was originally prepared in Japanese by Capt. (Navy) Atsushi Oi. For duty assignments of this officer, cf. n. 1. Chapter X. All source materials cited in this chapter are located in G-2 Historical Section Files, GHQ FEC.

2 The *jushin* (literally, "principal subjects") comprised all living statesmen who at one time had held the post of Premier; it also included the Privy Council President. Like the *genro*, or "elder statesmen" whose function they in part inherited, the *jushin* had no constitutional position. The Emperor's consultations with the *jushin* were usually carried out by the Lord Keeper of the Privy Seal, traditionally the Emperor's closest political advisor. Baron Reijiro Wakatsuki, Admiral Keisuke Okada, Koki Hirota, Prince Ayamaro Konoye, Baron Kiichiro Hiranuma, General Nobuyuki Abe, Admiral Mitsumasa Yonai and Baron Yoshimichi Hara, Privy Council President, made up the *jushin* group at that time. Wakatsuki, Okada and Konoye were more active in the move toward peace.

3 Statement by Admiral (ret.) Keisuke Okada, Premier, Jul 34–Feb 36.

4 General Hideki Tojo already held the post of War Minister in addition to the premiership. The action whereby he and Navy Minister Shimada installed themselves simultaneously as Chiefs of the Army and Navy General Staffs was criticized in some Army and Navy circles as reminiscent of the usurpation of Imperial prerogatives by the Tokugawa Shogunate. (Cf. Chapter X, n. 25.)

press for the resignation of the Tojo Government. Tojo realizing the graveness of the situation tried to settle it by reorganizing his Cabinet but this attempt failed and found himself without the support of the Emperor. This blow was fatal.[5]

On 18 July simultaneously with the public announcement of the fall of Saipan, General Tojo resigned with his entire Cabinet to end 33 months' continuous tenure of the premiership. The collapse of his Government marked an important political turning point of the war. The *jushin* had made their influence felt.[6]

The Tojo Ministry was succeeded by a government headed jointly by General (ret.) Kuniaki Koiso, and Admiral (ret.) Mitsumasa Yonai, a former Premier and well-known Navy moderate. Despite the cabinet change, however, a fundamental political shake-up such as was required to pave the way for any real peace endeavor had not been effected.

In the first place, although the Emperor had charged Koiso and Yonai with joint leadership of the new Government, Koiso was given the premiership. The hope was that he would succeed in establishing a national Cabinet, coordinate effectively political and military leaders, and maintain close liaison with the key fighting service. Furthermore, the Army consented to place its nominee, Field Marshal Sugiyama, in the Cabinet as War Minister on the condition that the new Government would continue the vigorous and determined prosecution of the war.[7]

Nevertheless, an initial success had been secured. In Admiral Yonai, who was himself a member of their group, the *jushin* now had a strong sympathizer in the Cabinet. Yonai, in accordance with the Emperor's mandate, took the post of Deputy Premier. More important, he assumed the concurrent post of Navy Minister, which automatically entitled him to first-hand information regarding the war situation and the plans and decisions of the Navy High Command.[8]

Although Yonai was the *jushin's* greatest hope in the new Cabinet, Premier Koiso himself was not always optimistic about the war. He believed that the wisest course lay in first attempting to wean China away from the Allied camp by means of a separate peace, and then seeking the mediation of the Soviet Union or some other neutral with a view to a general peace settlement on favorable terms. Before initiating any such move, however, he felt that it was essential to score one substantial military success in the field in order to improve Japan's bargaining position.[9]

With this end in view, Koiso devoted his first efforts to reinforcing the national war effort.

5 (1) *Kyokuto Kokusai Gunji Saiban ni okeru Hikokunin Kido Koichi no Sensei Kokyosho* 極東國際軍事裁判に於ける被告人木戸幸一の宣誓口供書 (Affidavit of Defendant Koichi Kido in International Military Tribunal for the Far East) For English translation, see IMTFE Record, pp. 31,076–104. (2) *Shusen Oboegaki* 終戰覺書, (Notes on the Termination of the War) Kobun-do, Tokyo, 1948, pp. 7–12.

6 General Tojo himself recognized that the *jushin* were instrumental in causing his downfall. When he went to the Imperial Palace to tender the Cabinet's resignation, he was asked by Marquis Kido if he desired to recommend a successor to the premiership. His reply was, "I think that the responsibility of the *jushin* for the present cabinet change is heavy. In consequence, they may have already formulated their own plan. Therefore, I will not express my views." In fact, the *jushin*, in response to an Imperial command, met at 1600 the same day (18 July) and decided their recommendations to the Throne.; Affidavit of Kido, op. cit., IMTFE Record, p. 31,082.

7 In accordance with standard practice, Marshal Sugiyama was selected for the War Minister's post by the Army "Big Three": the outgoing War Minister, General Tojo; the Chief of Army General Staff, General Yoshijiro Umezu; and the Inspectorate General of Military Training, Marshal Sugiyama himself. General Tojo had also resigned as Chief of Army General Staff on 18 July. General Umezu immediately succeeded him.

8 Since the Navy Minister had to be an admiral or vice admiral in active service, Yonai was hastily restored to the active list while the formation of the Cabinet was in progress. This essential move had been planned well in advance of the downfall of the Tojo Cabinet.

9 Direct Testimony of Defendant Kuniaki Koiso. IMTFE Record, pp. 32,251–4.

Soon after assuming office, he supplanted the old Liaison Conference between Imperial General Headquarters and the Government by a new Supreme War Direction Council.[10] The purpose of this body, in which key members of the Cabinet sat together with representatives of the High Command, was to assure full coordination of the civil and military branches of the Government. The Council, however, failed to achieve the results which Premier Koiso expected, and he discovered that he was unable to obtain all the military information he desired.

Despite the expectations of the *jushin*, the Koiso Cabinet's official policy continued to be one for all-out prosecution of the war. An Imperial conference held on 19 August formally adopted a basic policy declaration which emphatically stated that the war would be prosecuted to the end regardless of continued enemy successes or of unfavorable international developments. Foreign Minister Mamoru Shigemitsu prevailed upon the conference not to exclude diplomatic action designed to improve Japan's positions, notably by strengthening relations with the Soviet Union.[11] However, the military success which Premier Koiso deemed a prerequisite to any broad peace maneuver was still unrealized, and the situation remained at a stalemate.

In the fall of 1944, Koiso hoped that the *Sho-Go* Operations might bring the victory needed to pave the way for a peace attempt. His hope was soon shattered by events. The all-out effort of the sea and air forces to smash the enemy invasion of Leyte in October ended in failure, and in less than two months General MacArthur's forces had won a secure foothold in the central Philippines. On the diplomatic front, a proffer to dispatch a special mission to Moscow to begin the task of cementing Japanese-Soviet relations had meanwhile been rebuffed by the Soviet Government.[12]

Japan therefore faced a darker situation than ever before as the year 1945 began. So gloomy were the prospects that the Emperor himself felt impelled to reveal to the *jushin* his anxiety for an early peace. Early in January, when the enemy invaded Luzon and American carrier forces boldly swept Japanese shipping in the South China Sea, the Emperor told Marquis Kido, Lord Keeper of the Privy Seal, to arrange audiences with the *jushin*.[13]

Early in February these confidential meetings got under way. Because of fear that any inkling of their purpose might arouse opposition, extreme precautions were taken. No formal audience was granted to the *jushin* collectively. Instead, they were individually summoned to the Palace at different times, ostensibly for the purpose of paying their normal mid-winter respects to the Throne.[14]

---

10 The Supreme War Direction Council, formed on 5 August 1944, consisted of six members: the Premier, the Foreign, Navy, and War Ministers, and the Chiefs of the Army and Navy General Staffs.

11 Shigemitsu, in line with Army and Navy policy, inserted in the basic declaration a clause which provided that "in the event Germany collapses or makes a separate peace, Japan will attempt to improve her world position through utilizing the Soviet Union." According to post-war statements by Shigemitsu, his purpose in inserting this proviso was "to obtain a certain amount of freedom of diplomatic action, which could be gradually developed into a peace move." (Statement by Mamoru Shigemitsu, Foreign Minister, Apr 43–Apr 45.)

12 (1) Ibid. (2) Affidavit of Koiso, op. cit., IMTFE Record, p. 23,253.

13 Affidavit of Kido, op. cit., IMTFE Record, p. 31,110.

14 "Since it was the common belief that the *jushin* were advocates of peace (Konoye, Okada, Wakatsuki and Yonai), it was almost certain that, if an audience were granted, they would be suspected of advising the Emperor to end the war. Consequently, a very grave problem would arise." (Statement by Marquis Yasumasa Matsudaira, Chief Secretary to the Lord Keeper of the Privy Seal, Jun 36–Nov 45.)

Between 7 and 26 February, the Emperor conferred with seven senior statesmen, including ex-Premier Tojo and former Lord Privy Seal Count Nobuyuki Makino. The senior statesmen were unanimous in sharing the Emperor's deep concern over the trend of the war. Tojo, however, was the most optimistic, while Prince Konoye and Admiral Okada were the most pessimistic. Prince Konoye bluntly stated that Japan faced certain defeat and urged the Emperor to take resolute steps toward peace.[15]

The audiences marked a significant milestone on the road toward peace. For the first time since the outbreak of the war, the Emperor had held a free exchange of views with the majority of the *jushin*. More important, the advocates of peace now had a clear indication that the Emperor himself was strongly sympathetic towards their views. With this encouragement, they quietly waited for a fresh opportunity to advance the cause of peace.

During the next two months the Koiso Cabinet continued to fare badly on both the military and diplomatic fronts. While General MacArthur's forces made rapid headway on Luzon, the Pacific prong of the enemy's offensive swiftly overwhelmed Iwo Jima and then, on 1 April, thrust at Okinawa in the Ryukyus. On the diplomatic front, relations with Soviet Russia tended to worsen rather than improve, and Premier Koiso's hopes of winning China away from the Allies through peace overtures to Chungking remained unrealized.[16]

In spite of serious efforts, the Koiso Cabinet was unable to show any positive achievement. The Cabinet as a whole was unpopular. The war situation failed to improve and the Premier himself was unable to coordinate the political, diplomatic and military fronts. By the latter part of March, he had decided that it was useless to remain in office unless this situation was promptly remedied.

Early in April, the Army proposed to relieve Field Marshal Sugiyama of his post as War Minister and to replace him by General Korechika Anami. Koiso promptly decided upon a final effort to strengthen his position and proposed that, instead of substituting Anami, the Army agree to his own assumption of the war portfolio concurrently with the premiership. On 3 April Koiso intimated his hopes to Sugiyama, but the answer of the Army "Big Three" was a blunt refusal.[17]

Convinced that his position was no longer tenable, Premier Koiso tendered the collective resignation of the Cabinet on 5 April. Even as he did so, the complete failure of the diplomatic policies pursued by his Government was being underlined in Moscow by a blunt notification to the Japanese Ambassador, Naotake Sato, of the Soviet Government's intention to abrogate the Soviet-Japanese neutrality pact because it had "lost its significance."[18]

In relinquishing the premiership, Koiso made it plain that, under the circumstances confronting the nation, he considered it absolutely essential to install "an Imperial General Headquarters Cabinet." To all practical intents and purposes, this meant "a Cabinet in which the Premier assumes an influential

15 Affidavit of Kido, op. cit., IMTFE Record, pp. 31,113–14.

16 In March Koiso proposed to launch such overtures through the intermediary of Miao Pin, a Chinese who was believed to have close connections with General Ho Ying-chin, Chiang Kai-shek's minister of war. Koiso invited Miao Pin to Tokyo, but the plan collapsed due to the opposition of the Foreign, War and Navy Ministers, who strongly doubted that Miao Pin was a trustworthy and qualified intermediary. (1) Ibid., pp. 31,115–16. (2) Statement by General Koiso, previously cited.

17 Affidavit of Koiso, op. cit., IMTFE Record, pp. 32,254–6.

18 The five-year neutrality pact had been concluded on 13 April 1941 and was not to expire until April 1946. The Foreign Office revealed the Soviet Government's abrogation notice without comment in an announcement issued at 1400 on 6 April. *Nippon Times*, Tokyo, 7 Apr 45.

position in the Imperial General Headquarters to direct the conduct of the war."[19]

Obviously, however, the idea of installing an "Imperial General Headquarters Cabinet" did not conform to the ideas of the peace proponents among the *jushin*, nor did it conform to the wishes of all elements in the Japanese Army and Navy. Some of the *jushin*, notably Okada and Konoye, had already laid their plans for an attempt to install a cabinet which would offer greater possibilities for the realization of peace. In this delicate undertaking, they had the powerful support of Marquis Kido, the Emperor's closest advisor, who henceforth was to play a leading role in the struggle to attain the peace objective.

Formation of the Suzuki Cabinet

At 1700 on 5 April, less than seven hours after Premier Koiso had formally tendered the Cabinet's resignation to the Emperor, the *jushin* met in conference with Marquis Kido to deliberate on whom they should recommend to the Throne as head of the next Government. Six former Premiers, among them General Tojo, assembled at the Imperial Palace to attend this crucial meeting. The conference also included the president of the Privy Council, Admiral (ret.) Baron Kantaro Suzuki.[20]

The choice before the conference was both grave and delicate. With Okinawa already invaded and German resistance in Europe nearing complete collapse, it was certain that the Allies would soon begin massing their concerted forces for the final assault upon Japan itself. It was imperative in view of the complex and serious military and political problems that faced Japan that the proper man be chosen. Since the collapse of the Koiso Cabinet was foreseen in advance, the leaders of the peace group had ample time to consider all aspects of the question.

By 5 April, when the conference of senior statesmen convened, they had not only reached an understanding among themselves concerning their choice for Premier but they had also obtained assurance of the approval of Kido himself. Their unanimous decision was that only one man filled the exacting requirements essential to the success of their plans. That man was Admiral Suzuki, the 77 year old president of the Privy Council.

Admiral Suzuki's background unquestionably fitted him for the role which the peace group wanted him to play. His active naval career had ended in 1929 when, at the specific request of the Emperor, he had stepped down as Chief of Navy General Staff to become Grand Chamberlain, a post of advisorship to the Throne second only to that of Lord Privy Seal. After seven years in this position, he had barely escaped death by assassins' bullets in the abortive military uprising of 26 February 1936, one of the aims of which was the elimination of moderate influences around the Throne.[21]

---

19 (1) Affidavit of Kido, op. cit., IMTFE Record, pp. 31,118–19. (2) *Kido Nikki* 木戸日記 (Kido Diary) Entry for 4 Apr 45.

20 The six former Premiers attending the conference were: Prince Ayamaro Konoye, Baron Kiichiro Hiranuma, Baron Reijiro Wakatsuki, Admiral Keisuke Okada, Mr. Koki Hirota, and General Hideki Tojo. Affidavit of Kido, op. cit., IMTFE Record, p. 31,122. (Two *jushin* were absent: Admiral Mitsumasa Yonai, who was barred from participating by reason of his tenure of the deputy premiership and naval portfolio in the outgoing Cabinet; and General Hiroyuki Abe, currently serving as Governor-General of Korea.)

21 Suzuki was among those marked for assassination because he was believed to have played an important part in engineering acceptance of the London naval disarmament treaty over the opposition of the Navy General Staff, and also because he opposed military encroachment in political affairs. Admiral Viscount Makoto Saito, then Lord Privy Seal, Viscount Korekiyo Takahashi, then Finance Minister, both liberals, and General Jotaro Watanabe, then Army Inspectorate General of Military Training, were those actually killed. Admiral Okada, then Premier, eacaped assassination only because his brother-in-law was mistaken for him. Suzuki's life was saved by his wife, who shielded his body with her own after he had been wounded, preventing the assassins from completing the job. Both Prince Saionji and Count Makino, the Emperor's most trusted statesmen, were saved thanks to a timely police warning.

The Emperor had then relieved him of his duties as Grand Chamberlain, and Suzuki had subsequently held no post other than membership of the non-political and largely inactive Privy Council. He had risen to its presidency late in 1944.

The important thing about Suzuki's record was that, while it generally stamped him as a man of moderate ideas, it contained nothing which put him in the camp of the peace advocates. Thus, although some extremist elements were bound to regard him with suspicion, they could find few concrete charges to level against him. There was no doubt, moreover, that Suzuki possessed the full confidence and trust of the Emperor.[22] The fact alone would go far to assure his Cabinet of popular support.

In spite of this understanding among the members of the peace group and Marquis Kido, the conference of the senior statesmen on 5 April lasted three hours and was marked by weighty discussion of the problem of peace or war as well as the premiership. At the very outset, General Tojo demanded that the conference first decide on the basic issue of peace or continuation of the war: "The next Cabinet must be the last Cabinet in this war," he said. "There prevails within the country today, two views—one, that we would continue the war to the bitter end, and the other, that we should meekly accept the terms of unconditional surrender and sue for peace as soon as possible. It is necessary that we settle this question first."[23] Baron Wakatsuki did not entirely agree with Tojo: "The Council of Senior Statesmen is consulted only in the selection of the premier-designate. The suggestion brought up by Tojo is beyond the scope of this conference." At the same time, he warned however: "The new Cabinet must win the war. If we should propose peace now, it is clear that our ultimate fate will be unconditional surrender."[24]

Hiranuma on the other hand agreed with Tojo in principle: "The selection of the new Premier had an intimate and important bearing on the question of peace or continued resistance. The new Premier must be a man who will fight to the end. We cannot nominate an end-the-war peace advocate."[25] The peace faction, thus, refused to disclose their hand, either side-stepping the issue on the ground that it lay outside the province of the *jushin*, or unctuously supporting Tojo's contention that the new Cabinet must carry on the war to the bitter end. Suzuki, himself, unaware that he was being considered by the peace group, agreed, before his own name had been placed in nomination, that the next Premier must possess the "will to fight to the last."[26]

Underlying the deliberations of the conference, however, was the well-phrased remark voiced earlier by Admiral Okada, that the next Cabinet might have to "shoulder the nation's destiny to the end."[27]

This remark by a leading member of the *jushin* peace group was a veiled clue to the group's real intent, the installation of a cabinet oriented toward the earliest possible termination of the war. Admiral Okada and his associates

22 Marquis Matsudaira, who had long been a prominent Court official and who served as chief secretary to Lord Privy Seal, Marquis Kido for ten years, describes the relations between the Emperor and Admiral Suzuki as follows: "The Emperor embraced an especially warm feeling for the Admiral......He always felt, 'I can tell Suzuki what I really want to say without hesitation." (Statement by Marquis Matsudaira, previously cited.)

23 Affidavit of Kido, op. cit., IMTFE Record, p. 31,123.

24 Ibid., pp. 31,131–136.

25 Ibid., pp. 31,124–134.

26 Ibid., p. 31,132.

27 Ibid., p. 31,123.

felt that now, at last, they had reasonable hopes of attaining this difficult objective. The main opposition they had to face within the conference of senior statesmen itself was that of General Tojo. Far more important, Army popularity, under the impact of successive military reverses overseas and mounting war suffering at home, had reached its lowest ebb since the outbreak of hostilities. There was consequently less probability that the Army would insist upon installing a candidate of its own for premiership.[28]

Still, the leaders of the peace group realized that they must play their hand with the utmost caution. No one could yet advocate peace openly. Moreover, if the peace group aggressively advocated a Premier who was obviously from their own camp, there was danger that some extremist elements would either block the formation of such a cabinet or sabotage its efforts once it was installed. The problem of the peace advocates, therefore, lay in finding a candidate who would fulfill all the difficult requirements for the premiership. From their own viewpoint, he must be essentially a moderate whom they could influence toward accepting their thesis that the nation's best interest lay in seeking an early peace. On the other hand, he must be a person who would minimize the possibilities of inviting suspicion from the extremist elements; lastly and very important, he must be someone free of political ties or past commitments which would brand him as an opponent of the war.

These three basic requirements were absolutely essential, but there were others almost equally as vital. It was important too that the next Premier possess the stature necessary to restore public confidence in the national leadership and secure united obedience to the decisions of the Government. On this, perhaps, the civilian peace group and the military were in full agreement, with different objectives, however, in mind. It was also of particular importance to the plans of the peace faction that the Premier be someone whose loyalty to the Throne was outstanding and in whom the Emperor reposed implicit trust. The reason for this lay in the fact that the peace group clearly saw the necessity of making the Emperor's will for peace a decisive factor in the realization of their objective.

It was not until the discussions were well along that Baron Hiranuma finally proposed Suzuki. The latter protested that he did not want the nomination, but the peace group unitedly backed Hiranuma's suggestion, as did Marquis Kido. General Tojo, supported only by Hirota, strongly insisted that the fighting of the homeland battle required an Army man as Premier and stated that he considered Marshal Hata the best choice. Unless the senior statesmen took care, he warned, the Army might turn its back upon the new Cabinet.[29]

---

28 At the time of the Koiso cabinet's resignation, considerable sentiment existed among certain elements in the War Ministry and Army High Command in favor of a succeeding government headed by an active Army General, mainly with a view to fighting the decisive homeland battle. The most favored candidate of these elements was General Korechika Anami, Inspectorate General of Army Aviation and former commander of the Second Area Army in Western New Guinea and Halmahera, while some quarters favored Marshal Shunroku Hata, and others General Umezu, Chief of the Army General Staff. Nevertheless, when Marquis Kido, immediately following Koiso's resignation, asked for the views of the War and Navy Ministers and Chiefs of the Army and Navy General Staff regarding the next Cabinet, neither War Minister Sugiyama nor General Umezu positively expressed themselves in favor of a cabinet under Army leadership. (1) Statement by Lt. Col. Masao Inaba, Military Affairs Bureau, War Ministry. (2) Kido Diary, op. cit. Entry for 5 Apr 45, appended pages relative to talks with Marshal Sugiyama and General Umezu.

29 Affidavit of Kido, op. cit. IMTFE Record, pp. 31,138–141.

Tojo's blunt words precipitated the only heated exchange of the three-hour conference. Marquis Kido retorted that, were an Army man to take the premiership in disregard of the strong anti-militarist sentiment in the country, it might be the people that would "turn its back." Admiral Okada, siding with Kido, pointedly asked Tojo what he meant by saying that the Army, whose duty was to defend the nation, might refuse to support a Premier who had duly received the Emperor's mandate.[30]

Despite these arguments the conference had served its purpose. Marquis Kido had heard the views of everyone present, and it was clear that Suzuki was the choice of the majority. The only obstacle lay in persuading Suzuki to accept the nomination.

Marquis Kido convinced Suzuki to become Premier in a private talk with him after the *jushin* meeting. Their discussion was of the highest importance, for in it Kido referred obliquely to the necessity of a "*volte-face* in policy" in view of the grave war situation. Suzuki replied that, if commanded to assume the premiership, he would do his best to fulfill the Emperor's wishes.[31] Kido then submitted his recommendation to the Throne, and at 2200 the Emperor received Admiral Suzuki in audience and gave him the Imperial mandate to form a cabinet. His Majesty seized the occasion to emphasize his deep concern over the trend of the war, but he judiciously refrained from binding the hands of the premier-designate by an explicit request for early peace.[32]

Admiral Suzuki thus embarked on the formation of his Cabinet free from any rigid policy commitment but at the same time well aware of the over-all mission which the Emperor desired the Cabinet to carry out.[33] He now faced the task of obtaining the Army's cooperation. Failure in this would mean the prompt collapse of his efforts to form a government.

As General Tojo's remarks at the *jushin*

---

30 Ibid., p. 31,142.

31 No written record of this conversation exists, and there is some question as to the explicitness of the language used by Marquis Kido to convey to Admiral Suzuki the Emperor's desire for early peace. Replying to a questionnaire submitted by the author in May 1949, Kido stated that he explicitly told Suzuki that it was the Emperor's desire "to have the war terminated during your premiership." However, when interrogated further on this point on 20 December 1949, Kido admitted that he "did not explicitly say that the mission of the next government should be to end the war quickly", but that he used "such words as could be taken that way" and that his impression at the time was that "Suzuki understood what I meant." It appears most probable, therefore, that Kido did not go beyond stressing the Emperor's grave concern over the trend of the war and his desire for a drastic about-face in policy. Post-war statements by Suzuki confirm, however, that he understood from his talks with Kido and the Emperor that the mission of his Cabinet was to be the conclusion of an early peace. (Cf. n. 33) (1) Statements by Marquis Kido, previously cited. (2) Affidavit of Kido, op. cit., IMTFE Record, p. 31,143.

32 According to Marquis Matsudaira, who was at this time Chief Secretary to Marquis Kido, the Emperor refrained from any explicit mention of peace in his initial talk with Admiral Suzuki because he already "knew the general course in which the Admiral's thinking was being directed" and felt that a hasty approach at this stage would be unwise. Matsudaira states that the Emperor told him after the war, "I was aware of Suzuki's mind from the very first. . . .and was convinced that Suzuki understood what I really had in mind. Consequently, I was not in a hurry to communicate to him my desire for peace in express words." (Statement by Marquis Matsudaira, previously cited.)

33 (1) Interrogated by the United States Strategic Bombing Survey in December 1945, Admiral Suzuki stated, "At the time I became Premier, I did not receive any direct order from the Emperor, but I understood clearly from what the Emperor said to me at that time that he was very much concerned over the situation that Japan faced in the war, . . .over the death of civilians due to bombing, the general sickness and great number of civilian casualties, and the great losses in the field of battle. U. S. Strategic Bombing Survey (Pacific), *Interrogations*, No. 531. (2) Statement by Vice Adm. Seizo Sakonji, State Minister without portfolio, Suzuki Cabinet.

conference had forewarned, there was indeed an undercurrent of opposition to Suzuki's appointment among the extremist elements in the military. The rumor was that Suzuki would establish a Japanese "Badoglio Government." Actually, some desired a cabinet headed by General Korechika Anami, Inspectorate General of Army Aviation and ex-commander of the Second Area Army, who had already been slated to succeed Marshal Sugiyama as War Minister just prior to the fall of the Koiso Cabinet. The Cabinet's resignation, however, occurred before the change-over had been effected.

Following established practice, the Army imposed certain conditions for its cooperation in forming a cabinet. The Army's conditions were, in fact, drawn up in the War Ministry's Military Affairs Bureau, some of the members of which shared the extremists' strong suspicion of Suzuki's intentions. The conditions laid down were:[34]

1. *Prosecution of the war to the finish.*
2. *Formation of a cabinet such as to assure the earliest possible unification of the Army and Navy.*
3. *Prompt execution of measures desired by the Army to assure victory in the decisive battle of the homeland.*

All three stipulations had the same essential purpose: to prevent any peace designs on the part of the premier-designate. The outgoing War Minister, Field Marshal Sugiyama, approved the conditions *in toto* as drafted in the Military Affairs Bureau, and presented them to Admiral Suzuki on 6 April, when the latter called to request the Army's cooperation in forming the new cabinet.[35] In this talk with Marshal Sugiyama, Admiral Suzuki agreed to the Army's terms.[36] The premier-designate could not ignore the fact that if he took issue with the Army at this stage it would undoubtedly result in his failure to form a cabinet. Suzuki's sole hope of success lay in accepting the Army's conditions.

As it was previously decided, the Army promptly recommended General Anami as War Minister in the new Cabinet. This choice actually coincided with Admiral Suzuki's own desires. Concurrently with Suzuki's tenure as Grand Chamberlain, General Anami had served as Military Aide-de-Camp to the Emperor, and the two men respected and liked each other. Suzuki felt that he would have less difficulty in cooperating with Anami than with any other nominee the Army might have advanced to fill the vital war portfolio.

The next problem was the choice of a Navy Minister. Admiral Suzuki who was leaning heavily on the advice of Admiral Okada in regard to the formation of the Cabinet,[37] strongly desired Admiral Yonai to retain the naval portfolio. The Navy's choice was Admiral

34 Statement by Lt. Col. Masatoki Shirai, Military Affairs Bureau, War Ministry. (The drafting of the Army's conditions was assigned to Lt. Col. Shirai.)

35 (1) Ibid. (2) War Ministry Radio, Vice-Minister of War to Commanding Generals of all Armies, 6 Apr 45.

36 The Army and Navy required that War and Navy Ministers be chosen from general or flag officers in active service, which enabled them to block the formation of a cabinet by refusing to nominate a minister. Admiral Suzuki had to reckon with the possibility that the Army would resort to the device if he declined to accept the stipulated conditions.

37 According to Admiral Okada's son-in-law, Mr. Hisatsune Sakomizu, Admiral Suzuki contacted Okada by telephone early on 6 April and requested his assistance in selecting the Cabinet. Mr. Sakomizu, who himself was given the post of Chief Cabinet Secretary on Admiral Okada's recommendation, saw a significant indication of Admiral Suzuki's intentions in the fact that he relied almost exclusively on Okada's advice rather than on that of Baron Hiranuma, also a member of the *jushin* group which secured Suzuki's nomination, but a man with far less pronounced peace convictions than Okada. Only one person, Kozo Ota, recommended by Hiranuma, was given a cabinet post as Education Minister. (Statement by Mr. Hisatsune Sakomizu, Chief Cabinet Secretary, Suzuki Cabinet.)

Yonai, though Yonai himself wished to stay out of the Cabinet. Admiral Suzuki, however, wanted to sound out the Army's attitude regarding Yonai in the light of the second of the three conditions already accepted by him since Yonai was strongly opposed to the Army-Navy unification plan. There was also the Army's attitude regarding the reappointment of Yonai, who had been the Deputy Premier concurrently holding the position of Navy Minister in the out-going Koiso Cabinet. When the Army's opposition to Yonai did not appear so inflexible as to foreshadow outright rejection of his appointment, Suzuki decided to carry it out.[38] Yonai himself was persuaded to accept despite his earlier intention to remain out of the Cabinet.[39]

The most vital post still remaining to be filled was that of Foreign Minister. Marquis Kido had urged Admiral Suzuki on the night of 5 April to reinstate the out-going Foreign Minister, Mr. Shigemitsu.[40] Suzuki, however, received contrary advice from the retiring Premier, General Koiso, and therefore decided to offer the post to Mr. Shigenori Togo, who had been Foreign Minister at the outbreak of war but had resigned in September 1942 as a result of a disagreement with General Tojo's policies regarding the establishment of the Greater East Asia Ministry. Suzuki's decision won the concurrence of Admiral Okada and Marquis Kido, who agreed that Togo could be counted upon to work for the early termination of the war.[41]

Meanwhile, the composition of the rest of the Cabinet had been determined.[42] Rather than await the acceptance of the Foreign Ministry by Togo, who was then residing in Karuizawa, Admiral Suzuki decided that the new government should assume office immediately. The installation ceremony was therefore held at the Imperial Palace on 7 April, the Premier temporarily assuming the foreign affairs portfolio himself. The officially-controlled press

---

38 Mr. Sakomizu, already picked by Suzuki as Chief Cabinet Secretary, conferred with Lt. Gen. Masao Yoshizumi, Chief of the Military Affairs Bureau of the War Ministry, regarding the Army's attitude toward Yonai. Lt. Gen. Yoshizumi informed Sakomizu that the Army considered Yonai unsuitable for several reasons but did not go so far as to reject him. (Statement by Lt. Gen. Masao Yoshizumi, Chief, Military Affairs Bureau, War Ministry.)

39 Admiral Yonai felt that it was improper to enter the new Cabinet since he had been Deputy Premier of the preceding Government and shared responsibility for its failure. He wished to be succeeded as Navy Minister by Admiral Shigeyoshi Inouye, then Vice-Minister and a strong liberal, but the latter declined, and Yonai instead recommended Admiral Kiyoshi Hasegawa. Kido Diary, op. cit. (Entry for 5 Apr 45, appended pages covering talk with Admiral Yonai regarding next cabinet.)

40 Kido made this recommendation because he already knew Shigemitsu's views and felt that "his retention would facilitate the early realization of peace." Suzuki at this time expressed concurrence. (Statement by Marquis Kido, previously cited.)

41 (1) Affidavit of Admiral Baron Kantaro Suzuki. IMTFE Record, pp. 35,590–91. (2) Statement by Marquis Kido, previously cited.

42 The Cabinet list covering posts other than the top four already mentioned was as follows: Home Affairs, Genki Abe; Finance, Toyosaku Hirose; Munitions, Admiral Teijiro Toyoda; Agriculture and Commerce, Tadaatsu Ishiguro; Transportation, Naoto Kohiyama; Education, Kozo Ota; Justice, Hiromasa Matsuzaka; Welfare, Tadahiko Okada. Also included in the Cabinet were four State Ministers without portfolio: Vice Adm. Seizo Sakonji, Lt. Gen. Toji Yasui, Hyogoro Sakurai, and Hiroshi (Kainan) Shimomura (the last serving concurrently as President of the Cabinet Information Board). Three others who were not Cabinet members but sat in on its discussions were: Lt. Gen. Tsukizo Akinaga, Chief of the Cabinet Coordination and Planning Bureau; Mr. Hisatsune Sakomizu, Chief Cabinet Secretary; and Naokai Murase, Chief of the Legal Bureau.

promptly hailed the new cabinet as one for the vigorous prosecution of the war,[43] and Premier Suzuki himself furthered this impression by a strongly-worded radio broadcast to the nation on the night of the 8th.[44]

The Cabinet was now safely installed, but the vital question of who would become Foreign Minister remained undecided. Premier Suzuki conferred with Togo immediately upon the latter's arrival in Tokyo on 8 April and urged him to take the post. This meeting confirmed that Togo was strongly for ending the war. In fact, as a condition for entering the Cabinet, he demanded such clear-cut peace commitments on the part of the Premier that Suzuki, still uncertain in his own mind as to how soon peace should be attempted, was unable to satisfy Togo completely. The talk therefore ended without an agreement.[45]

This unexpected hitch spurred the peace forces into action. Chief Cabinet Secretary Sakomizu called on Togo the following day and assured him that the Premier's goal was an early peace, although his position made it impossible for him to give a clear-cut commitment. Marquis Kido, Admiral Okada, Marquis Matsudaira and others also appealed to Togo to accept. They emphasized that Suzuki's unduly optimistic estimate of how much longer Japan could continue fighting—the only point of disagreement—was not inflexible, and that it was essential for Togo to enter the Cabinet in order to influence the Premier. Yielding to these concerted pleas, Togo again met Suzuki on 9 April and agreed to join the Government.[46]

The completion of the Cabinet with Togo's installation as Foreign Minister marked a signal achievement for the advocates of peace. Of the four principal Cabinet figures, two—Yonai and Togo—were definitely in the peace camp. Premier Suzuki himself was basically convinced that the war must be ended and knew that this was the Emperor's desire, although it was evident that he did not, at this stage, grasp the full gravity of the war situation and the consequent urgency of initiating peace without delay.

Still, the formation of the Cabinet was only the beginning, and the most difficult part of the road remained to be travelled. To win and retain the Army's support, thus keeping his Cabinet in being, Premier Suzuki had pledged his Government to a policy of fighting the war to the finish and realized further that he would

---

43 The *Mainichi* dubbed the new Cabinet as a "Cabinet to repulse the enemy" and saw its primary mission in the "consolidation of the fighting strength of the entire nation." Although this was the general theme, the more liberal *Asahi* cautiously referred to the importance of diplomacy as the "other half of war policy" and hinted that the new Cabinet might be obliged to give its diplomacy "a drastic turn in a new direction." (1) *Mainichi Shimbun* 毎日新聞 (Mainichi Newspaper) Tokyo, 8 and 9 Apr 45. (2) *Asahi Shimbun* 朝日新聞 (Asahi Newspaper) Tokyo, 7 and 11 Apr 45.

44 In this broadcast, Suzuki exhorted the nation "to march on over my dead body when I die for my country." Raikichi Kono, *Shusen no Hyojo: Suzuki Kantaro Jutsu* 終戦の表情： 鈴木貫太郎述 (Features of the Termination of the War; as told by Kantaro Suzuki) Rodo Bunka Sha, Tokyo, 1946, pp. 10–11.

45 Togo told Marquis Matsudaira, Kido's chief secretary, following this meeting that he (Togo) and Premier Suzuki were agreed in principle regarding the necessity of ending the war. Whereas, Togo, however, held that it must be terminated as soon as possible, Premier Suzuki had expressed the opinion that Japan might be able to continue fighting for "two or three years more, depending on the circumstances". In view of this difference of opinion, Togo told Matsudaira, he felt that he and Premier Suzuki would have difficulty in agreeing on high policy regarding the war, and he therefore did not wish to accept the Foreign Ministry until he was certain of Suzuki's attitude. Affidavit of Marquis Matsudaira, previously cited. IMTFE Record, p. 35,596.

46 (1) Affidavit of Hisatsune Sakomizu. IMTFE Record, pp. 35,604–5. (2) Affidavit of Marquis Matsudaira, op. cit., IMTFE Record, pp. 35,596–7. (3) Affidavit of Shigenori Togo. IMTFE Record, pp. 35,779–80.

have to adhere faithfully to this policy until a more auspicious moment when the necessity for peace was generally recognized.[47]

Premier Suzuki had now entered upon a delicate and complex task—one fraught with danger, personal as well as national. He realized not only that he risked his own life, but that any mis-step or ill-considered action might cause the newly formed Cabinet to collapse, thus not fulfilling the mission entrusted to it by the Emperor.[48]

Initial Peace Discussions

In view of these highly complex circumstances, it was natural that the Suzuki Cabinet, during the first month of its existence, showed no visible sign of harboring peace intentions. Indeed, Premier Suzuki and even the more strongly peace-minded of his ministers fully realized that it was essential to make the strongest possible defense of Okinawa and to intensify homeland battle preparations to the maximum in order to place Japan in a better position for an eventual peace move.[49]

The Government therefore concentrated its initial attention on moves to expand war production and ameliorate the increasingly serious situations in regard to food and transportation. At the same time, acting through the Supreme War Direction Council, Premier Suzuki requested a comprehensive survey by the Council secretariat of all aspects of the national fighting strength.[50] This survey was of vital importance, for it was intended to provide a sound basis for judgement as to whether and how long Japan could continue the war.

Meanwhile, as the rapid convergence of Allied armies on Berlin heralded the imminent end of hostilities in Europe, the Army High Command became more alarmed over the possibility that Soviet Russia might forsake

---

47 The hostile attitude of some of the extremist elements toward the Suzuki Government was not entirely allayed by Suzuki's acceptance of the Army's policy conditions. One evidence of this was the fact that Col. Kenji Ogoshi, Chief of the General Affairs Section of *Kempeitai* Headquarters, called on the Premier a week after his appointment and warned him that there were elements who suspected that he was "conspiring for surrender". To this Suzuki cryptically replied that his greatest concern was "to remain loyal to the Emperor." *Jo-o* 女王 (Queen) Magazine, Tokyo, Sep 49: Fall of the Japanese Army.

48 In a statement made to American interrogators after the surrender, Admiral Suzuki described his difficult position upon becoming Premier as follows: "On the one hand, I had to carry out to the best of my ability the mission given me by the Emperor, to arrange for the conclusion of the war, whereas if anyone heard of this, I naturally would have been attacked and probably killed by people opposed to such a policy." USSBS Interrogation No. 531, op. cit.

49 In talks with the War and Navy Ministers and top leaders of the Army and Navy General Staffs soon after the formation of the Suzuki Cabinet, Foreign Minister Togo strongly emphasized the necessity of expending the greatest possible strength in order to win the Okinawa battle. Victory, he pointed out, would afford Japanese diplomacy a "basis for positive action," whereas defeat would virtually rule out such action. These talks, referred to elsewhere, had to do with proposed moves to assure continued Soviet neutrality toward Japan and did not go so far as to touch on diplomatic action for peace, but Togo at least had both objectives in mind. Togo Memoirs (Unpublished Memoirs of Foreign Minister Shigenori Togo, covering the period Apr–Aug 45, pp. 7–8.)

50 The secretariat of the Supreme War Direction Council consisted of four secretaries and seven assistant secretaries. The secretaries were: Chief Cabinet Secretary Sakomizu; Lt. Gen. Masao Yoshizumi, Chief of the Military Affairs Bureau of the War Ministry; Vice Adm. Zenshiro Hoshina, Chief of the Naval Affairs Bureau of the Navy Ministry; and Lt. Gen. Tsukizo Akinaga, Chief of the Cabinet Coordination and Planning Bureau. The assistant secretaries included subordinate officials of the Military and Naval Affairs Bureaus and Cabinet Coordination and Planning Bureau; of the Army and Navy General Staffs, and of the Foreign Office and Greater East Asia Ministry. The assistant secretaries usually conducted such surveys. (Statement by Chief Cabinet Secretary Sakomizu, previously cited.)

neutrality toward Japan and join the Allies as a belligerent in the Far East.[51] Since such action by Russia would seriously prejudice hopes of successfully defending the homeland against invasion, Lt. Gen. Torashiro Kawabe, Deputy-Chief of the Army General Staff, when he made a courtesy call on the new Foreign Minister in the latter part of April, expressed his views and urged Togo to launch diplomatic efforts designed to keep the Soviet Union out of the Far Eastern war.[52] Togo, in fact, had already instructed the Japanese Ambassador in Moscow to seek assurances regarding Soviet neutrality. To this *démarché* the Soviet Government, on 27 April, replied evasively that its attitude remained unchanged. Togo held that it was futile to press for a more concrete pledge unelss Japan's bargaining position was improved by a favorable turn in the Okinawa battle and expressed this opinion to the leaders of the High Command. However, since the High Command maintained that there was still ground for optimism regarding the Okinawa fighting, action was deferred pending further developments.[53] Almost simultaneously, the unconditional surrender of Germany on 8 May brought an end to the war in Europe. These developments, coupled with the mounting devastation of the homeland by enemy air attacks, convinced Foreign Minister Togo that even a temporary brightening of the military picture could no longer be expected. It was therefore imperative, in his opinion, to launch peace efforts without delay regardless of the obvious weakness of Japan's diplomatic position. Waiting, Togo felt, would only weaken that position further through the gradual exhaustion of Japan's fighting strength, thus jeopardizing the chances of obtaining a negotiated peace short of unconditional surrender.

The surrender of Germany, in fact, removed an important obstacle to broaching the question of peace, for Japan was thereby liberated from her treaty obligation not to come to separate terms with the Allies.[54] Reporting to the Emperor on the Nazi collapse, Foreign Minister Togo voiced the opinion that it was now necessary to consider the future policy of the Government in view of the unfavorable developments in Europe. The Emperor still refrained on this occasion from making any strong expression of his views but nevertheless made it clear that he favored an early peace.[55]

---

51 This fear had been sharply increased by the Soviet Government's notification on 5 April that it did not intend to renew the five-year Soviet-Japanese neutrality pact. In addition, there were ominous reports that Soviet troops were being shifted back from Europe to the Soviet-Manchurian border.

52 (1) Togo Memoirs, op. cit., p. 4. (2) This matter was first broached to Foreign Minister Togo by Lt. Gen. Kawabe, Deputy-Chiet of the Army General Staff, on 22 April. Coincidentally, this was the same day that Soviet troops entered Berlin. Statement by Lt. Gen. Torashiro Kawabe, Deputy-Chief, Army General Staff.

53 Togo Memoirs, op, cit., pp. 4 and 8.

54 In his affidavit to the International Military Tribunal for the Far East, Marquis Kido stressed the shackling effect of Japan's treaty commitments to Germany in the following words: "Up to Germany's surrender, it was thought that it would be open to criticism from the viewpoint of international faith to make peace proposals on the part of the Emperor even if the time had virtually come for the purpose.... Now that Germany was gone, I believed that it would be desirable for the Emperor to make up his mind at this juncture, as peace would be restored to the world according to Japan's attitude." Affidavit of Kido, op. cit., IMTFE Record, pp. 31, 143–4.

55 In his report to the Emperor, Foreign Minister Togo analyzed the causes of Germany's defeat, attributing major importance to the crushing effect of Allied aerial warfare. He pointed out in this connection that enemy air attacks on Japan were steadily mounting in severity, and that as a result production had slumped so alarmingly as to render continuation of the war a virtual impossibility. Future policy, he said, would have to be formulated in the light of these conditions. (1) Statement by Foreign Minister Togo, previously cited. (2) Togo Memoirs, op. cit., p. 6. (3) Affidavit of Togo, op. cit., IMTFE Record, pp. 35, 781–2.

Introduction of the volatile issue of peace into top-level discussions between the Government and the High Command, however, was a delicate problem. Foreign Minister Togo decided to take advantage of the military's wishes for a diplomatic move toward the Soviet Union He actually believed that little chance existed of obtaining from Russia what the military desired. However, he hoped that, if the leaders of the Government and High Command were brought together for highly confidential talks on this topic, it might be possible to swing the discussions around to the fundamental issue of terminating the war.[56]

Togo therefore proposed to General Umezu, Chief of the Army General Staff, that the Supreme War Direction Council meet for a comprehensive discussion of policy toward Soviet Russia and Japan's over-all war situation. Since full meetings of the council, including the secretaries, would make it impossible "to get down to fundamentals" because of the fear of leakage, he urged that the sessions be restricted to the six regular members only: the Premier, Foreign Minister, War and Navy Ministers, and Chiefs of the Army and Navy General Staffs.[57]

General Umezu fully concurred in these proposals and agreed to take them up with War Minister Anami, whose consent he successfully obtained. Togo meanwhile explained his ideas to Premier Suzuki, Suzuki gave his approval, and with the final concurrence of the Navy Minister, Admiral Yonai, and the Chief of Navy General Staff, Admiral Koshiro Oikawa, the way was paved for the Council to open its discussions.[58]

The six members convened for their first meeting on 11 May, continuing the discussions at further meetings on the 12th and 14th. The secrecy already obtained by the exclusion of the secretaries was further strengthened by an agreement among the members not to disclose the substance of the meetings even to their immediate subordinates.[59] Thus was created the necessary atmosphere of confidence to enable the members to speak their real minds on the subject of peace without fear of untoward consequences.

At the initial meeting, the six leaders immediately grappled with the issue of moves toward Soviet Russia. Pointing to evidence of large-scale Soviet troop tranfers from Europe to the Far East, General Umezu declared that it was vitally necessary to take diplomatic steps with a view to preventing Russia's entry into

56 Togo was encouraged to adopt this course of action by his impression, based on confidential talks with War Minister Anami and General Umezu, Chief of the Army General Staff, that neither of these two top Army leaders were personally opposed to the principle of terminating the war. From the viewpoint of obtaining peace conditions which the Army could accept, they insisted that a big military victory must precede any approach to the Allies. In an exchange of views which Togo had with Anami toward the end of April, Anami expressed general agreement that the situation was serious and indicated that he favored continued efforts to reach a separate peace with China as a prelude to a general peace attempt. (1) Statement by Foreign Minister Togo, previously cited. (2) Togo Memoirs, op. cit., p. 14.

57 Togo was convinced that only by restricting the meetings of the Supreme War Direction Council in this manner could an opportunity be created for the leaders of the Government, on the one hand, and the High Command, on the other, to engage in frank discussions on the question of terminating the war. (It is highly significant that, subsequent to the May meetings on the Soviet problem, the principle of restricted discussions was followed up to the final surrender decision, with the noteworthy exception of the Council meeting on 6 June, dealt within the following section of this chapter.) (1) Statement by Foreign Minister Togo, previously cited. (2) Togo Memoirs, op. cit., p. 6.

58 Togo Memoirs, op. cit., p. 6.

59 According to Togo, this agreement was adopted at the insistence of one of the military members of the Council and was strongly supported by the Foreign Minister himself. His recollection is that it was Admiral Koshiro Oikawa, Chief of Navy General Staff, who insisted that the agreement be adopted. (Statement by Foreign Minister Togo, previously cited.)

the war against Japan. This was unanimously agreed to by the other members, and this discussion turned to a proposal by Navy Minister Yonai that negotiations with the Soviet Union should also aim at obtaining positive Russian aid in the form of war materials, supplies, and especially oil.

The latter proposal precipitated a clash between Yonai and Foreign Minister Togo. The former maintained that it was not too late for such an effort, whereas Togo flatly stated that it would be " hoping against hope " to think that Russia could be won over to Japan's side in view of the over-all military situation and the probable strengthening of Allied-Soviet ties at the Yalta conference. In fact, declared the Foreign Minister, the deterioration of Japan's position was such that earnest consideration must be given to steps for the termination of the war itself.

Premier Suzuki intervened at this point to advance a new proposal. He acknowledged that Togo was possibly correct in his estimate that it was too late to win positive Soviet friendship, but he maintained that it would be unwise not to try to utilize Russia in some way advantageous to Japan. Since there was vital need of finding a way to terminate the war, he proposed that the negotiations with the Soviet Union encompass the additional objective of securing Soviet mediation for a general peace settlement with the Allies.[60] No opposition to the principle of Soviet mediation was voiced by the military members at this time.

Near the close of the three-day meetings, however, after general agreement had been reached on the concessions which Japan would offer as a price for renewed Russian friendship,[61] Foreign Minister Togo again raised the mediation question in the hope of obtaining a more concrete understanding. He pointed out that negotiations on this matter would be expedited if an agreement were reached immediately regarding the general peace terms which Japan would be willing to accept.

War Minister Anami promptly stated that the conditions of peace must be fixed on the basis of the existing war situation. Japan, he insisted, was not yet defeated and still held enemy territory of far larger area than the Japanese territory occupied by enemy forces. Togo retorted that such an attitude on Japan's part would mean the prompt collapse of any mediation effort. The peace conditions, he stated, must take into account not only the existing status but the probable future trend of the war, as far as it could be reasonably foreseen.

It was readily apparent that there was no hope of agreement on peace terms and that forcing the issue to a showdown at this stage

60 Suzuki, in advancing this proposal, expressed the opinion that Stalin was a man who could be trusted. Togo took a contrary view, arguing that Russia's natural desire must be to reduce Japan to a position of weakness, and that only by extensive concessions could she be induced to modify her attitude. In fact, Togo's outlook regarding the use of Russia as a peace intermediary was extremely skeptical, but it was generally recognized at this time that the Soviet Union was the only neutral power which, if it could be persuaded to do so, might influence the United States and Britain to relax their insistence upon Japan's unconditional surrender. Ibid.

61 It was agreed that Japan must be prepared to scrap the Portsmouth Peace Treaty and other basic treaties with Russia, and to make the following concessions: 1. Recession of southern Karafuto (Sakhalin); 2. Relinquishment of fishing rights in Soviet waters; 3. Freedom of the Tsugaru Strait; 4. Transfer of North Manchurian railways; 5. Recognition of Soviet sphere of influence in Inner Mongolia; 6. Lease to Russia of Port Arthur and Dairen; 7. If necessary, cession to Russia of the northern half of the Chishima (Kurile) Islands. However, Korea was to remain Japanese, and a maximum degree of independence was to be sought for Manchuria by means of establishing South Manchuria as a neutral zone and other means. Official *Précis* Agreements reached by Supreme War Direction Council, 11, 12 and 14 May 45.

might compromise future discussions. Navy Minister Yonai therefore proposed that further consideration of the mediation proposal be deferred, and the Council ended its deliberations.[62] An official *précis* drawn up by Foreign Minister Togo after the final meeting on 14 May stated, in part, that the Supreme Council had agreed on "the speedy opening of talks with Soviet Russia" for the purpose of achieving the following three objectives:[63]

1. Prevention of Soviet entry into the war against Japan;

2. If possible induce the Soviet Government to observe a benevolent neutrality toward Japan; and

3. Eventually obtain Soviet mediation of a nature advantageous to Japan for the purpose of ending the war.

In order to clear the decks for this vital diplomatic effort, the Cabinet took formal action on 15 May to denounce all treaties with Germany and Italy, including the Three-Power Anti-Comintern Pact of 1937, which Russia had always regarded as directed against her.[64] Togo further decided that it would be advisable, before launching formal negotiations, to sound out the Soviet attitude through unofficial preliminary talks. This mission was entrusted in the latter part of May to Mr. Koki Hirota, a senior statesman, who had at various times served as Premier, Foreign Minister, and Ambassador to Moscow.[65]

Owing to difficult wartime conditions and the evacuation of the foreign diplomatic missions from Tokyo, Hirota was unable to meet the Soviet Ambassador, Mr. Jacob A. Malik, until 3 June. Conversations took place on both 3 and 4 June at the mountain hotspring resort of Gora. Hirota intimated that the Japanese Government desired a long-term agreement with the Soviet Union stabilizing friendly relations between the two countries and asked Malik for his views on how this might be accomplished. The Soviet envoy cautiously replied that it would require time to consider the Japanese proposal, and the talks were therefore discontinued for the time being.[66]

Since the Government's policy had been decided, at least in principle, in favor of seeking Soviet mediation, Foreign Minister Togo, during the latter part of May, abandoned a tentative move initiated by the preceding Foreign Minister, Mr. Shigemitsu, looking toward the possible utilization of Sweden as a peace intermediary. The Japanese Minister at Stockholm, Mr. Suemasa Okamoto, was instructed that no further efforts should be made along this line.[67]

---

62 Account of the discussions which took place at the Supreme War Direction Council meetings on 11, 12 and 14 May is based on the following sources in the absence of any official minutes: Statements by Admiral Koshiro Oikawa, Chief of Navy General Staff, and Foreign Minister Togo, previously cited. (Togo and Oikawa are, at the date of this writing, the only surviving participants in these discussions.)

63 Immediately after the three-day meeting Togo drew up a *précis* and it was initialled by the participants. This original document was destroyed when the Foreign Ministry was bombed on 25 May, but Togo made a replica which was initialled, this time, only by Premier Suzuki. This replica was given to Admiral Soemu Toyoda when he succeeded Admiral Oikawa as Chief of Navy General Staff, and this is the only copy of the *précis* remaining. (1) Official *Précis*, op. cit. (2) Statement by Foreign Minister Togo, previously cited.

64 Foreign Minister Togo hoped that this action would help toward "removing, if possible, the distrust and ill-feeling existing between Russia and Japan." Togo Memoirs, op. cit., p. 6.

65 The object of the preliminary talks was mainly to sound out how far the aims agreed upon could be attempted and what would be the best way to persuade the Soviet Government. Ibid., p. 7.

66 Foreign Office Radios: No. 843, Foreign Minister to Japanese Ambassador at Moscow, 28 Jun 45.

67 During his tenure of office under the Koiso Cabinet, Foreign Minister Shigemitsu had held private talks with Mr. Widar Bagge, Swedish Minister to Japan, regarding possible Swedish mediation between Japan and the Allies. Togo was not informed of these talks when he became Foreign Minister, but on 11 April Bagge, then about to return to

Parallel with these developments, the Navy learned around the middle of May of an American move to ascertain Japanese peace intentions. Comdr. Yoshikazu Fujimura, special naval attaché at Berne, Switzerland, reported to the Navy General Staff and the Navy Ministry that he had been approached by Mr. Allen O. Dulles, a high-ranking American special agent in Europe. Dulles, Fujimura reported, proposed the dispatch of a Japanese naval representative of admiral's rank to enter into such talks.

Fujimura's messages, for the time being kept strictly secret within the Navy, met with a cautious reception. There was considerable suspicion of the sincerity of the American feeler, but despite this some high officials in both the Navy Ministry and the Navy General Staff were inclined to favor further exploration of the Dulles offer in the hope, at least, of ascertaining the American terms. However, Admiral Soemu Toyoda and Vice Adm. Takijiro Onishi, who took over as Chief and Vice-Chief, respectively, of the Navy General Staff in the latter part of May, both opposed any follow-up action on the ground that the move was nothing but a stratagem to weaken Japan's fighting unity. Consequently, by early June, nothing had yet been done even to bring up the Dulles offer for serious discussion between the Government and the High Command.[68]

Basic Policy of 8 June

While these highly tentative and cautious approaches were being made to the problem of peace, the military picture continued to darken throughout May. In the Philippines, Fourteenth Area Army resistance to General MacArthur's forces was rapidly disintegrating. On Okinawa, the ground battle had already become hopeless by early May, and thereafter the Japanese defense became principally an effort by the air forces to hamper and delay consolidation of the enemy's hold on this strategic offensive base for a thrust toward the homeland. The enemy air offensive against the home islands meanwhile continued to mount in severity.

The clear portent of these developments was that the margin of time before Japan Proper might be subjected to enemy assault was fast decreasing. Moreover, in view of the doubtful prospects for Soviet mediation, the absence of any official change in the enemy's stand on unconditional surrender, and the military's evident reluctance to accept unconditional

Stockholm, approached Togo through a third person to ascertain whether Togo also was agreeable to Swedish mediation. Togo expressed interest but was unable to pursue the conversations due to Bagge's departure. Back in Stockholm, Bagge contacted the Japanese Minister, Mr. Okamoto, sometime in May in order to discuss the matter further. Okamoto requested instructions from Tokyo and was notified that he should take no further action for the time being. According to postwar statements by Togo, he thought at first that Bagge had in mind an offer of mediation by the Swedish Government on its own initiative. However, when Okamoto's telegrams later clarified that Sweden would only act as a transmitting agent for Japanese peace proposals, Togo decided that such an approach was unlikely to produce anything better than unconditional surrender and that Soviet mediation was the only alternative. (1) Affidavit of Widar Bagge, former Swedish Minister to Japan. IMTFE Record, pp. 34,559–64. (2) Statements by Suemasa Okamoto, Japanese Minister to Sweden and Foreign Minister Togo, previously cited.

68 Sources covering the Dulles-Fujimura negotiations and the Tokyo reaction thereto are as follows: (1) Memorandum notes of Rear Adm. Sokichi Takagi, political adviser to Chief of Navy General Staff and Navy Minister. (2) Statements by Rear Adm. Sadatoshi Tomioka, First Bureau (Operations), Imperial General Headquarters, Navy Section; Admiral Soemu Toyoda, Chief of Navy General Staff; Vice Adm. Zenshiro Hoshina, Chief, Naval Affairs Bureau, Navy Ministry; Capt. Yoshimasa Suezawa, Chief, Naval Affairs Section, Navy Ministry; and Comdr. Yoshikazu Fujimura, Special Naval Attaché, Japanese Legation, Switzerland.

surrender before fighting the homeland battle, the chances were slim indeed that invasion might be averted by an intervening peace settlement. The Suzuki Cabinet, therefore, even while it strove to find a path toward peace, necessarily had to proceed until peace prospects became more definite, on the assumption that the homeland battle would be fought. This led, early in June, to two developments.

The first of these developments was a decision to convoke the Diet in special session from 9 June for the purpose of voting the Government sweeping emergency powers to prepare for the homeland battle. The decision was taken only after lengthy dispute within the Cabinet. Navy Minister Yonai argued that the Government could act by Imperial Decree without a prior legislative grant of authority and that it was an inopportune moment for convoking the Diet. Premier Suzuki at first inclined toward Yonai's view but finally swung over to the opinion of the majority of the Cabinet, including War Minister Anami and Hiroshi Shimomura, the Chief of the Cabinet Information Board, both of whom maintained that a Diet session was necessary to bolster national morale.[69]

The second development was the adoption by the Supreme War Direction Council on 8 June of a basic war policy which unequivocally called for prosecuting the war to the end. This was only one phase of the policy because the Government and even the top military leaders had already agreed at the secret mid-May conferences to seek eventual Soviet mediation. The one-sided nature of this policy, however, was inevitable in the light of the circumstances which surrounded its origin, drafting, and final adoption by the Council in plenary session.

Earlier, in mid-April, soon after the formation of the Suzuki Cabinet, a basic war policy draft had been prepared within the Army as was the customary practice. The policy draft was intended for submission to the Government in an effort to coordinate the future policy of the Government and the High Command. The draft was completed by Col. Sako Tanemura of the Army General Staff and was approved by Chief and Deputy-Chief of the Army General Staff and by the War Minister.[70]

During the meetings of the assistant secretaries of the Supreme War Direction Council, who had no knowledge that the six Council leaders had already begun secret deliberations on peace, the idea developed and was generally agreed that adoption by the Council of a new basic war policy was necessary to meet the ever-increasing threat of an invasion of the Homeland. The military element in the assistant secretaries' group vigorously pushed this idea and since the discussions were similar in substance to the Army policy draft prepared earlier by Tanemura, the Army draft bearing official approval was submitted to the Cabinet representatives

69 Since the early part of May, Diet leaders had been insistently pressing the Government to convoke a special session. Certain members of the Cabinet felt, in view of these demands, that it would have an injurious effect on public morale if the Government by-passed the Diet and enforced severe emergency war measures by decree. Yonai, on the contrary, felt that a Diet session was highly inopportune since it would oblige the Government to take an outwardly strong stand for continuation of the war, which would interfere with behind-the-scenes efforts to promote an agreement on peace. Marquis Kido likewise feared that the strong war sentiment usually expressed at Diet sessions would "hamper peace moves." (1) Statements by Chief Cabinet Secretary Sakomizu and Vice Adm. Sakonji, both previously cited. (2) Memorandum notes of Rear Adm. Takagi, op. cit. (3) Affidavit of Kido, op. cit., IMTFE Record, p. 31, 146.

70 Col. Tanemura submitted his draft to the Chief and Deputy-Chief of the Army General Staff and to War Minister Anami and obtained their approval in mid-April. He notified the Cabinet representatives of this fact when he transmitted the draft to them in latter part of May. (Statement by Col. Sako Tanemura, Military Affairs Bureau, War Ministry.)

in the Council secretariat in the latter part of May for further deliberations by the Council.[71]

It was a foregone conclusion that a policy declaration emanating from the lower level of the Army leadership would make no mention of efforts for peace. The Army's draft clearly reflected the dominant thinking of the military who were adamantly against acknowledging defeat and demanded the waging of a final all-out battle in defense of the homeland.

The draft was then taken up in preliminary discussions with the Cabinet representatives in the Council secretariat. The atmosphere in these discussions rendered any expression of peace ideas completely out of the question. Nevertheless an attempt was made to modify the terms of the policy to give it an interpretation which eventually might be exploited for peace purposes and also to limit the war objectives to the preservation of the homeland and the national polity. However, the final wording remained extremely warlike.[72] The text of the basic policy draft read as follows:[73]

General Policy:

On the strength of its advantageous geographical position and the undying loyalty and solidarity of its people, (the Empire) will prosecute the war to the end in order to preserve the national polity and protect the Imperial Homeland,[74] thereby securing the foundations for the future development of the race.

Measures:

1. Preparations for battle on the soil of the homeland will speedily be strengthened, and the main fighting strength of the Imperial forces will be concentrated there. In other areas, fighting strength will be disposed according to actual capabilities and with primary emphasis on combating the United States as the principal enemy, taking into account at the same time the possibility of a sudden change in the situation to the north.

2. Changing tendencies in the world situation will be exploited through the pursuit of an active and aggressive diplomacy, especially toward Soviet Russia and China, with a view to furthering the prosecution of the war.

3. On the home front, all necessary preparations to enable the entire nation to fight unitedly

71 The seven assistant secretaries of the Council, of whom four were representative of the War and Navy Ministries and Army and Navy General Staffs, met twice weekly to discuss matters of war policy. It was customary for these officials to exercise a large amount of initiative in proposing and preparing for regular Council meetings, including the drafting of documents and reports to be acted upon. It is not clear whether the Army representatives initiated the proposal for adoption of a new basic war policy in this instance, but they definitely played the major role in drafting the policy itself. (Statements by Col. Tanemura and Capt. Suezawa, both previously cited; Suezawa was Navy Ministry member and Tanemura War Ministry member of the assistant secretaries group.)

72 In the absence of any written records covering this preliminary phase of the discussions, post-war statements by surviving participants are the only available sources of information. These statements are contradictory on many points of fact, but a careful weighing of all the evidence supports the following conclusions, which are essentially stated in the text: 1. Both the wording and interpretation of the basic policy draft were discussed between the military and Cabinet representatives. 2. The Cabinet representatives did not attempt to insert in the draft any explicit peace provision. They did, however, try to soften the wording and to place major emphasis, in the interpretation of the policy, on the limitation of war objectives to "the preservation of the national polity and the protection of the Imperial Homeland." 3. Actually, only minor changes were made in the wording of the Army-proposed draft. (1) Statements by Chief Cabinet Secretary Sakomizu, Capt. Suezawa and Col. Tanemura, all previously cited.

73 *Saiko Senso Shido Kaigi Tsuzuri* 最高戰爭指導會議綴 (Proceedings of the Supreme War Direction Council, Part II: Basic Policy for Future Direction of the War.)

74 Throughout the basic policy draft, the term *Kodo* 皇土 (Imperial domain) was employed in place of *hondo*, 本土 (Homeland) the term customarily used to denote the Japanese Homeland. The former, in its strict and normal meaning, included Korea and Formosa, but as used in the basic policy, it was unquestionably intended to denote only the homeland, since decisive battle operations were planned only for the defense of that area.

in the decisive battle of the homeland will be rushed to completion in thorough conformity with the spirit of a people's war. In particular, the Government will lay stress in all its measures on the replenishment of the nation's material strength, with special attention to securing food supplies and augmenting the production of special weapons, as well as on boosting fighting morale and consolidating national unity, with the formation of people's volunteer units as the backbone of the program.

4. Swift action will be taken by the proper authorities to plan and execute concrete measures based on the present basic policy.

The stage was now set for action. On 6 June, the Supreme War Direction Council convened for a preliminary discussion of the draft, final action on which was to be taken at a formal meeting in the Emperor's presence on 8 June. Before the discussion began, the Council heard two special reports presented by the secretaries, one summarizing the current state of the nation's war strength and the other outlining the international situation. These reports supposedly formed the basis of the policy draft but were strikingly at variance with the decisions contained in it.

The analysis of the national war strength pointed to a sharp reduction of fighting potential due to air raid destruction and the severance of sea communications, particularly emphasizing decreased steel and aircraft production and the near-exhaustion of liquid fuel supplies.[75] In the report on the world situation, these gloomy findings were underlined by the estimate that the Allies would launch a massive assault on the homeland at an early date and that there was probability of Soviet entry into the war against Japan.[76]

Despite this dark picture, Lt. Gen. Torashiro Kawabe, Deputy-Chief of the Army General Staff,[77] strongly emphasized that, in defending the homeland itself, the armed forces would be favored by advantages which they had not enjoyed in previous island campaigns in the Pacific. For the first time, the Army would be able to throw its main strength into decisive battle, aided by short communication lines and the full support of the population. It would take the offensive with determination to fight to the death and with "faith in certain victory." He added, however, that continued Soviet neutrality was absolutely indispensable to enable Japan to withstand invasion.[78]

Admiral Soemu Toyoda, Chief of the Navy General Staff, spoke with less optimism. There was considerable possibility, he said, that the enemy might invade southern Kyushu in July or August, or might strike directly at

75 The conclusions stated in the report on national strength were based on data gathered by the Council secretariat as part of the comprehensive resurvey of the national war potential begun shortly after the formation of the Suzuki Cabinet. This was completed about the middle of May. Proceedings of the Supreme War Direction Council, op. cit. Part II: Present State of the National Strength.

76 The report noted incipient signs of dissension among the Allies but estimated that enemy unity would be maintained unless the war were prolonged. Although Soviet entry into the war against Japan was rated the greater probability, the report pointed to a possibility that Russia might press Japan to make peace with the United States. Ibid., Estimate of the World Situation.

77 Lt. Gen. Kawabe attended the Council meeting in place of General Umezu, who had left Tokyo for an important conference at Dairen with the commanding generals of the Kwantung Army and China Expeditionary Forces. Umezu carried with him an Imperial General Headquarters order calling for a revision of command dispositons on the Continent to meet a possible attack by Soviet Russia. (Statement by Lt. Gen. Kawabe, previously cited.)

78 Lt. Gen. Kawabe's statement on 6 June is summarized from an original document giving the substance of his identical statement made two days later at the Imperial conference which finally adopted the basic policy. Proceedings of the Supreme War Direction Council, op. cit. Part II: Summary of Statement by the Deputy-Chief of Army General Staff.

Honshu as early as September. In either case, the Navy would have difficulty in assembling sufficient strength, and it consequently was improbable that a large proportion of the invading force could be destroyed by attack operations prior to landing. This proportion would increase the later the invasion came, but only if the production of special attack weapons and especially of aviation fuel could be maintained in spite of enemy bombings.[79]

In the light of these High Command statements and the reports already presented to the Council, the ensuing discussion of the policy draft itself centered on two critical questions: first, whether there was hope of keeping Soviet Russia out of the war by diplomatic action; and second, whether it would be possible to keep war production at a sufficient level to assure reasonable prospects of successfully opposing an enemy invasion. On both these points, Foreign Minister Togo expressed serious doubt.[80]

Reminding the Council that in wartime strong diplomacy was contingent upon military victories, Togo declared that the war situation was so highly unfavorable to Japan that no success whatever could be expected in efforts to win positive Soviet friendship. He acknowledged that some hope remained of keeping Russia from actively entering the war against Japan, but he emphasized that even this hope would become tenuous unless Japan's fighting strength could be maintained at a sufficiently high level to convince the Russians that there was only a slight probability of an early and certain Allied victory.

Togo was no less pessimistic with regard to the possibility of keeping the nation's war strength up to the required level. Production, he stated, could hardly be augmented in the face of steadily intensifying enemy air operations against the homeland. On the contrary, there was grave danger that it would decline still further unless the enemy air superiority could be overcome.

Togo's stand, however, failed to gain any support from others. The Premier, in fact, displayed a firm attitude throughout the deliberations[81] and persuaded Togo to agree to the Council's adoption of the basic policy in view of prevailing circumstances. Admiral Yonai for the most part remained silent.[82] On the other hand, Munitions Minister Teijiro Toyoda, who had been specially summoned to attend the meeting, bolstered the case for adoption by stating that it might be possible to

79 Admiral Toyoda's remarks at the 6 June meeting were more pessimistic than his statement at the subsequent Imperial conference, which was redrafted between the two meetings. In the absence of any record of his 6 June statement as actually presented before the Council, its general contents are summarized from the draft materials prepared in the Navy General Staff and submitted to Toyoda to be used as a basis for his statement. Ibid., Summary of statement by the Chief of Navy General Staff.

80 According to statements made by Togo after the war, he had not been notified of the contents of the basic policy draft before attending the 6 June meeting, although it was customary for the Council secretaries to circulate such documents to all members of the Council in advance of the meeting at which they were to be discussed. In this case, it appears that Togo was the only member who had not seen the policy draft in advance. He states: "When the conference got under way, I was surprised to discover that a very strongly-worded resolution to continue the war was being brought up for debate. . . . The whole thing was a surprise to me." (Statement by Foreign Minister Togo, previously cited.)

81 The council deliberated on the question of evacuating the capital near the end of the conference. Premier Suzuki, on this point, strongly contended that the capital must be defended to the last. Proceedings of the Supreme War Direction Council, op. cit. Part II.

82 The draft policy had been submitted to Yonai before the 6 June meeting by Vice Adm. Hoshina, Chief of the Naval Affairs Bureau of the Navy Ministry and one of the secretaries of the Supreme War Direction Council. According to Hoshina, Yonai at this time paid little attention to the draft. (Statement by Vice Adm. Hoshina, previously cited.)

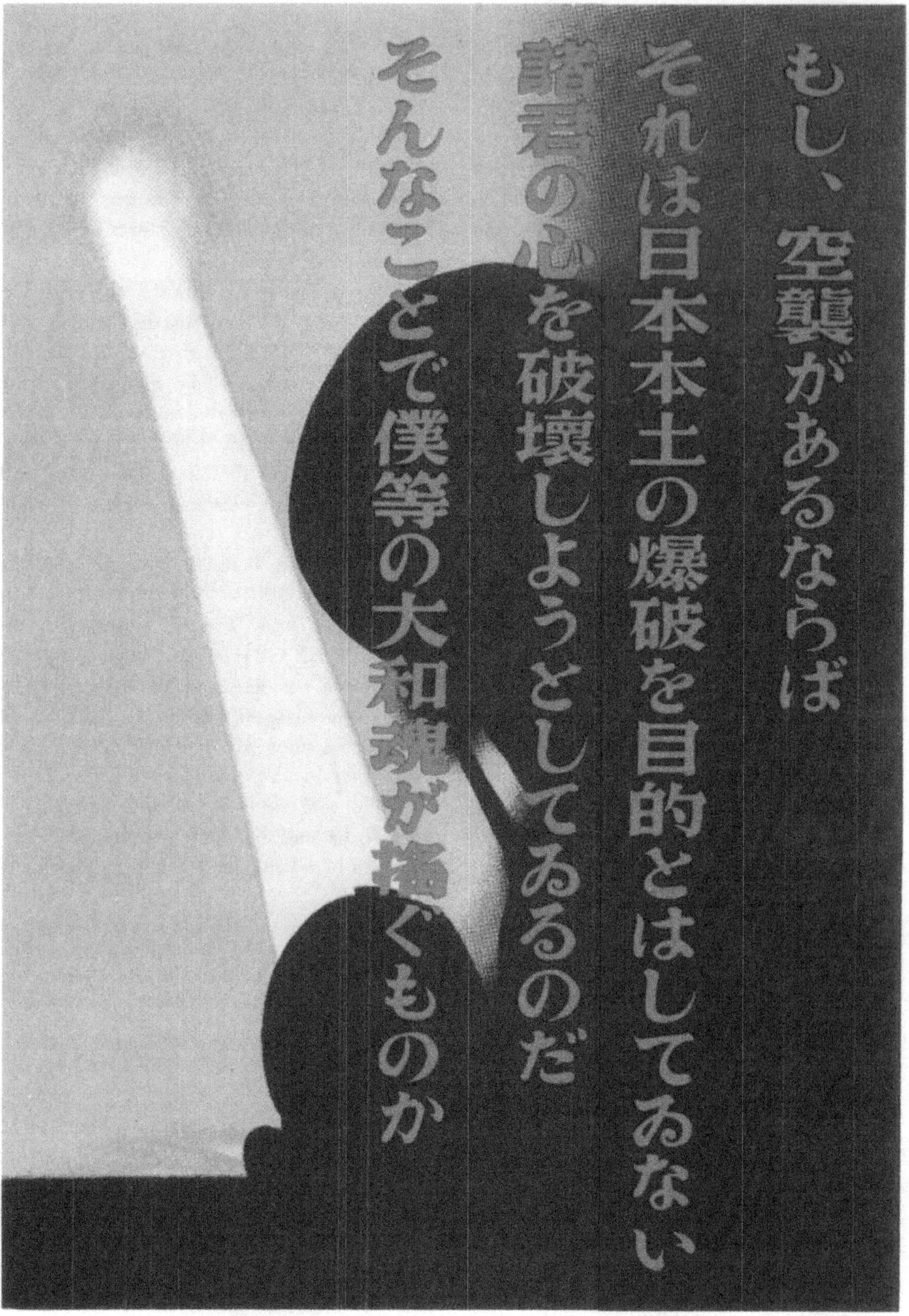

Should there be air raids,
They will not be intended for destroying our Homeland,
But will be aimed to strike at our morale.
Are we to let them demoralize our YAMATO fighting spirit?

PLATE NO. 164
Morale Poster

maintain war production at the required level if determined steps were taken to boost working morale, ensure greater coordination, and especially to strengthen the protection of land and sea transportation against air attack.[83]

No further opinions were put forward in regard to the fundamental issue. The Council, which had been in session the greater part of the day, approved a minor change in the wording of the basic policy and then adopted it unanimously to conclude the deliberations.[84] This decision was ratified by the Cabinet on 7 June, and on the 8th the Council reconvened in the presence of the Emperor to finalize its action.

The proceedings of the Imperial conference were rigidly formal. Whereas there had been some spontaneous discussion on 6 June, the meeting before the Emperor followed a predetermined pattern. Most of the statements were prepared in advance and designed to fit in with the decision which the conference would automatically take to adopt the basic policy.

As on 6 June, the proceedings began with presentation of the reports on the national strength and world situation. The body of these reports remained unchanged, except for a few revisions in the conclusions. The statements of the High Command representatives followed. Lt. Gen. Kawabe, again attending in place of General Umezu, read the same statement which he had made on 6 June. However, the remarks of the Chief of Navy General Staff, Admiral Toyoda, had been redrafted for the Imperial conference and were noticeably more optimistic. Although again emphasizing that the Navy's effectiveness depended upon the maintenance of war production, Toyoda expressed confidence that close to one-half of an enemy invasion force could be destroyed on the water in the event of an early assault on the Homeland, and that the rate of destruction would increase the later the assault came.[85]

Following statements by the Munitions and Agriculture Ministers, Foreign Minister Togo briefly addressed the conference. In contrast to his outspokenly critical attitude at the 6 June meeting, Togo on this occasion read a restrained statement which reiterated the pessimistic outlook for negotiations with Russia but avoided any clear-cut expression of opposition

83 In the absence of detailed official records covering the 6 June meeting of the Supreme War Direction Council, the discussions on the basic policy draft are summarized on the basis of the following sources: (1) Original Notes taken by Admiral Soemu Toyoda, Chief of Navy General Staff, at 6 June meeting. Proceedings of the Supreme War Direction Council, op. cit. Part I: Official Summary of Statement by the Munitions Minister at 8 June Council Meeting; and Part II. (The Munitions Minister's remarks at both 6 and 8 June meetings were substantially the same.) (2) Statements by Foreign Minister Togo, Chief Cabinet Secretary Sakomizu, Lt. Gen. Yoshizumi, Vice Adm. Hoshina, all previously cited.

84 The opening statement of general policy was amended to read: "On the strength of its advantageous geographical position and the undying loyalty and solidarity of its people, (the Empire) will prosecute the war to the end in order to preserve the national polity and protect the Imperial homeland, thereby *accomplishing the objectives of the war.*" The italicized passage was substituted for the words "securing the foundations for the future development of the race." According to the recollection of Foreign Minister Togo, the change was made because the original phrasing was thought susceptable to misinterpretation. (1) Proceedings of the Supreme War Direction Council, op. cit. Part I: Basic Policy for Future Direction of the War. (2) Statement by Foreign Minister Togo, previously cited.

85 As redrafted in the Navy General Staff, Admiral Toyoda's statement contained the estimate that 60 to 70 per cent of an enemy invasion force would be able to effect a landing if the invasion attempt were made at an early date. In delivering the statement at the Imperial conference, however, Toyoda modified this passage to state that close to half of the enemy invasion force could be destroyed before landing. (1) Summary of statement by the Chief of Navy General Staff, Imperial conference, 8 Jun 45, with pencilled corrections made by Toyoda. Proceedings of the Supreme War Direction Council, op. cit. Part II. (2) Statement by Admiral Toyoda, previously cited.

to the basic policy. Baron Hiranuma, President of the Privy Council, then spoke, after which the basic policy was formally read and adopted without further discussion.[86]

Concluding the conference, Premier Suzuki declared:

The Government and the High Command, acting in closest unity, will henceforth strive to carry into effect the terms of the Basic Policy for Future Direction of the War here adopted. However, it is clear from today's discussions that extraordinary effort on both sides is an absolute prerequisite for the effectuation of this policy.

For its part, the High Command is requested to ensure real unity between the Army and Navy so that maximum effectiveness may be realized through fully coordinated military operations. At the same time, on the Government's side, all members of the Cabinet pledge full determination literally to sacrifice their lives for the purpose of carrying out the policy, in particular Items 2 and 3. (Cf. pp. 640–1)

The present situation of the Empire is truly perilous. Indeed, I think it can be said that we as a nation are struggling for life in the very jaws of death. This situation cannot be met by intelligence or ingenuity alone. There is but one course, to drive straight ahead without faltering toward the objectives in which we believe. I hereby affirm the Government's resolve to do this.

Throughout the conference the Emperor had remained silent in strict accordance with constitutional custom. After Premier Suzuki's final words of exhortation, he quietly rose and left the conference chamber to bring the meeting to a close.[87] Those attending the conference, realizing that they were the cause of deep concern on the part of the Emperor, were overwhelmed with awe.

Emperor's Initiative

The basic policy decision evoked its strongest and most significant repercussion in the inner circle of the Court. Just before the Imperial conference met, Marquis Kido, the Emperor's closest advisor, obtained access to the conference documents and decided, in view of the basic policy's unequivocal reiteration of intent to fight on, that the only hope of swinging the High Command and Government toward peace lay in attaining direct Imperial intervention. On 8 June, the same day that the Imperial conference formally adopted the basic policy, Kido drew up for submission to the Emperor a vital plan of action for peace embodying this central idea.

The plan, which Kido cautiously entitled "A Draft Plan of Countermeasures to Meet the Situation", was prefaced by an explanation of the reasons why he considered it imperative to end the war without delay. The Okinawa battle, he pointed out, was almost certain to end soon in complete defeat. Furthermore, the report on the national strength submitted to the Imperial conference in conjunction with the basic policy clearly indicated that it would become virtually impossible in the near future for Japan to carry on the war.

Turning to the question of enemy strategy, Kido expressed fear that a continuation of mass

86 Baron Hiranuma had not participated in the 6 June Council meeting but was invited to attend the 8 June conference in accordance with the customary practice of including the President of the Privy Council in important Imperial conferences. In his statement at the meeting Hiranuma emphasized the importance of the nation's morale and concluded that every effort should be made to bring the war to a successful conclusion banning any idea of peace. The Agriculture and Commerce Minister also was specially included in the Imperial conference in order to make a statement concerning the food situation.

87 The account of the Imperial conference proceedings is based on an official summary, classified "National Top-Secret," which was drawn up for distribution to a highly restricted number of key officials. It includes the principal statements made at the conference but not those of the High Command. Proceedings of the Supreme War Direction Council, op. cit. Part I.

air attacks with incendiary bombs would "make a holocaust" of towns and villages throughout the country. Along with the destruction of their homes, people would lose their food stocks and clothing. Social unrest of alarming proportions would become inevitable with the advent of winter, and there was grave danger that the situation might get out of control. Kido's plan continued:[88]

From the viewpoint set forth above, I consider that it is essential for Japan to make a resolute move to terminate hostilities and restore peace. By what methods and steps, then, should this objective be attained? This problem calls for the most careful consideration.

In the light of various announcements, speeches and articles publicized by the enemy as part of a peace offensive, it is almost certain that the major enemy objective is the overthrow of the so-called **gumbatsu,** *or militarist clique, in this country. Consequently, I believe that the proper way to start peace negotiations would be for the fighting services to propose peace initially, and for the Government then to decide on a peace plan and open negotiations. However, in the light of actual conditions in Japan, such action is virtually impossible at this juncture. Furthermore, if we wait for an opportunity to ripen for such action, it may come too late, and Japan may share the fate of Germany, making it impossible to secure even our minimum demands: the safeguarding of the Imperial family and the preservation of the national polity.*

Although it will be a departure from past precedent and a course which we can contemplate only with awe and trepidation, I believe that there is no course left for us but to petition for Imperial intervention for the sake of the people and take steps to end hostilities and restore peace in accordance with the following policy:

1. Negotiations will be opened with an intermediary power in pursuance of a personal message by the Emperor. If it were possible to do so, it might be wise to open direct peace negotiations with the United States and Britain. However, it seems more advisable, in view of Soviet Russia's neutrality toward Japan, to ask that country to mediate between Japan and the Allies so as to allow some latitude to Japan.

2. The keynote of the Emperor's personal message will follow that of the Imperial Rescript on the declaration of war, emphasizing His Majesty's constant concern for peace and his decision to terminate hostilities on broad terms in view of war ravages and for the sake of world peace, by bearing the unbearable.

3. Definition of peace terms: Honorable peace (it may be inevitable that this will be the minimum term.) In view of the object of Japan's declaration of war, if a guarantee is obtained that the Pacific will be made true to its name, Japan will renounce her right of occupation and claim to leadership over those areas now under her occupation, provided only that the nations and peoples of these areas attain independence. Japan, on her own initiative, will withdraw her armed forces from the occupied areas. (In this case, some Japanese forces may be compelled to abandon their arms on the spot, but this question may be left to future negotiation.)

4. Japan must be prepared to carry out a heavy reduction of armaments, which will be demanded of her. There will be no alternative but to be content with the minimum armament required of national defense.

These are my personal opinions which I have candidly expressed. The plan contains only essential points. It goes without saying that the terms of negotiation and other matters will have to be adjusted on the basis of the views of experts in various quarters.

On 9 June Kido laid his plan before the Emperor. His Majesty had been profoundly distressed over the decision taken by the Imperial conference the preceding day and considered it illogical and inconsistent with the admittedly serious situation of the national strength.[89] After listening to Kido's plan, the Emperor gave his full approval and instructed Kido to enter into immediate consultations with the Premier and the War, Navy, and Foreign

88 Kido Diary, op. cit. Entry for 8 Jun 45, appended pages entitled *Jikyoku Shushu Taisaku Shian* 時局收拾對策試案 (Draft Plan of Countermeasures to Meet the Situation.)

89 Statement by Marquis Matsudaira, previously cited.

Ministers with a view to its realization. The Diet had already begun its special session, however, and Kido was unable to confer with the four Government leaders until its adjournment four days later.[90]

As Navy Minister Yonai had predicted, extreme war sentiment dominated the Diet session. In his opening speech, Premier Suzuki alluded to a statement he had made at San Francisco in 1918 to the effect that the United States and Japan would surely suffer "the punishment of Heaven" if they ever made the Pacific the scene of war.[91] Premier Suzuki's speech immediately led to hostile discussion among various members of the Lower House. Ultra-rightist members acting in collusion with certain Army elements seized upon Premier Suzuki's speech as a pretext for launching an attempt to discredit the Cabinet. The situation quieted after War Minister Anami had warned his subordinates against further involvement in anti-Government activity.[92] After this brief but significant flurry, the Diet passed the emergency war measures and adjourned on 13 June.

While the Diet was in session, the Emperor's concern for an early peace was heightened by two highly authoritative reports on the military situation. General Umezu, Chief of the Army General Staff, returned to Tokyo from his official mission to Dairen immediately after the Imperial conference of 8 June and, on the 9th, made a pessimistic report to the Emperor concerning the military outlook on the continent.[93] Three days later, an equally gloomy report was presented to His Majesty by Admiral Kiyoshi Hasegawa, Special Naval Inspector, whom the Emperor had entrusted with the inspection of special attack forces for the homeland battle.[94]

These reports, which the Emperor communicated in substance to Marquis Kido, spurred the latter to act swiftly on his peace plan as soon as the Diet adjourned on 13 June. Kido the same day had separate conversations with Premier Suzuki and Navy Minister Yonai, to both of whom he disclosed his plan. Yonai expressed full readiness to support the plan on the condition that Suzuki, as head of the Cabinet, would do likewise. The Premier, though asking for a little time to consider, also

90 Affidavit of Kido, op. cit., IMTFE Record, pp. 31,151–2.

91 *Kampo* 官報 (Official Gazette) Special Issue, 9 Jun 45, containing Stenographic Record of Proceedings of the House of Representatives, 87th Diet. (2) Premier Suzuki insisted upon retaining this particular passage in his speech despite the fact that most members of the Cabinet strongly advised him to leave it out. According to State Minister Sakonji, Suzuki wished to hint that "an unreasonable war is not good" and ought to be terminated speedily through the efforts of both sides. (Statement by Vice Adm. Sakonji, previously cited.)

92 In view of widespread rumors that the Army was behind the anti-Government move in the Diet, Chief Cabinet Secretary Sakomizu, after the close of the session, sought and obtained assurances from Lt. Gen. Yoshizumi, Chief of the Military Affairs Bureau of the War Ministry, that the Army intended to continue its support of the Cabinet. (Statement by Lt. Gen. Yoshizumi, previously cited.)

93 General Umezu acknowledged to the Emperor that the combined strength of the Japanese forces in Manchuria was barely adequate to meet an eventual attack on the Continent, and that ammunition reserves would be exhausted in one major action. This revelation, though applicable to the situation on the Continent, had a marked influence on the Emperor's estimate of the prospects for the homeland battle since he was convinced that the troops in Japan Proper were far inferior in training and equipment to those on the Continent. (1) Memorandum notes of Rear Adm. Takagi, op. cit., (Conversation with Marquis Matsudaira, Chief Secretary to the Lord Privy Seal, in which Matsudaira disclosed the nature of Umezu's report to the Emperor and His Majesty's reaction, 14 Jun 45.) (2) Matsudaira learned the details of Umezu's report from Marquis Kido shortly after the Emperor's audience with Umezu on 9 June and later obtained the same information directly from the Emperor. (Statement by Marquis Matsudaira, previously cited.)

94 Statement by Admiral Kiyoshi Hasegawa, Special Naval Inspector, Feb—Aug 45.

expressed general agreement and assured Kido that he would do his best to put the plan into effect.[95]

Kido next outlined his ideas to Foreign Minister Togo in an interview on 15 June. He explained that he had already obtained the Emperor's full approval and asked Togo to draw up without delay a definite and detailed plan. The Foreign Minister replied that he was in general accord with Kido's proposals, pointing out at the same time that efforts were already underway with a view to securing eventual Soviet mediation. Togo was surprised to discover that neither Kido nor the Emperor had been informed of the decisions reached at the six leaders' conferences in mid-May with respect to diplomatic action toward Russia.[96]

The Foreign Minister promptly brought Kido up to date on these developments and informed him that, although the six leaders had agreed to defer mediation until the Soviet attitude had first been sounded out and some agreement had been reached on the Japanese side regarding peace terms, he (Togo) and Navy Minister Yonai both felt that it was now high time to press for action. He told Kido that he and Yonai had agreed, in a talk on 13 June, to take up the matter with Premier Suzuki and War Minister Anami. Togo emphasized, however, that the Government was in an extremely awkward positon because of the one-sided basic policy adopted at the Imperial conference on 8 June. Kido replied that he was fully aware of this and would do what he could to facilitate matters.[97]

Immediately after his talk with Kido, Foreign Minister Togo again conferred with Yonai. The Navy Minister informed Togo that he had spoken to Premier Suzuki regarding steps to press the mediation issue and that the Premier had agreed. Suzuki had confided in him, Yonai told Togo, that he thought the Foreign Minister himself should be sent to Moscow to conduct negotiations with the Soviet Government.[98]

The Army's attitude, however, still remained unknown. War Minister Anami was out of Tokyo on an inspection tour, and it was not until 18 June, following his return, that Marquis Kido was able to sound out his views. The War Minister, after listening to Kido's proposals, agreed in general that the situation was serious but insisted that it was necessary to wage the homeland battle and inflict heavy losses on the enemy before initiating peace action. Kido, on the other hand, said he could see little hope of success in the defense of the homeland and strongly emphasized the Emperor's fear that such a course would finally result in laying waste the entire country and sacrificing the national polity itself. Anami, however, remained non-committal.[99]

Meanwhile, Premier Suzuki had decided to call a restricted meeting of the six Supreme

95 (1) Affidavit of Kido, op. cit., IMTFE Record, pp. 31,153–5. (2) Statement by Marquis Kido, previously cited. (3) Memorandum notes of Rear Adm. Takagi, op. cit., (Conversation with Admiral Yonai regarding consultation with Kido, dated 14 Jun 45.)

96 It had been agreed at the time of the mid-May conferences that Premier Suzuki would report to the Throne on the decisions taken. Asked by Togo following his conversation with Kido on 15 June whether he had made such a report, the Premier admitted that he had neglected to do so. (Statement by Foreign Minister Togo, previously cited.)

97 (1) Affidavit of Kido, op. cit., IMTFE Record, pp. 31,155–6. (2) Statement by Marquis Kido, previously cited. (3) Togo Memoirs, op. cit., p. 9.

98 Togo Memoirs, op. cit., pp. 9–10.

99 In the course of this conversation, Anami significantly cautioned Kido that certain Army elements were suspicious that Kido was working for peace and were talking of his removal. (1) Affidavit of Kido, op. cit. IMTFE Record, pp. 31,156–8. (2) Statement by Marquis Kido, previously cited.

War Direction Council members for the purpose of reopening discussions on peace. The meeting took place in the evening of 18 June. War Minister Anami and the Chiefs of the Army and Navy General Staffs unitedly maintained that Japan must keep on fighting as long as the enemy insisted upon unconditional surrender, and argued that there would be better chances of securing a moderation of the enemy's terms if the homeland battle were fought and heavy losses inflicted on the invading forces. Nevertheless, under advice rom Suzuki, Togo and Yonai, the War Minister and High Command leaders finally agreed to an immediate acceleration of the negotiations with Russia in an effort to obtain Soviet mediation for a peace settlement on acceptable terms.[100]

Foreign Minister Togo communicated this decision the following day to Mr. Hirota, who was still awaiting a response from Soviet Ambassador Malik to his initial proposals made on 3–4 June. Togo urged Hirota to seize the first opportunity to resume conversations with Malik and to ascertain, if possible, the Soviet attitude toward mediating between Japan and the Allies on the basis of terms more favorable than unconditional surrender.[101]

On 20 June Togo reported directly to the Emperor on the steps taken to pave the way for Soviet mediation, while Premier Suzuki informed Kido of the outcome of the six leaders' conference on the 18th.[102] Although a good start had been made, Kido felt that, to clinch the reorientation of policy toward peace, it was advisable for the Emperor to summon the six leaders to an Imperial conference and directly command them to speed diplomatic peace efforts. Kido submitted his opinion to the Emperor on the afternoon of 20 June, and the conference was called two days later.[103]

At the Imperial conference of 22 June, the Emperor boldly took the initiative. Pointing out that Japan was already experiencing extreme difficulty in carrying on the war owing to the critical military and domestic situation, he expressed fear that these difficulties would steadily become greater under intensified enemy air attacks. He therefore urged the Government and the High Command not to limit their efforts to carrying out military preparations for the defense of the homeland as provided in the basic policy of 8 June, but also to exert all possible effort to bring the war to an acceptable conclusion by diplomatic means. His Majesty then invited the opinions of the six leaders.

Premier Suzuki promptly acknowledged the extreme gravity of Japan's war situation and agreed with the Emperor's view that every effort should be made to bring about a diplomatic peace settlement at the same time that preparations were continued to defend the homeland. Navy Minister Yonai and Foreign Minister Togo seconded the Premier's remarks. Yonai also insisted that every step be taken to induce Soviet Russia to act as a mediator between Japan and the Allies for a negotiated peace. Togo admitted that an approach to Russia involved certain dangers as well as the necessity of making heavy concessions, but he emphasized that Soviet mediation offered the sole hope of avoiding unconditional surrender.

Asked to state the views of the Army High Command, General Umezu pointed out that the launching of peace proposals would have

---

100 (1) Togo Memoirs, op. cit., p. 10. (2) Affidavit of Kido, op. cit., IMTFE Record, pp. 31,159–60.

101 Togo Memoirs, op. cit., p. 10.

102 (1) Ibid. (2) Statement by Foreign Minister Togo, previously cited. (3) Affidavit of Kido, op. cit., IMTFE Record, pp. 31,159–60.

103 (1) Affidavit of Kido, op. cit., IMTFE Record, pp. 31,160–62. (2) Statement by Marquis Kido, previously cited.

profound repercussions at home as well as abroad, and that it was consequently imperative to act with the utmost caution and only on the basis of a most thorough appraisal of all aspects of the situation. The Emperor pointedly queried whether Umezu meant by his statement that he opposed peace action. To this Umezu replied that he fully recognized the necessity for prompt diplomatic measures without awaiting the invasion of the homeland and only meant to stress the need for caution. No further opinions were offered either by War Minister Anami or Admiral Toyoda, Chief of the Navy General Staff, and the Emperor retired to end the conference.[104]

The Emperor's clear-cut expression of his will at the 22 June Imperial conference marked a decisive stride forward on the thorny road toward peace. For the first time since the formation of his Cabinet, Premier Suzuki felt confident that he could succeed in the mission entrusted to him by the Emperor.[105] The Government's policy, hitherto vacillating and indecisive, was now set definitely toward terminating the war by diplomatic negotiation, and the Army's top leaders had committed themselves before the Emperor to support that policy. Despite this commitment, there was some concern in the peace group as to whether the Army could control its extremist elements.

Negotiations with Russia

Imperial intervention resulted in the immediate renewal of the Hirota-Malik talks, suspended since early June. Events, however, rapidly bore out Foreign Minister Togo's judgment that it would be extremely difficult to induce a reversal of the hostile trend in Soviet policy toward Japan.

Meeting Malik on 24 June, Hirota vigorously pressed for a reply to his earlier overtures for a Soviet-Japanese rapprochement but was unable to get from the Soviet envoy anything more than a promise to refer to his Government any concrete proposals which might be advanced from the Japanese side. In the course of the talk, Hirota found occasion to express hope for the early restoration of peace in the Far East. Malik's reply was so curt, however, that any approach to the subject of mediation was impossible.[106]

Since continuation of the talks now required specific proposals on the part of the Japanese Government, Foreign Minister Togo handed Hirota on 28 June a written statement to transmit to Malik, proposing the conclusion

---

104 The proceedings of the Imperial conference of 22 June are summarized on the basis of the following sources: (1) Kido Diary, op. cit. (Appended pages dated 22 Jun 45, recording the Emperor's account of the Imperial conference as orally given to Kido immediately following the conference.) (2) Memorandum notes by Rear Adm. Takagi, op. cit. (Conversation with Navy Minister Mitsumasa Yonai, 23 Jun 45, in which Yonai gave an account of the Imperial conference proceedings.) (3) Togo Memoirs, op. cit., pp. 10–1. (4) Statements by Foreign Minister Togo and Admiral Toyoda, both previously cited.

105 Premier Suzuki told Chief Cabinet Secretary Sakomizu following the 22 June Imperial conference that he now felt confident that he could terminate the war. "Today," he said to Sakomizu, "the Emperor said what everyone has wanted to say, but yet did not dare to say." Article entitled *Kofukuji no Shinso* 降伏時の眞相 (Truth of the Surrender Period) by Hisatsune Sakomizu. Asahi Newspaper, Tokyo, 13 Jan 46.

106 Hirota took advantage of Malik's reference to the return of peace between Russia and Germany to state that he hoped Japan might follow the Soviet example and restore peace at an early date. Malik replied that, since the Soviet Union was not a belligerent in the Far East, it had nothing to do with the question of peace there. Foreign Office Radios, op. cit., No. 843, Foreign Minister to Japanese Ambassador at Moscow, 28 Jun 45.

of a long-term Soviet-Japanese pact of mutual assistance and non-aggression.[107] Hirota had earlier told Togo that he feared pressing the negotiations too strongly might give away Japan's weakness. Togo nevertheless reiterated that haste was vital under the circumstances and urged Hirota to transmit the Japanese proposal without delay.[108]

Acting on these instructions, Hirota handed the Japanese Government's written statement to Malik on 29 June and requested a reply at the earliest possible date. On 30 June Togo telegraphed this information to the Japanese Ambassador in Moscow, instructing him to " do the utmost " at that end to speed the reply of the Soviet Government.[109] The negotiations were now on an official government-to-government basis, but the vital question of Soviet mediation still had not been broached.

The first week of July passed without any answer from Moscow to the preliminary Japanese proposals. The Emperor became increasingly concerned over the delay and finally, acting on the advice of Marquis Kido, summoned Premier Suzuki to the Palace on 7 July to urge that prompt steps be taken to bring mediation into the negotiations Stating that it would not do to miss the opportunity for peace by spending too much time in sounding out Soviet intentions, His Majesty proposed that mediation be requested without further delay and that a special envoy be dispatched to Moscow for this purpose, bearing a personal message from the Emperor.[110]

The dispatch of a special enovy was, in fact, already part of the six leaders' contemplated plan of action. Premier Suzuki and Foreign Minister Togo had tentatively agreed that the mission should be entrusted to Prince Konoye, and on 7 July, prior to Suzuki's audience with the Emperor, Foreign Minister Togo left Tokyo for Karuizawa to ascertain unofficially whether Konoye would accept the assignment. In a talk with Togo on 8 July, Konoye indicated that he would undertake the mission if the Emperor commanded him to do so. He and Togo agreed that it would be best if he could go to Moscow with wide discretion as regards peace terms.

Following his return to Tokyo, Foreign Minister Togo learned from the Premier on 9 July of the Emperor's desire to speed the dispatch of a special emissary. It was promptly decided to seek the approval of the Supreme War Direction Council, and Premier Suzuki summoned the six members to a restricted meeting on the 10th. No objection was encountered at the meeting. After the Premier communicated the Emperor's wish, the six leaders swiftly agreed that immediate steps be taken to arrange for the special mission, including the dispatch of a message transmitting to the Soviet Government the Imperial desire to end the war.[112]

Meanwhile, Moscow's silence on the Japanese proposal for a nonaggression pact remained unbroken. Naotake Sato, Japanese Ambassador at Moscow, was not aware that the proposal

107 The statement said that Japan was prepared, under the proposed pact, to neutralize Manchuria upon conclusion of the East Asia war, and to relinquish fishing rights in Siberian waters in exchange for Soviet oil. It further stated that Japan would discuss any other terms or conditions which the Soviet Government desired. Ibid., No. 853, 30 Jun 45.

108 (1) Togo Memoirs, op. cit., p. 12, (2) Statement by Foreign Minister Togo, previously cited.

109 Foreign Office Radios, op. cit., No. 852, 30 Jun 45.

110 Affidavit of Kido, op. cit., IMTFE Record, pp. 31,164–5.

111 Statement by Foreign Minister Togo, previously cited.

112 Togo Memoirs, op. cit., p. 13.

was intended to pave the way for a mediation request, and he had refrained from pressing it on the ground that circumstances rendered rejection a practical certainty.[113] For this failure to obey instructions, Sato was criticized by Foreign Minister Togo on 9 July. Togo cabled the Ambassador to carry out orders "apart from your own opinion" and reiterated that he must obtain an interview with Foreign Commissar Molotov before the latter's impending departure for the Potsdam conference and "try to induce the Soviet side to fall in with our project."[114]

Since Molotov was currently engaged in top-level talks with the visiting Chinese Premier, T. V. Soong, Sato had difficulty in arranging a meeting. However, after a preliminary exchange of views with Foreign Vice-Commissar Lozovsky on 10 July, Sato succeeded in obtaining a 20 minute interview with Molotov on the 11th. Both conversations got nowhere. Ambassador Sato insistently pressed for an indication of the Soviet attitude regarding the Japanese pact proposal, but Molotov replied that no answer could be given pending further careful study.[115]

Even while Sato was conferring with Molotov, new instructions were en route from Tokyo bringing the Ambassador his first knowledge that Japan was seeking to end the war through utilization of Soviet Russia. The instructions directed Sato to sound out Molotov speedily with this objective in view, and a further message dispatched by Foreign Minister Togo on the evening of the 11th authorized the Ambassador to state that Japan, in connection with the termination of the war, had "no thought of annexing or retaining the territories under her occupation."[116]

While Sato prepared to act on these new instructions, further developments took place in Tokyo. During the morning of the 12th, Premier Suzuki conveyed to the Emperor and Marquis Kido the Government's desire to send Prince Konoye to Moscow. Since the Prince was already on his way to Tokyo from Karuizawa, His Majesty decided to summon him to the Palace the same afternoon and command him to accept the mission if it materialized.[117]

Prince Konoye was received in audience by the Emperor at 1500 the same day. His Majesty, after obtaining a clear-cut assurance that Konoye agreed with him on the vital necessity of terminating the war, disclosed that he wished the Prince to go to Moscow as a special peace emissary. Konoye replied that he was prepared to sacrifice his life to carry out the Emperor's will.[118]

It still was necessary, however, to obtain Soviet consent to Konoye's mission. Foreign Minister Togo therefore dispatched an urgent message to Ambassador Sato on the evening of 12 July, instructing him to request prompt admittance for Prince Konoye and his suite to the Soviet Union and the provision of an official plane to transport the party from the Soviet-Manchurian border to Moscow. The Ambassador was further instructed to communicate directly to Molotov, prior to the impending three-Power conference, the following explanation of the Emperor's desire to

---

113 Foreign Office Radios, op. cit., No. 1330 and No. 1331, Japanese Ambassador at Moscow to Foreign Minister, 5 and 6 Jul 45.

114 Ibid., No. 875 and No. 884, Foreign Minister to Japanese Ambassador at Moscow, 5 and 9 Jul 45.

115 Ibid., No. 1364 and 1379, Japanese Ambassador at Moscow to Foreign Minister, 11 Jul 45.

116 Ibid., No. 890 and No. 891, Foreign Minister to Japanese Ambassador at Moscow, 11 Jul 45.

117 Kido Diary, op. cit., IMTFE Record, pp. 31,166–69.

118 Ibid., (Entry for 12 Jul 45, appended pages covering Emperor's audience with Prince Konoye.) IMTFE Record, pp. 31,170–71.

terminate the war:[119]

His Majesty the Emperor of Japan, being solicitous of the increasing hardships and war-suffering imposed on the belligerent nations, desires that the war should promptly be brought to an end. So long as Britain and the United States adhere to their demand for Japan's unconditional surrender in the Greater East Asia War, Japan has no alternative but to fight on at all costs for the honor and self-preservation of the nation. However, it is far from our desire to see further bloodshed by the belligerent nations on this account. His Majesty hopes to restore peace as speedily as possible for the sake of the happiness of mankind. . . .He intends to dispatch Prince Konoye as his special envoy to the Soviet Union, bearing a personal message which will convey his wishes to the above effect.

Ambassador Sato, immediately upon receiving this message on 13 July, requested a further interview with Molotov. The latter, however, replied that his impending departure for the three-Power conference made a meeting impossible, and he advised Sato to see Lozovsky instead. Sato therefore called on the Foreign Vice-Commissar at 1700 the same day and handed him a written statement of the Emperor's wish to terminate the war together with a confidential note for transmittal to Molotov, requesting the Soviet Government's assent to the Konoye mission. Lozovsky assured Sato that he would immediately transmit both documents to Molotov, but he stated that it would be practically impossible to make a reply prior to Molotov's departure.[120]

Stalin and Molotov left Moscow for the Potsdam conference on the afternoon of 14 July, nearly twenty-four hours after Ambassador Sato had handed the documents relative to Prince Konoye's mission to Foreign Vice-Commissar Lozovsky. Consequently, despite Sato's failure to see Molotov directly, it was considered certain in Tokyo that the Soviet leaders went to the tripartite talks with knowledge of the Emperor's desire to dispatch a peace mission to Moscow.[121]

On the 14th, a further meeting of only the six Supreme War Direction Council members was held. Premier Suzuki disclosed the Emperor's decision to entrust the Moscow mission to Prince Konoye, and Foreign Minister Togo explained the steps already taken to obtain the consent of the Soviet Government. It was swiftly agreed that Konoye should be accompanied by the Vice Minister of Foreign Affairs and by a general and an admiral representing the High Command, but on the question of peace terms the discussions again reached a deadlock. War Minister Anami repeated his contention that the terms must be based on the fact that Japan, from a territorial viewpoint, was still far from defeated. Togo and Navy Minister Yonai argued on the contrary that the probable future development of the military situation must be taken into account. Since no agreement seemed possible, the final decision on terms was deferred until Konoye actually reached Moscow and began the mediation parleys.[122]

The Soviet Government, however, appeared no more eager to respond to the proposal for Konoye's mission than to the earlier overture for nonaggression pact. For five days after the communication of this proposal to Lozovsky on the 13th, there was complete silence. Then, on the evening of 18 July, Lozovsky finally

119 Foreign Office Radios, op. cit., No. 893, Foreign Minister to Japanese Ambassador at Moscow, 12 Jul 45.

120 Ibid., No. 1385, Japanese Ambassador at Moscow to Foreign Minister, 13 Jul 45.

121 Foreign Minister Togo reported to the Emperor on 18 July that he was confident that His Majesty's peace intentions had been transmitted to Stalin and Molotov before their departure from Moscow. The Emperor expressed satisfaction. Togo Memoirs, op. cit., p. 16.

122 (1) Ibid., p. 15. (2) Statement by Foreign Minister Togo, previously cited.

handed Ambassador Sato a note which stated that the Soviet Government could give no definite reply because the statement of the Emperor's desire for peace contained no concrete proposal and because the nature of Prince Konoye's mission was not sufficiently clear.[123]

Ambassador Sato's radio reporting the contents of the Soviet note did not reach the Tokyo Foreign Office until the morning of 20 July. On the 21st, Foreign Minister Togo telegraphed back a reply for transmission to Lozovsky, stating that Prince Konoye's mission was to request the good offices of the Soviet Government for bringing the war to an end in accordance with the Emperor's wishes, to convey the concrete intentions of the Japanese Government in this connection, and to negotiate on matters relative to the establishment of cooperative relations between Japan and the Soviet Union. A separate message addressed to Sato explained that while Japan could not accept unconditional surrender under any circumstances, she desired, through Soviet good offices, to secure a peace settlement short of unconditional surrender. The message further stated that the Japanese Government could not immediately proffer concrete peace terms owing to the delicate situation, but that Prince Konoye would clarify Japan's intentions when he reached Moscow and would also give "full consideration to Russian demands in East Asia."[124]

The Japanese note clarifying Konoye's mission was communicated to Lozovsky by Ambassador Sato on 25 July. In dispatches to Foreign Minister Togo the same day, Sato stated that Lozovsky appeared deeply impressed by Japan's appeal for the good offices of the Soviet Government and promised to communicate the Russian reply as soon as it was available.[125]

The report of this interview was barely in the hands of the Tokyo Foreign Office when a startling development upset the Japanese plans. From Potsdam, where Stalin and Molotov were still conferring with the leaders of the British and American Governments, came the announcement of the Three-Power Anglo-American-Chinese Declaration of 26 July, stating the terms which Japan must accept in order to obtain peace.

The Potsdam Declaration

The Three-Power Declaration was picked up by official radio monitoring stations in Tokyo at about 0600 on 27 July. It stated:[126]

We, the President of the United States, the President of the National Government of the Republic of China, and the Prime Minister of Great Britain, representing the hundreds of millions of our countrymen, have conferred and agreed that Japan shall be given an opportunity to end this war.

The prodigious land, sea and air forces of the United States, the British Empire and of China, many times reinforced by their armies and air fleets from the west, are poised to strike the final blows upon Japan. This military power is sustained and inspired by the determination of all the Allied Nations to prosecute the war against Japan until she ceases to resist.

The result of the futile and senseless German resistance to the might of the aroused free peoples of the world stands forth in awful clarity as an example to the people of Japan. The might that now converges on Japan is immeasurably greater than that which, when applied to the resisting Nazis, necessarily

123 Foreign Office Radios, op. cit., No. 1417, Japanese Ambassador at Moscow to Foreign Minister, 19 Jul 45.

124 Ibid., No. 931 and No. 932, Foreign Minister to Japanese Ambassador at Moscow, 21 Jul 45.

125 Ibid., No. 1449 and No. 1450, Japanese Ambassador at Moscow to Foreign Minister, 25 Jul 45.

126 U. S. Department of State Publication No. 2423, *The Axis in Defeat*, pp. 27–9.

laid waste to the lands, the industry and the method of life of the whole German people. The full application of our military power, backed by our resolve, will mean the inevitable and complete destruction of the Japanese armed forces and just as inevitably the utter devastation of the Japanese homeland.

The time has come for Japan to decide whether she will continue to be controlled by these self-willed militaristic advisers whose unintelligent calculations have brought the Empire of Japan to the threshold of annihilation, or whether she will follow the path of reason.

Following are our terms. We will not deviate from them. There are no alternatives. We shall brook no delay.

There must be eliminated for all time the authority and influence of those who have deceived and misled the people of Japan into embarking on world conquest, for we insist that a new order of peace, security and justice will be impossible until irresponsible militarism is driven from the world.

Until such a new order is established and until there is convincing proof that Japan's war-making power is destroyed, points in Japanese territory to be designated by the Allies shall be occupied to secure the achievement of the basic objectives we are here setting forth.

The terms of the Cairo Declaration shall be carried out, and Japanese sovereignty shall be limited to the islands of Honshu, Hokkaido Kyushu, Shikoku and such minor islands as we determine.

The Japanese military forces, after being completely disarmed, shall be permitted to return to their homes with the opportunity to lead peaceful and productive lives.

We do not intend that the Japanese shall be enslaved as a race or destroyed as a nation, but stern justice shall be meted out to all war criminals, including those who have visited cruelties upon our prisoners. The Japanese Government shall remove all obstacles to the revival and strengthening of democratic tendencies among the Japanese people. Freedom of speech, of religion and of thought, as well as respect for the fundamental human rights, shall be established.

Japan shall be permitted to maintain such industries as will sustain her economy and permit the exaction of just reparations in kind, but not those which would enable her to re-arm for war. To this end, access to, as distinguished from control of, raw materials shall be permitted.

The occupying forces of the Allies shall be withdrawn from Japan as soon as these objectives have been accomplished and there has been established, in accordance with the freely expressed will of the Japanese people, a peacefully inclined and responsible government.

We call upon the Government of Japan to proclaim now the unconditional surrender of all the Japanese armed forces, and to provide proper and adequate assurance of their good faith in such action. The alternative for Japan is prompt and utter destruction.

Receipt of the Three-Power Declaration touched off a flurry of intense official activity. Foreign office experts immediately began translating and analyzing its contents, their attention promptly focussing on two highly significant facts. The first was that the Soviet Government, although almost certainly consulted at Potsdam regarding issuance of the declaration, remained non-party to it, thereby preserving Russia's legal neutrality in the Pacific war. The second was that the issuing Powers—the United States, Britain and China—abandoned their previous insistence upon unconditional surrender pure and simple and, instead, laid down eight specific conditions for a peace settlement with Japan. The term " unconditional surrender " appeared only once in the document and was specifically limited in application to " the Japanese armed forces." [127]

The fact that the Three-Powers, presumably

127 Foreign Office Radios, op. cit., *Bei-ei-shi Potsudamu Sengen no Kento* 米英支ポツダム宣言の檢討 Study of the Anglo-American-Chinese Declaration at Potsdam) Treaty Bureau, Foreign Ministry, 9 Aug 45. (Though dated 9 August, this study was begun immediately after receipt of the declaration on 27 July.)

informed by the Russians of Japan's desire to end the war, had proffered certain peace conditions instead of demanding blanket unconditional surrender immediately impressed Foreign Minister Togo as of paramount importance. The conditions were unquestionably severe, particularly with regard to the reduction of Japanese territories, but Togo was not surprised in view of Japan's war situation. Moreover, while the Allies stated that they would not deviate from these terms, Togo believed that if the Soviet Government finally agreed to Japan's mediation request, it would be possible at least to negotiate through Russia to assure that the Potsdam terms would be interpreted in the most favorable way for Japan. It was therefore essential, first, not to reject the Allied declaration, which would at once close the door to further peace negotiations, and second, to await Russia's final answer on the Konoye mission.[128] Premier Suzuki promptly concurred in these views.[129]

Having won the Premier's concurrence, Foreign Minister Togo proceeded to the Palace at about 1030, less than five hours after the receipt of the Three-Power Declaration, and reported on its contents directly to the Emperor and to Marquis Kido. In his report to the Emperor, Togo stated the opinion that rejection of the declaration would "invite serious consequences" and that, since an approach had already been made to the Soviet Government for the purpose of ending the war, no action on the declaration should be taken until the Soviet attitude was ascertained.[130]

Immediately following Togo's audience with the Emperor, the six leaders of the Supreme War Direction Council met in an urgent and restricted session to decide Japan's general policy toward the Three-Power pronouncement. Togo, supported by Premier Suzuki, advocated the same course that he had recommended to the Emperor. War Minister Anami and the Chiefs of the Army and Navy General Staffs, on the other hand, argued that the Potsdam terms were so close to out-and-out unconditional surrender that they allowed no basis for negotiation, and that prompt official rejection was necessary to obviate the serious effect which any display of indecision would have on the morale of the fighting services and people. On the latter point, however, the military members finally yielded, and it was agreed that no action should be taken on the declaration pending the outcome of the Russian negotiations.[131]

The Cabinet, meeting the same afternoon, debated the delicate question of how the Three-Power pronouncement should be handled domestically. Togo, who feared that immediate publication would put the Government in a position where it would be pressured into coming out openly against the Potsdam terms, urged withholding the declaration from the public for the time being. Other members of the Cabinet, however, argued that this would be unwise since the declaration had been broadcast throughout the world and would quickly leak out to the Japanese public. War Minister Anami urged that it was essential for the Government, in publishing the declaration, to repudiate it explicitly and even endeavor to exploit it for the purpose of boosting the fighting spirit of the nation.

Against the War Minister's stand, Premier Suzuki again came out strongly in support of Togo's thesis that the Government should

128 Statement by Foreign Minister Togo, previously cited.

129 Interrogation of Chief Cabinet Secretary Sakomizu, previously cited. USSBS Interrogation No. 609, p. 5.

130 (1) Togo Memoirs, op. cit., p. 17. (2) Kido Diary, op. cit., Entry for 27 Jul 45.

131 (1) Togo Memoirs, op. cit., p. 17. (2) Statement by Foreign Minister Togo, previously cited.

neither accept nor reject the Potsdam Declaration pending the outcome of the negotiation with Moscow. The Cabinet finally agreed, therefore, that although the declaration should be published, certain passages which seemed particularly aimed at lowering the nation's will to fight should be withheld, and the Government should rigidly refrain from any official comment whatsoever. The press also was to be instructed to play down the declaration as far as possible.[132]

In practice, this compromise arrangement worked out very badly. The Cabinet Information Board, within which the Army and Navy were represented, duly instructed the press to publish the declaration in abridged form, leaving out the passages which stated that the Allied did not intend " that the Japanese shall be enslaved as a race or destroyed as a nation " and that Japanese troops, after being disarmed, " shall be permitted to return to their homes with the opportunity to lead peaceful and productive lives." The Board further barred publication of any comment attributed to official sources, but since it was feared that no mention whatever of the Government's attitude might be construed publicly as an indication of wavering, the press was authorized to report, without crediting any official source, that the Government appeared to be ignoring the Allied Declaration.[133]

This guidance was obediently followed in the morning editions of 28 July. The declaration was published in abridged form, and the newspapers reported no official comment. The papers themselves refrained from editorial comment in compliance with the directive to play down the declaration, but in their news columns they reported briefly and without quoting any official source that the Government was paying no attention to the enemy pronoucement which appeared to be a propaganda device designed to drive a wedge between the military and the Japanese people.[134]

This halfway treatment was far from satisfactory to Togo, but it was even more unsatisfactory to the military. On the morning of 28 July, when the regular Saturday information conference of the Government and High Command took place at the Imperial Palace, the War and Navy Ministers and the Chiefs of the Army and Navy General Staffs closeted themselves and again urged that the Government come out openly against the Potsdam Declaration. Subjected to this insistent pressure, Suzuki agreed to make a brief statement at his scheduled press conference the same afternoon explicitly confirming the already published newspaper reports that the Government was ignoring the declaration. Foreign Minister Togo was absent from the Palace conference and knew nothing of this development.[135]

The Premier's statement to the press, phrased for home consumption, was extremely maladroit. He told reporters that he regarded the Three-Power pronouncement as a mere restatement of the Cairo Declaration, and that the Government attributed no significant value to it. The declaration, he added, would not be heeded, and the nation's only course was to go ahead with the all-out prosecution of the

132 (1) *Shusenki* 終戰記 (Record of the Termination of the War) by Hiroshi (Kainan) Shimomura, Oct 48, pp. 87–8. (2) Togo Memoirs, op. cit., pp. 17–8.

133 (1) Asahi and Mainichi Newspapers, Tokyo, 28 Jul 45. (2) Record of the Termination of the War, op. cit., pp. 87–9.

134 Asahi and Mainichi Newspapers, Tokyo, op. cit., 28 Jul 45.

135 (1) Record of the Termination of the War, op. cit., p. 90. (2) Statement by Foreign Minister Togo, previously cited.

war.[136]

Although some effort appears to have been made to caution the press against giving undue prominence to the Premier's remarks, the newspapers on 29 July played up his assertion that the Government would take no notice of the Potsdam Declaration.[137] Moreover, the statement was promptly disseminated to the world through Japanese news broadcasting facilities. Foreign Minister Togo lodged a strong protest with the Premier on the ground that his statement violated the decisions reached by the six leaders and the Cabinet on 27 July. A retraction, however, was obviously impossible, and it was decided to do nothing further until Russia's attitude had been ascertained.[138]

The negotiations with Moscow had been at a standstill since Ambassador Sato's interview with Lozovsky on 25 July. Stalin and Molotov were still at Potsdam and had authorized no reply to the Japanese proposal for the Konoye mission. On 30 July, Sato again pressed Lozovsky for an answer. He told the Foreign Vice-Commissar that, provided the formula of unconditional surrender could be avoided, Japan desired "to end the war on broad terms of compromise, so long as its honor and existence are guaranteed." He asked Lozovsky to transmit this information to the Soviet leaders at Potsdam, so that they might consider taking steps to remove such obstacles as the Anglo-American-Chinese declaration placed in the way of Soviet good offices for peace. Lozovsky promised compliance with this request.[139]

In Tokyo, meanwhile, the publication of the Potsdam terms—even in abridged form—was beginning to produce important effects. The military in general remained adamantly opposed to acceptance, but on the other hand many senior diplomats and influential persons in business, politics and press circles began privately urging Marquis Kido and members of the Cabinet that it was necessary to take advantage of the Three-Power offer to bring a prompt end to the war.[140]

These were encouraging signs from the viewpoint of the Cabinet peace group. Yet it was readily apparent that the Potsdam Declaration had rendered the attitude of the military leaders, with the exception of Navy Minister Yonai, even more intransigent than it had been hitherto. The peace group therefore clung to the fast-vanishing hope of Soviet acceptance of the role of mediator as the sole means of finding a way out of this impasse.

The decision to await Russia's answer was thus dictated not only by the exigencies of the tenuous internal situation but with slim hope of Soviet acceptance of the mediator's role. It was also influenced, however, by the fact that at no time had the Potsdam Declaration, despite its threat of "prompt and utter destruction", been construed as an ultimatum, non-acceptance of which by Japan would result in anything worse than a gradual intensification of enemy action against the homeland, culminating in the final invasion itself. There was certainly no suspicion that the declaration constituted forewarning of immediate and more terrible reprisal if Japan did not yield.

On 6 August, just ten days after the receipt of the Potsdam Declaration, the people of Hiroshima tragically learned otherwise.

136 Record of the Termination of the War, op. cit., p. 90.

137 Statement by Chief Cabinet Secretary Sakomizu, previously cited.

138 Statement by Foreign Minister Togo, previously cited.

139 Foreign Office Radios, op. cit., No. 1484, Japanese Ambassador at Moscow to Foreign Minister, 30 Jul 45.

140 (1) Kido Diary, op. cit., (Entries dated 28 Jul to 3 Aug 45.) (2) Togo Memoirs, op. cit., p. 18.

The Atomic Bomb

Hiroshima, situated in southwestern Honshu, was a city of 343,000. Its inhabitants considered themselves fortunate in that the city, thanks to its relative lack of war industry, had so far been over-looked by the enemy's incendiary-laden B-29's. However, it seemed problematical how long this good fortune would continue. Located in the city since April 1945 was the headquarters of the Second General Army, the top operational headquarters for southwestern Honshu, Shikoku and Kyushu. The next-door Inland Sea port of Ujina had lost its original importance as a military port for the shipment of troops and supplies overseas, but in the southeastern part of Hiroshima there remained a considerable number of military supply dumps and warehouses.

The 6th of August dawned sultry but fairly clear. At 0709 the early-warning radar net detected a small number of aircraft headed toward southern Japan, and a preliminary alert was sounded throughout the Hiroshima area. Radio stations promptly went off the air. Shortly thereafter, enemy planes circled over Hiroshima and withdrew without bombing. An "all clear" was sounded at 0730. People began their daily activities, thinking that the danger was past.

At 0800, however, two enemy B-29's were spotted heading toward Hiroshima. Radio broadcasting stations again issued a warning, telling the people that it would be advisable to take shelter if the enemy craft appeared over the city but that their mission appeared to be reconnaissance rather than bombing. Most factory and office workers had already reported to their places of employment, and those who were en route confidently continued on their way. The city's school children and some industrial workers were busily engaged in building firebreaks and evacuating valuables to the country.

Two B-29's soon appeared over the city, flying at high altitude. Since no bombing was expected, many people declined to take shelter and curiously watched the enemy planes as they came over. Observers later reported that they saw a parachute descend from one of the planes. Then, at approximately 0815, there was a terrific exp'osion over the central section of the city. A white flash of blinding intensity was all that many saw before they were struck down by the wall of searing heat that pushed outward from the explosion. As a huge cloud of smoke and dust spiralled up over the city, Hiroshima was cloaked in a ghastly pall of darkness. Then hundreds of fires, breaking out almost simultaneously, transformed the city into a blazing inferno.

By nightfall, Hiroshima had become a waste of ashes and smouldering ruins. The city writhed with the agony of its dazed, injured and dying. Those who had escaped with their lives, including many who had been horribly burned, streamed out of the city in confusion. Only the outlying southern and eastern sections of the city, where ironically most of the military supply installations were located, had escaped destruction.

Of the total civilian population of 343,000, approximately 78,150 were killed, with an additional 51,408 injured or missing. Military casualties, not included in these figures, were light. Nearly 48,000 of the total 76,327 buildings in Hiroshima had been completely destroyed, and 22,178 half-demolished or badly damaged. More than 176,987 persons were left homeless.[141]

141 *Taiheiyo Senso ni yoru Waga Kuni no Higai Sogo Hokokusho* 太平洋戰爭による我國の被害綜合報告書 (Consolidated Report of Losses and Damage Suffered in the Pacific War) Economic Stabilization Board, 1949, pp. 177–80.

Because of the complete disruption of communications in the stricken area, the first news of the catastrophe did not reach the Government in Tokyo until the afternoon of the 6th. The initial reports were meager and fragmentary, but they clearly indicated that the enemy had unleashed on Hiroshima a new-type bomb of unprecedented destructive power.

Early on 7 August, the Government and the High Command learned from American and British radio broadcasts that the destruction of Hiroshima had been wrought by an atomic bomb. President Truman's announcement to this effect shocked the Japanese authorities.[142] Shortly thereafter, the Army and Navy in Tokyo received direct reports from Hiroshima and Kure telling of the immense havoc wrought by the enemy weapon.

At the Cabinet meeting which took place the same morning, Foreign Minister Togo gave a lengthy summation of the information contained in foreign broadcasts with regard to the atomic bomb and indirectly hinted that the Potsdam Declaration should be accepted. The service ministers presented preliminary reports on the effects of the Hiroshima bombing, but the Cabinet withheld its judgment pending further investigation.

The general public was not informed of the bombing officially until the following day. The newspapers on 8 August carried an Imperial General Headquarters communique, dated 7 August which merely stated that Hiroshima had been "considerably" damaged by a "new-type bomb." After considerable discussion it had been decided not to use the term "atomic bomb" in the communiqué until the facts had been established by an official investigation on the spot.

The investigating committee dispatched by the High Command reached Hiroshima on the 8th and swiftly confirmed that the extent and nature of the destruction pointed to the use of an atomic bomb.[143] But even before official reports from this party reached Tokyo, Premier Suzuki and Foreign Minister Togo, aroused by the unusual importance attached to Hiroshima by the enemy radio, conferred and decided that Togo should communicate the substance of the Allied reports to the Emperor, recommending prompt acceptance of the Potsdam terms.

Received in audience at the Palace on the afternoon of the 8th, Togo conveyed to the Emperor his opinion that the atomic bomb would revolutionize modern warfare and warned that more bombs would probably be dropped on Japanese cities unless resolute action were taken to accept the Potsdam Declaration. The Emperor, in reply, agreed that Japan should no longer delay its decision to end the war and requested Togo to convey his wishes to the Premier. Suzuki then tried to convoke an immediate meeting of the Supreme War

142 Many Japanese authorities were aware that scientists abroad were engaged in atomic research for war purposes, but it was not believed that success would be achieved during the present war. Following successive reports from the scene of disaster, the authorities grew more convinced of the truth of the American broadcast, though there were some who at first were inclined to discredit the Truman announcement as propaganda intended to frighten the Japanese. (1) Record of the Termination of the War, op. cit., p. 98. (2) Atomic research had been conducted in great secrecy by both the Army and Navy for some time during the war, but experiments were dropped early in 1945 because of the prodigious expense and the small likelihood of achieving success in the immediate future. (Statement by Lt. Gen. Kawabe, previously cited.)

143 Included in the mission led by Lt. Gen. Seizo Arisue, were Dr. Yoshio Nishina, foremost atomic energy authority in Japan, the only civilian; and a number of experts from the Army Aeronautical Department and the Army Medical School. Delayed by an accident en route, Nishina's party did not reach Hiroshima until the afternoon of 8 August, but it quickly established the facts. (Statements by Dr. Yoshio Nishina and Lt. Gen. Seizo Arisue, Chief, Second Bureau (Intelligence), Imperial General Headquarters, Army Section.)

Direction Council but was unable to do so because of the inability of some members to be present.[144]

Soviet Entry Into War

The Cabinet meanwhile hopefully awaited the long-delayed interview which Ambassador Sato was scheduled to have with Foreign Commissar Molotov at 2300 8 August (1700 hours, 8 August, Moscow Time.) Stalin and Molotov had returned to Moscow on 6 August, much later than the Japanese Government anticipated. Sato had requested an immediate interview, but not until 7 August, a day after the United States revealed to the world it had dropped its first atomic missile, did Molotov agree to a meeting.

Russia's stunning answer was a declaration of war. Within the hour that Molotov was delivering the abrupt reply, Red Army troops were attacking at various points along the long Manchurian frontier and the North Korean border, with the greatest concentration at points bordering Mongolia. South Sakhalin was also invaded simultaneously. At dawn Soviet planes were attacking cities in Manchuria and North Korea as well as Japanese convoys in the Japan Sea. Although the Soviet offensive was sluggish at the outset, the Kwantung Army, weakened by diversion initially of much of its strength to the Southern Regions and more recently for the defense of Japan Proper, was unable to check the enemy offensive.[145]

Neither the Government nor the High Command in Tokyo was aware of the Soviet action until 0400 hours, 9 August, when the Domei News Agency intercepted the Tass message which read as follows:[146]

With the defeat and capitulation of Hitlerite Germany, Japan remains the only great Axis power continuing the war.

With the rejection by the Japanese Government of the 26 July demand of the United States, Great Britain and China, for the unconditional surrender of the Japanese armed forces, the proposal of the Japanese Government to the Soviet Union with regard to mediation in the Far East loses its basis. In view of Japan's refusal to surrender, the Allied nations have submitted to the Soviet Government a proposal that it join the war against Japanese aggression, in order to hasten the end of the conflict, reduce the number of victims, and contribute to the early restoration of universal peace.

The Soviet Government, fulfilling its obligations to its Allies, has accepted this proposal by the Allied nations and has joined in their declaration of July 26. The Soviet Government considers that a policy such as it has adopted is the only means of expediting the return of peace, freeing the several peoples from further sacrifices and suffering and helping the Japanese nation avoid the dangers and destruction suffered by Germany after her refusal to surrender unconditionally.

For the foregoing reasons, the Soviet Government declares that as of tomorrow, 9 August, a state of war will exist between the Soviet Union and Japan.

The immediate reaction to the sudden Soviet attack was one of pained disappointment and

---

144 Statement by Foreign Minister Togo, previously cited.

145 Suspicious of Soviet intentions, Imperial General Headquarters had already ordered the Kwantung Army to prepare against Soviet aggression. Hard pressed on the Pacific front, almost all the efficient weapons, such as aircraft and antiaircraft guns, as well as the better-trained troops, were withdrawn to Japan for the anticipated homeland battle, leaving the once-mighty Kwantung Army with nothing but poorly equipped troops. Accordingly, Imperial General Headquarters stipulated in May 1945 that the main line of defense was to be limited to south Manchuria and north Korea.

146 (1) The Japanese Government was not officially informed of the Russian declaration of war until 1115 hours, 10 August, when the Soviet envoy Malik, called on Foreign Minister Togo to deliver his Government's note. (2) *Soren Sansen ni tomonau Tsuzuri* 蘇聯參戰に伴ふ綴 (File on Soviet Entry into War) Part I, 9–20 Aug 45.

indignation. The Kremlin's declaration of war had shattered the last remaining hope of the Japanese to end the war through Soviet good offices. Premier Suzuki, upon hearing the news, remarked that the "inevitable has finally occurred."[147]

The Soviet entry into the war, coming quickly on the heels of the atom bomb, multiplied the urgency of arriving at a decision. The Government no longer could brook delays. When the atomic nature of the Hiroshima bombing had been finally confirmed, Suzuki, upon Togo's advice, had attempted to convoke the "Big Six" council that evening to secure a decision, but was forced to postpone the meeting until the morning of the 9th. Such being the case, the Emperor, Suzuki and Togo were ready to accept the Potsdam Declaration when they heard the Tass announcement of war.

At 0840, 9 August, Togo conferred with Suzuki on the unprecedentedly grave situation at the latter's home and reaffirmed their agreement to place the prompt-surrender problem before the Big Six conference. En route back to his office, the Foreign Minister stopped at the Navy Ministry where he obtained Yonai's prompt and willing consent to go along with Suzuki and Togo on accepting the Potsdam Declaration.

At 1000 hours, the Premier was at the Imperial Palace. There he was advised by Marquis Kido that the Emperor, who only a short time before had been notified by the Foreign Office of the Soviet act of war, had ordered him to point out to Suzuki the urgency of accepting the Declaration immediately.[148] Suzuki, fearing it would be exceedingly difficult to secure swift unanimity on this grave and delicate issue, then requested and obtained the Emperor's assurance of an Imperial decision in the event of a deadlock.[149]

At 1030, following the Palace interviews, the six leaders of the Supreme War Direction Council were gathered at the Imperial Palace, at the summons of Suzuki, for a critical discussion of the surrender question, the first of a series of crucial parleys which eventually decided the fate of the nation. By this time the six leaders were all agreed that the situation was serious, but they had not made up their minds what the final decision would be.

Suzuki proposed at the outset that Japan should promptly accept the Potsdam Declaration, which was the Emperor's wish. In the ensuing discussion the following four points were considered: (1) the preservation of the "national polity," (2) the extent and nature of the Allied occupation of the homeland, (3) the disarmament of the military forces, and (4) the disposition of war criminals.[150]

The conferees promptly agreed that the first and most important condition must be that acceptance of the Potsdam Declaration would not endanger the "national polity," or the prerogatives of the Imperial family. But Anami, the War Minister, and Umezu, Chief of Army General Staff, insisted that three additional conditions be approved:

1. That the Japanese military forces overseas be disarmed and demobilized by Japan itself;

147 "The Premier seemed quite unperturbed when told of the news of the Russian entry into the war and merely replied that the inevitable had finally occurred." The question to many in high Government quarters was not whether the Soviet Union would enter the war but when. Truth of the Surrender Period, Asahi Newspaper, Tokyo, 14 Jan 46.

148 Kido Diary, op. cit., IMTFE Record, p. 31,172.

149 Fearing that a complete agreement was impossible if the issue of continued war or surrender was left to the Big Six Council and the Cabinet, Suzuki, during his morning audience with the Emperor, asked him to express an Imperial opinion. The Emperor quickly agreed. (Statement by Chief Cabinet Secretary Sakomizu, previously cited.)

150 Statement by Admiral Toyoda, previously cited.

2. That all war criminals be prosecuted by the Japanese Government itself; and,

3. That there be no Allied occupation of Japan.[151]

The principle of self-disarmament was endorsed by Admiral Toyoda, Chief of Navy General Staff. He reluctantly supported Anami and Umezu on the two other conditions. Foreign Minister Togo, however, vigorously opposed any extra conditions. He warned that to press for terms other than the sole reservation governing the Imperial prerogatives would jeopardize the peace negotiations even before they started. Both Suzuki and Yonai supported Togo in his agruments, but after three hours of inconclusive deliberations, the Council was forced to adjourn in a deadlock.

First Cabinet Deliberations on Peace

The issue of immediate peace or continued war was then submitted by the Premier to a full dress Cabinet session at 1430, at which time Togo revealed to his Ministers for the first time the secret peace maneuvers which had thus far been the principal preoccupation of the Supreme War Direction Council.[152] Following a detailed explanation of the peace effort, Suzuki hoped to elicit the frank opinions of each Minister on the issue of peace before finally casting the die in the fateful deliberations. Suzuki emphasized both the gravity of the situation and the imperativeness of reaching an immediate decision, but the Cabinet wrangled indecisively for more than seven hours without nearing an agreement.

In the Cabinet considerations, as in the earlier Council deliberations, the two opposing views represented by Togo, Suzuki and Yonai on one side, and Anami on the other, were clearly delineated. The pro-surrender group saw no hope of victory. They were convinced that war meant only more destruction and ultimate extinction of the nation itself, and that the Potsdam Declaration offered acceptable terms. The opposing group, maintaining that the real test of Japan's strength was yet to come in the expected Homeland battle, argued that once severe casualties had been inflicted on the invasion force, it would be possible to terminate the war on more favorable terms.

Foreign Minister Togo, the outspoken peace advocate, reminded his colleagues that the atom bomb (a second atomic bomb had been dropped at 1103 hours that morning, 9 August)[153] and the Soviet war on Japan had further strengthened the hands of the Allies since the Potsdam ultimatum was issued and insisted that there was no way open for hard diplomatic bargaining in view of the uncompromising nature of the

---

151 A summary of what took place in this conference can be found in the following document: Memorandum Notes of Lt. Gen. Sumihisa Ikeda, Chief, Cabinet Coordination and Planning Bureau, 27 Jul–17 Aug 45.

152 The Cabinet had been ordered to assemble by noon by which time it appears Suzuki had hoped to secure the six leaders' approval of the Government's peace plan. The Cabinet did not meet until 1430 because of the cleavage in the Council and was closeted until 2230 in acrimonious debate. Record of the Termination of the War, op. cit., p. 116.

153 While the Nagasaki bombing does not appear to have had much influence on the peace discussions in the Cabinet, the fact that a second atom bomb had been dropped even before the Japanese had begun to recover from the horrible experiences at Hiroshima aroused widespread fear among the people that other cities might soon be deluged by this extremely destructive enemy weapon. A rumor was then circulating that what was left of Tokyo would be destroyed before 15 August. Nagasaki, like Hiroshima, had until 9 August escaped the general air raids experienced by other major cities. In proportion to the area hit, Nagasaki suffered much more heavily in dead and injured than Hiroshima. Out of a total population of 270,000, there were 23,753 killed and 43,020 injured. (1) Consolidated Report of Losses and Damage Suffered in the Pacific War, op. cit., pp. 179–180. (2) *Tokyo Sous Les Bombes*, an article by Father Joseph Flaujac, appearing in the Bethany Institute Bulletin No. 5, 1940–47, p. 31.

Declaration. Japan, he proposed, should accept the Allied terms with only the reservation governing the Imperial family attached.

War Minister Anami rejected Togo's arguments and while admitting that the Soviet Union could take over Manchuria, where the heaviest fighting was occurring, in two or three months time, he attempted to minimize the effectiveness of the atom bomb on the home front. Far from giving up, he concluded, Japan should insist on the four conditions discussed by the six leaders of the Supreme War Direction Council that morning, and continue hostilities in the event those conditions were rejected.

The protracted and grim debate was punctuated by heated argument between the War and Navy Ministers on Japan's ability to continue the war. Yonai, painting a definitely pessimistic picture, emphasized that the capacity to wage war should be viewed from over-all considerations and that since the problem of continuing the war was a matter to be decided on the basis of a total war, he suggested that the conditions in the munitions, food producing and transportation industries, as well as morale problems, should be discussed frankly. By so doing, he said, the Government would obtain a clear over-all appraisal of the real situation. The extremely depressing reports by the Ministers in charge on these questions bore out Yonai's deep concern.[154]

Suzuki then requested each Minister to state his views on Togo's proposal that Japan accept the Declaration subject to the one provision. At the conclusion of the Cabinet session, the lineup on the Togo proposal stood as follows: Against—War Minister Anami, Home Minister Abe and Justice Minister Matsuzaka: For—Foreign Minister Togo and Navy Minister Yonai. Although one or two remained non-committal, the others favored intermediate positions with most of them leaning toward Togo. A suggestion by Education Minister Ota that the Cabinet resign because of its inability to resolve the impasse was quickly scotched by Suzuki.

With time growing short and the conferees weary of discussion, Suzuki ordered a recess at 2230 and sought to settle the matter by resuming the morning deliberation of the Supreme War Direction Council in the presence of the Emperor. Suzuki directly informed the Emperor that he was unable to achieve the necessary unanimity required of a Cabinet decision and petitioned the Emperor for a conference. This meeting was arranged to invoke the Emperor's will.[155]

Showdown on Surrender Issue

At 2330, an hour after the end of the heated Cabinet session, the "Big Six" council sat down with the Emperor in an air raid shelter

---

154 Munitions Minister Teijiro Toyoda revealed that munitions production had fallen sharply since late spring; that the slump was due to transportation difficulties and failing morale of the workers who, he admitted, had lost "all sense of security and confidence in the Government and the armed forces." Agriculture and Commerce Minister Ishiguro reported that the crop year was the worst since 1931 and that in various areas famine conditions by winter were inevitable. Especially deplorable, he added, was the fact that soldiers were entering homes and taking whatever they needed to appease their hunger. Home Minister Abe stated bluntly that the people had lost confidence in winning the war; that the atomic bomb and the Russian invasion had aggravated war weariness, but said some elements of the people were determined to fight on. An equally grim outlook on transportation and communications prospects was given by Transportation Minister Kohiyama. Record of Termination of the War, op. cit., pp. 116–125.

155 Suzuki, never permitted either the Big Six Council or the Cabinet to break up their discussions. Whenever he found it impossible to secure the necessary unanimity, he called a recess. This policy proved successful because it enabled him to resume the meetings on a moment's notice, when time was so precious. (Statement by Chief Cabinet Secretary Sakomizu, previously cited.)

御前會議　寺內萬治郎

Original Painting by M. Terauchi

PLATE NO. 165

Imperial Conference of 9–10 August 1945

deep inside the Imperial Palace for a showdown on the surrender question. Present also were the president of the Privy Council, Baron Kiichiro Hiranuma and the Chief Aide-de-Camp, General Shigeru Hasunuma;[156] the Chief Cabinet Secretary Sakomizu and three other officials occupying important position in the over-all direction of the war.[157]

Sakomizu, on orders of Suzuki, read the Potsdam Declaration to the Emperor, and then the Premier himself placed on the agenda as the subject of consideration the following proposal, which had been drafted by the Foreign Minister:[158]

> *The Japanese Government is ready to end the war under terms of the Three Power Joint Declaration of 26 July 1945 with the understanding that the said Declaration does not comprise any demand prejudicial to the status of the Imperial family under the fundamental law of the nation.*

The Premier explained that this proposal had been submitted in preference to the other plan containing the four conditions because the majority of the Cabinet had favored, or at least desired, a minimum of reservations. This maneuver was a disappointment to Anami and the bitter-end resistance faction.[159] Backing up Suzuki, the Foreign Minister then strongly reiterated his position that the Allies could not be prevailed upon to ease surrender terms through negotiations. Japan, he insisted, should not propose any new conditions. To this, the Navy Minister expressed complete agreement.

The War Minister, however, remained adamant, stressing that acceptance of the Potsdam terms would result in national ruin. The four conditions were a minimum which Japan could consider, he asserted, adding that the very fact that Soviet Union had joined the Declaration had made acceptance even more unpalatable. Anami expressed the Army's confidence in Japan's ability to secure better terms and voiced the fear that if the Government ended the war without attempting further clarification, there would be chaos in the country.

Full concurrence with Anami's views was expressed by the Chiefs of the Army and Navy General Staffs.[160] Hiranuma made a number of pointed inquiries on the state of defense preparations and then agreed in principle with the Togo plan but suggested that the wording of the draft reply be amended to read that any Allied demand would not prejudice the *prerogatives of the Emperor as a sovereign ruler.* Hira-

156 Hiranuma attended the conference at the invitation of the Emperor. It was believed that the attendance of Hiranuma was necessary to help reach a final decision on the formal diplomatic note to be made to the Allies. (Statement by Chief Cabinet Secretary Sakomizu, previously cited.)

157 Two of them, Lt. Gen. Yoshizumi and Vice Adm. Hoshina, the Chiefs of the Military and Naval Affairs Bureaus of the War and Navy Ministries respectively, were ex-officio secretaries of the Supreme War Direction Council. The other was Lt. Gen. Sumihisa Ikeda, Chief of the Cabinet Coordination and Planning Bureau, which was also an ex-officio secretary of the Council since April 1945.

158 From a manuscript written by Vice Adm. Hoshina, who sat in discussions as ex-officio secretary.

159 War Minister Anami on 13 August intimated to his subordinates in the War Ministry that he had scented intrigue by the opposing faction when he found only the Foreign Minister's plan on the agenda. Anami instantly whispered to Umezu, Chief of the Army General Staff, who was seated next to him, that they should insist on continuation of the war instead of standing on their previous arguments for the four conditions. (Statement by Lt. Col. Inaba, previously cited.)

160 The Chief of the Army General Staff argued that the situation was not so bad that there was no other choice but surrender and that to lay down arms "we will do a great wrong to our war dead." The Chief of the Navy General Staff doubted that the Japanese people would agree to capitulate without trouble resulting. "The Japanese sense of honor is too great to endure the sight of their land being occupied or their arms being taken away," he declared. Unpublished Record of Japanese Navy General Staff, p. 3.

numa emphasized the point that Japan's ability to continue the war should be considered before negotiating for additional conditions. For this reason, he said, it was important to consider the condition of the civilian population as well as the strength of the armed forces. In the end, he felt, however, that the ultimate decision of surrender rested with the Emperor.

It was already past 0200 on 10 August; the cleavage in the council was as pronounced as ever. The Premier noted that despite several hours of continous discussion, the question remained unsettled; that in this extremely crucial hour there could be no further delays. Under the circumstances, he was proposing the Emperor's wish be made known. Suzuki, stepping forward in an unexampled act in Japan's constitutional history, requested the Emperor to state his opinion, and at the same time, pleaded with the group to accept the Emperor's resolve as final.[161]

The Emperor asserted at the start that he was in complete accord with the views of the Foreign Minister and then proceeded to give his reasons. To continue the war in the light of the world situation and Japan's internal conditions would be suicidal, he said, adding that to end the war on this occasion was the only way to save the nation from destruction.

He then pointed to the record of the military. The Emperor said it was apparent that their performance had fallen far short of the plans and promises expressed, pointing out that although he had been assured many times in the past that victory was certain, it had not been realized. Despite the War Minister's promise that the fortifications along the Kujukuri-Hama coast (in Chiba Prefecture) would be completed by the middle of August, they were far from completion. He said it had also been brought to his attention that new Army divisions had been organized, but that there were no arms with which to equip them.

In short, he said he could see no hope of victory over the highly mechanized Allied forces. He stressed, in conclusion, that he was deeply cognizant of the feelings of the sorrowed families of the war dead and the tremendous sacrifices made by the armed forces during the period of hostilities, but that for more important considerations he had decided that the war should be stopped.[162]

The Imperial wish had been made unmistakably clear. Suzuki then emphasized that this should be the conclusion of the conference.

The tense momentous meeting was finally adjourned at 0230, and half an hour later the Cabinet reconvened at the Premier's official residence where the Ministers had been waiting anxiously since the recess of the previous evening. Togo reported to his colleagues that his proposal had been adopted by the

161 Suzuki's decision to request Imperial intercession to solve an emergency issue was an extraordinary one in Japanese constitutional usage. In all previous conferences in the Emperor's presence, the agenda had been well prepared in advance and unanimously agreed upon among the authorities concerned. Suzuki's determination to pull Japan out of the war is perhaps best illustrated in his post-surrender reflections in Features of the Termination of the War, in which he defends his position in refusing to resign during the crucial period when his Cabinet was split on the surrender issue. Of significance are the following passages which read in substance: "In accordance with the established interpretation, the Emperor was exempt from political responsibilities, and his sanction was nothing but a legal formality. Even in national emergencies the Government decided policy. But it has been my long cherished idea that the Government ought to seek the Emperor's opinion in case they cannot reach an agreement among themselves on a national question of paramount importance." Features of the Termination of the War, op. cit., pp. 46–47.

162 The Emperor's statement is based on the unpublished manuscript of Vice Adm. Hoshina.

Imperial conference with a revision as suggested by Hiranuma. Both he and Suzuki were emphatic on the fact that the decision to surrender was the will of the Emperor. This time the full Cabinet unanimously agreed to Togo's proposal and signed the necessary document of approval.[163] The weary session broke up at 0400 hours, 10 August.

Initial Reply to Potsdam

At 0700 on 10 August a message including the following passages was transmitted to the Governments of the United States, Great Britain, the Soviet Union and China, through the intermediary of Switzerland and Sweden.[164]

In obedience to the gracious command of His Majesty the Emperor who, ever anxious to enhance the cause of world peace, desires earnestly to bring about a speedy termination of hostilities with a view to saving mankind from the calamities to be imposed upon them by further continuation of the war, the Japanese Government several weeks ago asked the Soviet Government, with which neutral relations then prevailed, to render good offices in restoring peace vis-a-vis the enemy powers. Unfortunately, these efforts in the interest of peace having failed, the Japanese Government in conformity with the august wish of His Majesty to restore the general peace and desiring to put an end to the untold suffering entailed by war as quickly as possible, have decided upon the following:

The Japanese Government is ready to accept the terms enumerated in the joint declaration which was issued at Potsdam on July 26th, 1945, by the heads of the Governments of the United States, Great Britain, and China, and later subscribed by the Soviet Government, with the understanding that the said declaration does not comprise any demand which prejudices the prerogatives of His Majesty as a Sovereign Ruler.

The Japanese Government sincerely hope that this understanding is warranted and desire keenly that an explicit indication to that effect will be speedily forthcoming.

But a real difficulty lay in the problem of how to prepare for surrender of a nation, which for months had been geared and groomed for war to the bitter finish. In the first place, it was virtually certain that the war faction would not be satisfied with this decision; in fact, there was danger of a *coup d'etat*. Moreover, it was almost impossible to keep such a drastic, *volte-face* in national policy hidden for long from the public, as well as the ever-watchful eye of the military. Indeed, as early as 10 August, officer elements in the Army, Navy and the High Command had become increasingly restive.

Thus, on the morning of 10 August, soon after the official reply had been sent to the Allies, General Anami felt it necessary to warn his subordinates in the War Ministry against any overt attempts to obstruct the Government's surrender decision.[165] At the same time, Admiral Toyoda, Chief of the Navy General Staff, noting the mounting restless atmosphere within his office, delivered a similar warning.[166]

The Cabinet was caught in a dilemma.

---

163 The entire Cabinet signed the ratification papers. (Statement by Chief Cabinet Secretary Sakomizu, previously cited.)

164 The Japanese Government's message to the four Allied Powers is contained in the Swiss Charge d'affaires note to the U. S. Secretary of State, James F. Byrnes, dated 10 August. *The Axis in Defeat*, op. cit., p. 29.

165 At 0900, Anami called together the officers of lieutenant colonel's rank and above to explain the decision taken at the Imperial conference of 9–10 August, declaring: "......though I am sorry for you, it can't be helped, because this is the Emperor's opinion. The important thing for us to do is to maintain order in the Army. No individual should resort to undisciplined action. Anyway, this decision is made on the condition that the national polity will be preserved. Consequently, the war has not ended with this decision. The Army must be prepared for both peace and war." (Statement by Lt. Col, Masahiko Takeshita, Military Affairs Bureau, War Ministry. Takeshita is the brother-in-law of General Anami.)

166 Statement by Adm. Toyoda, previously cited.

Unless the secret peace maneuvers were brought to fruition quickly, there were growing possibilities of an armed revolt. Yet, there were at the same time great uncertainties about the nature of the expected Allied reply. An unfavorable answer might force a fundamental reconsideration of the surrender question.

Under these circumstances, the Cabinet met again at 1400 to discuss what measures should be taken to make known the fateful decision, but here again there was a fresh clash of views. While some Ministers favored an immediate and frank explanation by the Government, others advised extreme caution. It was agreed in the end that no announcement of the negotiations would be made until an Imperial Rescript proclaiming Japan's acceptance of the Potsdam terms was issued. Instead, the nation would be directed obliquely toward peace in a piecemeal and gradual manner consistent with all developments.

The initial step in that course was taken through an official announcement made in the name of the President of the Cabinet Information Board. After drawing the nation's attention to the grave situation facing Japan as a result of two successive events—the employment by the United States of the extremely destructive atom bomb and the Soviet entry into the Pacific war, the statement read in conclusion:

> "*Despite the prompt action taken by our forces to check enemy aggression, we must admit our fortunes are now at their lowest ebb. The Government will do its utmost to defend the national polity and the honor of the nation. But at the same time, the Government expects you, the one hundred million people of Japan, to overcome all difficulties to uphold our national polity.*[167]

The statement, although pregnant with weighty implications, was vague and ambiguous. It also reflected the delicate situation which existed in the Government and military circles in the period just prior to the surrender. The original draft prepared by the President of the Cabinet Information Board Shimomura had been reworded after lengthy discussion before its release to the press. The Cabinet felt, however, that it should refrain from going beyond this line for the present. Hardly had the announcement been drafted when Shimomura was urgently notified that the newspapers had been requested by the Army to publish some "Instructions to Officers and Men" under the name of the War Minister, the essence of which was in direct contravention to the spirit and intent of the Cabinet statement. This proclamation, prepared by the Military Affairs Bureau of the War Ministry, had been sent to the Tokyo newspapers by junior officers of the organization with the request that it be published together with the Cabinet's announcement. Newspaper editors, recognizing the subtle discrepancies between the two statements, referred the matter to the Cabinet Information Board.

This development touched off a flurry of telephone conversations between the Cabinet Information Board, the Premier's secretariat, and the War Ministry on how to deal with the embarrassing situation. The Chief Cabinet Secretary urged Shimomura to suspend the War Minister's pronunciamento since it was sure to create confusion. However, on the ground that it was too late to suppress the Army release, it was finally decided between Anami and Shimomura to let the matter take its own course.[168]

---

167 Yomiuri-Hochi and Asahi Newspapers, op. cit., Tokyo, 11 Aug 45.

168 (1) On the morning of 10 August, on the recommendation of one of his subordinates, Anami approved a plan to issue instructions to the officers and men on the fields as necessary to maintain the Army's morale in the face of the Soviet entry into the war. Before the draft was formally approved, officers in charge of Army press releases

Thus, in the newspapers of 11 August, the Japanese public read with no little amazement the cryptic Cabinet statement, alongside of which was the War Minister's instructions, in large type, which read substantially as follows:[169]

Instructions to all officers and men of the Army:

The Soviet Union has finally taken up arms against this Empire. Try as she may to disguise the facts by rhetoric, her aspirations to conquer and dominate Greater East Asia are obvious. In the face of this reality, we shall waste no words. The only thing for us to do is to fight doggedly to the end in this holy war for the defense of our divine land.

It is our firm belief that though it may mean chewing grass, eating dirt, and sleeping in the field, a resolute fight will surely reveal a way out of a desperate situation....

10 August 1945
Minister of War

The Government's intention—to disclose by degrees its resolution to accept the Potsdam terms provided the national polity was preserved—was thus obscured by the non-committal phraseology of the Cabinet Information Board announcement and the simultaneous release of the War Minister's belligerent proclamation. The fundamental course had been determined, but the abrupt about-face from total war to unconditional surrender was no easy matter to achieve.

Final Imperial Conference on Surrender

Shortly before 0100 on 12 August, the Foreign Office intercepted a U. S. shortwave broadcast from San Francisco, which disclosed U. S. Secretary of State Byrnes' eagerly-awaited reply to the Japanese communication of 10 August. From the Japanese point of view, the Allied broadcast evaded a definite reply to Japan's requested reservation on the Emperor's prerogatives, around which the whole crucial surrender issue revolved. The Allied reply left much to interpretation and inference. The terms were stated in five brief paragraphs as follows:[170]

With regard to the Japanese Government's message accepting the terms of the Potsdam proclamation but containing the statement, 'with the understanding that the said declaration does not comprise any demand which prejudices the prerogatives of His Majesty as a sovereign ruler,' our position is as follows—

From the moment of surrender the authority of the Emperor and the Japanese Government to rule the State shall be subject to the Supreme Commander of the Allied Powers who will take such steps as he deems proper to effectuate the surrender terms.

The Emperor will be required to authorize and ensure the signature by the Government of Japan and the Japanese Imperial General Headquarters of the surrender terms necessary to carry out the provisions of the Potsdam Declaration, and shall issue his com-

---

prepared duplicates and handed them to newspapermen. The newspapers received the purportedly official Army release shortly before 1600, less than three hours before the printing deadline. Radio Tokyo broadcast the War Minister's instructions just preceding the Cabinet's announcement. (Statements by Lt. Col. Takeshita and Lt. Col. Inaba, previously cited.)

(2) Although the editorial staff of the Asahi Newspaper, which prominently carried both announcements, did not doubt the authenticity of the Army's request, it distinctly felt the sharp differences in the spirit of the Cabinet and War Ministry announcements. After a hot debate among the editors, it was decided to give the Army announcement as much prominence as the Cabinet statement. This action reflected the strong influence which the Army maintained over the press and propaganda organs. (Statement by Taihei Oshima, member of editorial staff of the Asahi Newspaper.)

169 Yomiuri-Hochi and Asahi Newspapers, op. cit,, Tokyo, 11 Aug 45.

170 *The Axis in Defeat*, op. cit., pp. 30–1.

mands to all the Japanese military, naval and air authorities and to all the forces under their control wherever located to cease active operations and to surrender their arms, and to issue such other orders as the Supreme Commander may require to give effect to the surrender terms.

Immediately upon the surrender the Japanese Government shall transport prisoners of war and civilian internees to places of safety, as directed, where they can quickly be placed aboard Allied transports.

The ultimate form of Government of Japan shall, in accordance with the Potsdam Declaration, be established by the freely expressed will of the Japanese people.

The armed forces of the Allied Powers will remain in Japan until the purposes set forth in the Potsdam Declaration are achieved.

An immediate study of the broadcast reply was undertaken by Foreign Vice-Minister Shunichi Matsumoto, together with Chief Cabinet Secretary Sakomizu. A hurried examination was also made by the High Command which had received the monitored text from the Army and Navy intelligence network. At 0800 as the Foreign Vice-Minister and Togo agreed to press for acceptance of the Allied terms, Sakomizu was advising the Premier to accept the Byrnes' reply.

The evasive Allied attitude toward the Emperor's ultimate position, however, aroused high military quarters and stiffened their attitude toward the whole surrender problem. At 0820, General Umezu and Admiral Toyoda, the respective Chiefs of the Army and Navy General Staffs, went to the Imperial Palace together and made a joint plea to the Emperor to reject the Allied conditions.

They declared that the terms subjecting the Emperor to the will of the Allied Supreme Commander was tantamount to reducing Japan to a mere vassalage; the Emperor would be placed in a subordinate position, a situation which neither the Army nor the Navy could countenance. They argued that acceptance of the terms would invite internal chaos and result in the ultimate removal of the Imperial family.[171]

The news of the Allied answer generated uneasiness in Army circles, and the threat of an organized Army *coup d'etat*, which had long been feared by the peace group, appeared to take concrete form. Soon after Toyoda and Umezu had returned from their unscheduled Palace visit, a group of young officers in the War Ministry appeared before Anami, suggesting that the Army undertake a *coup*, if necessary, to stop the peace movement. Anami, however, smothered the suggestion of rebellion by adroitly evading a definite commitment; yet, at the same time, he went to Suzuki to voice his opposition to the Allied reply.[172]

The text of the Allied broadcast and the Foreign Office's interpretation of the terms were submitted to the Emperor at 1100 by Foreign Minister Togo, shortly after he had conferred with Suzuki. Immediately thereafter, Togo met with Marquis Kido, the Emperor's personal advisor, and received assur-

171 Joint plea to the Emperor by Gen. Umezu and Adm. Toyoda, Chiefs of the Army and Navy General Staffs, respectively. The Unpublished Record of the Japanese Navy General Staff, p. 3.

172 (1) Anami was approached by a group of young officers as he was leaving his office for the outside. After listening to the pleas of his subordinates, he hurriedly left the room without committing himself. (Statement by Lt. Col. Takeshita, previously cited.) (2) Anami visited Suzuki's official residence at 1130, 12 August. *Dosei-Tsuzuri* 動静綴 (Record of the Movements of Cabinet Ministers.) This record was kept by the Political Affairs Section of the Tokyo Kempeitai Headquarters. (3) Foreign Minister Togo told Matsumoto late in the afternoon of 12 August that he feared Suzuki had been influenced by Anami and Hiranuma not to accept the broadcast Allied reply. (Statement by Shunichi Matsumoto, Foreign Vice-Minister, May-Aug 45.)

ances of his support in the event further opposition developed over the interpretations put on the Allied communication. Togo, while confident of his assessment of the Allied reply —that is, the Allies eventually would allow the Emperor to remain—feared that the opposition might seize upon Paragraph Four[173] as an argument against capitulation, and thus disrupt the existing unanimity in Government councils.

Meanwhile, Navy Minister Yonai, infuriated by reports that Toyoda had joined Umezu in the palace protestation, reprimanded him for his "thoughtless and irresponsible" act. It was to guard against such actions, he pointed out angrily, that the Navy only a day earlier had instructed local commanders to keep a strict watch over the activities of their subordinates.[174]

The repercussions of the Allied broadcasts became more manifest in Japanese political circles that afternoon. At 1250, Baron Hiranuma, President of the Privy Council who enjoyed a wide personal following in ultranationalistic groups, called on the Premier, and later, on the Lord Keeper of the Privy Seal, to register his opposition to the Allied reply. Hiranuma urged both Suzuki and Kido to press the United States for a clarification on its ultimate position *vis-a-vis* the Emperor disclosing at the same time that the worst possible interpretations were being put on the Allied broadcasts.[175]

The Emperor, undeterred by the impassioned pleas of the High Command representatives summoned the Imperial princes and the families of the princes of the blood to a Palace meeting at 1500 to solicit their support in his resolve for peace. The meeting was to last for nearly three hours, during the course of which he reaffirmed his determination to end the war. At that same hour, the Cabinet, without waiting for the formal Allied communication, was called into an extraordinary session at the Premier's official residence to exchange opinions on the broadcast reply.

Foreign Minister Togo, reporting on the broadcast text, admitted at the outset that the Allied terms could not be considered satisfactory in every respect, but that to make any request for clarification of the Emperor's status or to submit additional reservations would be dangerous. He said that such action could be interpreted by the Allies as an indication of Japan's intention to break off negotiations.

Togo argued at length that however odious and unpleasant were the Allied terms, especially those assigning the position of the Emperor, the fulfillment of the provisions embodied in Paragraphs One and Two, not to mention the all-important Paragraph Four, would, in effect, mean that the status of the Emperor would remain unchanged.[176]

173 Paragraph Four: "The ultimate form of Government of Japan shall, in accordance with the Potsdam Declaration, be established by the freely expressed will of the Japanese people." The military feared that this clause implied an ultimate change in the statutory position of the Emperor. Togo and his supporters argued that this paragraph should be interpreted to mean that the Japanese themselves would be permitted to decide what form of government they wanted and that such being the case it was inconceivable that the Japanese would desire any fundamental change in the Emperor System. Togo Memoirs, op. cit., pp. 23–4.

174 Statements by Vice Adm. Hoshina and Chief Cabinet Secretary Sakomizu, previously cited.

175 By this time the news of the secret peace negotiations were beginning to leak out to the public, and the Government was subject to both threats and encouragement. Handbills began appearing in the streets of Tokyo with the declaration: "Down with the Badoglio regime of Suzuki-Togo-Yonai."

176 The statement in the Allied answer, based on the Potsdam Declaration, which provided that the ultimate form of Government of Japan was to be determined by the freely expressed will of the Japanese people indicated that the Emperor system was not to be abolished, but rather, was clear to those who knew the Japanese people, that it was to be preserved. If the Emperor system were accepted, it naturally followed that the current position of the Emperor would likewise be accepted. (Statement by Foreign Minister Togo, previously cited.)

靖國の英靈に額く　寺内萬治郎

Original Painting by M. Terauchi

PLATE NO. 166
Paying Homage to Nation's War Dead, Yasukuni Shrine

War Minister Anami, together with Home Minister Abe and Justice Minister Matsuzaka, argued that no such interpretation could be placed on the Allied reply and insisted that further inquiries be made. Anami further argued for the attachment of two conditions governing disarmament and occupation, which had been shelved previously.

But more disconcerting to Togo and the peace bloc than the indefinitive arguments was the surprising, stiffening attitude of Suzuki as the deliberations progressed. Suzuki, revealing dissatisfaction with the Allied reply, stated bluntly that there was no alternative but to continue the war. Togo replied sharply that Japan should not continue the war irresponsibly and that unless there were some prospects of victory the Government should forge ahead with its peace negotiations.

Suzuki's unexpected wavering was a sharp disappointment to Togo, who now was virtually fighting a single-handed battle. It was past 1725; the Cabinet had been wrangling heatedly for more than two hours. Fearing that any further discussion would end in a victory for the continued-war faction, as symbolized by Anami, Togo suggested that the meeting adjourn until a formal communication had been received from the Allies.

Following the meeting, Togo upbraided Suzuki for his vacillation and warned that he would take the issue personally to the Emperor if the Premier persisted on talking of continued war.[177] Togo then returned to the Foreign Office and conferred with his aide, Matsumoto, at which time it was decided to withhold news of the expected Allied reply until the peace faction could induce Suzuki to recant. Both Togo and Matsumoto felt that to reopen the deliberations that same night under such adverse circumstances would be fatal.[178]

At 1830, Togo called at the Imperial Palace where he urged Kido to exercise his influence in swinging Suzuki back to the peace fold. The Foreign Minister expressed doubt that the peace moves could be successfully culminated unless immediate action was taken.

In the meantime, the formal Allied reply, dated 11 August and signed by U. S. Secretary of State James F. Byrnes on behalf of the Allied Governments, was received officially by the Foreign Office, almost 12 hours after its dispatch from Berne, Switzerland. The communication was found to be identical with the broadcast text from San Francisco, but its receipt was kept secret.

Later that evening, at 2130, Suzuki was prevailed upon by Kido to accept Togo's interpretation of the Allied terms and to go ahead resolutely on that basis. Suzuki was warned that continued resistance would only compound Japan's misery and that any hesitation or wavering at this point would be extremely hazardous. Suzuki concurred with Kido.[180]

At 0210, 13 August, the Foreign Office received an urgent cablegram from Suemasa Okamoto, the Japanese Minister to Sweden, which reported that the United States, in

177 Statement by Foreign Minister Togo, previously cited.

178 Statement by Foreign Vice-Minister Matsumoto, previously cited.

179 (1) The Allied reply was received by the Foreign Office message center at 1840. It had been dispatched from Berne by the Japanese Minister in Switzerland at 2324, 11 August (0724, 12 August, Tokyo Time.) Foreign Office Radios, op. cit., Japanese Minister Kase's Radiogram Relaying Byrnes Reply, 12 Aug 45. (2) Upon specific instructions from the Foreign Vice-Minister, Akira Oe, Chief of Radio Section, stamped the time of official receipt of the message as 0740, 13 August. News of the receipt was withheld so that the peace faction in the Foreign Office could martial their forces for the impending showdown on the surrender issue. (Statement by Foreign Vice-Minister Matsumoto, previously cited.)

180 Affidavit of Kido, op. cit., IMTFE Record, pp. 31, 185–186.

mapping the reply to the surrender offer, had stoutly resisted heavy pressures by the Soviet Union and China for outright removal of the Emperor.[181]

Armed with this telegram, Matsumoto, on his own initiative, hurriedly conferred with Sakomizu and then with the Premier, pointing out that the message, in effect, supported Togo's strong contention that any attempt by Japan to clarify the Emperor issue would result in a severance of the negotiations. He warned that to quibble over terms would only be playing into the hands of the intransigent elements in the Allied camp who were bitterly opposed to the retention of the Emperor system. Suzuki agreed promptly to accept the Allied terms.

At 0730, Anami paid an unexpected call on Kido and restated his objections to Paragraph Four. Kido pointed out that should Japan refuse to accept the Allied terms she would do so without any valid reason, and emphasized that the Allies would find it difficult to understand the change in attitude. This reaction, he concluded, would do more harm than good. They parted without agreement. In the meantime, the Emperor was informed by the Foreign Minister that the Allied note was identical in wording to the broadcast reply.

At 0900, Suzuki fortified with renewed determination from the night before, convoked an urgent meeting of the six leaders of the Supreme War Direction Council—its first on the Allied reply. As had been the case of the 9 August meeting, the deliberations were characterized by violent clashes of opinion, and the same cleavage persisted with the Premier, Navy and Foreign Ministers aligned on one side and the War Minister and the Chiefs of the Army and Navy General Staffs on the other.

Anami, Umezu and Toyoda argued it was inconceivable that the authority of the Emperor should be subjected to the restrictions imposed by the Allied supreme commander. They emphatically objected to Paragraph Four, expressing the fear that the ultimate expression of the people's will would be coercively channelled by the occupying powers. The negotiations, they insisted, must be continued until there was no doubt as to the Emperor's position.

Suzuki, Togo and Yonai explained that the Allied demands were reasonable since the limitations to be imposed by the Allied Occupation commander on the Emperor's prerogatives would only be within the sphere of the fulfillment of the surrender terms. They argued further that the clause governing the ultimate form of government was an implied guarantee that the Japanese people would not be coerced by the Allies. They cited the case of the Saar plebiscite, calling attention to the fact that the outcome of this issue would depend solely upon the Japanese people themselves. To attempt any further negotiations, they maintained adamantly, would only imperil Japan's final chance to sue for peace.

The cleavage widened rather than narrowed as the procrastinating debate continued. With the six leaders showing no sign of reaching an agreement, Suzuki called an adjournment of the session at 1500 and ordered a full Cabinet meeting for 1600 hours. The arena of discussion again shifted back to the Cabinet.

This time, Premier Suzuki took a more positive and aggressive role in the discussions. He asked each member of his Cabinet to express his real feelings on the Allied offer. The majority favored outright acceptance. Home Minister Abe and Justice Minister Matsuzaka, who earlier had lined up with Anami in opposing the Allied terms,

181 Foreign Office Radios, op. cit., copy of telegram received by Foreign Office at 0210, 13 August, from Suemasa Okamoto, Japanese Minister to Sweden.

agreed eventually to leave the final decision up to the Emperor.

Anami alone stubbornly refused to back down. Moreover, he insisted on appending the two conditions governing self-disarmament and the Allied occupation of the homeland. On this point, Togo again maintained that the conditions should be proposed separately as mere solicitations, a proposal which most of the Ministers supported. He warned repeatedly that insistence on further negotiations might allow time for opposition to gain strength in the Allied camp to the retention of any form of the Imperial family system. Thus, the Cabinet deliberated inconclusively again for more than three hours before adjournment.[182]

That same afternoon Japanese monitors heard U. S. broadcasts in which charges were renewed that the Japanese Government was deliberately delaying a reply. American carrier aircraft meanwhile struck viciously in increasingly bold sweeps throughout the Kanto area. Enemy carrier strikes were also made in other areas of northern Japan.

The surrender discussions had now been deadlocked for more than four days, and the hour of decision had arrived. The Premier, who had thus far failed by persuasion and arguments to steer the Cabinet and the Big Six council, urgently sought to settle the matter once and for all at what he hoped would be a conclusive meeting in the Emperor's presence. The War Minister and the Chiefs of the Army and Navy General Staffs, however, refused to sanction Suzuki's request for an Imperial conference.[183]

An hour after the Cabinet had adjourned at 1930, Umezu, Toyoda and Togo met at the Premier's official residence. This unscheduled meeting, urged by both Chiefs of General Staffs, continued for two hours and was merely a futile repetition of the old arguments. The two Chiefs of General Staff refused to accept Togo's interpretation of the Allied reply and pressed him to request the American Covernment to confirm his interpretation. Togo desisted, declaring that such a move would be construed as a refusal of the proffered terms and thus irreparably wreck Japan's last diplomatic chances.

Meanwhile, at the War Minister's official residence at Miyakezaka, Anami was closeted with several of his subordinates in a meeting beginning at 2000, 13 August, to hear a plan for a *coup d'etat*. According to this plan the *coup* was to be carried out after obtaining approval from both the War Minister and the Chief of Army General Staff. The officials, representing the bitter-end resistance faction, attempted in vain to secure a clear-cut commitment from Anami, and the meeting broke up with an assurance by the War Minister that a definite reply would be forthcoming "tomorrow."[184]

182 Record of the Termination of the War, op. cit., pp. 138144.

183 Chief Cabinet Secretary Sakomizu, while successful in his maneuver of the 9th to obtain prior consent by Umezu and Toyoda to the convocation of an Imperial conference, was unable to do so on the 13th. On the morning of the 13th, officials of the War Ministry and the Army and Navy General Staffs agreed not to sanction an Imperial meeting until the Big Six Council had agreed unanimously on the surrender issue. The War Minister, who had steadfastly resisted the peace moves in the Council and Cabinet debates, insisted that the surrender issue should be resolved by the members of the Council and the Cabinet and that invocation of the Emperor's decision should be avoided as far as possible. (Statement by Chief Cabinet Secretary Sakomizu, previously cited.)

184 A half-dozen officers holding key positions in the Military Affairs Bureau of the War Ministry, including Major Hatanaka (who later was to become one of the ringleaders in the Palace revolt), called on Anami at 2000 hours, 13 August, bearing plans for a large-scale *coup d'etat*. The plan roughly was as follows: (a) The aim to elicit from the Emperor sanction to continue negotiations until the desired peace conditions were met: (b) The forces to be used; the 1st Imperial Guards Division and the Twelfth Area Army: (c) Course of action; the bitter-end

Early on the morning of 14 August, American planes dropped over Japanese cities, including Tokyo, leaflets containing the Japanese Government's offer of 10 August to accept the Potsdam terms and the Allied reply written in Japanese. The secret negotiations, which the Government had tried to conceal from the public, were now revealed.

At 0830, Kido, carrying one of the leaflets, went directly to the Emperor and urgently advised him to take prompt action. Explaining he feared the leaflets would have a profound effect on the armed forces and on the people, he emphasized that every moment peace was delayed was dangerous. He pointed out the humiliating effect the enemy's disclosure of the Cabinet's secret maneuvers would have on the homeland troops who were being deployed against an invasion, stressing that their attitude would give the extremists a powerful weapon with which to oppose the Emperor's will to sue for peace. It was imperative, therefore, that the Emperor declare war be ended immediately lest he might lose control of the armed forces in the field. The Emperor readily agreed.

Soon after the audience, Kido conferred with Premier Suzuki who had come to the Palace to seek the Emperor's assistance in convoking an Imperial conference. Suzuki revealed the Government's helpless position resulting from the unyielding attitude of the Chiefs of Army and Navy General Staffs. Both he and Kido agreed it was useless to engage in further discussions and that in view of the increasingly perilous situation a final showdown was immediately necessary. Both men then held a joint audience with the Emperor shortly before 0900 and petitioned him to convoke on his own initiative an Imperial conference in defiance of a time-honored precedent.[185]

At 1000, the Emperor, in furtherance of his peace resolution, summoned Fleet Admiral Osami Nagano, Field Marshals Sugiyama and Shunroku Hata, the highest ranking military men present in the homeland at the time, and requested them to secure the obedient compliance of the armed forces to his surrender decision.[186]

Thus occurred the climactic development when the full Cabinet, the two Chiefs of the Army and Navy General Staffs, the President of the Privy Council, and the secretaries of the Supreme War Direction Council were urgently summoned to assemble at the Imperial Palace. The Cabinet Ministers had gathered at the Premier's official residence in a meeting scheduled for 1000, and when they received the Imperial summons they had no time to change to court clothing, so they were given special permission to proceed to the palace in their

resistance faction, taking advantage of a provision in the Imperial Ordinance concerning martial law, which authorizes local garrison commanders to proclaim martial law provisionally without Imperial sanction in the case of emergency, was to place the whole Tokyo area under martial law. By so doing, the Emperor would be isolated from the peace group and would be requested to issue such order as the Army might desire. (d) Prerequisites ; the War Minister, the Chief of the Army General Staff, the Commander of the Eastern District Army, and the Commander of the 1st Imperial Guards Division should all agree to the plan. Anami, while appearing favorably disposed to the plan, begged for time to make a final decision and said, " I will give you a definite reply tomorrow morning after I have thought it over." (Statement by Lt. Col. Inaba, previously cited. Inaba was one of six members who discussed the *coup d'etat* plan.)

185 In accordance with custom, more rigidly observed in recent years, an Imperial conference was convoked upon the joint request of the Premier and the Chiefs of the Army and Navy General Staffs and only after they had agreed upon the issues. This left the Emperor with virtually no necessity to express his own opinion. Under the circumstances, it was almost unthinkable that an Imperial conference would be convoked without the previous consent of the High Command.

186 The developments noted on this page are taken from material found in : (1) Affidavit of Kido, op. cit., pp. 31, 189–191. (2) Statement by Marquis Kido, previously cited.

regular business suits.

The final pre-surrender conference began at 1100 in the same air raid shelter in the Palace where, on 9–10 August, the Emperor had revealed unmistakably his desire for peace. Suzuki opened the meeting with a brief review of developments since 10 August. The Premier noted that most of the Cabinet members were in favor of accepting the Potsdam terms, but there had been no unanimity. He then proposed that the Emperor hear the opposing views before making a final decision. He named in turn the Chief of the Army General Staff, the Chief of the Navy General Staff, and the War Minister to express their views.

The three hold-outs voiced the same opinions and tearfully pleaded for further inquiries of the Allies, asserting that under the stated terms of the Byrnes' communication it would be difficult for Japan to preserve her national polity. If that were not possible, they concluded it would be better to continue the war. No others were called upon by the Premier to speak.

The Emperor then broke the grim silence and poignantly reiterated the Imperial wish, his voice trembling with emotion. He desired that all members agree with him. He said he was of the opinion that the Allied reply was satisfactory; that this opinion was unchanged since 10 August; and that his decision was based on a studied observation of the world and domestic situations. Continued war, he emphasized again, would mean ultimate destruction of Japan, whereas otherwise the nation's revival could be assured as long as the "seeds and source" existed. He pleaded with his audience to put aside personal sentiments.

The Emperor spoke with choked emotion as he referred to the plight of the war sufferers and the families of the war dead and said he would do everything he could for their consolation. While admitting fear that the shock of surrender would be especially far-reaching in the armed forces, the Emperor emphasized he was ready to appeal directly to his troops if necessary—he was willing to undertake any task which would facilitate the conclusion of hostilities.[187]

Finally, he offered to make a radio broadcast to the people of his decision to surrender and asked his Ministers to draw up a rescript ending the war. The attendants wept aloud at the Emperor's unprecedented plea. Suzuki then assured the Emperor a rescript would be prepared immediately. The historic Imperial conference was concluded at noon.[188]

Japan's Reply to Final Allied Offer

With the Imperial verdict unequivocally stated, the Cabinet members, upon whom lay the official responsibility for surrender, met at the Premier's official residence at 1300 to give the finishing touches to a draft Imperial Rescript which had been under preparation by the Cabinet secretariat since the Imperial conference of 10 August. Emphasis was placed on the Emperor's statements he had made at the Imperial conference on 10 August.

Although it had been feared by many Cabinet members that Anami might tender his resignation at this critical moment and thus compel the whole Cabinet to collapse, the War Minister chatted amiably with his colleagues,

187 Following the conference, the Emperor revealed to the War and Navy Ministers of his readiness to dispatch his aide-de-camps to Army and Navy headquarters to appeal to the armed forces. Navy Minister Yonai promptly declined the offer, saying he would do his utmost to secure the Navy's compliance with the Imperial wish for surrender. War Minister Anami followed suit. Notes on the Termination of the War, op. cit., p. 59.

188 (1) Ibid., pp. 58–9, (2) Truth of the Surrender Period, op. cit. Asahi Newspaper, Tokyo, 19 Jan 46.

and participated actively in the discussions on the Rescript. If the outcome had been a crushing defeat for the War Minister, he showed no outwards signs of it.[189]

At 2100, after nearly eight hours of continuous discussion, the Cabinet finished the wording of the draft rescript, after which it was taken to the Emperor immediately by the Premier for his seal and signature. The Ministers then counter-signed the Rescript to formalize the Emperor's surrender decision. It was already 2250. At that same moment, the Emperor was recording his impending broadcast message on a phonograph disc in the room of the Imperial Household Ministry. At 2300, the Imperial Rescript was ordered to be proclaimed but because of printing difficulties it was not until the following day that the official copies were distributed. Japan's surrender, however, had become effective at 2300, 14 August.[190]

With the promulgation of the Rescript, the following telegram, which had been prepared by Togo earlier in consultation with Foreign Office experts, was swiftly dispatched to the four Allied Governments through the facilities of the Swiss Government:[191]

With reference to the Japanese Government's note of August 10 regarding its acceptance of the provisions of the Potsdam Declaration and the reply of the Governments of the United States, Great Britain, the Soviet Union and China sent by the American Secretary of State Byrnes under the date of August 11, the Japanese Government has the honor to communicate to the Governments of the four Powers as follows:

1. His Majesty the Emperor has issued an Imperial Rescript regarding Japan's acceptance of the provisions of the Potsdam Declaration.

2. His Majesty the Emperor is prepared to authorize and ensure the signature by His Government and the Imperial General Headquarters of the necessary terms for carrying out the provisions of the Potsdam Declaration. His Majesty is also prepared to issue his commands to all the military, naval and air authorities of Japan and all forces under their control wherever located to cease active operations, to surrender arms and to issue such other orders as may be required by the Supreme Commander of the Allied Powers for the execution of the above-mentioned terms.

The long internal struggle was over. The Emperor's will had prevailed, and the die was cast for peace.

---

189 As the Army's top agent in the Cabinet, Anami was fully cognizant of both the direction being taken by the Government and the widespread desire within the Army for the continuation of the war. At the same time, while vigorously championing the cause of bitter-end resistance in the conference discussions, he did not disclose the Cabinet's peace intentions to the Army staff until 10 August, but took full responsibility for his own actions. Anami could have at any time wrecked the Cabinet by resigning or effecting a *coup d'etat,* and thus forced a fight to the finish, but he was reluctant to oppose what were the explicit wishes of the Emperor. (1) Statements by Marquis Kido, Chief Cabinet Secretary Sakomizu, Foreign Minister Togo and Lt. Col. Takeshita, all previously cited. (2) Record on the Termination of the War, op. cit., pp. 173–4. (3) Fall of the Japanese Army, op. cit., p. 51. (4) "Mitsumasa Yonai and Korechika Anami," by Seizo Sakonji, Maru magazine, Sep 49.

190 Statement by Chief Cabinet Secretary Sakomizu, previously cited.

191 *The Axis in Defeat,* op. cit., pp. 31–2.

CHAPTER XXI
RETURN TO PEACE

Imperial Announcement of Surrender

With the promulgation late on 14 August of the Emperor's Rescript accepting the Allied surrender terms, Japan's decision to capitulate had become final. But not until thirteen hours later did the nation at large learn the tragic and stunning news that the Empire had bowed in defeat for the first time in its recorded history.[1]

From early morning on 15 August, radio stations throughout the country alerted the people to stand by for an important broadcast at 12 o'clock noon. In Tokyo and other cities, special preparations were hastily carried out to ensure that a maximum number of people would hear the broadcast. Loudspeakers were set up in public places as well as in government and private offices where most employees stayed during the noon hour to eat meager box lunches.

As the time for the broadcast drew near, tense expectancy gripped the nation. In homes, offices and factories and on city streets throughout the land, people paused to gather near radios or loudspeakers. Few outside of limited official and press circles had any inkling of what was to come. The vast majority of the nation, even though feeling that defeat could not be far off, expected that they were about to hear a new exhortation to fight to the death or the announcement of an Imperial declaration of war on Soviet Russia.

The strident note of the noon time-signal was followed by the muted strains of the national anthem. Then, listeners heard State Minister Shimomura, President of the Cabinet Information Board, announce that the next voice would be that of His Majesty the Emperor. People caught their breaths in quick surprise for never before had the Emperor spoken directly to his subjects by radio. They listened in tense silence as the Emperor's voice came over the air, reading the solemn and fateful words of the Imperial Rescript:[2]

After pondering deeply the general trend of the world situation and the actual state of Our Empire, We have decided to effect a settlement of the present crisis by resort to an extraordinary measure. To Our good and loyal subjects, we hereby convey Our will.

We have commanded Our Government to communicate to the Governments of the United States, Great Britain, China and the Soviet Union that Our Empire accepts the terms of their Joint Declaration.

To strive for the common prosperity and happiness of all nations as well as the security and well-being of Our subjects is the solemn obligation handed down to Us by Our Imperial Ancestors, and We keep it close to heart. Indeed, We declared war on America and Britain out of Our sincere desire to ensure Japan's

---

1 This chapter was originally prepared in Japanese by Rear Adm. Katsuhei Nakamura, Imperial Japanese Navy. Duty assignments of this officer were as follows: Chief Secretary, Navy Ministry, 5 Nov 40–24 Feb 43; Commander, H. M. S. *Myoko*, 2 Mar–10 Dec 43; Naval Attaché, Japanese Embassy at Nanking, and concurrently Vice-Chief of Staff, China Area Fleet, 4 Jan–23 Jul 44; Directer, General Affairs Bureau, Navy Aeronautical Department, 7 Aug 44–5 Sep 45. All source materials cited in this chapter are located in G-2 Historical Section Files, GHQ FEC.

2 *Shosho* 詔書 (Imperial Rescript) 14 Aug 45. (Cf. Plate No. 167)

self-preservation and the stabilization of East Asia. It was not Our intention either to infringe upon the sovereignty of other nations or to seek territorial aggrandizement.

The hostilities have now continued for nearly four years. Despite the gallant fighting of the Officers and Men of Our Army and Navy, the diligence and assiduity of Our servants of State, and the devoted service of Our hundred million subjects—despite the best efforts of all—the war has not necessarily developed in Our favor, and the general world situation also is not to Japan's advantage.[3] *Furthermore, the enemy has begun to employ a new and cruel bomb which kills and maims the innocent and the power of which to wreak destruction is truly incalculable.*

Should We continue to fight, the ultimate result would be not only the obliteration of the race but the extinction of human civilization. Then, how should We be able to save the millions of Our subjects and make atonement to the hallowed spirits of Our Imperial Ancestors? That is why We have commanded the Imperial Government to comply with the terms of the Joint Declaration of the Powers.

To those nations which, as Our allies, have steadfastly cooperated with the Empire for the emancipation of East Asia, We cannot but express Our deep regret. Also, the thought of Our subjects who have fallen on the field of battle or met untimely death while performing their appointed tasks, and the thought of their bereaved families, rends Our heart, and We feel profound solicitude for the wounded and for all war-sufferers who have lost their homes and livelihood.

The suffering and hardship which Our nation yet must undergo will certainly be great. We are keenly aware of the innermost feelings of all ye, Our subjects. However, it is according to the dictates of time and fate that We have resolved, by enduring the unendurable and bearing the unbearable, to pave the way for a grand peace for all generations to come.

Since it has been possible to preserve the structure of the Imperial State, We shall always be with ye, Our good and loyal subjects, placing Our trust in your sincerity and integrity. Beware most strictly of any outburst of emotion which may engender needless complications, and refrain from fraternal contension and strife which may create confusion, lead ye astray and cause ye to lose the confidence of the world. Let the nation continue as one family from generation to generation with unwavering faith in the imperishability of Our divine land and ever mindful of its heavy burden of responsibility and the long road ahead. Turn your full strength to the task of building a new future. Cultivate the ways of rectitude, foster nobility of spirit, and work with resolution so that ye may enhance the innate glory of the Imperial State and keep pace with the progress of the world. We charge ye, Our loyal subjects, to carry out faithfully Our will.

The Government had taken special pains to phrase the Imperial message in such a way as to soften the impact of surrender on the nation at large. The word "surrender" itself had been studiously avoided. Nevertheless, the import of the Emperor's words was painfully clear to all who heard them. Even those who did not know what the Joint Declaration of the Allied Powers signified—and there were many such in the rural areas—swiftly understood that the Emperor was announcing the termination of hostilities on terms laid down by the enemy. Japan, in short, was accepting final defeat after more than three and a half years of costly fighting and hard sacrifice.[4]

Although the horrible destruction of Hiro-

3 This was one of the most controversial passages in the Rescript. As originally drafted for Cabinet approval, it read: "Despite the gallant fighting of the Officers and Men of Our Army and Navy. . . .*the war situation worsens from day to day*, and the general world situation also is not to Japan's advantage." During the Cabinet discussion of the text on the afternoon of 14 August, War Minister Anami objected to this wording as carrying too strong an implication of the total defeat of the Japanese armed forces. A compromise was finally reached whereby the italicized passage was amended to read, "*the war has not necessarily developed in Our favor.*" *Shusenki* 終戦記 (Record of the Termination of the War) by Hiroshi (Kainan) Shimomura, Oct 48, p. 161.

4 Owing to poor radio reception, there were some outlying districts in which the Emperor's broadcast was not understood and many thought that they were being urged to fight on. Press publication of the Imperial Rescript, however, promptly corrected this misunderstanding.

詔書

朕深ク世界ノ大勢ト帝國ノ現狀トニ鑑ミ非常ノ措置ヲ以テ時局ヲ收拾セムト欲シ茲ニ忠良ナル爾臣民ニ告ク

朕ハ帝國政府ヲシテ米英支蘇四國ニ對シ其ノ共同宣言ヲ受諾スル旨通告セシメタリ

抑帝國臣民ノ康寧ヲ圖リ萬邦共榮ノ樂ヲ偕ニスルハ皇祖皇宗ノ遺範ニシテ朕ノ拳々措カサル所曩ニ米英二國ニ宣戰セル所以モ亦實ニ帝國ノ自存ト東亞ノ安定トヲ庶幾スルニ出テ他國ノ主權ヲ排シ領土ヲ侵スカ如キハ固ヨリ朕カ志ニアラス然ルニ交戰已ニ四歳ヲ閲シ朕カ陸海將兵ノ勇戰朕カ百僚有司ノ勵精朕カ一億衆庶ノ奉公各々最善ヲ盡セルニ拘ラス戰局必スシモ好轉セス世界ノ大勢亦我ニ利アラス加之敵ハ新ニ殘虐ナル爆彈ヲ使用シテ頻ニ無辜ヲ殺傷シ慘害ノ及フ所眞ニ測ルヘカラサルニ至ル而モ尚交戰ヲ繼續セムカ終ニ我カ民族ノ滅亡ヲ招來スルノミナラス延テ人類ノ文明ヲモ破却スヘシ斯ノ如クムハ朕何ヲ以テカ億兆ノ赤子ヲ保シ皇祖皇宗ノ神靈ニ謝セムヤ是レ朕カ帝國政府ヲシテ共同宣言ニ應セシムルニ至レル所以ナリ

朕ハ帝國ト共ニ終始東亞ノ解放ニ協力セル諸盟邦ニ對シ遺憾ノ意ヲ表セサルヲ得ス帝國臣民ニシテ戰陣ニ死シ職域ニ殉シ非命ニ斃レタル者及其ノ遺族ニ想ヲ致セハ五内爲ニ裂ク且戰傷ヲ負ヒ災禍ヲ蒙リ家業ヲ失ヒタル者ノ厚生ニ至リテハ朕ノ深ク軫念スル所ナリ惟フニ今後帝國ノ受クヘキ苦難ハ固ヨリ尋常ニアラス爾臣民ノ衷情モ朕善ク之ヲ知ル然レトモ朕ハ時運ノ趨ク所堪ヘ難キヲ堪ヘ忍ヒ難キヲ忍ヒ以テ萬世ノ爲ニ太平ヲ開カムト欲ス

朕ハ茲ニ國體ヲ護持シ得テ忠良ナル爾臣民ノ赤誠ニ信倚シ常ニ爾臣民ト共ニ在リ若シ夫レ情ノ激スル所濫ニ事端ヲ滋クシ或ハ同胞排擠互ニ時局ヲ亂リ爲ニ大道ヲ誤リ信義ヲ世界ニ失フカ如キハ朕最モ之ヲ戒ム宜シク擧國一家子孫相傳ヘ確ク神州ノ不滅ヲ信シ任重クシテ道遠キヲ念ヒ總力ヲ將來ノ建設ニ傾ケ道義ヲ篤クシ志操ヲ鞏クシ誓テ國體ノ精華ヲ發揚シ世界ノ進運ニ後レサラムコトヲ期スヘシ爾臣民其レ克ク朕カ意ヲ體セヨ

御名御璽

昭和二十年八月十四日

各國務大臣副署

IMPERIAL RESCRIPT

To Our good and loyal subjects:

After pondering deeply the general trends of the world and the actual conditions obtaining in Our Empire today, We have decided to effect a settlement of the present situation by resorting to an extraordinary measure.

We have ordered Our Government to communicate to the Governments of the United States, Great Britain, China and the Soviet Union that Our Empire accepts the provisions of their Joint Declaration.

To strive for the common prosperity and happiness of all nations as well as the security and well-being of Our subjects is the solemn obligation which has been handed down by Our Imperial Ancestors, which We lay close to heart. Indeed, We declared war on America and Britain out of Our sincere desire to ensure Japan's self-preservation and the stabilization of East Asia, it being far from Our thought either to infringe upon the sovereignity of other nations or to embark upon territorial aggrandizement. But now the war has lasted for nearly four years. Despite the best that has been done by every one—gallant fighting of military and naval forces, the diligence and assiduity of Our servants of the State and the devoted service of Our one hundred million people, the war situation has not necessarily improved and the general trends of the world are also not to Japan's advantage. Moreover, the enemy has begun to employ a new and most cruel bomb, the power of which to do damage is indeed incalculable, taking the toll of many innocent lives. Should We continue to fight, it would not only result in an ultimate collapse and obliteration of the Japanese nation, but also it would lead to the total extinction of human civilization. Such being the case, how are We to save the millions of Our subjects; or to atone Ourselves before the hallowed spirits of Our Imperial Ancestors? This is the reason why We have ordered the acceptance of the provisions of the Joint Declaration of the Powers.

We cannot but express the deepest sense of regret to our Allied nations who have consistently co-operated with the Empire towards the emancipation of East Asia. The thought of those officers and men as well as others who have fallen in the fields of battle, those who died at their posts of duty, or those who met with untimely death and all their bereaved families, pains Our heart night and day. The welfare of the wounded and war-sufferers and of those who have lost their home and livelihood are the objects of Our profound solicitude. The hardships and sufferings to which Our nation is to be subjected hereafter will be certainly great. We are keenly aware of the inmost feelings of all ye, Our subjects. However, it is according to the dictate of time and fate that We have resolved to pave the way for a grand peace for all the generations to come by enduring the unendurable and suffering what is insufferable.

Having been able to safeguard and maintain the structure of the Imperial State, We are always with ye, Our good and loyal subjects, relying upon your sincerity and integrity. Beware most strictly of any outburst of emotion which may engender needless complications, or any fraternal contention and strife which may create confusion, lead ye astray and cause ye to lose the confidence of the world. Let the entire nation continue as one family from generation to generation, ever firm in its faith of the imperishableness of its divine land, and mindful of its heavy burden of responsibilities, and the long road before it. Unite your total strength to be devoted to the construction for the future. Cultivate the ways of rectitude; foster nobility of spirit; and work with resolution so as ye may enhance the innate glory of the Imperial State and keep pace with the progress of the world.

The 14th day of the 8th month of the 20th year of Showa

Signature	Office
SUZUKI, Kantaro, Baron	Prime Minister
YONAI, Mitsumasa	Minister of Navy
MATSUZAKA, Hiromasa	Minister of Justice
ANAMI, Korechika	Minister of War
TOYODA, Teijiro	Minister of Munition
OKADA, Tadahiko	Minister of Welfare
SAKURAI, Hyogoro	Minister of State
SAKONJI, Masazo	Minister of State
SHIMOMURA, Hiroshi	Minister of State
HIROSE, Toyosaku	Minister of Finance
OTA, Kozo	Minister of Education
ISHIGURO, Tadaatsu	Minister of Agriculture and Commerce
ABE, Genki	Minister of Home Affairs
TOGO, Shigenori	Minister of Foreign Affairs
YASUI, Toji	Minister of State
KOHIYAMA, Naoto	Minister of Transportation and Communication

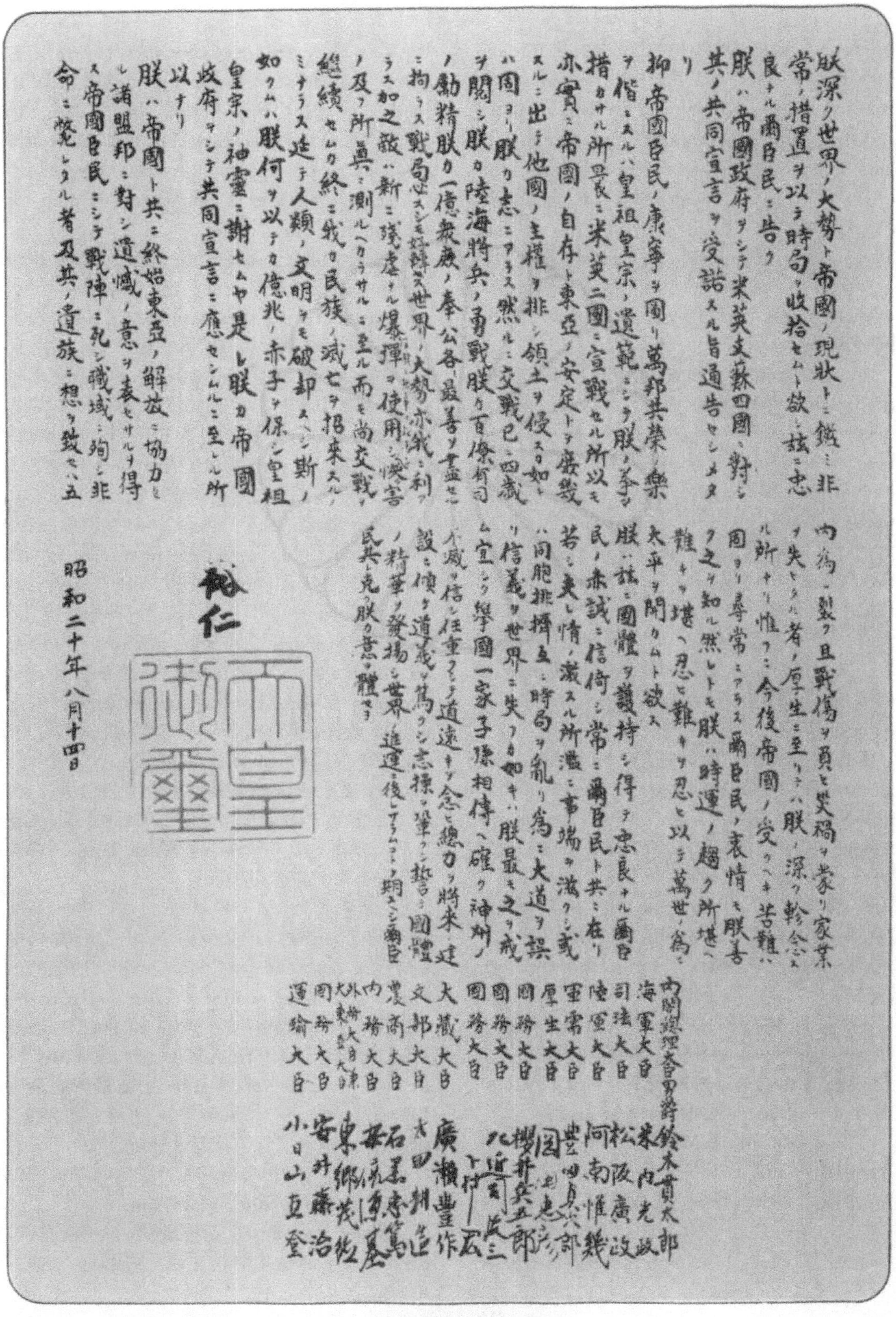

朕深ク世界ノ大勢ト帝國ノ現狀トニ鑑ミ非常ノ措置ヲ以テ時局ヲ收拾セムト欲シ茲ニ忠良ナル爾臣民ニ告ク

朕ハ帝國政府ヲシテ米英支蘇四國ニ對シ其ノ共同宣言ヲ受諾スル旨通告セシメタリ

抑々帝國臣民ノ康寧ヲ圖リ萬邦共榮ノ樂ヲ偕ニスルハ皇祖皇宗ノ遺範ニシテ朕ノ拳々措カサル所曩ニ米英二國ニ宣戰セル所以モ亦實ニ帝國ノ自存ト東亞ノ安定トヲ庶幾スルニ出テ他國ノ主權ヲ排シ領土ヲ侵スカ如キハ固ヨリ朕カ志ニアラス然ルニ交戰已ニ四歳ヲ閲シ朕カ陸海將兵ノ勇戰朕カ百僚有司ノ勵精朕カ一億衆庶ノ奉公各々最善ヲ盡セルニ拘ラス戰局必スシモ好轉セス世界ノ大勢亦我ニ利アラス加之敵ハ新ニ殘虐ナル爆彈ヲ使用シテ頻ニ無辜ヲ殺傷シ慘害ノ及フ所眞ニ測ルヘカラサルニ至ル而モ尚交戰ヲ繼續セムカ終ニ我カ民族ノ滅亡ヲ招來スルノミナラス延テ人類ノ文明ヲモ破却スヘシ斯ノ如クムハ朕何ヲ以テカ億兆ノ赤子ヲ保シ皇祖皇宗ノ神靈ニ謝セムヤ是レ朕カ帝國政府ヲシテ共同宣言ニ應セシムルニ至レル所以ナリ

朕ハ帝國ト共ニ終始東亞ノ解放ニ協力セル諸盟邦ニ對シ遺憾ノ意ヲ表セサルヲ得ス帝國臣民ニシテ戰陣ニ死シ職域ニ殉シ非命ニ斃レタル者及其ノ遺族ニ想ヲ致セハ五内爲ニ裂ク且戰傷ヲ負ヒ災禍ヲ蒙リ家業ヲ失ヒタル者ノ厚生ニ至リテハ朕ノ深ク軫念スル所ナリ惟フニ今後帝國ノ受クヘキ苦難ハ固ヨリ尋常ニアラス爾臣民ノ衷情モ朕善ク之ヲ知ル然レトモ朕ハ時運ノ趨ク所堪ヘ難キヲ堪ヘ忍ヒ難キヲ忍ヒ以テ萬世ノ爲ニ太平ヲ開カムト欲ス

朕ハ茲ニ國體ヲ護持シ得テ忠良ナル爾臣民ノ赤誠ニ信倚シ常ニ爾臣民ト共ニ在リ若シ夫レ情ノ激スル所濫ニ事端ヲ滋クシ或ハ同胞排擠互ニ時局ヲ亂リ爲ニ大道ヲ誤リ信義ヲ世界ニ失フカ如キハ朕最モ之ヲ戒ム宜シク擧國一家子孫相傳ヘ確ク神州ノ不滅ヲ信シ任重クシテ道遠キヲ念ヒ總力ヲ將來ノ建設ニ傾ケ道義ヲ篤クシ志操ヲ鞏クシ誓テ國體ノ精華ヲ發揚シ世界ノ進運ニ後レサラムコトヲ期スヘシ爾臣民其レ克ク朕カ意ヲ體セヨ

裕仁

天皇御璽

昭和二十年八月十四日

内閣總理大臣男爵 鈴木貫太郎
海軍大臣 米内光政
司法大臣 松阪廣政
陸軍大臣 阿南惟幾
軍需大臣 豊田貞次郎
厚生大臣 岡田忠彦
國務大臣 櫻井兵五郎
國務大臣 左近司政三
國務大臣 下村宏
大藏大臣 廣瀬豊作
文部大臣 太田耕造
農商大臣 石黒忠篤
内務大臣 安倍源基
外務大臣兼大東亞大臣 東郷茂徳
國務大臣 安井藤治
運輸大臣 小日山直登

PLATE NO. 167
Imperial Rescript Ending the War

shima and Nagasaki by the atom bomb, coupled with Soviet entry into the war, had increased popular realization that the war was lost, the sudden, grim news of capitulation caught the nation psychologically unprepared. So thorough had been the indoctrination for a last-ditch defense of the homeland that many at first could not comprehend that Japan had surrendered. Then, as the meaning of the Emperor's words sank in, the people were stunned and bewildered.

Mingled with the sense of shock were grief, despair and disillusionment. To a proud and intensely patriotic people who had never before been conquered in war, defeat was a bitter pill to swallow. Many wept openly as they listened to the Emperor's solemn announcement. Some felt swift anger against the nation's leaders and the fighting services for their failure to avert defeat; others blamed themselves for falling short in their war effort. Above all, there was a feeling of intense sympathy for the Emperor, who had been forced to take so tragic and painful a decision.

Following the Emperor's broadcast, war factories throughout the country dismissed their workers and closed their doors. The newspapers, which had been ordered to suspend their customary morning editions, came out in the afternoon, all carrying the text of the Imperial Rescript, an unabridged translation of the Potsdam Declaration, and texts of the notes exchanged with the Allied Powers. In Tokyo, throughout the afternoon, crowds of weeping citizens gathered in the vast plaza before the Imperial Palace and at the Meiji and Yasukuni Shrines to bow in reverence and prayer.

The shock and grief of the moment, combined with the dark uncertainty of the future, prevented any widespread feeling of relief at the termination of hostilities. Bombings and bloodshed were at an end, but defeat seemed likely to bring only a continuation of hardship and privation. Starvation already gripped the land. To this might now be added the breakdown of public discipline and order, acts of violence and oppression by the enemy forces of occupation, and a staggering burden of reparations.

Despite the dark outlook, the nation drew solace and courage from the Emperor's assurance that he would remain to lead the people through the difficult period that lay ahead. His appeal for strict compliance with the Imperial will made a profound impression, and "Reverent Obedience to the Rescript" became the rallying cry with which the nation prepared to face the harsh consequences of capitulation.

Three hours after the Emperor's broadcast, the Suzuki Cabinet, its difficult mission at last accomplished, tendered its collective resignation. The Premier stated that the Government, by resigning, meant to bear responsibility for its unprecedented action in appealing to the Emperor to make the final surrender decision, and he declared that it was vitally necessary for younger leaders to take over the stupendous task of national reconstruction.[5] The Emperor accepted the resignation but commanded the Government to remain in office until a new Cabinet could be formed.

Broadcasting on the evening of the 15th, Premier Suzuki appealed to the nation to unite in absolute loyalty to the Throne in face of the grave national crisis. Japan, he stated, had accepted the terms of the Potsdam Declaration "fully confident that there will be no change in the sovereignty of the Emperor." He emphasized further that His Majesty's decision to end the war had been taken out of compassion for his subjects and in careful consideration of existing circumstances.

Addressing himself specifically to the members of the armed forces, the Premier voiced

5 Letter of resignation submitted by Premier Suzuki to the Throne, 15 Aug 45.

full sympathy with their feelings regarding surrender. He stressed, however, that it was the primary duty of the Emperor's subjects to assist in fostering the glory and prosperity of the Throne and that this could be accomplished only through complete loyalty and obedience to the Imperial will. He concluded with this exhortation:[6]

I hope that the intrinsic sincerity of our people will be given full expression in the fulfillment of the terms of the Potsdam Declaration, and that thereby we shall speedily regain the confidence of the world. Our primary concern must be to restore the Empire to its proper place in the world as quickly as possible through conscientious conduct and the display of the moral strength of our nation under Imperial rule. To this end we and succeeding generations must labor mightily and indomitably, uniting as one family in faithful observance of the Imperial Rescript.

Military Reaction

The Emperor's personal radio message of the carefully phrased Rescript concerning the surrender decision was couched in almost the exact words used by him in the last Imperial conference. As the members of the Government had hoped, the message had the desired effect. The people at large, though shocked and grief-stricken, were impressed that this decision represented the actual will of the Emperor himself rather than an act of the Government, which the sovereign had merely ratified. As the momentous day of 15 August came to a close, there was scarcely any doubt but that the nation as a whole would obediently accept the Imperial command.

The reaction of the armed forces, however, was more dangerously problematical. Although the top leaders of both fighting services had finally bowed to the Emperor's will at the Imperial conference of 14 August, they themselves were deeply concerned over the possibility of action by the extremist elements within the Army and Navy to defy the surrender decision. In fact, even before this decision was finally taken on the 14th, War Minister Anami was already aware that a coterie of young officers in the War Ministry and the Army General Staff was planning an armed *coup d'etat* for the purpose of forestalling surrender and establishing a military government which would continue the war.[7]

This was obviously a situation fraught with danger. The Army leaders had held out until the very last moment against acceptance of the Potsdam terms, and they still might have blocked or seriously impeded the surrender by giving their sanctions to the *coup d'etat* plan. Nevertheless, when the final decision was taken at the Emperor's command, they loyally decided to obey and took swift action to ensure discipline

6 *Yomiuri-Hochi Shimbun* 讀賣報知新聞 (Yomiuri-Hochi Newspaper), Tokyo, 17 Aug 45.

7 The ringleaders of the *coup d'etat* plot were principally subordinate officers of the Military Affairs Bureau of the War Ministry but also included two officers of the Army General Staff. On the night of 13 August, five of the conspirators called on General Anami at his official residence and secretly disclosed their plans to him, seeking his approval. (Cf. Chapter XX, p. 669) Anami was non-comittal at this time, but after conferring with General Umezu, Chief of the Army General Staff, the following morning, he told the plotters that the *coup d'etat* must be abandoned in view of Umezu's disapproval. Anami made no move to arrest the ringleaders, among whom was his own brother-in-law, but he ordered Lt. Gen. Sanji Okido, Commanding General, Army Military Police (Kempeitai), and Lt. Gen. Takeshi Mori, Commanding General, 1st Imperial Guards Division, to exercise precautionary surveillance over them lest they proceed to act in defiance of his orders. Cooperation of the Guards Division, assigned to protection of the Imperial Palace, was vital to the *coup d'etat* plan. Lt. Gen. Mori, on 12 August, had already rejected a request by the plotters for his support in executing the plan, but some of his staff officers had been independently won over by the conspirators. (Statements by Col. Saburo Hayashi, Private Secretary to War Minister Anami and Lt. Col. Masahiko Takeshita, Military Affairs Bureau, War Ministry.)

PLATE NO. 168
Scene in Front of Imperial Palace

and order throughout the Army.

Following the final Imperial conference on 14 August, the Army " Big Three "—War Minister Anami, Chief of the Army General Staff Umezu, and Inspectorate-General of Military Training General Kenji Doihara—chanced to meet at the War Ministry together with Field Marshals Hata and Sugiyama, the top operational commanders of the Army forces in the homeland. These five men affixed their seals to a joint resolution pledging that the Army would " conduct itself in accordance with the Imperial decision to the last." The resolution was endorsed immediately afterward by General Masakazu Kawabe, over-all commander of the Army air forces in the homeland.[8]

In accordance with this decision, General Anami and General Umezu both called meetings of their ranking subordinates during the afternoon of the 14th, notified them of the outcome of the final Imperial conference, and directed strict obedience to the Emperor's command.[9] A short while later, special instructions to the same effect were radioed to all top operational commanders (of General Army level) jointly in the names of the War Minister and Chief of Army General Staff. These instructions stated in part:[10]

1. Negotiations have taken place with the enemy on the basis of our conditions that the national polity be preserved and the Imperial domain maintained. The stipulations laid down by the enemy, however, rendered the realization of these conditions extremely difficult, and for that reason we vigorously and consistently maintained that these stipulations were absolutely unacceptable. Although we reported to the Throne to this effect on various occasions, His Imperial Majesty nevertheless has decided to accept the terms of the Potsdam Declaration. . . . . .

2. The Imperial decision has been handed down. Therefore, in accordance with the Imperial will, it is imperative that all forces act to the end in such a way that no dishonor shall be brought to their glorious traditions and splendid record of meritorious service, and that future generations of our race shall be deeply impressed. It is earnestly desired that every soldier, without exception, refrain absolutely from rash behavior and demonstrate at home and abroad the everlasting fame and glory of the Imperial Army.

Simultaneously with the action of the Army leaders, the Navy also took steps to ensure disciplined compliance with the surrender decision. At 1630 on 14 August, Navy Minister Yonai communicated this decision to his principal subordinate in the Navy Ministry and instructed that all necessary measures be taken to secure obedience. Orders were also issued to all major naval commands within Japan, directing them to dispatch their chiefs of staff to Tokyo immediately in order to receive instructions.[11]

8 (1) Field Marshal Hata was in Tokyo as the Emperor had summoned a meeting of all the Field Marshals and Fleet Admirals at 1000, 14 August. (2) The joint resolution was originally suggested to them by Lt. Gen. Torashiro Kawabe, Deputy-Chief of Army General Staff, the main purpose of which was to reaffirm the unwavering loyalty to the Imperial will under these extraordinary circumstances. (Statement by Lt. Gen. Torashiro Kawabe, Deputy-Chief, Army General Staff.)

9 General Anami declared in his address that the entire Army must act in complete accord with the Imperial decision. " Even if it means sleeping on grass and eating stones," he added, " I ask you all to do your utmost to assure the preservation of the national polity." (Statements by Lt. Gen. Masao Yoshizumi, Chief, Military Affairs Bureau, War Ministry; Lt. Col. Shiro Hara, Staff Officer (Operations), Imperial General Headquarters, Army Section; and Lt. Col. Takeshita, previously cited).

10 Army Top Secret Radio No. 68, 14 Aug 45, *Soren Sansen ni tomonau Tsuzuri* 蘇聯參戰に伴ふ綴 (File on Soviet Entry into War) compiled by Maj. Hideo Egawa of Military Affairs Bureau, War Ministry, Part I, Doc. No. 12

11 Daily Record of Navy Minister's Secretariat.

The prompt action taken by the Army and Navy authorities proved, for the most part, highly effective. In the Army, where the more threatening situation prevailed, the final, unequivocal decision of the top leaders to abide by the will of the Emperor dealt a crippling blow to the smouldering plot of the young officers for a *coup d'etat* to block the surrender. The conspirators had predicated their plans upon unified action by the Army as a whole. Since the decision of the Army leadership clearly rendered this impossible, most of the principal plotters reluctantly abandoned the *coup d'etat* scheme on the afternoon of 14 August.[12]

Not all of the conspirators, however, were willing to bow to the leaders' decision. Two of the most fanatical—Maj. Kenji Hatanaka and Lt. Col. Jiro Shiizaki, both of the Military Affairs Bureau in the War Ministry—were determined to make a last, desperate attempt to block the surrender and force a reversal of the Imperial decision to terminate the war. They talked Lt. Col. Masataka Ida, a colleague in the Military Affairs Bureau, into joining them and also secured the promised support of two staff officers of the 1st Imperial Guards Division.[13]

Hatanaka and Shiizaki made it their ultimate objective to prevail upon the Emperor to change his mind and forestall his scheduled broadcast on the 15th while galvanizing the Army into a united opposition to the surrender. To accomplish this, they had in mind a military government and planned first to deploy troops of the Guards Division during the night of 14–15 August to seize control of the Imperial Palace thereby insulating the Emperor. They also planned to usurp control of the central Tokyo radio station since the broadcast would end all possibility of contrary action by the Army. In case of failure, they were fully prepared for suicide.

The conspirators were ready to go into action late on the 14th. Between 2300 and midnight, Hatanaka, Shiizaki and Ida proceeded to the headquarters of the 1st Imperial Guards Division, situated close by the Northwest (*Inui*) gate of the Palace grounds (Plate No. 169), and met the two division staff officers who had been won over to the plot. Shiizaki and these two officers—Majs. Hidemasa Koga and Sadakichi Ishihara—immediately began planning the disposition of division troops. Shortly after midnight Shiizaki, Hatanaka and Ida gained admittance to the office of the Guards Division commander, Lt. Gen. Takeshi Mori, and

12 According to Lt. Col. Takeshita, brother-in-law of War Minister Anami and a key member of the original *coup d'etat* plot, the young officers considered it essential from the first that the coup be assured of unified Army support so that it would not become "a mere local attempt" opposed by sections of the Army forces. The approval of the War Minister and Chief of Army General Staff was therefore regarded as "an absolute prerequisite" for execution of the plan. When the Army leaders finally came out for rigid compliance with the Emperor's decision, most of the original plotters decided that no action could be undertaken. (Statement by Lt. Col. Takeshita, previously cited.)

13 Account of the attempted military uprising led by Maj. Hatanaka and Lt. Col. Shiizaki is based on the following sources: (1) *Chian Gunchitsu Kankei Hatsu Rai Kan Tsuzuri* 治安軍秩關係發來翰綴 (File of Incoming and Outgoing Communications on Public Peace and Military Discipline) Military Police Headquarters, Dec 45: a. Report Concerning August 15th Incident and Military Police Measures; b. Eastern District Army Headquarters Report; c. August 15th Incident. (2) *Kido Nikki* 木戸日記 (Kido Diary) Entries for 15 Aug 45. (3) *Shusen Hishi* 終戰秘史 (Untold History of the Termination of the War) by Hiroshi (Kainan) Shimomura, May 50, pp. 115–181. (4) Statements by Baron Yoshihiro Tokugawa, Chamberlain, Imperial Household Ministry; Lt. Col. Masao Inaba and Lt. Col. Masataka Ida, both of Military Affairs Bureau, War Ministry; Col. Hiroshi Fuwa, Staff Officer (Operations) and Maj. Kiyoshi Tsukamoto, Adjutant, both of Eastern District Army Command; Col. Hayashi and Lt. Col. Takeshita, previously cited.

demanded that he order the division to rise against the surrender decision. If Mori's troops acted, Hatanaka declared, the Army as a whole would follow.

Mori flatly refused. The Emperor's decision was already taken, he said, and the Guards Division, particularly charged with the mission of protecting the sovereign, would under no circumstances act contrary to his will. Hatanaka declined to accept this answer as final and pressed Mori to change his mind. Ida meanwhile went into the adjoining office of the division chief of staff, Col. Kazuo Mizutani, and tried to persuade him to support the plot.

For some time heated discussion continued between Hatanaka and Mori. The Guards Division commander was still standing firm in his refusal to act, when another conspirator, Capt. Shigetaro Uehara of the Army Air Academy, came in and joined Hatanaka. Uehara urged Hatanaka to act speedily lest the plot fail. Hatanaka then appealed to Mori once more to give his consent, and when Mori refused, Hatanaka drew his pistol and shot the Guards commander in cold blood. At the same time, Uehara drew his sword and cut down Mori's brother-in-law, a staff officer of the Second General Army, who had been with General Mori when the conspirators appeared at the division headquarters and who at this moment attempted to intervene.

The plot now moved into high gear. Shiizaki and the two staff majors of the Guards Division had completed planning the disposition of troops, and a false division order was drawn up over Lt. Gen. Mori's official seal to put the dispositions into effect. The order directed the bulk of the division strength to take up positions encircling the Palace grounds, while one company of the 1st Guards Infantry Regiment was to occupy the Radio Tokyo building. All communications between the Palace and the outside were to be severed, except those leading to division headquarters.[14]

Hatanaka, Shiizaki and Uehara did not wait for the false order to be completed. Leaving Koga and Ishihara to handle matters at division headquarters, they immediately proceeded to the Palace grounds at about 0200 and were admitted to the command post of the Palace guard detachment near the Nijubashi entrance. Two battalions of the 2d Guards Infantry Regiment under command of Col. Toyojiro Haga, the regimental commander, were currently standing guard duty. Hatanaka and Shiizaki, posing as staff officers specially assigned to the Guards Division by Imperial General Headquarters, told Haga that a division order was being issued to secure the Palace.

At 0200 the false order written by the conspirators was issued at division headquarters. Shortly thereafter an adjutant delivered it to Col. Haga at the Palace guard command post. Haga immediately issued implementing orders to his troops. All entry and egress to and from the Palace grounds were stopped at once.

This action immediately produced an apparent stroke of luck for Hatanaka and his fellow conspirators. Two cars attempting to leave through the Sakashita gate were stopped by sentries and found to be carrying State Minister Shimomura and a number of officials of the Japan Broadcasting Corporation, who had been engaged in recording the Emperor's surrender Rescript at the Imperial Household Ministry. Shimomura and those accompanying him were detained, searched and questioned.

14 The false order was issued as Kin (Guards Division) Operations Order No. A-584, 0200 15 Aug 45. File of Incoming and Outgoing Communications on Public Peace and Military Discipline, op. cit. Eastern District Army Headquarters Report.

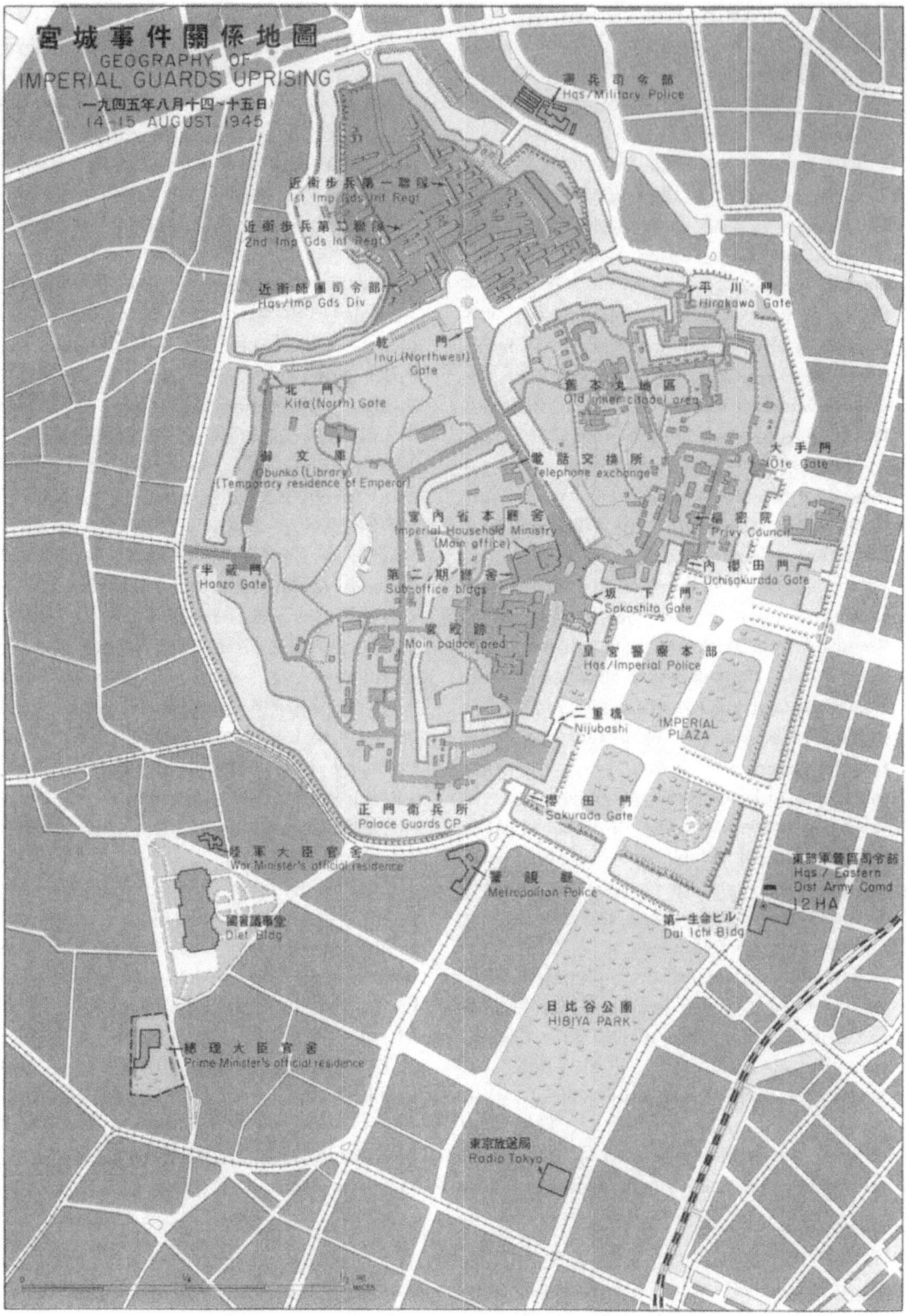

PLATE NO. 169
Geography of Imperial Guards Uprising

Since they did not have the recording with them, Hatanaka and Shiizaki were certain that it must still be in the Imperial Household Ministry.

It was now urgent for the conspirators to find the Emperor's recordings of the Rescript in order to stop the broadcast which was scheduled for noon on the 15th. Accompanied by squads of Haga's troops, they lost no time in carrying out a frenzied search of the Household Ministry building. General Shigeru Hasunuma, chief aide-de-camp to the Emperor, and several Imperial Household officials were taken into custody and grilled, but they steadfastly refused to disclose the whereabouts of the recording. The Lord Privy Seal, Marquis Kido, and Imperial Household Minister Sotaro Ishiwata managed to escape discovery by hiding in an underground air-raid shelter beneath the Ministry. The searchers repeatedly ransacked Kido's office but failed to locate the vital recording.

By this time also, the plot was going awry in another quarter. Immediately after the assassination of his commanding officer, the Guards Division chief of staff, Col. Mizutani, and Lt. Col. Ida, who was acting in collusion with Hatanaka and Shiizaki, had proceeded to the headquarters of General Shizuichi Tanaka, Commanding General, Twelfth Area Army and Eastern District Army. Mizutani's object was to report the plot so that the Area Army might take appropriate counter-action. Ida, however, went with some hope that General Tanaka might support the uprising.

General Tanaka and his staff swiftly decided to assume direct command of the 1st Imperial Guards Division and stop the revolt. Steps were immediately taken to contact the various regiments and instruct them to disregard the false division order.[15] Tanaka wished to proceed at once to the division headquarters and take command but was dissuaded from doing so by his chief of staff, who argued that he might be shot down in the darkness. It was consequently decided to wait until dawn.

Lt. Col. Ida, upon discovering that the Area Army command was entirely disinclined to support the uprising, left General Tanaka's headquarters for the Palace grounds. There, he briefly met Hatanaka and advised him that the situation was hopeless, urging that the plot forthwith be abandoned. Hatanaka, however, was not yet ready to admit failure. Ida then left the Palace grounds and proceeded to the official residence of the War Minister, where another part of the drama was unfolding.

About midnight of 14–15 August, Hatanaka temporarily suspending negotiation with Lt. Gen. Mori, had called on Lt. Col. Masahiko Takeshita, one of the leaders of the original *coup d'etat* plot and brother-in-law of War Minister Anami. Takeshita, with most of the other ring-leaders, had already renounced the plot, but Hatanaka now disclosed to him that he (Hatanaka) and Shiizaki intended to incite action by the Guards Division in a final effort to block the surrender. Hatanaka urged Takeshita to use his influence for the purpose of winning over War Minister Anami. Takeshita finally consented to do so if the Eastern District Army Command also supported the *coup*.

At about 0100, Takeshita went to the War Minister's official residence in order to report Hatanaka's plan. However, finding upon his

15 The Area Army was apprised of the false division order by the commander of the 7th Guards Infantry Regiment, Col. Minami. Because of the dubious nature of the order, Col. Minami personally went to General Tanaka's headquarters to report the matter. This report was made shortly after the Area Army learned from Col. Mizutani and Lt. Col. Ida of the assassination of the Guards Division commander. Ibid.

arrival that General Anami was preparing to commit suicide, he remained silent concerning the purpose of his visit until about an hour later when shots were heard from the direction of the Palace. Takeshita then told Anami that Hatanaka and Shiizaki were launching an armed *coup* with Guards Division troops in order to stop the Emperor's surrender broadcast and spur the Army to united action.

General Anami received this news with a curious and enigmatic calm. He manifested no intention of acting to put an end to the uprising or of changing his suicide plan. He merely remarked to Takeshita that General Tanaka's Twelfth Area Army would probably not join in the revolt, and that as long as the Area Army stayed quiet, the insurrection would collapse. Even when Takeshita, a short while later, relayed to Anami a telephone report that the rebels had assassinated the commander of the Imperial Guards Division, the War Minister showed no signs of taking positive action. His mind appeared absorbed with the thought of impending death.

It was at approximately 0340 that Lt. Col. Ida appeared at the War Minister's residence after conferring with Hatanaka at the Palace grounds. His intention was to lay bare the events which had transpired, but when he found General Anami ready for suicide, he too remained silent. Ida joined Anami and Takeshita in conversation and after an hour he departed.

At about 0445, Anami made his final preparations for death. Donning a white shirt given him by the Emperor, he knelt facing toward the Palace. A few moments later he plunged a dagger into his abdomen, then raised it and slashed his own throat. Col. Takeshita, who witnessed Anami's death, found two brief suicide messages. One was a 31-syllable verse expressing gratitude for the favors of the Emperor. The other read, "Confident of the everlastingness of our divine fatherland, I give my life in atonement for great wrongs."[16]

Meanwhile, as Anami had predicted before taking his own life, the Hatanaka-Shiizaki coup was rapidly collapsing. General Tanaka's headquarters had already succeeded in establishing telephonic contact with some units of the Guards Division, and these units promptly stopped compliance with the false division order. There was, however, the most important 2d Infantry Regiment commanded by Col. Haga still active in and around the Palace grounds, and also the 1st Infantry Regiment on reserve yet to be brought under control.

At about 0450, General Tanaka, accompanied by his aide and an officer of his staff, drove to the headquarters of the Imperial Guards Division. There, he found the 1st Guards Infantry Regiment still preparing to move in accordance with the false division order. He immediately ordered the regimental commander to dismiss his troops and placed under arrest Maj. Ishihara, one of the two division staff officers who had acted in collusion with the plotters.

General Tanaka was now ready to tackle the confused situation prevailing in the Palace grounds. At about 0600 he proceeded to the Northwest (Inui) gate, where Col. Haga, in response to telephone instructions, met him and reported all developments in detail. After cautiously admonishing Haga, Tanaka ordered him to return his troops immediately to their normal duties. Also at this time Lt. Col. Ida, accompanied by his superior at the War Ministry, drove to the Northwest gate but Gen. Tanaka stopped them from entering the Palace

16 "Great wrongs" in one of the suicide messages of War Minister Anami is interpreted as the repeated crimes committed by the Army particularly since the Manchurian incident of 1931. (Statement by Col. Hayashi, previously cited.)

grounds. Then, with one of Haga's battalion commanders as an escort, he personally proceeded to the Obunko (Emperor's private library and temporary residence) to ensure the safety of the Imperial family. The Eastern District Army commander then hurried to the Imperial Household Ministry to set at liberty Gen. Hasunuma and other Court officials whom the rebels had confined. Gen. Tanaka accompanied by Gen. Hasunuma reported to the Throne at 0735 what had taken place in the Palace grounds. He then carefully inspected the Imperial Household Ministry buildings and other key installations to assure that his orders were carried out. He left the Sakashita gate about 0830 and returned to his headquarters.

The ill-advised plot had collapsed completely. Hatanaka, Shiizaki, Koga and Uehara signified their intention of committing suicide and were therefore allowed by Haga to remain at liberty. Koga killed himself at the Imperial Guards Division headquarters at 1300 on the 15th, Hatanaka and Shiizaki in the Imperial Palace plaza at 1400 the same day, and Uehara at the Air Academy on the night of 17–18 August.[17]

By 0800 on 15 August, complete order had been restored in the capital. A small force of insurgent-led troops, which had occupied the Radio Tokyo building before dawn, had already withdrawn after failing in an attempt to put a false broadcast on the air. The reckless plan of Hatanaka and Shiizaki to touch off a general Army uprising against the surrender had ended in total failure.[18]

The Government rigidly banned from publication all news of the uprising attempt. A War Ministry communique issued on the 15th and published the next day briefly announced the suicide of General Anami but remained silent on the circumstances surrounding it. Vice Adm. Takijiro Onishi, Vice-Chief of the Navy General Staff and organizer of the first Kamikaze corps in the Philippines, followed Anami's example on the morning of 16 August. Suicides by other high-ranking military officials continued in late August and September.[19]

With the collapse of the Hatanaka-Shiizaki

17 Lt. Col. Ida, in view of the fact that he did not participate in the insurrection after the initial stage and tried to persuade Hatanaka, Shiizaki and the others to abandon the attempt, was let off by a military court with light punishment. Maj. Ishihara, Guards Division staff officer, also was exonerated in view of his change of heart. On 19 August, however, he was shot and killed in Ueno Park, Tokyo, while assisting in efforts to pacify a band of recalcitrant anti-surrender troops from the Army Air Signal Training Division at Mito. These troops, carrying arms and led by several officers, travelled to Tokyo by train on 17 August and encamped in the Ueno Art Museum. Prompt security measures by Twelfth Area Army headquarters and the Imperial Guards Division prevented any serious outbreak of disorder and the troops returned to Mito on 20 August. The officers involved committed suicide. File of Incoming and Outgoing Communications on Public Peace and Military Discipline, op. cit. August 15th Incident; and Activities in Tokyo of Officers and Troops of the Mito Army Air Signal Division.

18 Other acts of violence, entirely unrelated to the Hatanaka-Shiizaki plot but motivated by anti-surrender sentiment, were committed during the night of 14–15 August by a small group of soldiers and civilians from the Yokohama area, led by Capt. Takeo Sasaki. Driving to Tokyo in motor trucks in the early hours of the morning, this band fired machine-guns into and then set fire to, both the official and private residences of Premier Suzuki and the private residence of the President of the Privy Council, Baron Hiranuma. Capt. Sasaki and his followers were rounded up by military police and returned to their units for disciplinary measures on 16 August. File of Incoming and Outgoing Communications on Public Peace and MilitaryDiscipline, op. cit. August 15th Incident.

19 General Shizuichi Tanaka, Commanding General, Eastern District Army and Twelfth Area Army, committed suicide on 24 August. On 11 September, ex-Premier General Hideki Tojo unsuccessfully attempted to kill himself when American Military Police went to his home to place him under arrest. Field Marshal Sugiyama, First General Army

plot and the safe execution of the Emperor's surrender broadcast, the most critical moment had passed. There were still to be scattered instances in both the Army and Navy of action in defiance of the surrender edict, some of a fairly serious and potentially dangerous nature. However, the firm resolve of the top military leaders and the discipline of the armed forces had already proved equal to their gravest test on the fateful night of 14–15 August.

Formation of New Cabinet

Japan now faced the many complex and difficult problems involved in the practical effectuation of the surrender. But before these could be tackled, it was first necessary to install a new government in place of the weary, struggle-worn Suzuki Cabinet.

Immediately after Suzuki's resignation on the afternoon of 15 August, the Emperor commanded Marquis Kido to recommend a new Premier. Speed was essential since General MacArthur's headquarters in the Philippines was already dispatching orders to the Japanese Government with regard to surrender arrangements. Marquis Kido therefore decided to dispense with a full advisory conference of the *jushin* and summoned only Baron Hiranuma, President of the Privy Council, for consultation.

Kido and Hiranuma swiftly agreed in view of the troubled domestic situation that a member of the Imperial family should be called upon to head the Cabinet. Since it was particularly vital that the nominee had the confidence of the armed orces, the choice fell upon General Prince Naruhiko Higashikuni, uncle-in-law of the Emperor, member of the Supreme War Direction Council, and former commander of the General Defense Army. The Emperor approved Kido's recommendation on the evening of the 15th, and at 1000 on the 16th Prince Higashikuni formally received the Imperial command to orm a new government.[20] This marked the first time that an Imperial Prince had been appointed Premier since the promulgation of the Meiji Constitution.

Marquis Kido strongly emphasized to the premier-designate the urgency of forming a cabinet which would be capable of effecting the difficult transition to peace swiftly and without incident. He pointed in particular to the nervousness of the people in the face of impending enemy occupation and the ever-present threat of rebellious action on the part of Army and Navy elements opposed to surrender. General MacArthur, he informed the Prince, was already demanding the prompt dispatch of a surrender mission to Manila.[21]

Messages had begun arriving from MacArthur's headquarters during the night of the 15th. The Japanese Government and Imperial General Headquarters, however, took no action in compliance until formal notification was received from the Allies at 1030 on 16 August that the final Japanese surrender offer had been accepted. The Allied note, signed by the United States Secretary of State, James F. Byrnes, directed the Japanese authorities to

---

Commander, killed himself the following day. Other general and flag officers who committed suicide later included General Shigeru Honjo, General Teiichi Yoshimoto, Lt. Gen. Kumaichi Teramoto, Lt. Gen. Yoshio Shinozuka, and Vice Adm. Matsuo Morizumi. In addition to the suicides of military officers, there were a few instances of mass suicides by members of nationalistic and patriotic societies in the later part of August. (1) Untold History of the Termination of the War, op. cit., pp. 180, 224, and 248. (2) Foreign Office Memorandum No. 1734 (LCR), 24 Aug 50, and No. 1711 (LCR), 21 Aug 50.

20 Kido Diary, op. cit. (Entries for 15–16 Aug 45.) IMTFE Record, pp. 31, 198–202.

21 Prince Naruhiko Higashikuni, *Watakushi no Kiroku* 私の記錄 (My Record) Toho Shobo, Mar 46, pp, 119–22.

order the prompt cessation of hostilities and stated that General MacArthur had been named Supreme Commander for the Allied Powers to effect the surrender.[22] The note read:

You are to proceed as follows:

1. Direct prompt cessation of hostilities by Japanese forces, informing the Supreme Commander for the Allied Powers of the effective date and hour of such cessation.

2. Send emissaries at once to the Supreme Commander for the Allied Powers with information of the disposition of the Japanese forces and commanders, and fully empowered to make any arrangements directed by the Supreme Commander for the Allied Powers to enable him and his accompanying forces to arrive at the place designated by him to receive the formal surrender.

3. For the purpose of receiving such surrender and carrying it into effect, General of the Army Douglas MacArthur has been designated as the Supreme Commander for the Allied Powers, and he will notify the Japanese Government of the time, place and other details of the formal surrender.

The earlier radios direct from General MacArthur had specified that representatives, accompanied by competent advisors representing the Army, Navy and Air forces, should be sent to Manila to receive instructions regarding effectuation of the surrender. The delegation was directed to travel aboard an aircraft of specified type and markings and to proceed via a designated route, leaving Sata-Misaki, on Kyushu, between 0900 and 1100 on 17 August.[23] This allowed barely 24 hours in which to make preparations for the mission.

Execution of the Allied instructions was hampered almost immediately by disagreement in Imperial General Headquarters and the Foreign Office as to the exact nature of the mission. Some officials interpreted the instructions to mean that the delegates must be armed with full powers to receive and agree to the actual terms of surrender and consequently must be top representatives of the Government and High Command. Others understood them to mean that the mission would be purely preparatory for the purpose of working out technical surrender arrangements and procedures.[24]

Late in the afternoon of 16 August, a message was dispatched to General MacArthur's headquarters asking for a clarification of this matter and requesting more time in which to organize and prepare the mission. At the same time the Supreme Commander was notified that the Emperor, at 1600 on 16 August, had issued a "cease hostilities" order to all Japanese armed forces and was prepared to dispatch members of the Imperial family to the various combat areas to assure obedience by the field commanders. The message pointed out, however, that it might take some time for knowledge of the order to reach the front-line troops before resulting in full cessation of hostilities.[25]

A reply clarifying the purpose of the mission to Manila was received from the Allied Supreme Commander on the morning of 17 August. The Government and Imperial General Headquarters promptly acted to hasten the necessary preparations, but appointment of the head of the mission was further held up pending the installation of Higashikuni Cabinet.

The premier-designate had pushed the

22 Foreign Office Memorandum No. 23 (MA), 10 Jan 50.

23 Radio No. 2501, SCAP to Japanese Government, 15 Aug 45. Contained in File on Soviet Entry into War, op. cit. Part I, Doc. No. 37.

24 Statement by Mr. Shunichi Matsumoto, Vice-Minister for Foreign Affairs.

25 Radio No. 1, Japanese Government to SCAP, 16 Aug 45. Contained in File on Soviet Entry into War, op. cit. Part I, Doc. No. 38.

formation of the new government at top speed, and on the afternoon of the 17th the official ceremony of installation took place in the Emperor's presence. Until General Shimomura could be summoned to Tokyo from his North China Area Army command, Prince Higashikuni himself assumed the portfolio of War Minister concurrently with the premiership Admiral Mitsumasa Yonai remained in the critical post of Navy Minister. Prince Ayamaro Konoye, in accordance with a special recommendation by Marquis Kido, entered the Cabinet as Minister without Portfolio to act as Prince Higashikuni's closest advisor. The post of Foreign Minister was given to Mr. Mamoru Shigemitsu, who had previously served in the Koiso Cabinet.[26]

With the new government duly installed in office, Prince Higashikuni broadcast to the nation on the evening of 17 August. He declared that his policies as Premier would conform to the desires of the Emperor as expressed in the Imperial mandate to form a Cabinet. These policies, he said, were to control the armed forces, maintain public order and surmount the national crisis, with scrupulous respect for the Constitution and the Imperial Rescript terminating the war.[27]

Cessation of Hostilities

The first steps toward actual cessation of hostilities had already been taken by Imperial General Headquarters in advance of the formal order received from the Allied Powers on 16 August. On the 14th, the Navy Section of Imperial General Headquarters had ordered immediate suspension of all attack operations against the United States, Britain, China and the Soviet Union.[28] The Army Section followed suit on the 15th with an order which stated:[29]

1. It is the intention of Imperial General Headquarters to comply with the provisions of the Imperial Rescript of 14 August 1945.

2. Until further orders are received, all Armies will continue performance of present duties, but all active attack operations will be suspended. Steps will be taken to maintain military discipline and solidarity of action. In the homeland, Korea, Sakhalin and Formosa, precautions will be taken against possible disturbances of the public peace.

Following receipt of the Allied order on 16 August, immediate steps were taken to extend the suspension of attack operations to all hostile activities. Orders sent out the same day by the Army and Navy Sections of Imperial General Headquarters directed all Army and Fleet commands to order the forces under their control to "cease hostilities forthwith." Army and Fleet commanders were ordered to report back to Imperial General Headquarters the effective dates and hours set by them for the cessation of hostilities in their respective operational areas.[30]

26 The full Cabinet list was as follows: Premier and War Minister, Prince Naruhiko Higashikuni; Navy, Admiral Mitsumasa Yonai; Foreign Affairs, Mamoru Shigemitsu; Home Affairs, Iwao Yamazaki; Finance, Juichi Tsushima; Justice, Chuso Iwata; Welfare, concurrently Education, Kenzo Matsumura; Agriculture and Commerce, Kotaro Sengoku; War Provisions, Chikuhei Nakajima; Transportation, Naoto Kohiyama; Ministers without Portfolio, Prince Ayamaro Konoye and Taketora Ogata. A few changes were made in the original Cabinet prior to the formal signing of the surrender on 2 September, General Sadamu Shimomura becoming War Minister, Tamon Maeda Education Minister, and Binshiro Obata Minister without Portfolio.

27 (1) Yomiuri-Hochi Newspaper, Tokyo, op. cit., 18 Aug 45. (2) My Record, op. cit., pp. 138–41.

28 *Daikairei Dai Shijushichi-go* 大海令第四十七號 (Imperial General Headquarters Navy Order No. 47) 14 Aug 45.

29 *Dairikumei Dai Sensambyakuhachijuichi-go* 大陸命千三百八十一號 (Imperial General Headquarters Army Order No. 1381) 15 Aug 45.

30 (1) *Daikairei Dai Shijuhachi-go* 大海令第四十八號 (Imperial General Headquarters Navy Order No. 48) 16 Aug 45. (2) *Dairikumei Dai Sensambyakuhachijuni-go* 大陸命第千三百八十二號 (Imperial General Headquarters Army Order No. 1382) 16 Aug 45.

On 17 August, the Emperor personally backed up these orders with a special Rescript to the armed services. His Majesty's message, carefully worded to assuage military aversion to surrender, paid tribute to the gallantry, sacrifice and undiminished fighting spirit of the armed forces, but exhorted them to comply with the Imperial decision to conclude peace since this was necessary to assure the preservation of the national polity. The Rescript read:[31]

More than three years and eight months have elapsed since We declared war against the United States and Great Britain. We are deeply grateful to Our beloved soldiers and sailors for fighting gallantly in pestilential wastelands and over raging tropical seas, undeterred by every hardship.

Recently, the Soviet Union entered the war against Us, and various other factors, domestic and international, have led Us to conclude that any further continuation of hostilities would only result in increased disaster and ultimately bring destruction of the very basis of the Empire's existence. Therefore, although the fighting spirit of the Army and Navy is in no way diminished, it is Our intent to conclude peace with the United States, Britain, the Soviet Union and the Chungking Government in order to assure the preservation of Our glorious national polity.

We are sincerely grieved at the loss of many of Our loyal and brave officers and men, who have fallen in battle or died of illness. At the same time, We believe that the distinguished deeds and true loyalty of Our fighting men will embody the soul of the Japanese people for all generations.

We charge you, the members of the armed forces, to comply faithfully with Our intentions, to preserve strong solidarity, to be straightforward in your actions, to overcome every hardship, and to bear the unbearable in order to lay the foundations for the enduring life of Our nation.

On the same day that the Rescript to the armed forces was issued, three Imperial Princes left Tokyo by air as personal representatives of the Emperor to urge compliance with the surrender decision upon the major overseas commands. The envoys chosen all held military rank as officers of the Army, and they had been guaranteed safety of movement by General MacArthur's headquarters. General Prince Yasuhiko Asaka was dispatched as envoy to the headquarters of the expeditionary forces in China, Maj. Gen. Prince Haruhiko Kanin to the Southern Army, and Lt. Col. Prince Tsuneyoshi Takeda to the Kwantung Army in Manchuria. By 20 August, they had completed their missions.[32]

Meanwhile, on 19 August, Imperial General Headquarters had issued orders fixing 22 August as the final deadline for the cessation of all acts of hostility by Army and Navy forces in the homeland area.[33] A Navy Section order issued 22 August directed the Southeast, Southwest and China Area Fleets to cease hostile acts with the least possible delay after 22 August. The same day, an Army Section order set 25 August as the final date for cessation of hostilities by all overseas Army forces.[34]

The Army High Command had been par-

31 *Asahi Shimbun* 朝日新聞 (Asahi Newspaper), Tokyo, 18 Aug 45.

32 Imperial General Headquarters Communique, 22 and 23 Aug 45. Yomiuri-Hochi Newspaper, Tokyo, op. cit., 23 and 24 Aug 45. (2) Radios exchanged between SCAP and Imperial General Headquarters. File on Soviet Entry into War, op. cit. Part I, Docs. No. 38, 39 and 62; Part II, Doc. No. 26.

33 (1) *Daikairei Dai Goju-go* 大海令第五十號 (Imperial General Headquarters Navy Order No. 50) 19 Aug 45. (2) *Dairikumei Dai Sensambyakuhachijuroku-go* 大陸命第千三百八十六號 (Imperial General Headquarters Army Order No. 1386) 19 Aug 45.

34 (1) *Daikairei Dai Gojushi-go* 大海令第五十四號 (Imperial General Headquarters Navy Order No. 54) 22 Aug 45. (2) *Dairikumei Dai Sensambyakuhachijuhachi-go* 大陸命第千三百八十八號 (Imperial General Headquarters Army Order No. 1388) 22 Aug 45.

軍旗奉焼式　向井潤吉

Original Painting by J. Mukai

PLATE NO. 170
Burning Regimental Flag

ticularly apprehensive with regard to the reaction of the forces in China to the surrender. The China Expeditionary Army, with a strength of about one million men, had never suffered a decisive military defeat in eight years of fighting. On about 11 August, after learning that Japan was suing for peace, General Yasutsugu Okamura, China theater commander, had addressed a strong protest against acceptance of unconditional surrender to both War Minister Anami and the Chief of the Army General Staff.[35]

The concern felt by the Army authorities, however, proved needless. Following the promulgation of the Imperial Rescript terminating the war, General Okamura promptly ordered the forces under his command to comply with the surrender decision. With the exception of minor elements which were under attack by Chinese Communist troops, all components of the China Expeditionary Army had ceased hostilities by 21 August.

Although delays had been anticipated, the cessation of hostilities was actually accomplished in most areas with remarkable speed. By midnight on 17 August, all Army and Navy forces in Japan and Southern Korea and all Navy forces in China and the Rabaul area had reported the end of hostile action. By 19 August, similar reports had been received from Northern Korea, the Kuriles, most sections of Manchuria, the Central Pacific islands and New Guinea. Army forces in the Rabaul area and, with some exceptions, the forces in Southeast Asia had reported by 22 August. No report, however, was received from General Yamashita's beleaguered headquarters in the mountains of northern Luzon.[36]

Mission to Manila

Parallel with the steps to bring about cessation of hostilities, preparations were hastily completed for the dispatch of the surrender mission to Manila. The installation of the Higashikuni Cabinet on 17 August removed one cause of delay, and in the afternoon of the same day a message from General MacArthur's headquarters clarified the nature and purpose of the mission. On the basis of this clarification, it was promptly decided that Lt. Gen. Torashiro Kawabe, Deputy-Chief of the Army General Staff, should head a delegation of sixteen members mainly representing the Army and Navy General Staffs.[37]

Lt. Gen. Kawabe was formally appointed by the Emperor on 18 August. By late afternoon of the same day, assembly of the data required by the Allied Supreme Commander was largely completed, and a message was dispatched to Manila informing General MacArthur's headquarters that the mission was prepared to leave the following morning.[38] The itinerary received prompt approval from the Supreme Commander.

At 0611 on 19 August, the surrender mis-

---

35 Statement by Lt. Gen. Kawabe, previously cited.

36 Imperial General Headquarters received no reports confirming the cessation of hostilities by units in the Philippines until early in October. Dates given for the other areas are based on special reports prepared by the First Demobilization Bureau, 1 Dec 47 and Second Demobilization Bureau, 24 Nov 47.

37 (1) File on Soviet Entry into War, op. cit. Part I, Docs. No. 42, 52 and 53. (2) Full composition of the delegation was as follows: Chief delegate, Lt. Gen. Torashiro Kawabe; Army representatives, Maj. Gen. Masakazu Amano, Col. Arata Yamamoto, Lt. Col. Masao Matsuda, Lt. Col. Kiyoshi Minami, Lt. Col. Mamoru Takakura, Lts. Sadao Otake and Harumi Takeuchi; Navy representatives, Rear Adm. Ichiro Yokoyama, Capt. Toshikazu Ohmae, Capt. Hidemi Yoshida, Comdr. Yoshimori Terai, Mr. Shuichi Mizota and Mr. Kazuma Sugita; Foreign Office, Mr. Katsuo Okazaki and Mr. Morio Yukawa.

38 Radio No. 9, Imperial General Headquarters to SCAP, 18 Aug 45. File on Soviet Entry into War, op, cit. Part I, Doc. No. 63.

sion took off from Haneda Airport, outside Tokyo, and first proceeded to the naval air base at Kisarazu on Boso Peninsula in Chiba Prefecture. There, they boarded two disarmed Navy medium bombers displaying the distinctive markings specified by General MacArthur's headquarters. The planes took off at 0707 for Ie-Shima in the Ryukyus, where the delegates were to transfer to an American aircraft for the rest of the journey to Manila.

Special precautions were taken during the initial part of the flight owing to rumors that anti-surrender elements in the Air forces might attempt to intercept and shoot the planes carrying the mission. The two aircraft took a circuitous route and flew with radios silenced.[39] No mishap occurred, and they put down safely at Ie-Shima at 1240. The mission then boarded an American Army transport which reached Nichols Field near Manila at 1800 the same day.

At the airport the delegation was officially met by Maj. Gen. Charles A. Willoughby, Assistant Chief of Staff, G-2. Jeering, hostile crowds of Filipinos lined the streets to watch as a procession of military cars drove the members of the mission to their hotel. At 2100 the first conference began at the Supreme Commander's headquarters in the Manila City Hall, with Lt. Gen. Richard K. Sutherland, General MacArthur's Chief of Staff, presiding.[40]

The initial discussions continued for four hours, during which speedy progresss was made toward the settlement of major problems. Data required by Allied headquarters with regard to the strength and disposition of Japanese Air forces, locations and condition of remaining naval vessels, condition of airfields, and other matters were submitted. Lt. Gen. Kawabe affirmed the Japanese intention to carry out faithfully all requirements fixed by the Allied Supreme Commander and expressed hope that the entry of the occupation forces into Japan could be effected smoothly and without incident.[41]

The conference now tackled the vital problem of the occupation time schedule. From a document handed to them at the start of the discussions, the Japanese delegates learned that General MacArthur's headquarters had already drawn up a tentative schedule which left an unexpectedly short time before the entry of the first occupation forces into Japan. This schedule called for the landing of an advance party at Atsugi Airfield on 23 August, the entry of naval forces into Tokyo Bay on the 24th, the arrival of General MacArthur at Atsugi on the 25th together with the start of the main landings of airborne troops and naval and marine forces, signature of the formal surrender instrument aboard an American battleship in Tokyo Bay on the 28th, and initial troop landings in southern Kyushu on 29–30 August.[42]

---

39 Joint statement regarding the surrender mission to Manila by Rear Adm. Ichiro Yokoyama, Capt. Toshikazu Ohmae, Capt. Hidemi Yoshida, and Comdr. Yoshimori Terai.

40 Those participating in the first conference were: For the Allied Supreme Commander, Lt. Gen. Richard K. Sutherland, Maj. Gen. Charles A. Willoughby, Maj. Gen, A. J. Chamberlin, Maj. Gen L. G. Whitlock, Brig. Gen. D. R. Hutchinson, and Rear Adm. Forrest P. Sherman. For the Japanese, Lt. Gen. Torashiro Kawabe, Maj. Gen. Masakazu Amano, Col. Arata Yamamoto, Lt. Cols. Matsuda, Minami, and Takakura, Rear Adm. Ichiro Yokoyama, Capts. (Navy) Ohmae and Yoshida, and Comdr. Terai. *Manira Haken Teikoku Daihyo no Kaigungawa Shuseki Zuiin Nimmu Hokokusho* マニラ派遣帝國代表の海軍側首席隨員任務報告書 (Duty Report of the Senior Naval Member, Manila Delegation) Aug 45.

41 (1) Statement by Lt. Gen. Kawabe, previously cited. (2) Duty Report of Senior Naval Member, op. cit.

42 Duty Report of Senior Naval Member, op. cit. Appendix A.

The Japanese delegates felt that this stringent schedule left far too little time for effecting the evacuation and disarmament of Army and Navy forces in the areas of initial occupation and for taking other precautionary measures against the occurrence of untoward incidents. Lt. Gen. Kawabe and the senior Japanese naval delegate, Rear Adm. Yokoyama, frankly stated this view to the American representatives and requested that a period of at least ten days be allowed for preparations before the entry of Allied troops. Lt. Gen. Sutherland replied that such an extended delay could not be granted, but in order to ease the difficulties on the Japanese side, he consented on behalf of the Supreme Commander to defer the arrival of the advance party at Atsugi for three days until 26 August, with the first major troop landings to be similarly held up until the 28th.[43]

At the second and concluding conference, which began at 1030 the following day, the Japanese delegation was handed several documents relating to the occupation and surrender procedures. The first of these was a statement of requirements covering the surrender of Japanese military forces, the entry of occupation units into the Tokyo and Kanoya areas, and logistic matters. The others were the texts of the formal surrender instrument to be signed in Tokyo Bay, of an Imperial Rescript to be issued following the surrender ceremony, and of a general order to be issued by Imperial General Headquarters simultaneously with the Rescript, commanding Japanese forces in the field to surrender to designated Allied commanders.[44]

The surrender parleys ended at 1215 on the 20th. Shortly thereafter, the Japanese delegation was escorted to Nichols Field for the return flight to Ie-Shima. Just before departure, however, Rear Adm. Yokoyama handed Maj. Gen. Willoughby of the American staff, a solicitation appealing to the Supreme Commander for special consideration in the enforcement of the occupation, to the traditional veneration of the Japanese for their shrines and ancestral tombs and the privacy of their homes.[45] This appeal was received in the same spirit of fairness which had marked the American attitude throughout the parleys.[46]

In contrast to their uneventful journey to Manila, the Japanese delegates ran into bad luck on the way back to Tokyo. At Ie-Shima, where they were to transfer back to their own planes, last-minute mechanical trouble with its brakes prevented one of the two aircraft from taking off. Since, however, it was extremely vital to carry the conference data back to Tokyo without a moment's delay, Lt. Gen. Kawabe and the other principal members of the delegation decided to take off immediately aboard the remaining aircraft. The plane left Ie-Shima at 1840 on the 20th.

Through an error made in refueling at Ie-Shima, the plane suddenly ran short of gasoline as it neared its home base at Kisarazu just before midnight. The pilot quickly decided upon a forced landing in shallow water just offshore near Hamamatsu and succeeded in

43 Ibid.

44 File on Soviet Entry into War, op. cit. Part II, Docs. No. 4, 5 and 13; Part III, Docs. No 89, 90 and 91.

45 Duty Report of Senior Naval Member, op. cit., Appendix E.

46 The Japanese delegation was unanimously impressed by the firm but fair attitude of the American staff officers with whom they dealt. Lt. Gen. Kawabe later stated that "the Americans listened carefully to what we had to say, expressed sympathy, and studied our suggestions with a cooperative attitude." Mr. Katsuo Okazaki, Foreign Office member of the delegation, similarly remarked that the American attitude "was stern, but they were not arrogant nor did they mock the vanquished." (Statements by Mr. Katsuo Okazaki, Vice-Minister of Foreign Affairs, and Lt. Gen. Kawabe, previously cited.)

bringing the plane down without serious injury to anyone aboard and without loss of the vital documents carried by the delegation. Early on the morning of 21 August, a special Army plane picked up the delegates at Hamamatsu airfield and flew them the short remaining distance to Tokyo.[47]

Immediately after reaching the capital, Lt. Gen. Kawabe made a full report on the results of the surrender mission to the Emperor, the Premier, and Imperial General Headquarters. With the scheduled arrival of the advance party of the Allied occupation forces only five days away, it was clear that only by superhuman effort could the necessary preparations and precautions be carried out in time to comply with the requirements fixed by General MacArthur's headquarters.

Preparations for Allied Occupation

The most pressing task facing the Japanese military authorities under the Allied directives received at Manila was the disarmament of combat units stationed in the areas of initial occupation and their evacuation outsides these areas. Only a few hours after Lt. Gen. Kawabe reported the outcome of his mission, the Army and Navy Sections of Imperial General Headquarters, on the afternoon of 21 August, issued basic orders for the execution of this requirement. Detailed implementing directives were issued the same day and on the 22nd.

The basic orders stated that Allied forces would begin occupying the homeland on 26 August and reaffirmed the intention of Imperial General Headquarters "to insure absolute obedience to the Imperial Rescript of 14 August, to prevent the occurrence of trouble with the occupying forces, and thus to demonstrate Japan's sincerity to the world." With respect to the disarmament and evacuation of Japanese forces in the initial occupation areas, the orders established the following deadlines:

Army Forces[48]

1. All forces except units specially authorized will be evacuated, leaving behind their equipment, from the "area of initial evacuation" along both sides of Tokyo and Sagami Bays (Plate No. 171) by 1200 on 25 August.[49]

2. All forces except units specially authorized will be evacuated, leaving behind their equipment, from the Kanoya area in southern Kyushu, by 1200 on 30 August.[50]

Navy Forces[51]

1. All combat forces except the minimum required to maintain order and guard weapons dumps, supplies and military installations, will be evacuated from the Atsugi area by 1800 on 24 August, and from all other parts of the "area of initial evacuation" by 1800 on 27 August.

2. All combat forces except those required for maintenance of order and guard duty will, after disarming, be evacuated from the Kanoya area, in southern Kyushu, by 1800 on 30 August.

The implementing directives subsequently issued by Imperial General Headquarters Speci-

47 The second plane, carrying the remaining members of the surrender mission, left Ie-Shima at 0830 on 21 August after undergoing repairs. It arrived safely at Kisarazu in the afternoon of the same day.

48 *Dairikumei Dai Sensambyakuhachijushichi-go* 大陸命第千三百八十七號 (Imperial General Headquarters Army Order No. 1387) 21 Aug 45.

49 The "area of initial evacuation" had been specifically defined by General MacArthur's headquarters in the documents handed to the Japanese surrender mission at Manila. These documents also defined a larger "Tokyo Bay area" inclusive of the "area of initial evacuation." (Plate No. 171) Certain requirements, such as the disarming of all coastal batteries, anti-aircraft guns and artillery, applied to the whole "Tokyo Bay area."

50 For the purpose of the evacuation of armed forces the Kanoya area was defined as "the Satsuma and Osumi peninsulas south of a line through Kushikino, Kajiki, Sueyoshi and Aburatsu."

51 *Daikairei Dai Gojuni-go* 大海令五十二號 (Imperial General Headquarters Navy Order No. 52) 21 Aug 45.

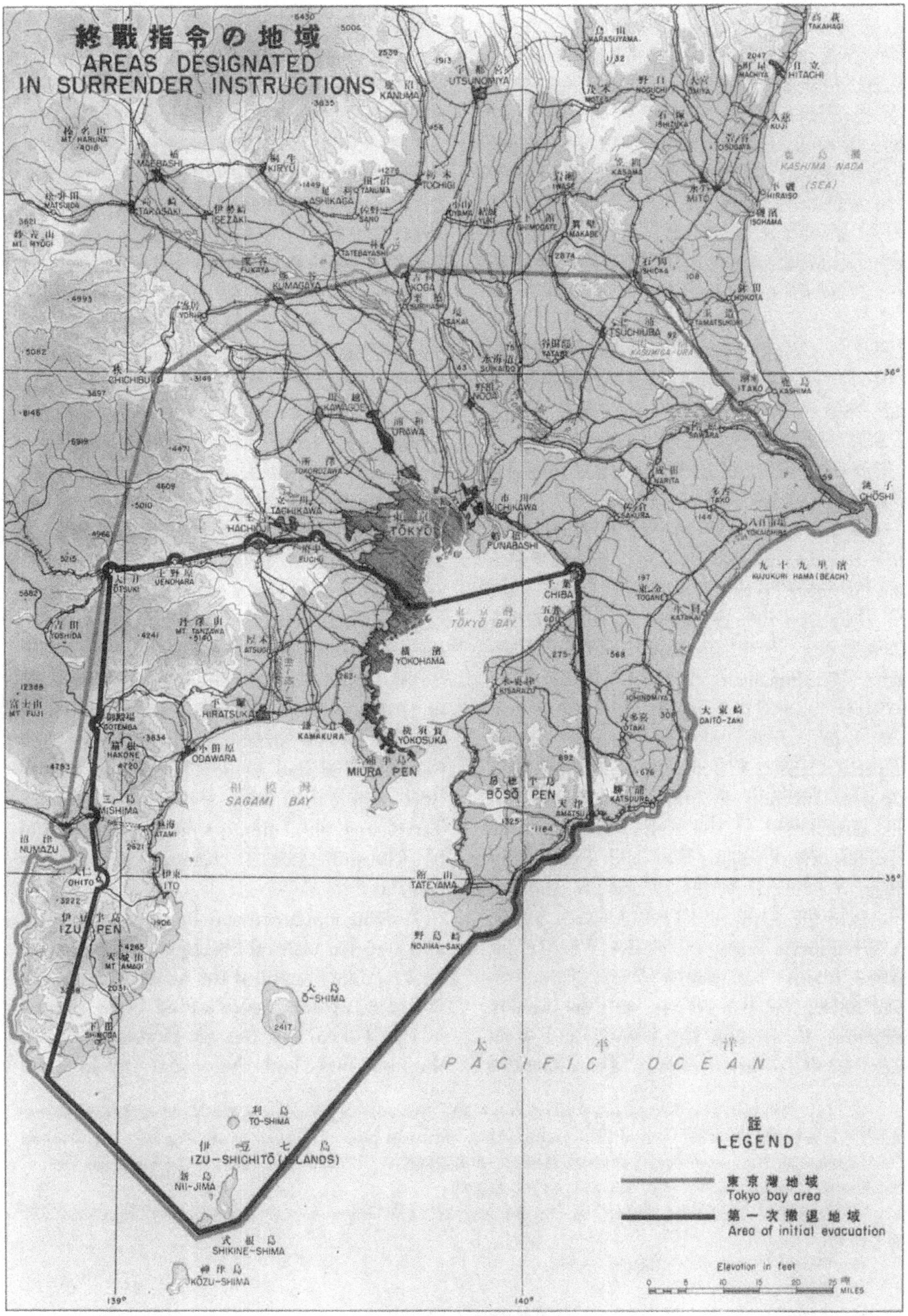

PLATE NO. 171

Areas Designated in Surrender Instructions

fied that Army forces in the areas to be evacuated should prepare to begin movement out of these areas at 1200 on 23 August, and that all personnel left behind should be disarmed and placed under reliable commanders. The evacuated troops were to be assembled as far as possible at points convenient for subsequent demobilization. Procedures were also laid down for the disarmament of troops and the disposition of weapons and explosives. All flights by military aircraft were prohibited after 1800 on 24 August. Movements of submarines and naval craft were similarly restricted. Steps to prepare the Atsugi and Kanoya airfields, as well as waterways and anchorages, for the arrival of the occupation forces were directed.[52]

The Government meanwhile acted to transform the wartime organizational structure into one for handling the transition to peace. On 22 August, the Supreme War Direction Coucil was abolished, and a War Termination Arrangements Council (*Shusen Shori Kaigi*) established in its place. The new council was a joint organ of the Government and Imperial General Headquarters, invested with top-level authority in matters connected with the termination of the war. On it sat the Premier, the Foreign, War and Navy Ministers, a Minister without Portfolio, and the Chiefs of the Army and Navy General Staffs. A subordinate organ called the War Termination Liaison Committee (*Shusen Jimu Renraku Iinkai*) was also set up, with the primary functions of assuring the prompt and exact execution of Allied directives. The committee was composed of officials of various ministries under the chairmanship of the Foreign Minister.[53]

Later, on 26 August, another organ was established in accordance with Allied directives. This was designated the Central Liaison Office, attached to the Foreign Ministry, and was charged with liaison between the Japanese Government and the Allied forces of occupation. Mr. Katsuo Okazaki, who had represented the Foreign Office on the surrender mission to Manila, received appointment as Chief of the Central Liaison Office. The main office of the agency was established in Tokyo, and this organization gradually supplanted the original War Termination Liaison Committee. The local liaison offices were later set up at such places as Atsugi, Tachikawa, Yokohama, Kyoto and other places as the occasion demanded.[54]

In addition to these agencies, the Government and Imperial General Headquarters established joint reception committee at the points of arrival of the occupation forces in order to expedite local preparations and cooperate with the command staff of these forces upon arrival. Such committees were stationed at Atsugi, Yokohama and Tateyama on the main island of Honshu, and at Kanoya in southern Kyushu.[55]

A joint announcement by the Government and Imperial General Headquarters, published on 22 August, notified the nation at large that Allied occupation troops would begin arriving in the Tokyo Bay area on 26 August.[56] At the same time, both the central and local au-

---

52 (1) *Dairikushi Dai Nisengohyakushijuku-go naishi Dai Nisengohyakugojuichi-go* 大陸指第二千五百四十九號乃至第二千五百五十一號 (Imperial General Headquarters Army Directives Nos. 2549–2551) 21–22 Aug 45. (2) *Daikaishi Dai Gohyakusanjusan-go naishi Dai Gohyakusanjuhachi-go* 大海指第五百三十三號乃至第五百三十八號 (Imperial General Headquarters Navy Directives Nos. 533–538) 22–24 Aug 45.

53 (1) Asahi Newspaper, Tokyo, op. cit., 24 Aug 45. (2) Central Liaison Office Report No. 9887 (GG), 29 Dec 47.

54 Central Liaison Office Report, op. cit.

55 Ibid.

56 Asahi Newspaper, Tokyo, op. cit., 22 Aug 45.

thorities took steps to calm public disquiet and apprehension caused by wild rumors that the occupying forces would commit acts of violence and brutality against the civilian population. A hasty evacuation of the coastal areas was already beginning as a result of these exaggerated fears.

To ease the public nervousness and combat rumors, the press and radio were fully employed. On 22 August, the Home Ministry announced that "all phases of the occupation by Allied troops will be peaceable" and urged the public not to "become needlessly panic-stricken" and, under no circumstances, to resort to physical force against the occupying troops.[57] Governor Hirose of Tokyo Metropolis similarly urged the people to shun irresponsible rumors and maintain a discipline worthy of the nation.[58] Residents of the Yokosuka area were assured that the evacuation of women and children due to fears concerning the conduct of occupation troops was totally unnecessary and that "no unpleasant incidents are expected to occur."[59]

While the authorities thus strove to instill confidence in the public with regard to the conduct of Allied troops, they themselves continued to be apprehensive lest hostile action by die-hard anti-surrender elements in the Japanese armed forces wreck the peaceful effectuation of the occupation. There had been ominous indications of trouble both from Kyushu, where the bulk of sea and air special-attack units were poised to meet an enemy invasion, and from Atsugi itself, which was to be the main point of entry for Allied airborne troops into the Tokyo Bay area.

The threat in these instances came from the Navy Air forces. Although the Navy Section of Imperial General Headquarters, as early as 14 August, had ordered the suspension of attack operations, Vice Adm. Matome Ugaki, Commander of the Fifth Air Fleet based on Kyushu, decided to defy the Imperial surrender decision in a final suicide attack on the enemy. Leading a formation of eleven *tokko* planes, he took off from Oita airfield on the afternoon of 15 August, within a few hours after the Emperor's surrender broadcast, and headed for Okinawa to attack enemy shipping. None of the aircraft returned. Ugaki, who had sent hundreds of his Kamikaze pilots into battle now realized that the only way left for him was to follow in the footsteps of his gallant airmen.[60]

Vice Adm. Ryunosuke Kusaka, appointed to succeed Ugaki, acted speedily to prevent further trouble by hastening the separation of units from their weapons. This device obviated a repetition of acts as serious as Ugaki's, but there was a temporary breakdown of discipline and control which hampered the orderly disarmament and evacuation of units from the Kanoya area in preparation for the arrival of occupation forces.

At Atsugi, an even more threatening situation developed in the Navy's 302d Air Group. Immediately after the announcement of the surrender, extremist elements in the Group led by Capt. Yasuna Kozono, flew over Atsugi and the surrounding area in their aircraft and scattered leaflets urging continuation of the war on the ground that the surrender edict was not the true will of the Emperor but the machination of "traitors around the Throne". The extremists, numbering 83 junior officers and noncommissioned officers, did not resort to hostile acts but refused to obey orders from their superior commanders.

On 19 August Prince Takamatsu, brother

57 Yomiuri-Hochi Newspaper, Tokyo, op. cit., 23 Aug 45.

58 Asahi Newspaper, Tokyo, op. cit., 26 Aug 45.

59 Ibid., 27 Aug 45.

60 *Shusenzen no Sakusen Gaikyo narabini Teisen* 終戰前の作戰概況並に停戰 (Outline of Operations Prior to Termination of the War and Activities Connected with the Cessation of Hostilities) 2d Demobilization Bureau, Feb 49, p. 16.

of the Emperor and a captain in the Navy, telephoned to Atsugi and personally appealed to Capt. Kozono and his followers to obey the Imperial decision. This intervention did not end the incident, however, for on 21 August the extremists seized a number of aircraft and flew them to Army airfields in Saitama Prefecture where they hoped to gain support from Army air units. They failed in this attempt, but it was not until 25 August that all members of the group had surrendered.[61]

As a result of the Atsugi incident, the Emperor decided on 22 August to dispatch Capt. Prince Takamatsu and Vice Adm. Prince Kuni to various naval commands on Honshu and Kyushu to reiterate the necessity of strict obedience to the surrender decision. Both princes immdiately left Tokyo to carry out this mission, but the situation so improved during the next two days that they were recalled before their scheduled tours had been completed.[62] The Emperor's action was nevertheless effective in helping to restore full order and discipline in the naval forces.

Meanwhile, physical preparations for the entry of the Allied occupation forces were being rushed ahead in the face of serious difficulties. On the night of 22–23 August, a typhoon struck the Kanto area, inflicting heavy damage and interrupting communications and transport vitally needed for the evacuation of troops from the occupation zone. It now appeared highly unlikely that the preparations to receive the Allied advance party could be completed by 26 August, the scheduled date of arrival. On the 24th, therefore, a message to this effect was dispatched to General MacArthur's headquarters.[63]

In response to the Japanese message, the Supreme Commander notified the Government and Imperial General Headquarters on 25 August that the occupation schedule would be set back an additional two days.[64] This 48-hour postponement was rendered even more necessary by a second typhoon which struck western Honshu during the 25th. The delay, however, proved sufficient to enable most of the disorganization and confusion to be overcome before the entry of the occupation troops. Preparation of troop billets, requisitioning of motor vehicles, and other required steps were carried out despite a serious lack of materials and facilities.

Disarmament and Demobilization

From the moment that Japan's surrender had become final, the military authorities had recognized that the key to assuring a smooth transition from war to peace lay in the speediest possible disarmament and demobilization of the armed forces. Imperial General Headquarters had therefore begun laying the groundwork for these vital processes even before the departure of the Japanese surrender mission to Manila.

The disarmament and demobilization of such large forces, scattered over a wide area of the Pacific and the Asiatic mainland, was obviously a formidable task. In the homeland alone, there were approximately 3,655,000 men under arms, both Army and Navy.[65] The

61 Record of Court Martial of Capt. Yasuna Kozono, Imperial Japanese Navy, 16 Oct 45.

62 Foreign Office Memorandum No. 1017 (LCR), 4 Oct 49.

63 Radio No. 31, Japanese Government to SCAP, 24 Aug 45. File on Soviet Entry into War, op. cit., Part II, Doc. No. 50.

64 Radio No. 2559, SCAP to Japanese Government and Imperial General Headquarters, 25 Aug 45. Ibid. Part II, Doc. No. 60.

65 Special Report, Military Intelligence Section, GHQ SCAP, *Final Progress of Demobilization of Japanese Armed Forces*, 31 Dec 46. pp. 12 and 79. (Report based principally on figures provided by the 1st and 2d Demobilization Bureaus).

strength of Army and Navy forces overseas, including the scattered remnants left behind and isolated in such places as New Guinea, the Bismarcks and the Solomons and in the mountains of the Philippines, could only be very roughly estimated owing to the lack of up-to-date information for many areas. Actually, later investigations indicated an approximate total of 3,575,323, with the largest concentrations in China and Manchuria.[66]

For the homeland area, disarmament procedures could be fairly well standardized and controlled. For the overseas theaters, however, they had to be left flexible to meet varying conditions, and local commanders were allowed a wide measure of discretion.[67] As far as possible, Imperial General Headquarters sought to apply the principle of self-disarmament by Japanese troops by order of their own commanders, so that the stigma of surrender would be less keenly felt by the individual soldier. This concern was manifested in similar Army and Navy Section orders issued respectively on 18 and 19 August, which stated in part:[68]

Military personnel and civilians attached to the armed forces, who come under the control of enemy troops following the promulgation of the Imperial Rescript, will not be regarded as prisoners of war. Nor will the surrender of weapons or any other act performed in accordance with enemy directives handed down by Japanese superiors be considered surrender.

In the homeland, full resort was made to the inactivation or transfer of military units from one place to another as a means of speedily separating personnel from their weapons. While generally effective, overhasty measures of this nature led to some confusion in local areas, and on 19 August the Navy Section of Imperial General Headquarters issued the following directive calculated to assure orderly disarmament procedures:[69]

In order to eliminate all possible sources of trouble between Japanese and American forces when the enemy occupation begins, to maintain effective control over personnel, and to facilitate further arrangements of various kinds, the Commander-in-Chief, General Navy Command, will make preparations in general conformity with the following outline:

1. Permission is granted to transfer and station forces as required, without regard to duties and stations assigned by existing orders.

2. Separation of forces into small units will be avoided, and so far as possible all units will be assembled promptly in suitable locations to facilitate direction and control.

3. Weapons and munitions which cannot be moved with units will not be destroyed, scattered or lost, but will be placed under the custody of necessary guards to be left behind.

4. Vessels will be anchored at their assigned mooring places; surface and undersea special-attack craft will be kept at their regular bases or close to them; and aircraft in general will be stored at their present locations. Necessary guards will be assigned.

By 26 August, two days before the scheduled arrival of the Allied advance occupation

66 Foreign Office Memorandum No. 1281 (GR), 26 Jun 50.

67 As an example of the latitude given to local commanders, an Imperial General Headquarters, Navy Section directive issued 19 August stated that "when necessary to facilitate smooth procedure, agreements may be entered into with the senior local commander of the Allied forces." It further gave permission "to take expedient measures in accordance with the actual situation." *Daikaishi Dai Gohyakusanjuichi-go* 大海指第五百三十一號 (Imperial General Headquarters Navy Directive No. 531) 19 Aug 45.

68 (1) *Dairikumei Dai Sensambyakuhachijugo-go* 大陸命第千三百八十五號 (Imperial General Headquarters Army Order No. 1385) 18 Aug 45. (2) *Daikairei Dai Goju-go* 大海令第五十號 (Imperial General Headquarters Navy Order No. 50) 19 Aug 45.

69 (1) *Daikaishi Dai Gohyakusanju-go* 大海指第五百三十號 (Imperial General Headquarters Navy Directive No. 530) 19 Aug 45. (2) Similar directive was issued by the Army Section. *Dairikushi Dai Nisengohyakushijuku-go* 大陸指第二千五百四十九號 (Imperial General Headquarters Army Directive No. 2549) 21 Aug 45.

party at Atsugi, the disarmament of Army and Navy forces in the homeland and in some overseas areas was well under way. On that date, Imperial General Headquarters, in accordance with directives from General MacArthur's headquarters, notified Army and Navy commands of procedures to be followed in negotiating the cessation of hostilities and in effecting the final transfer of weapons and equipment to Allied control.[70]

Parallel with the disarmament of the fighting forces, War and Navy Ministries and Imperial General Headquarters also tackled the broader problem of general demobilization. This was to be carried out in accordance with Allied directives, but even before such directives were issued, the Army and Navy authorities proceeded to draw up and disseminate basic demobilization plans so that no time would be lost. On 18 August the War Ministry issued both a general demobilization outline, approved by the Emperor, and an implementing order containing detailed basic regulation.[71] Three days later, on 21 August, the Navy Ministry radioed to subordinate commands a general plan for the demobilization of naval personnel.[72]

Actual execution of the plans began shortly thereafter. As the program got under way, the Emperor on 25 August issued a special message to all Army and Navy personnel expressing the Imperial desi e to see the demobilization carried out swiftly and in orderly fashion. The message read:[73]

To Our trusted soldiers and sailors on the occasion of the demobilization of the Imperial Army and Navy:

Upon due consideration of the situation, We have decided to lay down Our arms and abolish military preparations. We are overcome with emotion when We think of the precept of Our Imperial Forefathers and the loyalty so long given by Our valiant servicemen. Especially, Our grief is unbounded for the many who have fallen in battle or died of sickness.

On the occasion of demobilization, it is Our fervent wish that the program be carried out rapidly and systematically under orderly supervision, so as to give a crowning example of the perfection of the Imperial Army and Navy.

We charge you members of the armed forces, to comply with Our wishes. Turn to civilian occupations as good and loyal subjects, and by enduring hardships and overcoming difficulties, exert your full energies in the task of postwar reconstruction.

In the homeland, demobilization progressed rapidly, with some 1,500,000 men being discharged and returned to their homes by the end of August.[74] Demobilization of the forces overseas, however, could be accomplished only little by little as a long and difficult program

70 (1) *Daikaishi Dai Gohyakushiju-go* 大海指第五百四十號 (Imperial General Headquarters Navy Directive No. 540) 26 Aug 45. (2) *Dairikushi Dai Nisengohyakugojuhachi-go* 大陸指第二千五百五十八號 (Imperial General Headquarters Army Directive No. 2558) 27 Aug 45. (3) *Dairikushi Dai Nisengohyakugojuku-go* 大陸指第二千五百五十九號 (Imperial General Headquarters Army Directive No. 2559) 31 Aug 45.

71 *Meirei Tsuzuri* 命令綴 (File of Orders) Military Police Headquarters, Aug 45: Essentials of Imperial Army Domobilization; and Detailed Regulations Concerning Essentials of Imperial Army Demobilization

72 Secretariat Personnel Confidential Radio No. 210,110, Navy Ministry, 21 Aug 45. File of Incoming and Outgoing Communications on Public Peace and Military Discipline, op. cit. Navy Gazette No. 5176, 6 Sep 45.

73 *Meirei Shiji Shorui Tsuzuri* 命令指示書類綴 (File of Documents Pertaining to Orders and Directives) Military Police Headquarters, Aug 45.

74 Demobilization of Army and Navy personnel in the homeland was completed by the end of November 1945. At this time the War and Navy Ministries were dissolved, being replaced respectively, by the 1st and 2d Demobilization Ministries which continued to handle the demobilization of personnel repatriated from overseas. The ministries later became bureaus under the Welfare Ministry.

of repatriation was carried out. The program got under way soon after the formal signing of the surrender when the Navy hospital ship *Takasago Maru* was dispatched to Meleon in the Caroline Islands to bring back the Japanese personnel on the island.[75]

Entry of Occupation Forces

At 1030 on 27 August, elements of the United States Third Fleet entered Sagami Bay in execution of first step of the delayed occupation schedule. Preparations for the reception of the Allied forces had been completed as far as was humanly possible, and in a radio address on the 26th, the newly-appointed War Minister, General Sadamu Shimomura,[76] had made a final and urgent appeal to the Army to submit to the occupation quietly and without incident. He said:[77]

The course we must follow has clearly been indicated in the Imperial Rescript. We in the armed services have no choice but to obey reverently the Imperial will. ....Any action against or contrary to the Imperial decision will not be condoned even though it springs from the pure Bushido spirit. ....I strictly admonish you not to let yourselves be agitated by untoward incidents and not to assume a hostile attitude toward the foreigners or indulge in disorderly speech or activity. ....We must behave prudently, taking into consideration the future of the nation. Do not become excited and conceal or destroy arms which are to be surrendered. This is our opportunity to display the magnificence of our nation.

Between 0828 and 1100 on 28 August, the advance party of the Allied occupation forces, consisting of about 150 men under Col. C.T. Tench, arrived at Atsugi airfield. After hearing a report by Lt. Gen. Seizo Arisue, chairman of the official reception committee, the advance party completed technical arrangements for the arrival of the main forces and issued supplementary instructions to the Japanese authorities. Col. Tench expressed satisfaction with the preliminary preparations which had been carried out in compliance with the directives of the Supreme Commander.[78]

On 30 August the main body of the airborne occupation forces began streaming into Atsugi in a steady procession of transport planes, while naval and marine forces simultaneously began landing at Yokosuka on the south shore of Tokyo Bay. At 1405 on 30 August, the Supreme Commander himself arrived at Atsugi to set up his headquarters on Japanese soil. The Tokyo newspaper *Yomiuri-Hochi*, in a story published the following day, described the scene as follows:[79]

As the four-engined plane landed, the waiting officers, men and newspaper reporters surged forward. The door of the plane opened, and the tall, khaki-clad figure of General MacArthur appeared. He wore dark green sun-glasses under his large, visored military cap and held a corn-cob pipe in his mouth. Calmly, he paused with one foot on the debarkation ladder to

---

75 (1) SCAP Directive to Imperial General Headquarters, 1 Sep 45. File on Soviet Entry into War, op. cit. Part III, Doc No. 79. (2) It was initially estimated, in view of the almost complete destruction of Japanese shipping, that more than four years would be required to complete the repatriation program. Progress was more rapid than expected, however, due to the loan of over 200 ships—104 Liberty ships and 104 LST's—by SCAP for repatriation purposes between February 1946 and February 1947.

76 General Shimomura had been relieved on 23 August of command of the North China Area Army in order to assume the portfolio of War Minister in the Higashikuni Cabinet. Until Shimomura's appointment, Prince Higashikuni had filled the post of War Minister concurrently with the premiership.

77 Asahi Newspaper, Tokyo, op. cit., 27 Aug 45.

78 Statement by Lt. Gen. Seizo Arisue, Chairman, Atsugi District Reception Committee.

79 Yomiuri-Hochi Newspaper, Tokyo, op. cit., 31 Aug 45.

look around at the welcoming party.

In his first statement to Allied and Japanese reporters, General MacArthur declared:

From Melbourne to Tokyo is a long road. It has been a long and hard road, but this looks like the pay-off. The surrender plans are going splendidly and completely according to previons arrangements. In all outlying areas, fighting has practically ceased.

In this area a week ago, there were 300,000 troops which have been disarmed and demobilized. The Japanese seem to be acting in complete good faith. There is every hope of the success of the capitulation without undue friction and without unnecessary bloodshed.

Every newspaper in Japan printed illustrated accounts of the General's arrival at Atsugi. The man who for more than three years had waged a relentless counteroffensive against the Japanese armies in the Southwest Pacific and who now was to accept and conclude the surrender, thus made his first appearance before the Japanese people.

Immediately after his arrival, MacArthur proceeded to Yokohama, accompanied by Lt. Gen. Robert L. Eichelberger, Commander of the United States Eighth Army, who had arrived earlier in the day. The party drove some 20 miles eastward through peaceful, farming country and silent, war-devastated towns. The Supreme Commander set up his headquarters provisionally in the Yokohama Customs House. The headquarters of the American Eighth Army and the Far East Air Force also had been established in the same city, and representatives of the United States Pacific Fleet were attached to Supreme Commander's headquarters.[80]

In order to facilitate rapid liaison, various Japanese agencies were established in the Kanagawa Government Building adjacent to the headquarters of the Supreme Commander. They included the Yokohama District Reception Committee (later the Yokohama Liaison Office), the Imperial General Headquarters Liaison Committee and the office of the Governor of Kanagawa Prefecture.

Occupation forces arrived in ever-increasing numbers by land, sea and air until important cities and major military installations were under Allied control. The occupation of Tokyo itself was not carried out until 8 September, when some 8,000 troops of the First Cavalry Division moved into the capital in orderly fashion. Up to this time, traces of uneasiness had lingered in the minds of the general populace, but the occupation of the capital without incident did much to ease tension. The nation as a whole was deeply impressed when it discovered that the occupation troops, from whom ill-treatment and excesses had been feared, were well-disciplined and under enlightened leadership. In particular, Lt. Gen. Eichelberger's order of 9 September placing the Imperial Palace, temples, shrines and private dwellings "off limits" to occupation forces was gratefully welcomed by the citizens.[81]

Scattered incidents of a minor character were reported as contact between the occupation forces and the Japanese Government and people became more frequent. On the whole, however, the general public was submissive, and its trust in the occupation gradually deepened. Cooperation in all quarters improved as the Japanese came to know American soldiers as individuals and recognized the efficiency and discipline with which their work was carried

80 The headquarters of both the United States Eighth Army and the Far Eastern Air Force were set up in the New Grand Hotel near the Yokohama Customs House. Rear Adm. J. J. Ballentine acted as the U. S. Fleet liaison officer with SCAP.

81 Yomiuri-Hochi Newspaper, Tokyo, op. cit., 12 Sep 45. Lt. Gen. Eichelberger's order coincided with the "solicitations" which the Japanese mission to Manila left with Maj. Gen. Charles A. Willoughby, General MacArthur's G-2, at the conclusion of the surrender parleys. (Cf. p. 694).

out. The words of General MacArthur at Atsugi were proving prophetic.

Signing of the Surrender

The day set for the formal signature of the surrender instrument was a peaceful, cloudy Sunday, the second of September. Shortly after 0600, Foreign Minister Shigemitsu representing the Emperor and the Japanese Government, General Umezu representing Imperial General Headquarters, and nine accompanying officials of the Foreign Ministry, Army and Navy assembled at the Kanagawa Government Building in Yokohama. From there they drove to the South Pier and boarded an American destroyer which took them out to the battleship *U. S. S. Missouri* resting at anchor in the middle of Tokyo Bay.[82]

The Japanese delegation boarded the Missouri at 0855. At 0900 the surrender formalities began, with General MacArthur presiding. Not a word was spoken or a sign of greeting exchanged as the Japanese representatives walked to their places. At 0904, Foreign Minister Shigemitsu affixed his signature to the surrender document. He was followed by General Umezu, who signed for both the Army and the Navy.[83] Four minutes later, General MacArthur affixed his signature of acceptance, followed by representatives of the nine Allied Powers which had been at war with Japan.

The ceremony lasted 20 minutes. After receiving their copy of the surrender instrument, the Japanese delegates returned the way they had come, proceeding immediately to Tokyo to report to the Emperor.

The same afternoon, the Government and Imperial General Headquarters made public the contents of the instrument of surrender. The text of the instrument follows:[84]

We, acting by command of and in behalf of the Emperor of Japan, the Japanese Government and the Japanese Imperial General Headquarters, hereby accept the provisions set forth in the declaration issued by the heads of the Governments of the United States, China and Great Britain on 26 July 1945, at Potsdam, and subsequently adhered to by the Union of Soviet Socialist Republics, which four powers are hereafter referred to as the Allied Powers.

We hereby proclaim the unconditional surrender to the Allied Powers of the Japanese Imperial General Headquarters and of all Japanese armed forces and all armed forces under Japanese control, wherever situated.

We hereby command all Japanese forces, wherever situated, and the Japanese people to cease hostilities forthwith, to preserve and save from damage all ships, aircraft, and military and civil property and to comply with all requirements which may be imposed by the Supreme Commander for the Allied Powers or by agencies of the Japanese Government of his direction.

We hereby command the Japanese Imperial General Headquarters to issue at once orders to the Commanders of all Japanese forces and all forces under Japanese control, wherever situated, to surrender unconditionally themselves and all forces under their control.

We hereby command all civil, military and naval officials to obey and enforce all proclamations, orders and directives deemed by the Supreme Commander for the Allied Powers to be proper to effectuate this surrender and issued by him or under his authority, and we direct all such officials to remain at their posts and to continue to perform their non-combatant duties

82 The nine attendants were: Messrs. Katsuo Okazaki, Toshikazu Kase and Saburo Ota, of the Foreign Office; Lt. Gen. Shuichi Miyazaki, Maj. Gen. Yatsuji Nagai, and Col. Ichiji Sugita, of the Army; Rear Adms. Sadatoshi Tomioka and Ichiro Yokoyama, and Capt. Katsuo Shiba of the Navy.

83 File on Soviet Entry into War, op. cit. Part III, Doc. No. 90.

84 The nine powers whose representatives signed the instrument of surrender were the United States, Republic of China, United Kingdom, Union of Soviet Socialist Republic, Commonwealth of Australia, Dominion of Canada, Provisional Government of the French Republic, Kingdom of Netherlands and the Dominion of New Zealand. Ibid.

unless specifically relieved by him or under his authority.

We hereby undertake for the Emperor, the Japanese Government and their successors to carry out the provisions of the Potsdam Declaration in good faith, and to issue whatever order and take whatever action may be required by the Supreme Commander for the Allied Powers or by any other designated representative of the Allied Powers for the purpose of giving effect to that Declaration.

We hereby command the Japanese Imperial Government and the Japanese Imperial General Headquarters at once to liberate all Allied prisoners of war and civilian internees now under Japanese control and to provide for their protection, care, maintenance and immediate transportation to places as directed.

The authority of the Emperor and the Japanese Government to rule the state shall be subject to the Supreme Commander for the Allied Powers who will take such steps as he deems proper to effectuate these terms of surrender.

Made public at the same time was the text of a new Imperial Rescript announcing the formal conclusion of the surrender to the Japanese people. The draft of this rescript had been drawn up by General MacArthur's headquarters and had been delivered to the Japanese surrender mission at Manila. It read as follows:[85]

Accepting the terms set forth in the Declaration issued by the heads of the Governments of the United States, Great Britain and China on 26 July 1945 at Potsdam and subsequently adhered to by the Union of Soviet Socialist Republics, We have commanded the Japanese Imperial Government and the Japanese Imperial General Headquarters to sign on Our behalf, the instrument of surrender presented by the Supreme Commander for the Allied Powers and to issue general orders to the Military and Naval forces in accordance with the direction of the Supreme Commander for the Allied Powers. We command all Our people forthwith to cease hostilities, to lay down their arms, and faithfully to carry out all the provisions of the instrument of surrender and the General Orders issued by the Japanese Imperial Government and the Japanese Imperial General Headquarters hereunder.

Also in compliance with Allied directives received during the Manila parleys, Imperial General Headquarters issued Army-Navy General Order No. 1. This order contained specific stipulations for the surrender of local Japanese commanders over the entire Pacific war area, for the demobilization of the armed forces and the general disarmament of Japan, and the furnishing of information required by the occupation authorities.[86]

Simultaneously with and subsequent to the surrender aboard the Missouri, Japanese forces overseas formally surrendered to designated Allied commanders. At Truk, the surrender formalities took place the same day as the signing in Tokyo Bay. At Baguio on Luzon and at Chichi-Jima in the Bonin Islands, they were carried out the following day; at Batavia, on 5 September; at Rabaul, on 6 September; at Nanking and Seoul on 9 September, and at Singapore on 12 September.[87]

The words with which General MacArthur opened the ceremony aboard the Missouri stood as solemn testimony to the hopes of the entire human race, both victor and vanquished, as represented by the men there assembled.[88]

But rather it is for us, both victors and vanquished,

85 A specially prepared copy of this Rescript, signed by the Emperor and countersigned by Premier Prince Higashikuni and all other Cabinet Ministers, was handed to the Supreme Commander for the Allied Powers by the Japanese delegate aboard the Missouri prior to their signing of the surrender instrument. Ibid, Doc. No. 91.

86 Ibid., Doc. No. 89.

87 (1) *Kyu Nihongun Kofuku no Nichiji Basho* 舊日本軍降伏の日時場所 (Time and Place of Formal Surrender of Japanese Armed Forces) 1st and 2d Demobilization Bureaus, 24 Dec 48. (2) After the surrender of arms in all areas had been completed, the Imperial General Headquarters was first abolished on 13 September, and the Army and Navy General Staffs were dissolved on 15 October 1945.

88 Asahi Newspaper, Tokyo, op. cit., 3 Sep 45.

to rise to that higher dignity which alone benefits the sacred purposes we are about to serve. It is my earnest hope—indeed the hope of all mankind—that from this solemn occasion a better world shall emerge out of the blood and carnage of the past, a world founded upon faith and understanding, a world dedicated to the dignity of man and the fulfillment of his most cherished wish for freedom, tolerance and justice.

Glossary of Japanese Military Signs, Symbols, and Abbreviations Used in this Volume

ARMY GROUND FORCES

	Imperial General Headquarters
	General Army Headquarters
	Area Army Headquarters
	Army Headquarters
	Division Headquarters
	Brigade or Infantry Group Headquarters
	Regimental Headquarters
SA	General Army
HA	Area Army
A	Army
D	Division
MBs	Independent Mixed Brigade
B	Brigade
iB	Infantry Group or Brigade
i	Infantry Regiment
is	Independent Infantry Regiment
bs	Independent Infantry Battalion
TKD	Armored Division
TKB	Tank Brigade
TK	Tank unit
	Tank
	Artillery position
AA	Antiaircraft artillery

NAVY

	Combined Fleet or Area Fleet Headquarters
	Fleet Headquarters
	Division, Squadron, Flotilla or Base Force Headquarters
GF	Combined Fleet
HF	Area Fleet
F	Fleet
AF	Air Fleet
S	Battleship or Cruiser Division
Sf	Carrier Division or Air Flotilla
Sd	Destroyer Squadron
fg	Air Group
Bg	Base Force
aBg	Special Base Force
CV	Aircraft carrier
CVL	Light aircraft carrier
BB	Battleship
CA	Heavy cruiser
CL	Light cruiser
DD	Destroyer
AP	Transport
AO	Tanker
	Aircraft carrier
	Submarine

ARMY AIR FORCES

	Air Army Headquarters
	Air Division Headquarters
	Air Brigade Headquarters
FA	Air Army
FD	Air Division
FB	Air Brigade
FR	Air Regiment

GENERAL

-xxxxx-	General Army or Combined Fleet operational boundary
-xxxx-	Area Army or Area Fleet operational boundary
-xxx-	Army or Fleet operational boundary
-xx-	Division operational boundary
-x-	Brigade or Base Force operational boundary
	Regimental or lower unit operational boundary
	Airfield
	Aircraft
	Airborne landing
	Assembly area
	Defensive position
	Offensive position
	Advance or attack
	Troop movement
	Withdrawal

NOTE

1. Unless otherwise indicated, blue represents Japanese Forces and red represents Allied Forces.
2. Battalions are designated by Roman numerals; i. e. 2nd Battalion of 41st Infantry Regiment is designated II/41 i.
3. When the Japanese designation of a geographical location differs from that accepted by Department of the Army or SWPA, the Japanese designation is followed by (J) and the accepted American designation is given below it in parentheses. Where no accepted American designation can be ascertained, only the Japanese designation is given.

PLATE NO. 172

Glossary of Japanese Military Signs, Symbols and Abbreviations

APPENDIX

THE JAPANESE EMPEROR AND THE WAR

By

Yasumasa Matsudaira

Grand Master of Ceremonies,
Ministry of Imperial Household

Formerly Chief Secretary to
Marquis Koichi Kido, Lord
Keeper of the Privy Seal.

APPENDIX

THE JAPANESE EMPEROR AND THE WAR

The Emperor's Role at the Time of Surrender.

The Pacific War broke out despite all efforts exerted by the Emperor to prevent the calamity. Nevertheless, His Majesty's devotion to international concord was such that from the very moment hostilities began his mind was directed constantly toward the restoration of peace.[1]

As early as February 1942, when our forces were about to capture Manila, the Emperor instructed Premier Tojo to lose no opportunity in bringing the war to an end.

I believe it was in February 1944 that His Majesty began to grapple with the problem of peace as the most urgent issue confronting the Empire. I recall one day after the war ended the Emperor, speaking to us, his entourage, reminisced: "From the time our line along the Stanley Mountain Range in New Guinea was penetrated, I was anxious for peace, but, we had a treaty with Germany against concluding a separate peace; and we could not violate an international commitment. This was a dilemma that tormented me."

In February 1945, after the fall of Manila, the Emperor summoned individually, and on different days, Baron Wakatsuki and several other ex-Premiers and also Count Makino, former Lord Keeper of the Privy Seal, and asked for their views. Though they all favored peace, I understand none had any concrete suggestion.

It was toward the end of May, 1945 when the Yamato, our newest and biggest battleship dispatched to save Okinawa, was sunk and all hope was gone in naval warfare that General Umezu, Chief of the Army General Staff, informed the Throne of the impossibility of land operations to recoup our reverses at the Yunnan and Burma fronts. His Majesty was then resolved to seek peace at all costs.

In order to bring hostilities to an end, however, it was necessary to pursuade the officers and men, who were determined to fight to death, to lay down their arms. Here was a most delicate task. One false step would bring domestic turmoil to the country threatened with an imminent enemy invasion. The fall of Saipan, defeat on Leyte, and the loss of Okinawa had convinced the military leaders, as well as all the Cabinet members, of the pressing need for terminating the war—still they were at loss as to how to go about it. In the face of the nation's volcanic temper at that time, no one dared to voice his honest opinion or do anything that might arouse the suspicion of his entertaining such an opinion. At this juncture it was His Majesty who urged the Government to sue for peace.

The Supreme War Direction Council meeting be-

1 The Emperor, as the sovereign ruler of Japan, played an important role in the termination of the Pacific War. His influence in bringing hostilities to an end can be seen in the concluding chapters of this volume. It was felt however, that a more detailed report of the Emperor's feelings and actions would further aid in understanding his relation to the Pacific War. Therefore, this unpublished manuscript of Marquis Yasumasa Matsudaira, one of the Emperor's trusted advisors, has been included as an appendix to this volume. Marquis Matsudaira, currently the Grand Master of Ceremonies, Ministry of Imperial Household, served as Chief Secretary to the Lord Privy Seal, Marquis Koichi Kido, from June 1936 to November 1945. These positions has provided him with many opportunities to learn and understand the Emperor's views regarding numerous events of great historical significance. This manuscript was prepared in December 1949 by the Marquis in response to a request from the G–2 Historical Section, GHQ FEC, and has been published just as it was translated from the Japanese.

fore the Emperor on June 8, 1945, passed on the "Basic Policy Draft for Future Direction of War" which declared the official decision of the Government to continue the war. His Majesty was deeply shocked when this decision, in utter disregard of the developments at the front and situation at home, was adopted by the Supreme Council presided over by Premier Suzuki, who enjoyed Imperial confidence to the fullest degree. The Emperor wished to demand the Government as well as the military leaders to reconsider. But extraordinary caution was required for this. The Emperor was most reluctant to give out his personal opinion in the form of a command, for that would contravene the practice of our constitutional government and might produce grave repercussions. Herein lay the difficulty. The draft of the Basic Policy Draft for Future Direction of War contained an accurate description of both the war situation and the domestic conditions of the country, of which the impossibility of continuing the war was an obvious conclusion. Yet the document, by an amazing twist of logic, ended in advocating the continuation of war. At the June 8 conference the Emperor, though he made no comment, sensed the inconsistency at once. He desired to leave the matter to be threshed out at another meeting. Thus, with the consent of the Premier he called the Supreme War Direction Council meeting on June 22, at which he proposed peace.

"This is a critical moment," said His Majesty, "permitting no hesitation, nor delay. Despite the Basic Policy Draft adopted at the Supreme War Direction Council meeting on June 8, you will consider the question of ending the war as quickly as possible." To advocate peace, as His Majesty did in those days of frenzied chauvinism, was an act which required an extraordinary resolution involving a grave risk even on the august person of an Emperor.

What impressed those of us who were close to the Emperor was the steadfast and determined way His Majesty followed his own conviction. Thanks to the Imperial admonition, the Government was now moving toward the termination of hostilities. But as to the time and method and also the terms of peace, there was a wide divergence of views between the Cabinet and the military. The Premier and Navy Minister Yonai and Foreign Minister Togo were conspicuously active in working for peace, but the role of the Emperor, who openly and covertly gave constant encouragement and support to these men, should not be forgotten.

The government wavered until the very last minute. It was at the Imperial conference of the Supreme War Direction Council of August 14 that the Emperor himself pronounced his decision in favor of immediate acceptance of the Potsdam Declaration. Thus, did our nation emerge from eight long years of warfare to hail the dawn of peace. At both conferences, August 9 and 14, His Majesty, summarizing in a lucid fashion the general situation relating to the war developments on land and sea, the extent of air raid damage, the defense of the home islands, the munition production, and food supply, expounded why Japan should stop fighting and surrender. He was most deeply concerned not only over the sufferings of his subjects, which were being intensified each day, but also over the enormous waste in lives and property our futile resistance was causing our adversaries. These sentiments are partly revealed in the Imperial Rescript issued on the termination of war. I know well how heavily these matters weighed upon the mind of the Emperor throughout the war years.

Emperor's Efforts for Preservation of Peace.

The Pacific War was terminated by the decision of the Emperor. Why then had it not also been possible to prevent war by Imperial judgement?

This question is asked frequently abroad and even among our own people at home. The question arises out of ignorance of His Majesty's ceaseless endeavors toward peace and also from misunderstanding the position of the Emperor in the Japanese government.

The first part (1926–36) of the Showa Era witnessed great changes in Japanese thought and economy. The economic depression that followed the First World War oppressed heavily the lives of our people and engendered extremist movements both to the right and to the left. It was an era of political instability when the military clique raised its head and our foreign policy took on a markedly aggressive character.

His Majesty is a man of conviction who believes firmly in peace and it is no exaggeration to say that all his efforts throughout the 20 years of his reign were devoted to the preservation of peace. In order to bring out this fact concretely it is necessary to explain the following points:

(1) *The military—especially the Army—was extremely high-handed and intractable, and ignored the wishes of the Emperor.*
(2) *The Manchurian Incident (1931) and the China Affair (1937) broke out through arbitrary actions of the military on the spot.*
(3) *The Government and the Diet, and public opinion itself were all powerless before the military.*
(4) *The Emperor always adhered scrupulously to constitutional procedures.*

In the following pages I venture to state the facts as I remember them with respect to these points.

The Government and the Military

His Majesty's endeavors for peace had never been completely successful and to explain why, it is necessary to describe the relationship between the military and the Government.

From the beginning of the Showa Era (1925) the economic depression, and especially the sorry plight of the farming communities had brought on confusion and unrest. Radical ideologies affected young officers in the Army and Navy, breeding a spirit of defiance and revolt against their superiors. In 1931, the year of the Manchurian Incident, there occured the so-called "March Incident" and "November Incident," in which they made unsuccessful attempts to stage a coup d'etat to seize the Government. In the March Incident, officers at higher levels were involved, revealing how deeply the canker of recalcitrance had eaten into the heart of the Army. Both incidents were carefully hushed up, all information being withheld from the public, and even from the Government. They were not reported to the Emperor. None of the culprits were punished. The situation grew from bad to worse. On May 15, 1932 Premier Inukai was shot by young Navy officers. On the 26th of February, 1936, a mutiny was staged, in which the Lord Keeper of the Privy Seal, the Grand Chamberlain, the Premier, the Finance Minister and other important personages were either killed or wounded. Throughout these turbulent years His Majesty did everything to restore order and discipline in the armed services. To cite a few instances as I recollect them; in July 1929 the Tanaka Cabinet resigned under Imperial reprimand for its failure to carry out the Emperor's special instructions to track down and punish the Army men responsible for the murder, in the preceding year, of the Manchurian war lord, Chang Tsolin. In 1931 the Emperor warned the War and Navy Ministers against the interference of the military in government and diplomacy. War Minister Minami, in pursuance of the Imperial wishes, dispatched a special emissary to Manchuria. The emissary arrived in Mukden on the very eve of the outbreak of the Manchurian Incident. Before and after the May 15, 1932 Incident, and immediately prior to the February 26, 1936 mutiny, the Emperor warned the military leaders against interference by soldiers in government affairs, and called for strict enforcement of discipline.

At the time of the February mutiny it was the Emperor who in the midst of unspeakable confusion ordered the suppression of the uprising, and saved the country from an impending catastrophe. When the Army was showing a sympathetic attitude toward the mutineers and trying to make patriots of them, the Emperor pronounced them to be "rebels" and issued a strict command to the War Minister to suppress them. When Home Minister Goto, who was then acting as Premier, came to present the letters of resignation of his Cabinet members, the Emperor urged him to "use all efforts to suppress the rioters and to unite the Cabinet for the restoration of order." Unfortunately, the Cabinet was too weak and helpless. To the Deputy-Chief of Army General Staff, Sugiyama, who reported on the partial promulgation of martial law, the Emperor inquired if he had completed plans for suppressing the rebel force, and cautioned him against abuse of martial law by the military. The War Minister recommended to His Majesty, through His Majesty's Chief Aide-de-Camp, the formation of a temporary cabinet calculated to humor the rebel officers, who were then demanding the establishment of a military government under the War Minister. The Emperor rejected the proposi-

tion, saying " I have already ordered the Government to suppress the rebels." Despite repeated demands from the Throne, the War Minister hesitated to act. The Supreme Military Council, which was ordered to take the matter in hand, could reach no decision. Finally, the Emperor summoned his Chief Aide-de-Camp. " If nobody can stop the rebels," said His Majesty, " I will go myself. Fetch the Imperial standard, and follow me." Overawed, the Army leaders decided upon a punitive action, and the rebellion was put down completely on the 29th of February.

Notwithstanding the complaint voiced among young Army officers, the Emperor always treated the political affairs and military affairs on an equal footing. For the disposition of the former he relied on the advice and assistance of the ministers of state; and for the latter, on the counsel of his chiefs of staff—namely the Chief of the Army General Staff and the Chief of Navy General Staff. But whenever the recommendation of the military advisers ran counter to the policy of the Government, he reserved his approval. For instance, when at the outbreak of the Manchurian Incident in September 1931, a contingent of the Japanese Army in Korea crossed the border into Manchuria to reinforce the Kwantung Army without an Imperial Command and against the Government's policy of non-extension of hostilities, the Emperor did not give post facto approval on the arbitrary action of the Commander of our Korean forces until the Government had sanctioned an appropriation for this troop movement. Again, on the occasion of a Soviet-Japanese clash at Chang-ku-feng on the Soviet-Manchurian border, when the Chief of the General Staff recommended to the Throne the dispatch of reinforcements, His Majesty who had been advised by the Foreign Minister of the Government's opposition, disapproved the measure.

It may be noted that the Imperial Rescript on Japan's withdrawal from the League of Nations contains the sentence:

" We command that all public servants, whether civilian or military, shall faithfully perform his appointed duty. . . . ."

This was inserted at the express wish of the Emperor. It differs somewhat from the passage in the Imperial Rescript issued on the occasion of his accession to the Throne, which reads " with the civil service as the warp and the military as the woof." In the light of the political background as described above, the intention of the Rescript is clear and was to warn the military against trespassing upon the sphere of civil government.

Some time after the war's end His Majesty spoke to his entourage of his difficult position in those years as follows:

" I was greatly worried over the possible consequences of the reckless behavior on the part of the young Army hot heads. I seized every opportunity to caution the military leaders, but because of the spirit of insubordination and defiance pervading the ranks, their words were of no avail. But I could not very well speak directly to officers on the lower echelons. That would destroy the very structure of control and direction. Here was my problem. Among my trusted advisers there were many who strove day and night to curb the excesses of the military, however, these men had either to lose positions of influence, or even to sacrifice their very lives.

The Manchurian Incident and China Affair.

The Manchurian Incident of 1931 was aggravated and expanded in scope by the Japanese forces on the spot against the policy of the Government.

The Emperor was alarmed at the talk of the so-called strong foreign policy of solving by force the problems of Manchuria and Inner Mongolia prior to the outbreak of the Incident. He took up the matter with the then Premier Baron Wakatsuki.

" You are settling these problems," remarked His Majesty, " of Manchuria and Mongolia on the basis of Sino-Japanese amity, I hope."

The Premier at once relayed the Imperial words to his Cabinet, and the Government's policy for amicable solution was reaffirmed.

But scarcely a month had elapsed before the Incident started on September 18, 1931. The conflagration spread to Shanghai where a violent anti-Japanese movement flared out. In February 1932, the newly formed Inukai Cabinet was obliged to dispatch an Army contingent to rescue the Japanese marine corps which was threatened with annihilation by a superior Chinese force.

The consensus of both Government and private circles in Tokyo was: "Our armed forces, once on foreign soil, are beyond the control of the home Government, and hostilities are likely to expand into an all-out war between Japan and China."

However, contrary to these gloomy forebodings, General Shirakawa, Commander of the Expeditionary Force, after a series of fierce battles had succeeded in driving the enemy from around the International Settlement, proclaimed a cease fire order to all his forces, much to the chagrin of his staff officers who wanted to pursue the fleeing Chinese. It was May 3rd, the very day on which the League of Nations was scheduled to deliberate on the Shanghai incident.

This dramatic event had a background which was then not known to the world. General Shirakawa, appointed Commander of the Expeditionary Force, went to the palace to take leave of His Majesty. Calling the General to his side the Emperor discoursed on the international situation.

"This is entirely my personal wish," said the Emperor, "But when you have achieved the purpose of restoring peace to the Settlement, I hope you will hold back your troops and never let them march into the interior."

It was this Imperial wish which the General, as a loyal warrior, carried out to the letter, although by so doing he incurred the displeasure of the Army.

When the next March 3rd (1933) came around, the General was dead. The day being the Dolls Festival Day, His Majesty sent to Mrs. Shirakawa the following poem:

This day when maidens celebrate
The Feast of Dolls,
I remember the glorious deed
Of him who stopped war.

Japan's withdrawal from the League of Nations was, needless to say, not His Majesty's desire. With the return of Lord Lytton and his party to Geneva in October 1932, the Commission Report was made public. The Emperor, being opposed to the Army's scheme for an independent Manchuria, considered that the Lytton Report should be accepted inasmuch as it recognized Japan's legitimate rights and claims in Manchuria, while sustaining China's suzerainty over the territory. He would have therefore conveyed his view to Premier Saito and instructed him to reconsider the Government's unyielding attitude, but for the advice of Prince Saionji and Count Makino. It was the carefully considered opinion of these senior statesmen that it would be improper for His Majesty as a constitutional monarch to intervene in matters of foreign policy which had been decided upon by the Government and duly reported to the Throne by the Premier.

On March 27, 1933, when the Japanese Government notified the League of Nations of Japan's withdrawal from the League, an Imperial Rescript was issued. Considering the wish of the Emperor, the Rescript laid special emphasis on His Majesty's unchanging devotion to the cause of international peace and harmony. This was intended as a warning to the nation against the tide of anti-foreignism arising with its withdrawal from the League.

Matsuoka, Japan's delegate to Geneva, returned to Japan after the break with the League of Nations, and said: "From the moment I had received the Imperial Command and left Japan, I was always aware of what was expected of me by His Majesty, which certainly was not our withdrawal from the League. I do not understand why I, who have failed in this important mission, should be welcomed back like a victorious general by a segment of the public." The Emperor's sentiments are well indicated in this remark of Matsuoka.

The Jehol expedition of 1932–33 became a subject of discussion by the League of Nations. The Emperor thought of a plan of calling an Imperial conference. It failed to materialize because Prince Saionji held the view: "From the behavior of young officers these days, there is little hope for the success of such a plan. After all, it would be best to let the Premier assume the entire responsibility for coping with the current situation."

The China Affair started in 1937 by the Army on the spot as was in the case of the Manchurian Incident. The Emperor, deeply concerned over the trend of the conflict spreading from North China to Central China, contrary to the Government's avowed policy of non-extension of hostilities, conveyed his wish to Premier Konoye to call an Imperial conference and establish a basic national policy for the restoration of peace. The Premier, however, hesitated for various reasons to carry out this Imperial wish.

From the first conference which was called under the First Konoye Cabinet on January 11, 1938, to discuss China policy following the failure of the peace negotiations through the intermediation of the German Government to the last conference of December 1, 1941, which decided upon war with America and Britain, Imperial conferences were held eight times. But these conferences were purely ceremonial, as they did nothing but adopt formally, in the Imperial presence, the decisions that had already been reached at the Liaison Conference between the Government and High Command, independently of the intentions of the Emperor. Such being the case, the conferences, instead of providing an occasion to restrain the military, served rather to advance the cause of their aggressive policy.

The Pacific War

From the beginning to the end the Emperor was determined to prevent war. Words fail me to describe His Majesty's anxieties and endeavors in this respect. As one who personally attended him, I find it painful to recollect the momentous developments of those days leading to the Pacific War.

It was at the Imperial conference of September 6, 1941, that the Emperor counselled for peace, by reciting the famous poem of the Emperor Meiji:

> *When all on the four seas*
> *Are, I believe, brethren,*
> *Why this tumult of winds and waves—*
> *O world!*

This episode, which was mentioned in the policy statement of Premier Higashikuni before the post-surrender Diet, is minutely related in the memoirs of the late Prince Konoye.

On the eve of the September 6th conference Premier Konoye reported to the Throne on the draft resolution on the agenda. The Emperor was shocked to discover that the draft resolution contained a paragraph calling for a decision to declare war with the time limit set at the first part of October. He demanded a revision, but the Premier could not accede. Therefore, His Majesty summoned the Chiefs of the Army and Navy General Staff and obtained assurance to the effect that the High Command considered the resolution to give diplomatic negotiations priority over war. His Majesty caused both the Government and High Command to confirm the same assurance.

As stated before, an Imperial conference was strictly a formal affair, and it proceeded like a ceremony with the Premier presiding and the Emperor taking no part whatsoever in its deliberations. However, on this occasion His Majesty, breaking the custom, took the floor and made clear what he wanted. Nevertheless, the draft resolution was adopted without a word of amendment.

Now why did the Emperor approve this resolution to which he had openly declared his objection? From the days of Meiji it had been an unwritten law that the Emperor would approve any decision of the Government, never would he veto it. His Majesty adhered firmly to this constitutional usage as his own personal creed. As a matter of fact this practice was, I am sure, the most appropriate means of safeguarding monarchy from falling into dictatorship and fitting it into the system of a constitutional government.

With the stagnation of the Japanese-American conversation and the approach of the date set for decision to fight, the Konoye Cabinet resigned en bloc on October 18, 1941. War Minister Tojo, who wrecked the Konoye Cabinet by his war-like stand, received the Imperial command to form the successor cabinet. This has invited criticism from various quarters. Let me explain!

Under the old constitution of Japan the selection of a Premier was made, as a matter of established custom, through the recommendation of the Lord Keeper of the Privy Seal who was constantly in attendance upon the Throne to tender advice and assistance. Marquis Kido, Lord Keeper of the Privy Seal, who at this fateful juncture had to recommend a new Premier, took, needless to say, every possible care and caution. As Chief Secretary to the Lord

Keeper of the Privy Seal I know well the circumstance of General Tojo's selection. Marquis Kido believed that a man who could successfully cope with the situation at that critical moment had to meet the following two conditions:

1. *Instead of plunging immediately into war, he must seek to bring the Japanese-American negotiation to a successful conclusion.*
2. *In case of a settlement with the United States, he must command enough influence to hold down the malcontents in the armed services—especially in the Army.*

After careful consideration the Marquis came to the conclusion that no one but General Tojo could meet these qualifications. So, with the approval of the senior statesmen he recommended Tojo as Premier.

As for the first condition Marquis Kido had been assured of General Tojo's view that war with America and Britain must be averted as long as the Navy Minister lacked confidence in the war (the fact that he did lack it had been ascertained by Kido from a reliable source).

With respect to the second condition Tojo himself had told His Majesty that in case the Japanese-American negotiations succeeded, he would suppress even at the risk of his life any dissatisfaction within the Army. Moreover, the War Minister had demonstrated his power when he forced the resignation of the Army Commander, a Division Commander, and a Brigade Commander for the forcible landing of troops in French Indo-China, contrary to the instructions from Tokyo. It was these concrete facts which convinced the Marquis of the reliability of General Tojo with regard to his control over the Army.

On the day War Minister Tojo was commanded to form a new cabinet, Marquis Kido conveyed an Imperial message to both War and Navy Ministers stating: "In deciding upon the fundamental line of national policy, you will, without being bound by the decision of the September 9 Imperial conference decision, conduct a wider and deeper study of the situation at home and abroad and exercise utmost care and prudence." This message, in order to ensure accuracy, was drafted first in writing (I myself wrote the draft), and was read aloud to the ministers by the Marquis. His Majesty's intention was evident: "Give up your war decision, and work single-heartedly for peace!"

On October 18, 1941, the Tojo Cabinet came into being. The new Premier in conformity with the Imperial message made earnest endeavors toward the success of the Japanese-American negotiations. Unfortunately his efforts failed and war was declared, frustrating the hopes of His Majesty. But the Emperor strove to the last minute to prevent the war.

It was early in November 1941, as I recall, that the Emperor happened to hear about the Premier turning cool toward the Japanese-American negotiations. He asked Tojo if he was conducting in earnest the negotiations with the United States. Tojo replied in the affirmative.

On November 29, 1941, His Majesty summoned to the palace ex-Premiers, political leaders, the President of the Privy Council, as well as the Premier and principal ministers of states for an informal discussion. His Majesty had first suggested to the Government to call such a meeting, but when the Cabinet declined, the Cabinet members and the senior statesmen were called separately, to meet at the sametime, and talk matters over in an informal fashion. The meeting was designed by the Emperor to serve as an opportunity for the senior statesmen to speak their minds and moderate the stiff attitude of the Government.

The Emperor has since told me directly, "Most of the senior statesmen were opposed to war. Unfortunately, their opinion was abstruse and could not hold down the Cabinet members who cited figures to support their contentions."

On the morning of November 30, 1941, the day after the above-mentioned palace meeting, the Emperor heard from his Brother, Prince Takamatsu, attached to the Navy General Staff, that the Navy lacked confidence in a war with America and Britain. If true, that would be an absolute reason not to fight, or at least it would serve as a starting point for a move against war. The Emperor immediately asked the Premier to postpone the Imperial conference, scheduled for the following day. It proved to be the last of such conferences preceding the declaration of war. At the same time he summoned the Navy

Minister and the Chief of the Navy General Staff and inquired about the prospects of the naval operation. They replied that as Navy leaders they had full confidence in their plans for naval warfare. Thus on December 1, 1941, the Imperial conference attended by all members of the Cabinet, the Chiefs of Army and Navy General Staffs and the President of the Privy Council, was held, presided over by Premier Tojo. War with America and Britain was the final decision.

The message of President Roosevelt, his last attempt at preserving peace, reached Tokyo on December 7, 1941. But it was 15 minutes past midnight when a copy was delivered by Ambassador Grew to Foreign Minister Togo, who reported its contents to His Majesty at 0300, December 8. At that time it was 0730, December 7 in Honolulu, or some 20 minutes before Japanese naval aircraft reached Pearl Harbor. In Washington it was 1300, just the hour of appointment for Ambassador Nomura to deliver the last memorandum of the Japanese Government to Secretary of State Hull. In point of time it was no longer possible to call off the movements of Japanese land and sea forces which had already started on a gigantic scale.

Foreign Minister Togo, who had apparently conferred with Premier Tojo before coming to the palace, advised His Majesty that no reply was necessary. He departed after obtaining the permission to transmit a brief Imperial message to the American ambassador. This, I believe, was done.

Concerning the above episode His Majesty said to me later, "I wished to answer the personal cablegram from the American President, but I was persuaded not to do so by Foreign Minister Togo who told me I had better not send a reply because hostilities would start on the 8th, changing the situation completely."

According to what I was told by Ambassador Kurusu, who was invited to a banquet given by Premier Tojo soon after his return on the exchange ship in the summer of 1942, the conversation at the table had turned to the subject of President Roosevelt's message, and the Premier remarked, "If the American note of November 26 had not been as harsh as it was, and if the President's message had arrived earlier, war would have been impossible."

If, indeed, there had been a few days' time, as Premier Tojo said, the Emperor would not have desisted from cooperating in the peace proposal of the President.

The Emperor's Legal Status

The Emperor once told his entourage: "Before doing anything I have made it my rule to rely on the advice of my counsellors, and never to reject such advice. Only twice in my life have I acted on my own initiative, once at the time of the February 26 Incident, 1936, and again at the time of surrender."

The Emperor personally ordered the suppression of the rebellion attending the February 26 Incident because, with the Premier beseiged by rebels, the Cabinet had ceased to function. The Army leaders were all so upset that they did not know what to do. His Majesty's close advisers, the Lord Keeper of the Privy Seal had been killed, the Grand Chamberlain seriously wounded, and there were none near him to shoulder the responsibility for assisting and advising the Throne. The Emperor had no alternative but to assume the direction for handling the emergency. There can be no question of the constitutionality of his action in that case.

At the time of the termination of the war there existed both the Government and the High Command. But both had lost their functions as the supreme organs of state to advise and assist the Throne. In this respect the situation was similar to that of the February 26 Incident. That is to say, on the question of accepting or rejecting the Potsdam Declaration, opinion was so hopelessly divided among the leaders of the Government and the High Command that the Premier asked the Throne for a decision.

The Emperor's attitude was consistent throughout and the fault lay with the Premier and other leaders of the Government and the High Command who were found wanting in the discharge of their duty to advise and assist the Throne in the affairs of state. In fact, it was for exactly this reason that the Suzuki Cabinet resigned en bloc immediately following the Emperor's radio address on the termination of war.

Afterwards, the Emperor told the Minister of the Imperial Household Ishiwata: "At that time I was convinced of the necessity of ending the war. What worried me was that if the Government and the High Command had come to an agreement and the continuation of the war had been decided upon as the policy of state, I would have had to approve the decision against my personal conviction. To my great relief, the decision was left to me."

Now at the outbreak of the Pacific War the situation was totally different. In this case the Government and the High Command, which were responsible for tendering advice and assistance in these respective spheres of political affairs and military matters, had agreed. The policy of the state was decided upon and the Emperor could do no other than approve the decision in accordance with the constitutional usage. Prior to that fatal moment His Majesty, as described already, had done everything he could as a constitutional monarch to avert the war.

Adherence to the constitution was set down by the Emperor as the first principle of government. At every change of cabinet he would enjoin upon the new Premier strict observance of the constitution. This might appear all too proper for comment. It had however, a great significance at the time of his reign, when radical ideologies were sweeping over Japan. We owe it to His Majesty that we were able to preserve, somehow, our constitutional government through those years during which dictatorship and force prevailed in some quarters of the world. Years in which Japan itself saw the political parties become powerless and open threats to brush aside the parliamentary government itself.

"The King reigns but does not rule." This English proverb was literally translated into action by His Majesty. The principle under the old constitution that the Emperor shall always rely on the advice and assistance of his ministers in all matters of state, had established a rule that the Emperor was not to exercise the veto as an Imperial prerogative. His Majesty had an occasion early in his reign to confirm his faith in this unwritten law.

In July 1929 Premier Tanaka, as I said before, resigned after having received an Imperial reprimand. On recollecting the episode years afterwards, His Majesty said to us: "There I went too far; I was young. It gave currency to such annoying rumors as 'palace intrigue' or 'senior statesmen's block, which were invented by the military and the rightists, and which had a bad effect upon government. This bitter experience has taught me to be always on guard against exceeding my limits." His Majesty's attitude toward the constitution was born of a profound reflection on his responsibility as sovereign and his ceaseless self-discipline and self-restraint from wilful and arbitrary action.

In his Memoires Prince Konoye appears to blame the passive attitude of the Emperor for the inability of the government to stop the war. This is a most regrettable mis-statement on the part of the Prince who had failed to live up to his own responsibility as Premier.

As I have already said, His Majesty expressed his opposition to the view of the military at the Imperial conference of September 6, 1941, and prodded the Government on to peace efforts. If Prince Konoye hads hown a greater zeal for peace he would have been able to obtain His Majesty's support and hold down the military.

Again, if, at the Imperial conference of December 1, 1941, which decided upon war, there had been a single member of the Cabinet standing against war, it might have given the Emperor an opportunity to do something to avert war. But that conference attended by all ministers of state as well as the Chiefs of the Army and Navy commands unanimously decided on war. As head of the state having a constitution and parliamentary system it was unthinkable of the Emperor to veto this decision.

By respecting the constitutional function of the Government the Emperor did not mean to evade his responsibility as sovereign. Nor was he blindly bound by the letter of the constitution. He only conformed to the tenet of constitutional government, which rejects dictatorship and absolutism.

REPORTS OF GENERAL MacARTHUR

VOL I: The Campaigns of MacArthur in the Pacific
VOL I: Supplement: MacArthur in Japan: The Occupation, Military Phase
VOL II: Japanese Operations in the Southwest Pacific Area

". . . This report has been prepared by the General Staff to serve as a background for, and introduction to the detailed operational histories of the various tactical commands involved.

The pressure of other duties having prevented my personal participation in its preparation, it has been entrusted by me to that magnificent staff group which actually conducted the staff work during the progress of the campaigns. They speak with that sincere and accurate knowledge which is possessed only by those who have personally participated in the operations which they record . . ."

Preface by General Douglas MacArthur.

Senior Commanders: Southwest Pacific Areas

Gen W. Krueger: Sixth Army; Lt Gen R. L. Eichelberger: Eighth Army; Gen Sir Thomas Blamey: Aust. Imp. Forces; Lt Gen G. C. Kenny: AAF; Adm T. H. Kinkaid, USN: Seventh Fleet

The General Staff: GHQ: Southwest Pacific Area

Lt Gen. R. K. Sutherland, CofS; Maj Gen R. J. Marshall, D CofS; Maj Gen C. P. Stivers, G–1: Maj Gen C. A. Willoughby, G–2; Maj Gen S. J. Chamberlin, G–3; Maj Gen L. J. Whitlock, G–4; Maj Gen S. B. Akin, CSO; Maj Gen W. F. Marquat, AAO; Maj Gen H. J. Casey, CE; Brig Gen B. M. Fitch, AG; Brig Gen L. A. Diller, PRO.

Editor in Chief

Maj Gen Charles A. Willoughby, G–2

Senior Editors

Col E. H. F. Svensson, G–2; Gordon W. Prange PhD; Mr. Stewart Thorn

Associate and Contributing Editors

Brig Gen H. E. Eastwood, G–4; Col F. H. Wilson, G–2; Col R. L. Ring, G–2; Col W. J. Niederpruem, G–3; Lt Col M. K. Schiffman, G–2; Maj J. M. Roberts, G–3; Capt J. C. Bateman, G–2; Capt Mary Guyette, G–2; Capt John L. Moore, G–2; Lt Stanley Falk, G–2; Mr. Jerome Forrest; Mr. Kenneth W. Myers; Miss Joan Corrigan.

Translation–Interrogation–Production

Lt Col W. H. Brown, G–2; Louis W. Doll, PhD; Capt E. B. Ryckaert, G–2; Capt K. J. Knapp, Jr., G–2; Lt Y. G. Kanegai, G–2; Mr. James J. Wickel; Mr. John Shelton, ATIS; Mr. Norman Sparnon, ATIS; SFC H. Y. Uno, G–2; Mr. K. Takeuchi; Mr. S. Wada.

INDEX

☆ U.S. GOVERNMENT PRINTING OFFICE: 1994 360-580

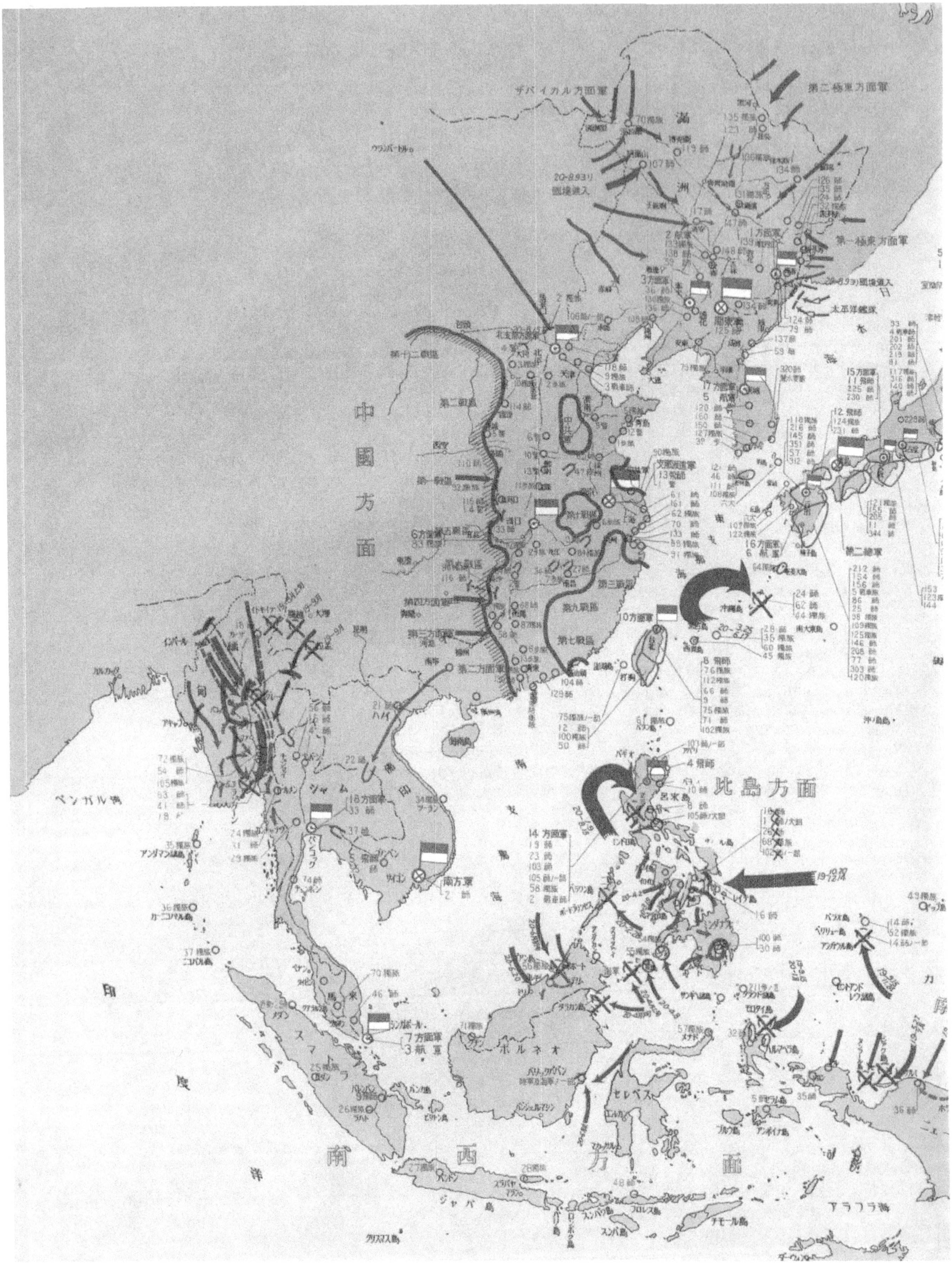

ザバイカル方面軍
第二極東方面軍
第一極東方面軍
満
洲
太平洋艦隊
中國方面
第十二戦區
第二戦區
第一戦區
第三戦區
第七戦區
第九戦區
支那派遣軍
6方面軍
第四方面軍
第三方面軍
第二方面軍
10方面軍
16方面軍
6 航軍
15方面軍
12方面軍
17方面軍
第二總軍
沖縄島
比島方面
14 方面軍
呂宋島
ベンガル湾
シャム
18方面軍
バンコク
サイゴン
南方軍
アンダマン諸島
カーニコバル島
ニコバル島
馬來
シンガポール
7方面軍
3航軍
スマトラ
ボルネオ
セレベス
南
西
方
面
印
度
洋
ジャバ島
アラフラ海
ロンボク島
スンバ島
フロレス島
チモール島
クリスマス島

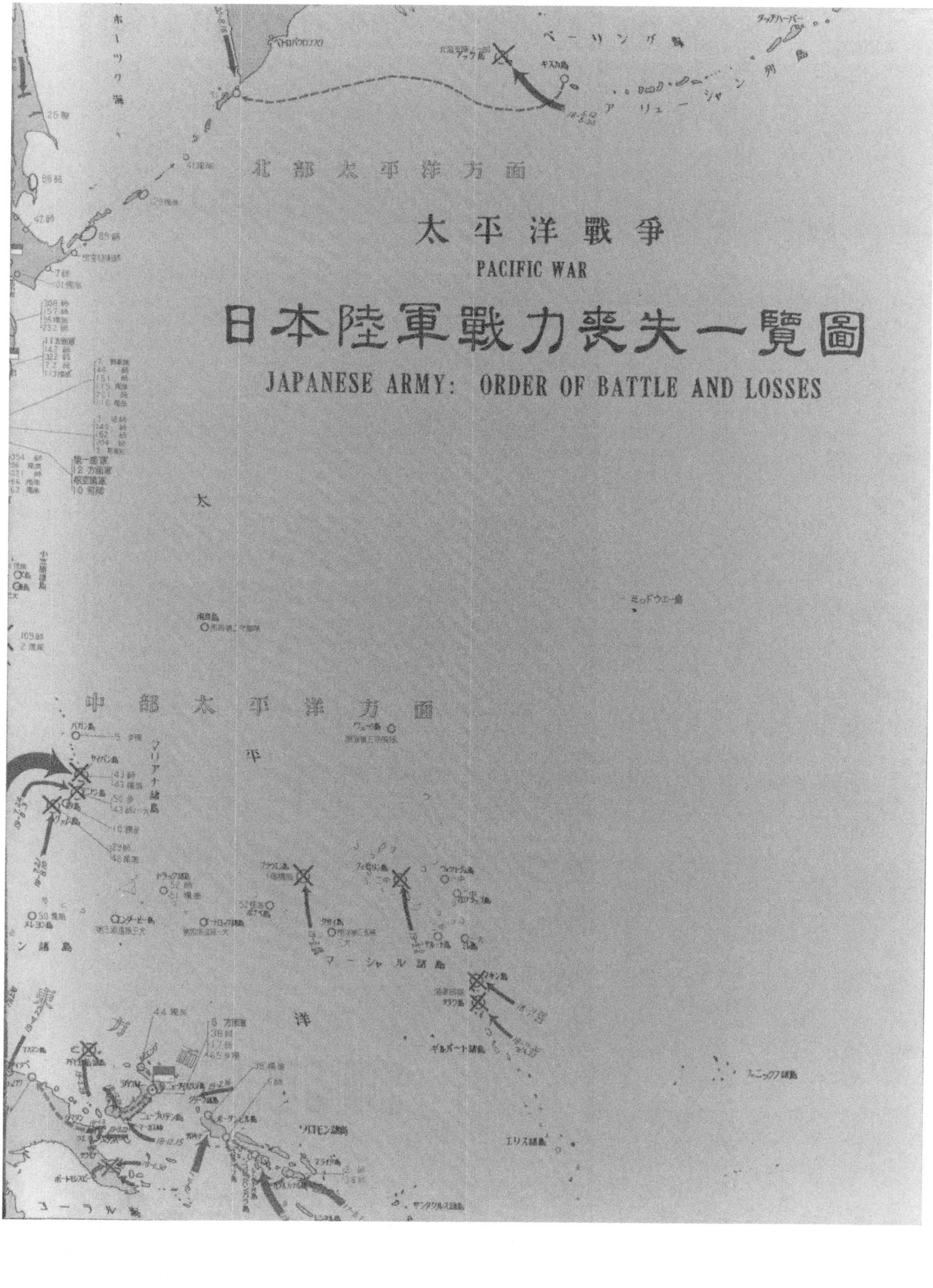
太平洋戰爭
PACIFIC WAR
日本陸軍戰力喪失一覽圖
JAPANESE ARMY: ORDER OF BATTLE AND LOSSES
北部太平洋方面
中部太平洋方面
ベーリング海
アリューシャン列島
オホーツク海
マリアナ諸島
マーシャル諸島
ギルバート諸島
ソロモン諸島
エリス諸島
ミッドウエー島
サンタクルス諸島
フェニックス諸島

Made in United States
North Haven, CT
11 July 2025

70597668R00252